Sustainable Transportation: Emerging Technologies

Sustainable Transportation: Emerging Technologies

Edited by
Jay Rhodes

MURPHY & MOORE
www.murphy-moorepublishing.com

Published by Murphy & Moore Publishing,
1 Rockefeller Plaza,
New York City, NY 10020, USA

ISBN: 978-1-63987-513-9

Cataloging-in-Publication Data

Sustainable transportation : emerging technologies / edited by Jay Rhodes.
p. cm.
Includes bibliographical references and index.
ISBN 978-1-63987-513-9
1. Transportation--Environmental aspects. 2. Sustainable engineering.
3. Transportation--Technological innovations. I. Rhodes, Jay.
HE147.65 .S87 2022
388.049--dc23

For information on all Murphy & Moore Publications
visit our website at www.murphy-moorepublishing.com

MURPHY & MOORE

Contents

Preface

This book has been an outcome of determined endeavour from a group of educationists in the field. The primary objective was to involve a broad spectrum of professionals from diverse cultural background involved in the field for developing new researches. The book not only targets students but also scholars pursuing higher research for further enhancement of the theoretical and practical applications of the subject.

Sustainable transportation is a significant feature of sustainable development. It makes use of renewable energy that is sustainable and has low impact on the health of humans and ecosystems. Sustainable transportation is commonly known as green transportation. It is affordable, allows basic access and has lower rate of emission and waste product generation. Cycling, walking, public transport and sailing are some examples of sustainable transportation. It results in reduction of greenhouse gases, traffic congestion, air pollution, dependence on non-renewable sources of energy, and transportation costs. It also increases physical activity and social interaction leading to a healthier lifestyle. This book explores all the important aspects of sustainable transportation in the present day scenario. The various studies that are constantly contributing towards advancing technologies and evolution of this field are examined in detail. The readers would gain knowledge that would broaden their perspective about the emerging technologies in sustainable transportation.

It was an honour to edit such a profound book and also a challenging task to compile and examine all the relevant data for accuracy and originality. I wish to acknowledge the efforts of the contributors for submitting such brilliant and diverse chapters in the field and for endlessly working for the completion of the book. Last, but not the least; I thank my family for being a constant source of support in all my research endeavours.

Editor

1

Exploring the Energy Efficiency of Electric Vehicles with Driving Behavioral Data from a Field Test and Questionnaire

Kezhen Hu,[1,2] Jianping Wu,[1] and Mingyu Liu[1]

[1] *Department of Civil Engineering, Tsinghua University, Beijing 100084, China*
[2] *Transport Studies Unit, School of Geography and the Environment, University of Oxford, Oxford, UK*

Correspondence should be addressed to Jianping Wu; jianpingwu@tsinghua.edu.cn

Academic Editor: Shamsunnahar Yasmin

With increasing concerns about urban air quality and carbon emissions, electric vehicles (EVs) have gained popularity in megacities, especially in Europe and Asia. The energy consumption of EVs has subsequently caught researchers' attention. However, the exploration of energy consumption of EVs has largely focused on people's revealed driving behavior and rarely touched on their self-perception of driving styles. In this paper, we developed a more human-centric approach, aiming to investigate how the energy efficiency of EVs is shaped by the driving behavior and driving style in the urban scenario from field test data and driving style questionnaires (DSQs). Field tests were carried out on a designated route for a total of 13 drivers in the city of Beijing, where vehicle operation parameters were recorded under both congested and smooth traffic conditions. DSQs were collected from a larger pool of drivers including the field test drivers to be applied to driving style factor analysis. The results of a correlation analysis demonstrate the dynamic interaction between drivers' revealed behavior and stated driving style under different traffic conditions. We also proposed an energy consumption prediction model with the fusion of collected driving parameters and DSQ data and the result is promising. We hope that this study would draw inspiration for future research on people's transitioning driving behavior in an electric-mobility era.

1. Introduction

To cope with the challenges of reducing greenhouse gas emissions and curbing urban pollution, electrification of the urban vehicle fleet has been a trend, particularly in densely populated urban areas, such as Paris in Europe and Beijing in Asia [1–3]. EVs have the advantages of zero tailpipe emissions, low noise, and higher propulsion efficiency. By the end of 2016, a fast-growing market has been fostered with an accumulated EV stock reaching two million globally with the support from central and municipal governments [4].

However, potential EV buyers are still concerned about limited range, which is referred to as "range anxiety." It is defined as anxiety or fear that one may not reach a destination before the battery is empty, which can occur while driving or prior to driving as the user worries about a later planned trip [5]. Even though EV motor efficiency is less sensitive to speed than are internal combustion engines, bidirectional energy flow resulting from regenerative braking and the dynamics of electric propulsion has caused substantial heterogeneity among drivers due to different driving behaviors, and this deserves further study [6].

On-road fuel consumption of vehicles can be influenced by drivers through three levels of decisions, including strategic decisions (vehicle selection and maintenance), tactical decisions (route selection and vehicle load), and operational decisions (driving behavior) [7]. Driving behaviors have been shown to result in fluctuations of vehicle energy efficiency with regard to conventional vehicles, where aggressive driving behavior can consume 30% more fuel [8]. In some EV pilot countries, researchers have pointed out the change of driving behaviors in the transition from conventional vehicles to EVs, which could be susceptible to the contexts and require further research [9, 10]. Hence, for progress to be made in the case of EVs, a comprehensive understanding of the interaction between people's driving style, driving behavior, and energy efficiency is needed.

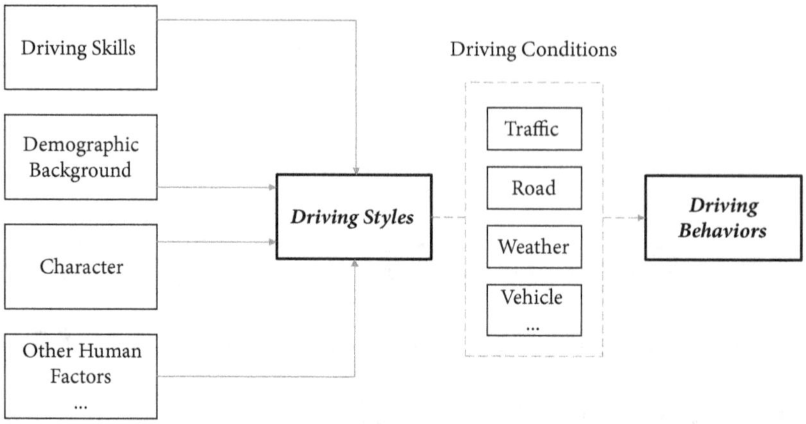

FIGURE 1: Schematic graph for the demonstration of the relationships between the key terminologies used in this paper.

Since terminologies such as "driving behavior" and "driving style" are used in this study and can sometimes be confounding, we would like to give a concise description beforehand. As understood by the authors, driving behaviors emerge out of the interplay of contextual conditions (traffic, road, weather, vehicle, et cetera) and person-specific driving styles [11, 12]. In the real-world tests, contextual conditions are more of a dynamic nature. Driving styles, dispositions to drive in particular ways that have been acquired over time, are more or less stable in the short term and can be measured by questionnaires. The relationships between these terms are illustrated in Figure 1.

Primarily, since aftermarket devices which can record multiple vehicle operation data sources on EVs (including speed, energy consumption) through onboard diagnostics (OBD) are widely available, the tech-centric approach has been adopted to evaluate driving behaviors from the abundant data collected in field tests [13]. Some research on the energy consumption of EVs and driving behaviors was based on experiments on test tracks without actual traffic flow, and the results elucidated the potential of good driving behavior on the improvement of energy efficiency [14–16]. Other authors conducted real-world tests to cluster drivers according to their speed and energy consumption profiles or calculate the average energy consumption for different road types [17–20]. Besides, there has been another series of studies, namely, more human-centric approaches, focusing on the use of interviews, questionnaires, and travel diaries to understand driving behavior or attitudes towards EVs [21]. Psychological methods were adopted to evaluate people's perception of EV display systems, portable driving feedback device, and their acceptance of EV itself [13, 22, 23]. Recently, Helman and Reed conducted an exploratory study on the correlation between driver behavior questionnaire (DBQ) scores and observed driving speed data [24], which inspired us to extend this comparative idea to the domain of EV energy consumption. These two strands of studies (field experiments and questionnaires) have remained parallel for most of the time, excluding the chance to understand driving behavior from people's underlying dispositions.

In summary, none of the existing work analyzed the impact of driving behaviors on EVs' energy consumption in the open driving environment from both Controller Area Network- (CAN-) based data and questionnaires. The overall EV energy consumption performance will depend on not only how well the vehicle works by itself but rather its interaction with the human driver. Analyzing behavior-related variables with multiple data sources allows for a comprehensive data triangulation [25], which could potentially yield a more accurate analysis result. In this study, compared with previous study that only featured measured field test data or questionnaire data separately [14, 22], a more "human-centric" experiment is proposed to explore how drivers' stated driving styles are reflected and correlated with their actual way of driving and to understand how they together contribute to the energy consumption.

The remainder of the paper is organized as follows: after the introduction of the EV field test in Section 2.1; Section 2.2 describes the driving style questionnaires (DSQs) in detail; then, an analysis of the results follows in Section 3; and finally, conclusions of the investigation are given in Section 4.

2. Experiment and Methodology

The goal of this research is to improve the understanding of the driving behavior of EV drivers through the integration of experiments and questionnaires and its impact on the energy efficiency created by EV users. It is noteworthy that there are a number of factors that can affect the energy efficiency of a vehicle in the real world [11, 13], including external factors like infrastructure design, vehicle design, road grade, weather conditions, congestion level, and internal factors of the driver such as driving behavior and driving style. Since this paper concentrates on studying driving behavior and driving style under different traffic conditions, other external factors remain as controlled variables. To this end, we recruited 13 EV drivers to collect their real-world field test data on the same route, and we also built a pool of driving style questionnaires (DSQs) with these drivers included. The DSQ data (more than 300 samples) was used to conduct a valid factor analysis so we could further score each of

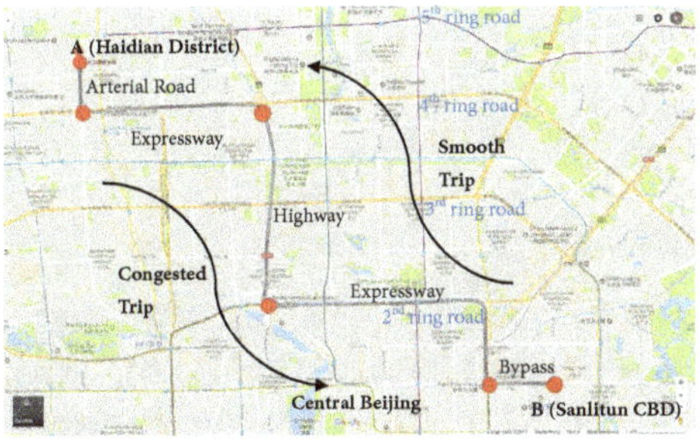

FIGURE 2: Map of the selected route.

the 13 drivers in terms of different dimensions of driving style. Observations on driving behaviors were conducted on a designated route in a "semicontrolled" condition to increase the comparability of the results across drivers. With this research we seek to (a) identify the relationship between self-stated driving style and observed driving behavior under different traffic conditions and (b) advance understanding of the energy efficiency created by EV users in a real-world driving scenario from the perspectives of driving behavior and driving style.

2.1. Field Test Data Used in This Study. Detail of the experimental design in which energy consumption and other driving behavior data were collected as described in our previous work [11]. In short, the data was collected from 13 participants who drove on a selected typical commuting route in Beijing in the same time of day, starting from the residential area of Haidian District (near the 5th ring road) and ending in Sanlitun CBD (between the 2nd and 3rd ring road) in Chaoyang District (shown in Figure 2). The drivers were selected based on a snowball sampling method to be representative of the current EV drivers in Beijing as shown in a survey, which indicated that 63.4% of EV drivers in Beijing were male, 79% of the drivers fell into the age group of 20-39, and the majority were well educated [26]. The majority of drivers in the experiment were well-educated young male adults who had driven an EV before. The drivers were instructed to drive as they normally would (without violating the speed limit) at the same time of the day (including both congested and smooth traffic conditions) to normalize the data against traffic conditions.

The vehicle used for the study was a battery electric vehicle Nissan LEAF model with an OBD data logger connected to the Electronic Control Unit (ECU) along the driving tests. A number of variables including vehicle speed, motor torque, motor speed, battery pack current, and voltage were recorded. The acceleration and energy consumption data could be calculated as derived values accordingly. All tests were carried out within two months after commencement in October 2015, excluding days of extreme weather (rainy, windy), and special events.

To ensure consistency, the vehicle started with the same state of charge (SOC) of battery for different participants during all the tests (80% for the congested trip and roughly 65% for the smooth trip), and no additional 'comfortable loads' (i.e., air-conditioning and heating or radio) were used. Other controlled variables included the identical test route and same departure time. Note that unlike the ideal conditions in a simulation platform or chassis test, the operation conditions could not be maintained in an identical fashion in our tests. We recognize the existence of uncertainty (traffic conditions were similar but not the same) in our tests but believe it is inevitable when the aim is to obtain real-world data for analysis.

Since the drivers were instructed to drive "naturally" as they normally would, we assume the field test data reflects participants' usual driving behavior under everyday traffic conditions. Each driver in the study also completed the DSQ, leaving us with the opportunity to examine how and to what extent their self-stated responses correspond to their driving behavior in the field test dataset.

2.2. Questionnaire Data Used in This Study. The Driving Style Questionnaire (DSQ) examines driving behaviors that are related to accidents and risks, including speed, headway (distance to the car in front), lane-changing behavior, violations of traffic signals, and human cognition and attitudes that are somehow associated with decision-making, for example, feelings of confidence, route planning, and en-route risk taking [27].

We designed a driving style questionnaire in Chinese according to previous research [28] and took into account the unique context in China to extract the driving characteristics. A total of 55 questions were included in the original version, and 40 questions were finally selected after eliminating unimportant questions. The questionnaire was mainly structured with each answer consisting of four ordinal measurements. For example, for question No. 30, "I can drive safely under complicated traffic conditions," there were four degrees of agreement: (A) No, (B) Seldom, (C) Sometimes, and (D) Always. Full details of the questionnaire are given in the Appendix.

In order to perform a valid DSQ analysis, there should be a minimum sample size of 5 times the number of questions in the questionnaire (to avoid the "curse of dimensionality"), which equals 200. Hence it is necessary to extend the questionnaire dataset to include a wider group of drivers. While all the 13 drivers were requested to complete the questionnaire, DSQs were also disseminated online and offline among valid driving license holders in the City of Beijing. Finally, more than 400 questionnaires were completed including the 13 drivers in the driving experiment. After excluding invalid samples with extreme values, discrepancies, or missing values (in terms of age, responding time, cross-validation questions, et cetera), the remaining 331 (80%) questionnaires were taken for further analysis. The demographic information is listed in Figure 3. The respondents are across all age groups (mean=38.09, Std Dev=9.075), with a balanced gender mixture and an average driving experience of 5 years.

Statistical Product and Service Solutions (SPSS) was used to carry out the calculation and analysis process. The Kaiser-Meyer-Olkin (KMO) Measure of Sampling Adequacy>0.85 and the Bartlett test of the sphericity (sig=0.000) of the

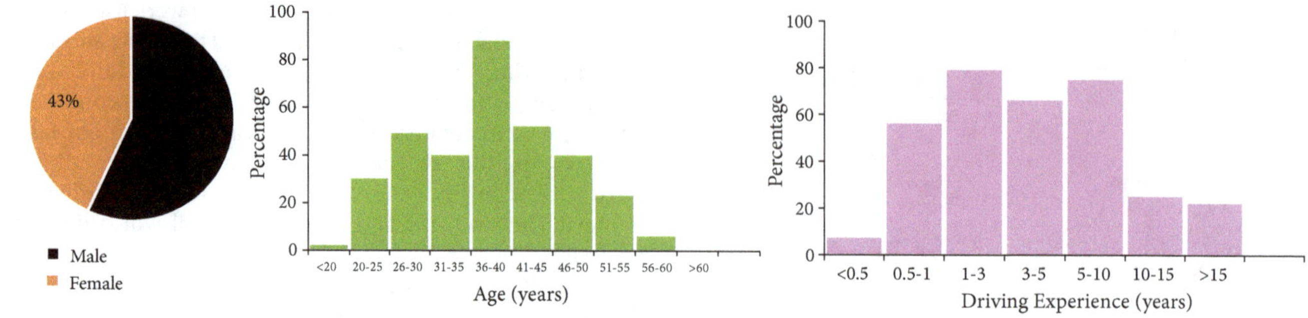

FIGURE 3: Demographic information of DSQ respondents.

TABLE 1: Variance explained by each factor.

Factor	Total	% of Variance	Cumulative %
Factor 1	5.216	13.040	13.040
Factor 2	3.631	9.077	22.116
Factor 3	2.297	5.743	27.859
Factor 4	2.094	5.236	33.095
Factor 5	1.945	4.862	37.958
Factor 6	1.882	4.704	42.662
Factor 7	1.739	4.347	47.009
Factor 8	1.664	4.159	51.169

questionnaires both indicate the suitability of a factor analysis (also called principal component analysis, PCA) [29]. Therefore, the factor analysis was applied to the DSQs and extracted eight components with eigenvalues greater than one (other factors with lower eigenvalues were ignored for interpretability purpose). Table 1 shows the contribution of the extracted factors to explain the variance after a varimax rotation. Together the eight factors accounted for 51.2% of the variance. The detailed interpretation of the extracted factors is given in Table 2.

3. Results

This section briefly summarizes the results from our experiments and continues with the analysis of the DSQs. It then explores the relationships between measured driving behavior and DSQ scores for the 13 recruited drivers under different traffic conditions. Finally, it proposes a concise multiple regression model to formulate the EV energy consumption using DSQ scores and the driving behavior data collected from the field tests.

3.1. Preliminary Results of the Field Tests and DSQ Scores. The results (shown in Figure 4) illustrated that the average energy consumption during congested traffic conditions is 15.6% higher than during smooth conditions (151.7 Wh/km and 131.3 Wh/km, respectively), which is statistically significant at the 95% confidence level in the Mann-Whitney test. The energy consumption variation among drivers is slightly larger during congested traffic condition than during smooth condition. The worst-performing versus the best-performing

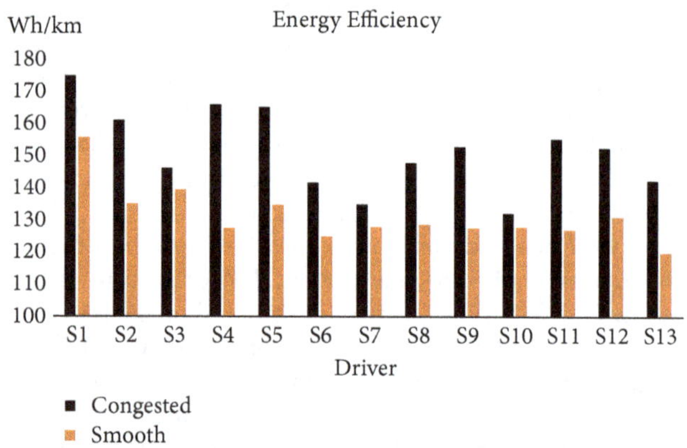

FIGURE 4: Energy efficiency for the drivers under different conditions.

driver is 32.4% during congested condition and 30.0% during smooth condition.

Previous research has suggested that fuel consumption and driving behavior variables such as speed and acceleration were correlated [30]. Corresponding to energy consumption, several driving behavior variables were recorded and calculated for each driver, respectively, including average speed, average acceleration, average deceleration, journey energy regeneration, and driving status distribution. Driving status distribution consists of acceleration, deceleration, constant speed, and idling. The idling mode means that the battery power is turned on to supply to the motor although the actual vehicle speed is 0; the acceleration mode is defined by the acceleration speed $a > 0.2m/s^2$ in the constant driving

TABLE 2: Question loading for the eight factors.

Factor	Loadings	Questions
Factor 1: Self-confidence in driving	0.79	I can drive safely under complicated traffic conditions.
	0.74	I can respond reasonably to unexpected events.
	0.72	I can maneuver the vehicle properly as I wish.
	0.71	I have confidence in teaching newcomers to drive.
	0.68	I am skillful in backing and parking.
	0.54	I can drive normally in harsh weather (windy, foggy, rainy, snowy).
	0.42	I enjoy driving.
	-0.57	I feel nervous when driving.
	-0.43	I am worried about a traffic accident when driving.
Factor 2: Distraction in driving	0.78	I answer the phone when driving.
	0.69	I talk to my passengers when driving.
	0.62	I check my phone/iPad during the red traffic signal period.
	0.47	I pay attention to the street scenery when driving.
	-0.56	I concentrate during driving and always ignore what others say.
Factor 3: Impatience in driving	0.52	I feel upset during congestion.
	0.50	I maintain a close distance to the vehicle in front.
	0.45	I always drive fast.
	0.43	I always change lanes during congestion.
	-0.59	I give way to other vehicles if they merge into my lane.
	-0.44	I obey and have faith in traffic rules.
Factor 4: Attachment to vehicles	0.67	I can repair a vehicle by myself.
	0.52	I always watch car racing TV shows.
	0.51	I feel uncomfortable when others want to cut into my lane.
	0.45	I use a vehicle even when there are alternatives.
Factor 5: Intersection acceleration behavior	0.80	I accelerate when approaching and crossing a yellow light when there is only one vehicle in front and it is accelerating.
	0.79	I accelerate when approaching and crossing a yellow light when there is no vehicle in front.
	0.41	I feel uncomfortable when others want to cut into my lane.
Factor 6: Proactive driving behavior	0.57	I change to the expected lane early in preparation for the next intersection.
	0.52	I observe the traffic light patterns from a distance.
	0.46	I adjust my vehicle in advance when entering into a queue.
Factor 7: Rude driving	0.52	I always honk the horn to the vehicle in front in congestion.
	0.44	I brake hard when I decelerate.
	-0.47	I find it is easy to remain a proper distance from the front vehicle.
Factor 8: Rigidity in driving	0.74	I stick to my accustomed route even when the road is congested.
	-0.50	I feel annoyed when following heavy-duty vehicles like buses and vans.
	0.44	I enjoy driving.

Table 3: Relationship of driving style factors with age and gender.

Factor	Correlation with Age	Male Mean	Female Mean	Mann-Whitney test p-value (2-tailed)
Factor 1	0.118*	0.337	-0.448	0.000*
Factor 2	-0.219**	0.098	-0.131	0.030*
Factor 3	-0.100	0.080	-0.106	0.078
Factor 4	0.082	0.254	-0.337	0.000 * *
Factor 5	-0.032	0.025	-0.033	0.704
Factor 6	0.188*	0.103	-0.136	0.023*
Factor 7	-0.086	-0.129	0.171	0.016*
Factor 8	0.035	-0.064	0.086	0.161

N=331
**. Correlation is significant at the 0.01 level (2-tailed).
*. Correlation is significant at the 0.05 level (2-tailed).

process; deceleration mode happens when acceleration goes below $a < -0.2m/s^2$ in the constant driving process; and the constant speed mode is defined as instantaneous acceleration $|a| < 0.2m/s^2$ while speed is above 0. Series of continuous moments of each status have been aggregated across individual trips as episodes. The share of these four types of driving status is calculated as the percentage of the total number of episodes. The idling share is significantly different for the two traffic conditions, that is, 3.8% for the departure trip in congested conditions versus 1.8% for the return trip in smooth conditions. This discrepancy is not yet significant for other statuses, which is normal since whether to maintain a constant speed or to accelerate/decelerate is the joint result of the traffic conditions and drivers' personal driving style. Full details of the field test results and spatial domain analysis can be found in our previous study [11].

As for the DSQs, the aforementioned extracted factors were further interpreted by examining the contents of the variable loadings accordingly. Loadings are the correlation coefficients between the variables and factor. A criterion of 0.4 was used as the minimum loading for a question item to be incorporated. Seven of the 40 question items were excluded, and the remaining 33 were used for further analysis. The loadings of the questions for each interpreted factor are shown in Table 2. According to the content of relevant question items, Factor 1 can be interpreted as the driver's self-assessment of confidence in driving. The positive loading (0.79) of the question "I can drive safely under complicated traffic conditions" indicates its positive correlation with Factor 1, while the negative loading (-0.53) "I feel nervous when driving" denotes the negative correlation (this result is intuitive since when you feel nervous during driving you are always less confident in driving). Eight factors regarding driving styles are characterized as *self-confidence in driving, distraction in driving, impatience in driving, attachment to vehicles, intersection acceleration behavior, proactive driving behavior, rude driving*, and *rigidity in driving*.

With the loadings for each factor, the factor score for each driver can be calculated as follows:

$$Score_{ik} = \frac{\sum D_{ij} L_{jk}}{\sqrt{E_k}} \quad (1)$$

where D_{ij} is the standardized value for driver i on question j and L_{jk} is the loading of variable j on factor k. E_k is the Eigenvalue of factor k. The positive and negative factor score means the associated driver is above or below average among the sample group, respectively.

We first check the results of the whole pool of samples rather than the recruited drivers only. The relationship between driving style factor, age, and gender were tested using Pearson's correlation to explore changes in factor scores with age and the Mann-Whitney test for gender difference. The results are shown in Table 3. Males had higher scores on six out of eight factors; among them, self-confidence in driving, distraction in driving, attachment to vehicles, and proactive driving behavior were significant.

We categorized the population by age into four groups and plotted the average score for the eight factors accordingly (shown in Figure 5). Self-confidence and proactive behavior in driving increased with age (R=0.118, significant at 0.05 level, and R=0.188, significant at 0.01 level, respectively), while distraction in driving decreased with age (R=-0.219, significant at 0.01 level). Impatience in driving was higher for people below 40 than people above 40, while attachment to vehicles was the opposite. Rude driving alleviated with age, though not significantly. As for rigidity in driving, people between 40 and 50 outperformed other groups. Other factors like impatience in driving, attachment to vehicles, intersection acceleration behavior, proactive driving behavior, and rigidity in driving did not show a prominent correlation with age.

3.2. Relationships between DSQ, Field Test Data and Energy Consumption. After extracting the factors and corresponding question loadings of the DSQ, factor scores for the 13 recruited drivers were calculated accordingly. Relationships between the DSQ factor scores and characteristics of the measured driving behaviors in both congested and smooth traffic conditions were investigated using Pearson's correlations as shown in Table 4. Factor 1 (self-confidence in driving) is correlated significantly (negatively) with the idling share for congested and smooth conditions, which indicates more confident drivers are less likely to stand still; confidence level is also positively correlated with average speed, showing the

Table 4: Correlations between DSQ and measured driving behaviors.

Smooth Traffic	Average Speed	Acc Share	Dec Share	Constant Share	Idling Share	Average Acc	Average Dec	Energy Regeneration	Variation of Steering Wheel Angle
Factor 1	0.50	0.15	-0.11	0.26	-0.76**	0.19	-0.15	0.16	-0.15
Factor 2	-0.37	-0.27	0.47	-0.37	0.53	-0.03	0.18	-0.09	0.13
Factor 3	-0.27	0.16	0.48	-0.58*	0.08	0.28	0.18	0.30	0.59*
Factor 4	0.66*	-0.35	-0.27	0.63*	-0.23	-0.44	-0.33	-0.34	-0.46
Factor 5	0.07	0.26	0.47	-0.57*	-0.15	0.48	0.35	0.60*	0.40
Factor 6	0.14	0.04	0.14	0.01	-0.40	-0.07	0.22	0.06	-0.22
Factor 7	-0.17	-0.05	0.50	-0.40	0.03	0.15	0.30	0.25	0.51
Factor 8	0.19	-0.27	-0.30	0.46	0.09	-0.48	-0.29	-0.32	0.17
Congested Traffic	**Average Speed**	**Acc Share**	**Dec Share**	**Constant Share**	**Idling Share**	**Average Acc**	**Average Dec**	**Energy Regeneration**	**Variation of Steering Wheel Angle**
Factor 1	0.51	0.02	-0.03	0.26	-0.58*	0.17	0.15	0.18	-0.23
Factor 2	0.05	-0.05	0.50	-0.33	0.21	0.03	0.22	0.02	0.16
Factor 3	0.10	0.29	0.48	-0.38	-0.08	0.30	0.30	0.30	0.48
Factor 4	0.31	-0.36	-0.21	0.47	-0.36	-0.20	-0.19	-0.15	-0.55
Factor 5	0.12	0.60*	0.50	-0.60*	0.01	0.62*	0.30	0.57*	0.62*
Factor 6	0.61*	0.07	0.42	0.00	-0.58*	-0.01	0.23	0.24	-0.17
Factor 7	-0.01	0.39	0.43	-0.46	0.04	0.29	0.47	0.42	0.62*
Factor 8	-0.39	0.14	-0.37	0.01	0.25	0.02	0.02	-0.04	0.06

N=13
**. Correlation is significant at the 0.01 level (2-tailed).
*. Correlation is significant at the 0.05 level (2-tailed).

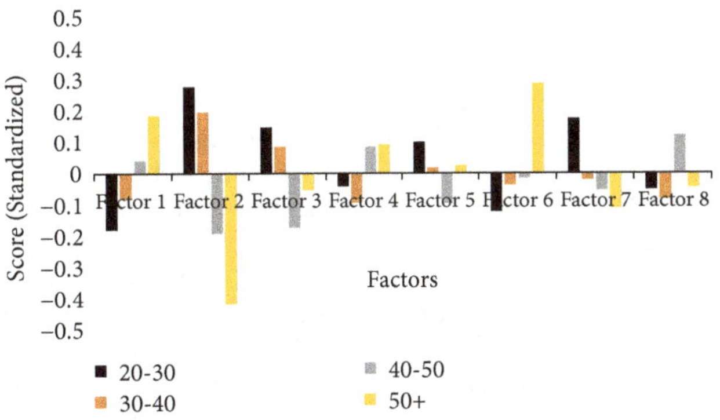

Figure 5: DSQ factor scores broken down by age groups.

speeding tendency among confident drivers. The constant speed share decreases with Factor 3 (impatience in driving) score, while the variation of steering wheel angle increases in smooth conditions, showing impatience potentially contributes to more speed variation and overtaking behavior. The pattern is similar in congested conditions, though less prominent due to the external constraint of traffic. Factor 4 (attachment to vehicles) is positively correlated with average speed and constant share in smooth conditions, showing people who are attached to vehicles tend to drive faster and also more stable. Factor 5 (intersection acceleration behavior) correlates negatively with constant speed share and positively with energy regeneration for both traffic conditions, and during congestion it shows positive correlations (0.60) with the average value and share of acceleration, as well as the variation of steering wheel angle, verifying the consistency and complementarity of these two data sources.

Further, we continued assessing the extent to which energy consumption is associated with DSQ factor scores and measured driving behavior. Pearson's correlations were calculated and the results are demonstrated accordingly in Figure 6. It can be concluded that generally the measured driving behaviors are correlated with energy consumption in the same direction regardless of the driving conditions, except idling share. This implies that this indicator (idling share) poses a bidirectional impact on energy consumption depending on external traffic conditions. During congestion, energy consumption is strongly correlated (>0.60, significant at the 5% level) with the average value and share of acceleration, and with variation of steering wheel angle (correlated with overtaking behavior). During smooth traffic conditions, energy consumption is only strongly correlated (>0.60, significant at the 5% level) with energy regeneration and variation of steering wheel angle. As for DSQ scores, Factor 3 (impatience in driving), Factor 5 (intersection acceleration behavior), and Factor 7 (rude driving) are strongly correlated (all significant at the 5% level) with energy consumption, but the pattern becomes less prominent in congested conditions. As a matter of fact, the correlations between energy consumption and all DSQ factor scores become weaker in congested conditions. It seems that the revealed driving

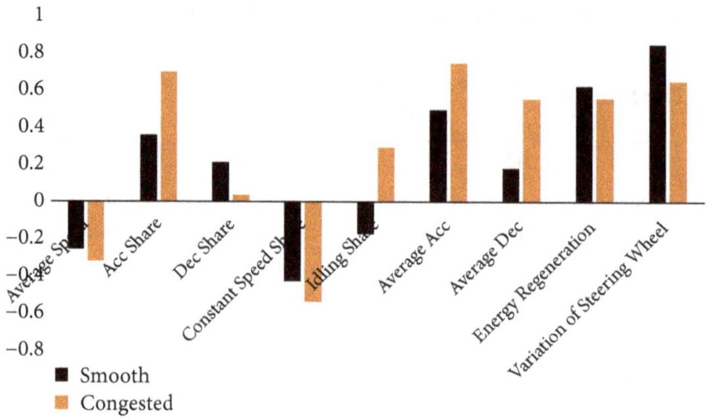

(a)

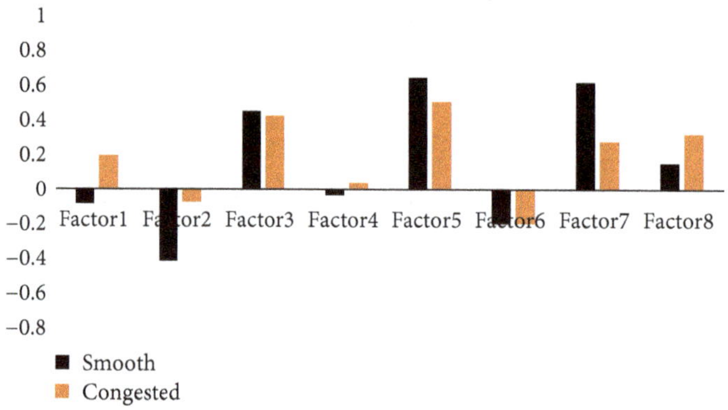

(b)

FIGURE 6: Correlations between journey energy consumption and DSQ (a), and measured driving behaviors (b).

parameters serve as better indicators of energy consumption during congested traffic conditions, while on the contrary DSQ scores perform better during smooth traffic conditions.

3.3. Multiple Regression Analysis. Using the correlation analysis conducted in Section 3.2, we adopted a concise regression formula to predict each driver's journey energy consumption for the urban driving condition based on field test data and DSQ scores. Compared with microscopic models for energy consumption developed by other researchers [31, 32], our macro model serves to predict the journey energy consumption from highly generalized input data, which also help to advance the understanding of the contribution of these two data sources. At the current stage we only consider linear relationship between the variables while further study might be able to elaborate a more complex model.

In order to ensure that all possible predictor variables were included and to prevent multicollinearity of the input variables, a stepwise multiple regression analysis was carried out in which all those variables that had shown significant bivariate correlations with energy consumption were entered as possible predictors. Each input variable needs to contribute significantly to the regression results according to F test otherwise it will be rejected, and multicollinearity is avoided in this process. Collinearity diagnostics are also used in later stage to double-check that there are no significant signs of multicollinearity of the final constructed models.

We have proposed three scenarios in terms of input variable combinations: (a) Scenario 1: measured driving behavior data only; (b) Scenario 2: DSQ factor scores only; and (c) Scenario 3: measured driving behavior combined with DSQ factor scores. The selected criteria for input variables are based on Adjusted R^2 in the stepwise regression process since it penalizes the statistic as extra variables enter into the model to prevent overfitting, while R^2 increases automatically when more variables are included. The best-performance regression models with the best Adjusted R^2 are listed as follows (see Table 5). In order to show the predictive ability of the regression model, Mean Absolute Percentage Error (MAPE) statistics are calculated from the leave-one-out cross-validation procedure (LOOCV). In LOOCV, the regression model is trained on all the data except for one point and a prediction is made for that point. MAPE is used to measure the deviation between the predicted and actual value.

If the driving behavior data was taken as input variables alone (Scenario 1), the model in smooth conditions outperforms that in congested conditions (MAPE 3.52% < MAPE 7.24%). When there is less traffic, a driver will display a more naturalistic and consistent driving style. Energy regeneration

TABLE 5: Performance of regression model.

		Best performance input indicators	MAPE	RMSE	Adjusted R^2	R^2
Smooth Traffic	Scenario 1: Driving behavior data	Energy Regeneration+ Variation of Steering Wheel	3.52%	4.68	0.72	0.77
	Coefficients	0.016, 0.85				
	Scenario 2: DSQ scores data	Factor 2+ Factor 3+ Factor 5+ Factor 6+ Factor 7	2.93%	3.44	0.85	0.91
	Coefficients	-4.34, 4.77, 1.94, -6.04, 6.68				
	Scenario 3: Driving behavior data +DSQ scores data	Variation of Steering Wheel+ Factor 2+Factor 5+ Factor 6+ Factor 7	2.57%	2.68	0.91	0.95
	Coefficients	0.56, -1.99, 2.20, -3.32, 4.55				
Congested Traffic	Scenario 1: Driving behavior data	Constant Speed Share+ Average Acc+ Average Dec+ Energy Regeneration+ Variation of Steering Wheel	7.24%	6.20	0.69	0.82
	Coefficients	-345.85, 67.42, 20.89, 0.07, 2.19				
	Scenario 2: DSQ scores data	Factor 1+ Factor 3+ Factor 5+ Factor 6+ Factor 7+ Factor 8	4.06%	5.92	0.72	0.86
	Coefficients	3.88, 6.79, 1.79, -9.97, 3.72, 6.52				
	Scenario 3: Driving behavior +DSQ scores data	Constant Speed Share+ Average Acc+ Variation of Steering Wheel+ Factor 3+ Factor 6+ Factor 8	3.17%	3.00	0.90	0.94
	Coefficients	-320.25, 53.88, 0.60, 5.84, -4.10, 4.85				

(more acceleration and deceleration behavior) and variation of steering wheel (more overtaking behavior) both increase energy consumption per kilometer. The situation is a bit more complicated in congested traffic, while average acceleration and deceleration (amplitude), constant speed share (frequency), energy regeneration, and variation of steering wheel jointly determine energy consumption per kilometer. Clearly, with only energy regeneration and variation of steering wheel data it is not enough to predict energy consumption in congested conditions.

The DSQ data alone (Scenario 2) also yields a better result in smooth conditions (MAPE 2.93% < MAPE 4.06%). Factor 3 (impatience), Factor 5 (intersection acceleration), Factor 6 (proactive driving), and Factor 7 (rude driving) appear in the final regression model in both traffic conditions with the same sign (only Factor 6, proactive driving, contributes negatively to energy consumption). Distraction appears in the smooth conditions as a negative input, which probably associates with familiarity with the road environment. And in the congested conditions, confidence and rigidity level serve as complements to the indicators mentioned above. More confident and rigid driving behavior seem to contribute negatively to energy efficiency in less smooth conditions. The MAPE and Adjusted R^2 denote that the driver's energy consumption is more predictable in Scenario 2 when the traffic is less congested.

The fusion of observed driving behavior data from the field test with the data from the DSQ scores (Scenario 3) yields the best result for the energy consumption prediction model in both traffic conditions (MAPE 2.57% vs MAPE 3.17%). In smooth conditions, variation of steering wheel together with Factor 2 (distraction in driving), Factor 5 (intersection acceleration), Factor 6 (proactive driving), and Factor 7 (rude driving) is able to build a quite accurate regression model. Energy regeneration and Factor 3 (impatience) which appear in Scenarios 1 and 2 are no longer necessary in the Scenario 3 model, indicating their information is somehow included in the final remaining indicators. This pattern is also seen in the model constructed in congested conditions, where average deceleration, energy regeneration, Factor 3 (impatience), and Factor 7 (rude driving) are excluded in the Scenario 3 final model. The remaining indicators of different dimensions are enough to characterize a subject's energy consumption.

This result makes us reflect on the representativeness of a traditional perfectly controlled driving test (on a simulator) and a fully naturalistic driving test (without controlling route and travel time). While DSQs alone might correlate well with simulated driving behaviors, the increased interaction with other traffic in real-world conditions confounds drivers' naturalistic driving styles, and therefore the energy consumption estimation requires the integration of different data sources.

While uncontrolled naturalist experiment results are not suitable for comparison study, the semi-naturalistic, real-world test in this paper compensates for the dynamic nature of the interplay between traffic conditions and person-specific driving styles, which is lost in fixed-condition traditional experiments (e.g., driving simulator). Our results imply that traditional DSQs can still add value to EV energy consumption estimation, completing the information not born in revealed driving behavior. The benefit is even more obvious considering that a DSQ is much more easily accessible than CAN-bus data. In this sense, potential EV drivers would be able to get an ex-ante assessment of their EV driving energy efficiency based on their DSQ results. In the near future, researchers and OEMs can make use of both data sources either to generate a person-specific ecodriving feedback or a user-friendly autonomous driving system that values personal driving styles.

4. Conclusions

This paper has innovatively designed a more "human-centric" experiment to explore how drivers' stated driving styles are reflected and correlated with their actual way of driving. The purpose of this study is to contribute to the research gap of the understanding of EV drivers' driving behaviors and energy efficiency through the integration of experimentation data and questionnaires.

While much of previous research has focused on safety implications of driving behaviors, in the current study we focus on energy efficiency, which has attracted increasing attention within the EV community. Some factors extracted from DSQs are strongly correlated with revealed driving behavior data regardless of external traffic conditions. For example, Factor 1 (self-confidence in driving) is negatively correlated with the idling share in the driving status distribution, and Factor 5 (intersection acceleration behavior) is positively correlated with the constant speed share in both traffic conditions.

During congestion, energy consumption is positively correlated with the average value and share of acceleration, and with variation of steering wheel angle. While in smooth conditions, energy consumption is only strongly correlated (>0.60, significant at the 5% level) with energy regeneration and variation of steering wheel angle. As for DSQ scores, Factor 3 (impatience in driving), Factor 5 (intersection acceleration behavior), and Factor 7 (rude driving) are strongly correlated (all significant at the 5% level) with energy consumption, although the pattern becomes less prominent in congested conditions. Thus, revealed driving parameters serve as better indicators of energy consumption during congested traffic conditions, while DSQ scores perform better during smooth traffic conditions.

Moreover, as demonstrated through our multiple regression results, the two complementary data sources, CAN-based field test data and questionnaires, could be combined to yield a more accurate energy consumption prediction model. Based on each driver's field test performance and DSQ scores, a personalized feedback for an ecodriving strategy depending on each driver's unique characteristics can be proposed for different driving scenarios. The framework of the present research opens up new ways that the field test data can be applied in real life. Additionally, our results show that traditional Driving Style Questionnaires could still be of value as more researchers now show interest in modeling the driving behavior of drivers. In the end, any predictor of EV driving energy that will not take into account the human driving style cannot be accurate. As a matter of fact, the prediction model of EV driving energy neglecting human driving style cannot be that accurate.

Nonetheless, it is also important to point out some limitations of this study: the number of participants (N=13) in the field test might not offer a rich enough variety of driving behaviors in comparison with our DSQ database (N=338). Additionally, care needs to be taken when deciphering our regression model since small sample size could potentially cause sampling errors. Further research could compensate for the defect by involving more drivers with sufficient demographic backgrounds as well as a wider range of EV models in the field tests.

In conclusion, this study has provided with preliminary evidence on the validity of the combinational strategy with both field test data and questionnaires in analyzing EV driving behavior, especially when focusing on the energy efficiency perspective. The relationship between self-stated driving style and observed driving behavior under different traffic conditions has been identified, and our understanding of EV users' energy efficiency in real-world driving scenarios has been advanced.

Appendix

Driving Style Questionnaire (Translated from Chinese)

Please fill in your basic information.

(1) Gender

______

(2) Age

______

(3) Do you have a driving license?

______

(4) How long have you had the driving license?

______

Please choose the answer that applies to you.

(1) I feel nervous when driving.

(A) No
(B) Seldom
(C) Sometimes
(D) Always

(2) I find it is easy to remain a proper distance with the front vehicle.

(A) No
(B) Seldom
(C) Sometimes
(D) Always

(3) I pay attention to the street scenery when driving.

(A) No
(B) Seldom
(C) Sometimes
(D) Always

(4) I talk to my passengers when driving.

(A) No
(B) Seldom

(C) Sometimes
(D) Always

(5) I accelerate when approaching and crossing a yellow light when there is only one vehicle in front and it is accelerating.

(A) No
(B) Seldom
(C) Sometimes
(D) Always

(6) I am worried about a traffic accident when driving.

(A) No
(B) Seldom
(C) Sometimes
(D) Always

(7) I change lanes during congestion.

(A) No
(B) Seldom
(C) Sometimes
(D) Always

(8) I follow the front vehicle closely in congested condition.

(A) Strongly disagree
(B) Disagree
(C) Agree
(D) Strongly agree

(9) I am skillful in backing and parking.

(A) Strongly disagree
(B) Disagree
(C) Agree
(D) Strongly agree

(10) I maintain a close distance to the vehicle in front.

(A) No
(B) Seldom
(C) Sometimes
(D) Always

(11) I feel annoyed when following heavy-duty vehicles like buses and vans.

(A) No
(B) Seldom
(C) Sometimes
(D) Always

(12) I adjust my vehicle in advance when entering into a queue.

(A) No
(B) Seldom
(C) Sometimes
(D) Always

(13) I feel uncomfortable when others want to cut into my lane.

(A) No
(B) Seldom
(C) Sometimes
(D) Always

(14) I can drive normally in harsh weather (windy, foggy, rainy, and snowy).

(A) No
(B) Seldom
(C) Sometimes
(D) Always

(15) I often clean my own vehicle.

(A) Strongly disagree
(B) Disagree
(C) Agree
(D) Strongly agree

(16) I accelerate when approaching and crossing a yellow light when there is no vehicle in front.

(A) No
(B) Seldom
(C) Sometimes
(D) Always

(17) I give way to other vehicles if they merge into my lane.

(A) No
(B) Seldom
(C) Sometimes
(D) Always

(18) I honk the horn to the vehicle in front in congestion.

(A) No
(B) Seldom
(C) Sometimes
(D) Always

(19) I check my phone/iPad during the red traffic signal period.

(A) No
(B) Seldom
(C) Sometimes
(D) Always

(20) I pay attention to my mobile phone during driving.

(A) No
(B) Seldom
(C) Sometimes
(D) Always

(21) I enjoy driving.

(A) No
(B) Seldom
(C) Sometimes
(D) Always

(22) I can repair a vehicle by myself.

(A) Strongly disagree
(B) Disagree
(C) Agree
(D) Strongly agree

(23) I drive fast.

(A) No
(B) Seldom
(C) Sometimes
(D) Always

(24) I brake hard when I decelerate

(A) No
(B) Seldom
(C) Sometimes
(D) Always

(25) I can respond reasonably to unexpected events.

(A) Strongly disagree
(B) Disagree
(C) Agree
(D) Strongly agree

(26) I obey and have faith in traffic rules.

(A) Strongly disagree
(B) Disagree
(C) Agree
(D) Strongly agree

(27) I can respond reasonably to unexpected events.

(A) Strongly disagree
(B) Disagree
(C) Agree
(D) Strongly agree

(28) I observe the traffic light patterns from a distance.

(A) Strongly disagree
(B) Disagree
(C) Agree
(D) Strongly agree

(29) I change to the expected lane early in preparation for the next intersection.

(A) No
(B) Seldom
(C) Sometimes
(D) Always

(30) I can drive safely under complicated traffic conditions.

(A) Strongly disagree
(B) Disagree
(C) Agree
(D) Strongly agree

(31) I pay attention to fuel consumption during driving.

(A) Strongly disagree
(B) Disagree
(C) Agree
(D) Strongly agree

(32) I have confidence in teaching newcomers to drive.

(A) Strongly disagree
(B) Disagree
(C) Agree
(D) Strongly agree

(33) I answer the phone when driving.

(A) No
(B) Seldom
(C) Sometimes
(D) Always

(34) I watch car racing TV shows.

(A) No
(B) Seldom
(C) Sometimes
(D) Always

(35) I use the vehicle even when there are other alternatives.

(A) No
(B) Seldom
(C) Sometimes
(D) Always

(36) I stick to my accustomed route even the road is congested.

(A) No
(B) Seldom
(C) Sometimes
(D) Always

(37) I concentrate during driving and always ignore what others say.

(A) No
(B) Seldom
(C) Sometimes
(D) Always

(38) I feel upset during congestion.

(A) No
(B) Seldom
(C) Sometimes
(D) Always

(39) You pay attention to the vehicles nearby when following the front vehicle.

(A) No
(B) Seldom
(C) Sometimes
(D) Always

(40) I have faith in myself to drive steadily.

(A) No
(B) Seldom
(C) Sometimes
(D) Always

Conflicts of Interest

The authors declare that there are no conflicts of interest regarding the publication of this article.

Acknowledgments

This research is supported in part by Tsinghua University Initiative Scientific Research Program (2015THZ01), National Natural Science Foundation of China (U1509205), and NSFC-Zhejiang Joint Fund for the Integration of Industrialization and Informatization (U1709212). The authors would like to acknowledge the support from Nissan (China) Investment Co., LTD, which kindly provided the test vehicle for this study. They also thank Dr. Geqi Qi from Beijing Jiaotong University for his assistance in DSQ data collection and also the EV drivers for their participation in this study. The first author would like to thank Dr. Tim Schwanen from Transport Studies Unit, University of Oxford, for his helpful insight in the early stage of this paper. The first author would also like to thank the China Scholarship Council, University of Oxford, and Tsinghua University for their financial support.

References

[1] M. Dijk, R. J. Orsato, and R. Kemp, “The emergence of an electric mobility trajectory,” *Energy Policy*, vol. 52, pp. 135–145, 2013.

[2] B. Singh and A. H. Strømman, “Environmental assessment of electrification of road transport in Norway: Scenarios and impacts,” *Transportation Research Part D: Transport and Environment*, vol. 25, pp. 106–111, 2013.

[3] T. Trigg, P. Telleen, R. Boyd et al., “Global EV Outlook 2016 Electric Vehicles Initiative,” *International Energy Agency*, pp. 1–40, 2013.

[4] K. B. Naceur and J. F. Gagné, “Global EV outlook 2017,” *International Energy Agency*, pp. 27–30, 2016.

[5] A. Lundström, C. Bogdan, F. Kis, I. Olsson, and L. Fahlén, “Enough power to move: Dimensions for representing energy availability,” in *Proceedings of the 2012 14th ACM International Conference on Human-Computer Interaction with Mobile Devices and Services, MobileHCI'12*, pp. 201–210, USA, September 2012.

[6] R. C. McIlroy, N. A. Stanton, and C. Harvey, “Getting drivers to do the right thing: A review of the potential for safely reducing energy consumption through design,” *IET Intelligent Transport Systems*, vol. 8, no. 4, pp. 388–397, 2014.

[7] M. Sivak and B. Schoettle, “Eco-driving: strategic, tactical, and operational decisions of the driver that influence vehicle fuel economy,” *Transport Policy*, vol. 22, pp. 96–99, 2012.

[8] D. L. Greene and S. E. Plotkin, *Reducing greenhouse gas emission from US transportation*, Pew Center on Global Climate Change, Arlington, Texas, USA, 2011.

[9] R. Hjorthol, “Attitudes, ownership and use of Electric Vehicles-a review of literature,” TOI Report, 2013.

[10] M. Helmbrecht, C. Olaverri-Monreal, K. Bengler, R. Vilimek, and A. Keinath, “How electric vehicles affect driving behavioral patterns,” *IEEE Intelligent Transportation Systems Magazine*, vol. 6, no. 3, pp. 22–32, 2014.

[11] K. Hu, J. Wu, and T. Schwanen, “Differences in energy consumption in electric vehicles: An exploratory real-world study in Beijing,” *Journal of Advanced Transportation*, vol. 2017, Article ID 4695975, 17 pages, 2017.

[12] C. Marina Martinez, M. Heucke, F.-Y. Wang, B. Gao, and D. Cao, “Driving Style Recognition for Intelligent Vehicle Control and Advanced Driver Assistance: A Survey,” *IEEE Transactions on Intelligent Transportation Systems*, vol. 19, no. 3, pp. 666–676, 2018.

[13] K. Boriboonsomsin, A. Vu, and M. Barth, *Eco-driving: pilot evaluation of driving behavior changes among us drivers*, University of California Transportation Center, 2010.

[14] S. Carroll and C. Walsh, “UK Electric Vehicle Case Studies , SAE Technical Paper,” Tech. Rep., 2011.

[15] C. Bingham, C. Walsh, and S. Carroll, “Impact of driving characteristics on electric vehicle energy consumption and range,” *IET Intelligent Transport Systems*, vol. 6, no. 1, pp. 29–35, 2012.

[16] M. Knowles, H. Scott, and D. Baglee, "The effect of driving style on electric vehicle performance, economy and perception," *International Journal of Electric and Hybrid Vehicles*, vol. 4, no. 3, pp. 228–247, 2012.

[17] C.-H. Lee and C.-H. Wu, "A Novel Big Data Modeling Method for Improving Driving Range Estimation of EVs," *IEEE Access*, vol. 3, pp. 1980–1993, 2015.

[18] W. Li, P. Stanula, P. Egede, S. Kara, and C. Herrmann, "Determining the main factors influencing the energy consumption of electric vehicles in the usage phase," in *Proceedings of the 23rd CIRP Conference on Life Cycle Engineering, LCE' 16*, pp. 352–357, Berlin, Germany, May 2016.

[19] H. Yu, F. Tseng, and R. McGee, "Driving pattern identification for EV range estimation," in *Proceedings of the 2012 IEEE International Electric Vehicle Conference, IEVC 2012*, pp. 1–4, 2012.

[20] E. Yao, Z. Yang, Y. Song, and T. Zuo, "Comparison of electric vehicles energy consumption factors for different road types," *Discrete Dynamics in Nature and Society*, vol. 2013, Article ID 328757, 7 pages, 2013.

[21] T. Franke, F. Bühler, P. Cocron, I. Neumann, J. F. Krems, and F. Bühler, "Enhancing sustainability of electric vehicles: A field study approach to understanding user acceptance and behavior," in *Advances in Traffic Psychology*, vol. 2012, pp. 295–306, 2012.

[22] P. Cocron, F. Bühler, I. Neumann et al., "Methods of evaluating electric vehicles from a user's perspective - The MINI e field trial in Berlin," *IET Intelligent Transport Systems*, vol. 5, no. 2, pp. 127–133, 2011.

[23] M. F. Jung, D. Sirkin, T. M. Gür, and M. Steinert, "Displayed uncertainty improves driving experience and behavior: The case of range anxiety in an electric car," in *Proceedings of the 33rd Annual CHI Conference on Human Factors in Computing Systems, CHI '15*, pp. 2201–2210, Seoul, Korea, April 2015.

[24] S. Helman and N. Reed, "Validation of the driver behaviour questionnaire using behavioural data from an instrumented vehicle and high-fidelity driving simulator," *Accident Analysis & Prevention*, vol. 75, pp. 245–251, 2015.

[25] T. Franke, M. Günther, M. Trantow et al., "Examining user-range interaction in battery electric vehiclesa field study approach," *Advances in Human Aspects of Transportation Part II*, pp. 334–344, 2014.

[26] Y. Xing, G. Tal, Y. Wang et al., "A Comparison of Plug-in Electric Vehicle Markets Between China and the U.S. Based on Surveys," *Blue Book of New Energy Vehicles*, pp. 314–338, 2016.

[27] O. Taubman-Ben-Ari, M. Mikulincer, and O. Gillath, "The multidimensional driving style inventory - Scale construct and validation," *Accident Analysis & Prevention*, vol. 36, no. 3, pp. 323–332, 2004.

[28] M. Ishibashi, M. Okuwa, S. Doi, and M. Akamatsu, "Indices for characterizing driving style and their relevance to car following behavior," in *Proceedings of the SICE(Society of Instrument and Control Engineers)Annual Conference, SICE 2007*, pp. 1132–1137, Japan, September 2007.

[29] H. F. Kaiser, "An index of factorial simplicity," *Psychometrika*, vol. 39, no. 1, pp. 31–36, 1974.

[30] T. Lee and J. Son, "Relationships between driving style and fuel consumption in highway driving," *SAE Technical Papers*, 2011.

[31] R. Zhang and E. Yao, "Electric vehicles' energy consumption estimation with real driving condition data," *Transportation Research Part D: Transport and Environment*, vol. 41, pp. 177–187, 2015.

[32] X. Wu, D. Freese, A. Cabrera, and W. A. Kitch, "Electric vehicles' energy consumption measurement and estimation," *Transportation Research Part D: Transport and Environment*, vol. 34, pp. 52–67, 2015.

2

Locating Station of One-Way Carsharing based on Spatial Demand Characteristics

Xiaohong Chen,[1] Jiaqi Cheng,[1] Jianhong Ye,[1] Yong Jin,[2] Xi Li,[2] and Fei Zhang[2]

[1]*Key Laboratory of Road and Traffic Engineering of the Ministry of Education, Tongji University, 4800 Cao'an Road, Shanghai 201804, China*
[2]*Global Carsharing & Rental Co., Ltd., 888 South Moyu Road, Shanghai 201805, China*

Correspondence should be addressed to Jianhong Ye; yjh1875@hotmail.com

Academic Editor: Emanuele Crisostomi

This study aims to explore factors that affect carsharing demand characteristics in different time periods based on EVCARD transaction data, which is the largest station-based one-way carsharing program in Shanghai, China. Monthly usage intensity and degree of usage imbalance are used as proxies of demand. This study uses three groups of independent variables: carsharing station attributes, built environment (density, diversity, design, and destination accessibility), and transportation facilities. The adaptive elastic net regression is developed to identify factors that influence carsharing usage intensity and degree of usage imbalance after factor selection using extra-randomized-tree algorithm. Finally, a station layout is proposed according to both usage intensity and degree of imbalance. The main results of this study are presented as follows: (1) different effects of built environment and transportation factors cause dynamic demand across different time periods; (2) factors with positive and negative effect on carsharing demand are divided clearly for guidance of the carsharing station layout; (3) public parking space leads to more personal vehicle trip compared to a carsharing trip; and (4) as public transportation, the relationship of the metro and carsharing is complementary. However, the bus stop and carsharing have a competitive relationship. This study provides a carsharing layout method based on both usage intensity and degree of imbalance. Furthermore, several policies concerning carsharing are proposed.

1. Introduction

Carsharing service allows users to avoid the money and time cost of car ownership, such as maintenance, insurance, and finding parking space. It offers more flexible car travel [1] and has been promptly developed in recent years. Many previous studies have shown that carsharing service can effectively reduce vehicle ownership [2, 3], reduce the vehicle miles traveled [4, 5], and control the greenhouse effect [6, 7], particularly when some carsharing uses electric vehicles [8]. Carsharing can improve social equality and provide a convenient trip option to people who cannot afford a car [9]. These benefits for the society rely on the scale of the carsharing program and the program sustainability, which are dependent on the profit and cost situation of carsharing.

The mode of operation is primarily considered based on profit and cost from the demand feature situations. Carsharing has two operation categories, namely, station-based carsharing (SBC) and free-floating carsharing (FFC). Moreover, there are two types of SBC. The first type requires returning the rented vehicle to the same station, which is called round-trip carsharing (RTC). The second type allows users to drop off the vehicle at any station after finishing their trip, which is called one-way carsharing (OWC). However, FFC allows users to pick up and drop off the vehicle at any location within a specified service area.

In RTC, users eventually drop off vehicles to the pick-up station, which does not require dispatch in normal operation. OWC improves the flexibility of using the vehicle compared to RTC while bringing extra operational burden. Given that dropping off at another station is allowed, there is a high probability of a spatial imbalance over the stations. The operator needs to arrange the vehicle relocation to maintain regular operation, which results in increased operation cost.

OWC and FFC allow one-way trip, which creates the problem of spatiotemporal imbalance of demand. This problem can be handled using different approaches. FFC can only maintain system balance by dispatching techniques. However, OWC is a station-based system; therefore, it can mitigate the problem of demand imbalance by appropriate location of stations. Supposing that there is a more demand-balanced area, the pick-up and drop-off at a station can be counteracted during a given temporal interval. Therefore, the layout of stations in the more balanced areas can strategically reduce vehicle relocation. In the long term, it can reduce dispatching cost. Therefore, it is essential to find a small-scale area with more balanced demand in OWC. However, the authors found that studies focusing on this issue are rare. Thus, this research will address this gap in literature.

Given that station usage intensity is related to the operator's operational profits and station usage imbalance is related to operational cost, finding areas with higher carsharing usage intensity but lower degree of station usage imbalance is significantly important to the station-based OWC program. Therefore, this research uses station usage intensity and degree of station usage imbalance as proxies to represent the demand characteristics to explore how factors affect the carsharing demand characteristics in different time period of the day.

Section 2 reviews the literature related to the identification of factors affecting the demand characteristics in the carsharing field. Section 3 describes the data profile and develops the monthly usage intensity model and usage imbalance model. Section 4 presents the result of the models and then discusses the factors affecting the demand characteristics of carsharing and provides result of station layout. Finally, Section 5 provides a summary and the limitations of this study.

2. Literature Review

The research on carsharing has increased rapidly since 2010 [10]. The current document mainly contains seven subfields: the users' attitude and behavior, demand analysis, depot location, vehicle relocation, benefit evaluation, business models, and policy. Two issues are included in the demand analysis: (1) user characteristics or potential demand and (2) identification of factors that influence the current demand. This work aims to address the second issue.

Table 1 shows the current references based on analysis units, methods, source of variable data, dependent variables, and the division stage according to the operating model.

The current references mainly focus on RTC [11–16] and FFC [17–22], but the study on OWC is insufficient. In basic analysis unit, the user is always selected for the RTC model [11–13]. The spatial scales of a single station, the station cluster to land parcel (including grid, census tracts, transportation analysis zone, and district), are in increasing order [14–24]. The smaller the spatial scale of the analysis unit is, the greater the model can reflect the demand characteristics of the local space. However, the negative effect is the larger proportion of unexplained variance [23]. When the amount of data can ensure that the proportion of random change cannot obviously affect the model fitting, the analysis should be conducted using smaller space scale. The study using vehicle as an analysis unit mainly focuses on analyzing the usage rate of different types of vehicles [9].

Given the privacy protection and business sensitivity, early researchers cannot obtain operational data but discontinuously collect some sparse data [14, 15]. These research studies assume that operators can dispatch a vehicle fully according to demand, where the vehicle number within the space unit can be used as proxy of the demand in that spatial unit [23]. Studies based on transaction data and operation record log increased. However, given that the demand is restrained by insufficient supply, the study on the demand can only be conducted using actual usage as proxy of the demand, which is more reliable than studies using supply as proxy for demand.

Dependent variables can be divided into usage frequency and usage duration, which are common dimensions of demand characteristics. Most studies take usage frequency and its various variants (such as vehicle usage rate) as dependent variables. These studies take booking quantity as usage frequency [9, 11–13, 15, 16, 19, 21] but do not consider pick-up and drop-off mainly because the pick-up and drop-off of RTC are in the same station. On the contrary, there is no station in FFC. In OWC, the station not only generates carsharing trip but also is attracting it as well. Completing a carsharing trip based on two stations requires a separate pick-up and drop-off. In addition, vehicles of RTC belong to one station; therefore, the effects of supply on demand can be eliminated by controlling the variable of the vehicle number. The vehicle usage rate can be used as dependent variable [14]. Without station in FFC, one method to obtain usage frequency is by calculating the bookings in area unit, which makes booking density the most common variable in FFC [17, 20–22] including usage duration. However, no study has been found on station usage imbalance.

Only few research studies use a simple method of descriptive statistics [11] similar to the method used in this study. Most studies directly use multiple linear models [12, 14–17, 19, 23, 24], while other studies exchange natural logarithms for dependent variables to make them in normal distribution [23]. Given that dependent variables meet with other exponential family distribution, the generalized linear model is used [12]. Moreover, given that an independent variable has large zero value, the zero-inflated model [13, 20] is used. The Dynamic Ordered Probit Model is used because decisions of multiple frequencies of usage are cumulatively related to each other [13].

In summary, the research on station-based OWC is rare, and this work will focus on this operational mode. Moreover, for the dependent variable, the station usage intensity used in this study comprehensively considers the demands for pick-up and drop-off. However, the degree of station usage imbalance has not yet been studied. Finally, the analysis of demand characteristics across different time period is not sufficient in the current literature. In this study, time is divided into several parts, and each time section of the day is modeled individually.

TABLE 1: Summary of document on identification of factors affecting demand characteristics.

Item	Category	Literature
Operational mode	RTC	[11–16]
	FFC	[17–22]
	NEM (not explicitly mentioned)	[9, 23, 24]
Basic analysis unit	User (RTC)	[11–13]
	Single station (RTC)	[14]
	Station cluster (NEM)	[23]
	Location (FFC)	[19, 20]
	Land parcel (RTC, FFC, and NEM)	[15–18, 21, 22, 24]
	Vehicle (NEM)	[9]
Data sources of dependent variable	Collected by authors	[14, 15]
	Transaction data or operation log	[9, 11–13, 16–24]
Dependent variable	Usage frequency [RTC and FFC]	[11–13, 16, 19, 21]
	Total vehicle hour travel [RTC]	[12]
	Monthly hours per vehicle [RTC]	[14]
	Vehicle usage rate [RTC and NEM]	[9, 15]
	Booking density [FFC]	[17, 20–22]
	Number of vehicles [NEM]	[24]
	Vehicle unused duration [FFC]	[18]
	Usage hours of station [NEM]	[23]
Method	Descriptive statistics	[11]
	Generalized linear model	[9, 12, 20, 21]
	Multiple linear model	[12, 14–17, 19, 23, 24]
	Dynamic ordered probit model	[13]
	Duration model technique	[18]
	Hot spots by ArcGIS	[22]
Develop individually models for several time periods of the day		[15, 21, 22]

3. Data and Methodology

3.1. Data Preparation. The four data sources used in this study are shown in Figure 3. The first data source is provided by the Shanghai EVCARD, which includes station information and transaction data starting from January 2015 to December 2016. EVCARD is the largest carsharing program in China. It is a station-based one-way system, which allows users to drop off car at any station after a trip. All vehicles in operation are Pure Battery Electric Vehicles (PBEVs), and the price is only based on the trip duration. The usage procedure is completely self-service. Station information includes station coordinates, station age (number of months from the start of operation until December 2016), parking space, if a station is located at an underground garage (this is related to whether or not the station is easy to find), and if a station has limited accessibility because it is located at an internal location of companies or government department where nonemployees are not allowed to enter for a specific time period or permanently. Other stations nearby can be calculated based on the coordinates data using Geographic Information System (GIS) software. A station includes several parking lots (from two to dozens). Transaction data includes membership number, vehicle license plate number, time and station name for pick-up and drop-off, trip duration, fee, deductions of the booking fee, and the reason for the reduction. Given that the data are commercially sensitive, these data are only permitted to display directly a limited part of the operator request.

The second data source is composed of Point of Interest (POI) data in Shanghai, which was collected through AMAP API in November 2016. AMAP is a Google-Maps-like E-map, which is popular and widely used in China. The total collected POIs reached 427210, which are categorized into eleven types based on land use attributes: residential area, culture places, business, medical service, public authority, social and recreation, shopping, college and university, industry, transportation, and intersections. Among these types, college and university is used instead of a larger type of education because, in addition to being found in preresearch, other educational areas, such as middle school, primary school, kindergarten, and training institution, have no statistically significant effect on the demand characteristics. However, college and university plays an important role; therefore, it replaces the educational type. The transportation type is very important; therefore, it is separately categorized into more specific subtypes, including airport, train station, intercity coach station, metro station, bus stop, and traditional car rental station. Intersection POIs are used to represent the street design. Moreover, POI-mixed entropy is calculated based on POIs, which refers to the entropy form of research [25].

The third data source is part of the latest census data of Shanghai, which was collected in 2010 by the Shanghai Municipal Bureau of Statistics. Resident density, job density, and average trip distance of each block were used in this research. Average trip distance is used to represent the accessibility of a destination.

Finally, the fourth data source is the information of road network in Shanghai, which includes freeway, arterial road, secondary road, local street, one-way road, and two-way road. We counted the length of each type of road in spatial unit to represent characteristics of local road network.

The four data sources are reorganized into three groups of features. The first feature is the station's attributes (S). The second feature is the transportation facilities. The number of each type of transportation facilities located within the buffer of the carsharing station is used to present the transportation environment within the spatial unit. The third group of features refers to the 5D principle of the built environment [26, 27]. However, the fifth "D" (distance to transit) is presented as the distance between a residential place or work place to the nearest public transportation station, which is not appropriate for this research. Therefore, we use the transportation facilities instead of the "distance to transit." Thus, independent variables shall be categorized and grouped as "$S+T+4D$." The descriptive statistics of features and targets are summarized in Table 2.

For sample selection, stable operational stations are selected in this study. Many stations have short operation period with data that ended on 10 December 2016. These stations are not well known to users; therefore, they are not included in this research because of their unstable performance. Data indicate that the use of vehicles becomes relatively stable when the station is operated into the fourth month (the booking changing rate is about 0.5. The booking changing rate reaches 7.1 in the second month and 0.78 in the third month). In this research, we only include stations with more than three months of operation, while the booking changing rate is less than 0.3 in three consecutive months from October to December 2016. These stations are judged as relatively stable. Eventually, 551 stations participated in this research. Figure 1 shows the spatial distribution of all stations operating in Shanghai as well as the stations participating in this research. The proportion of stations participating in this research is the same as the proportion of the whole station at each area. The central areas have less than 4% stations and parking spaces, while 95% of the stations are located at the outskirts area. We divided Shanghai into a hexagonal grid, where the area of the hexagonal spatial unit is 2 km^2 (similar to a circle with 800-meter radius in the area). The stations participating in this research are assigned to this hexagon unit based on spatial location. The hexagonal spatial unit is used as an analysis unit in this research.

Since the EVCARD station, as basic object in this research, is selected according to the fact that bookings changing rate is relatively stable during the three months from November to December 2016, the dependent variable in this research also requires the relative stability of station operation performance, where the collecting time of POI data shall be within this time range. Therefore, only the three-month transaction data from October to December 2016 is investigated, which covers about 40% of the total bookings.

3.1.1. Temporal Dynamics of Demand. To investigate the temporal dynamics of demand, the data is divided into two subsets: workday and nonworkday. In China, given that adjustment is made when national holidays overlap weekdays, some weekends become workdays and some workdays become holidays. Therefore, workday and nonworkday are used to separate time instead of weekday and weekend.

Based on these conditions, the two subsets of data are divided into five time sections of the day, which are as follows according to the trip temporal distribution of the EVCARD (Figure 2): early morning (EM) (2:00–6:00, 4 h), morning peak (MP) (6:00–10:00, 4 h), off-peak (OP) (10:00–16:00, 6 h), evening peak (EP) (16:00–20:00, 4 h), and night (NT) (20:00–2:00, 6 h). Given the few bookings in EM which are quite random, the later four time sections are included in this research. Finally, eight time sections were obtained, and these time sections are mathematically expressed as follows:

$$T = \left\{ T_a^b \mid a \in A, b \in B \right\}, \tag{1}$$

where A = {workday, nonworkday} and B = {MP, OP, EP, NT}.

3.2. Methodology. Two dependent variables are used as proxies representing demand: monthly usage intensity and degree of usage imbalance. Two machine-learning models are established in this research. The monthly usage intensity model is used to evaluate the effect of factors influencing usage intensity of exclusive parking space in the analysis spatial units per month. The usage imbalance model is developed to estimate the factors affecting the degree of imbalance of pick-up and drop-off at spatial units. The research framework is shown in Figure 3.

3.2.1. Features Selection. Extremely randomized trees algorithm (ET) is used to select the important features to avoid overfitting. The extra-trees algorithm builds an ensemble regression trees. Its two main differences from other tree-based ensemble methods, such as random forest [28], are that it divides the nodes by choosing the cut-points completely at random and that it uses the complete learning sample instead of a bootstrap replica to grow the trees. The complete extra-trees algorithm is described in [29]. Relative variance reduction is used to denote the goodness of point splitting. For a sample S and a splitting point s (a feature selected randomly), the goodness of point splitting is expressed as follows:

$$G(s,S) = \frac{\operatorname{var}\{y \mid S\} - (|S_l| / |S|) \operatorname{var}\{y \mid S_l\} - (|S_r| / |S|) \operatorname{var}\{y \mid S_r\}}{\operatorname{var}\{y \mid S\}}, \tag{2}$$

where var$\{y \mid S\}$ is the mean squared error of output y in the sample S. S_l and S_r are two subsets of sample S.

Table 2: Descriptive statistics of variables.

Variable type	Variable name	Abbr.	Variable type	Mean	Std.	Min.	Med.	Max.
Usage intensity	Workday 06–09 h	I_W0609	Numerical	190.13	260.16	0	105	2879
	Workday 10–15 h	I_W1015	Numerical	271.26	419.04	2	152	5287
	Workday 16–19 h	I_W1619	Numerical	280.09	480.29	2	140	6214
	Workday 20–01 h	I_W2001	Numerical	285.28	430.63	0	114	3987
	Nonworkday 06–09 h	I_NW0609	Numerical	61.13	86.56	0	32	948
	Nonworkday 10–15 h	I_NW1015	Numerical	149.21	231.39	0	80	2962
	Nonworkday 16–19 h	I_NW1619	Numerical	111.31	210.55	0	54	2925
	Nonworkday 20–01 h	I_NW2001	Numerical	110.7	173.35	0	44	1860
Usage imbalance	Workday 06–09 h	IB_W0609	Numerical	0.52	0.18	0.34	0.78	1
	Workday 10–15 h	IB_W1015	Numerical	0.46	0.26	0.31	0.71	1
	Workday 16–19 h	IB_W1619	Numerical	0.47	0.18	0.28	0.72	1
	Workday 20–01 h	IB_W2001	Numerical	0.49	0.22	0.32	0.75	1
	Nonworkday 6–09 h	IB_NW0609	Numerical	0.48	0.23	0	0.73	1
	Nonworkday 10–15 h	IB_NW1015	Numerical	0.41	0.20	0.19	0.65	1
	Nonworkday 16–19 h	IB_NW1619	Numerical	0.43	0.22	0.27	0.68	1
	Nonworkday 20–01 h	IB_NW2001	Numerical	0.45	0.27	0	0.72	1
Operational area attributes	Operational age	OA	Numerical	9.82	4.01	3	10	44
	Limited-access parking space	LA	Numerical	1.02	3.28	0	1	8
	Underground parking space	UG	Numerical	0.96	2.14	0	0	6
	Exclusive parking space	EPS	Numerical	4.73	1.83	0	4	20
Transportation	Metro station	MS	Binary	0.15	0.24	0	0	1
	Bus stop	BS	Numerical	10.28	4.96	0	9	44
	Car rental	CR	Numerical	0.63	0.84	0	0	9
	Intercity coach	IC	Binary	0.08	0.28	0	0	1
	Train station	TS	Binary	0.06	0.31	0	0	1
Built environment	Residential	RS	Numerical	35.27	42.08	0	15	484
	Public authority	PA	Numerical	8.93	11.58	0	2	133
	Medical hygiene	MH	Numerical	8.47	13.37	0	2	235
	Recreation and social	RC	Numerical	149.32	161.33	0	53	1697
	Culture	CU	Numerical	3.21	4.11	0	1	53
	Business	BU	Numerical	28.6	32.69	0	12	471
	University and college	UN	Binary	0.12	0.22	0	0	1
	Industry	IN	Numerical	3.88	2.79	0	3	30
	Public parking	PP	Numerical	43.19	39.9	0	20	454
	Shopping	SH	Numerical	17.67	14.16	0	10	126
	Average trip distance	ATD	Numerical	9.23	3.73	2.37	7.93	61.01
	Intersection density	ID	Numerical	23.86	8.45	7.22	21.98	69.47
	POI-mixed entropy	PME	Numerical	0.78	0.1	0.193	0.82	0.98
	Freeway (meter)	FW	Numerical	486.61	2049.96	0	0	15500.97
	Arterial road (meter)	AR	Numerical	740.64	1356.05	0	0	6557.62
	Secondary road (meter)	SR	Numerical	620.74	1104.11	0	0	6390.57
	Local street (meter)	LS	Numerical	5611.37	3894.72	0	5760.07	16209.97
	One-way road (meter)	OR	Numerical	3577.73	3708.48	0	2640.8	23501.03
	Two-way road (meter)	TR	Numerical	6657.71	3077.18	19.84	6418.64	17526.87

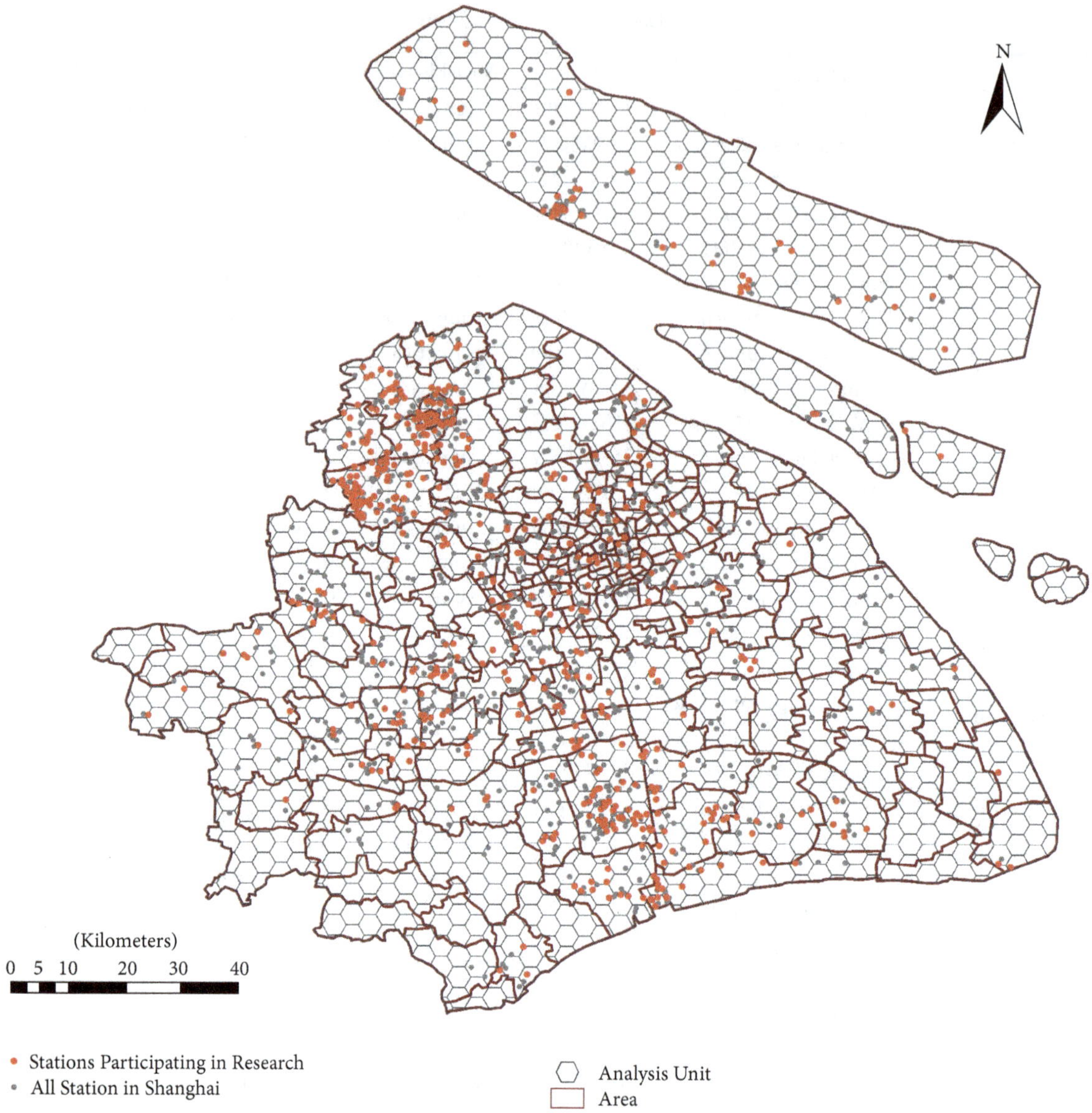

FIGURE 1: Spatial distribution of stations in Shanghai.

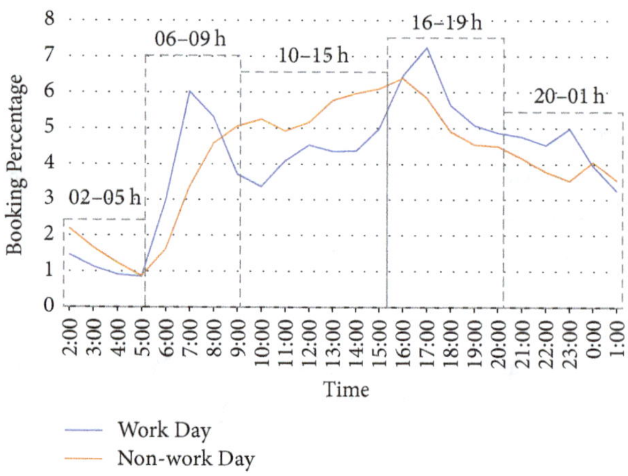

FIGURE 2: EVCARD trip temporal distribution and time division.

Let $G^i(s, S)$ be the goodness of ith ET; the feature importance is the average goodness of each ET, which can be expressed as follows:

$$\mathrm{FI} = \frac{1}{N}\sum_{i=1}^{N} G^i(s, S), \tag{3}$$

where FI denotes the feature importance and N is the number of ET.

Unimportant features have little to no effect on the mean squared error model, while important features should significantly decrease it.

3.2.2. Monthly Usage Intensity Model and Usage Imbalance Model. Considering the unequal duration of time sections, τ_T can be the duration of each T, where the unit is hour.

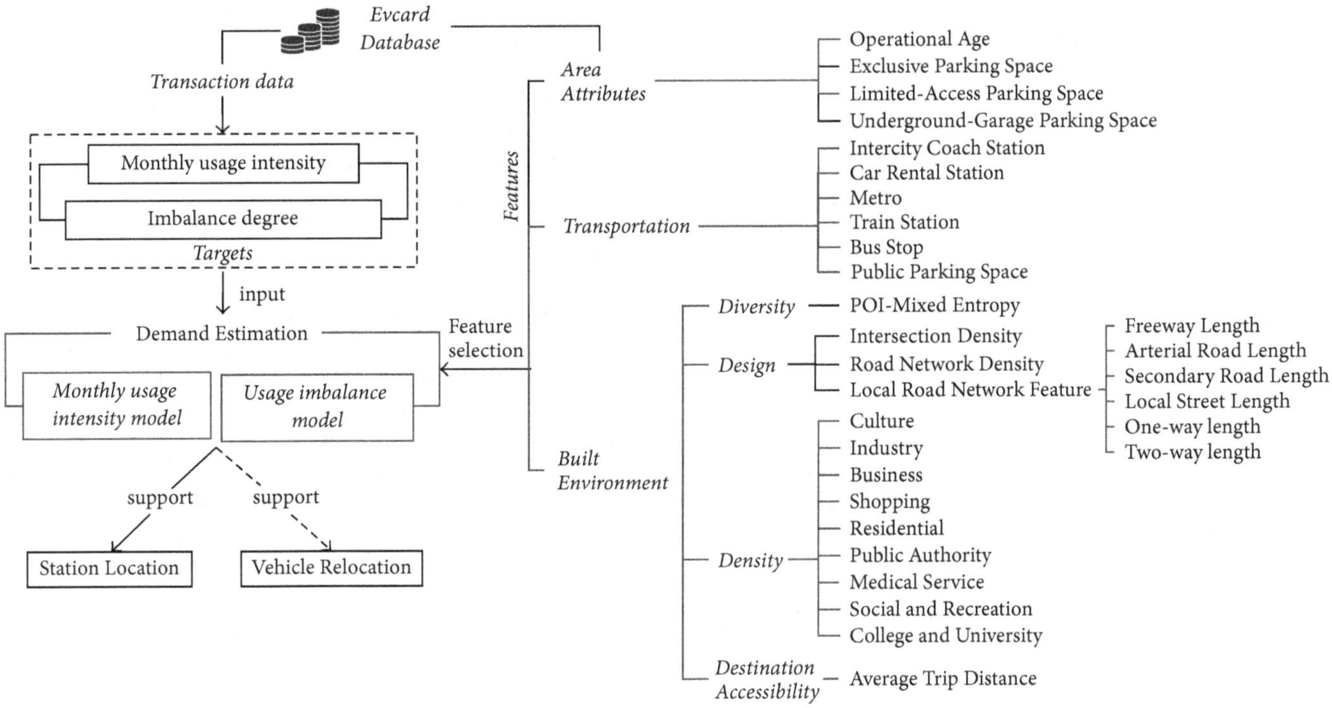

FIGURE 3: Research framework.

Usage intensity is defined as average value per hour of the total amount for pick-up and drop-off in specific time sections of the spatial unit. The calculation formula is shown as follows:

$$I_T = \frac{p_T + d_T}{l\tau_T}, \tag{4}$$

where I represents the usage intensity and p_T and d_T are the amount of pick-up and drop-off at specific time-interval. l means l months' transaction data is involved in this research; l is 3 here.

We use usage intensity as a proxy of demand rather than the number of bookings because the station is viewed as both origin and destination for the carsharing trip. It not only generates carsharing trip but also attracts it. Therefore, usage intensity is an appropriate index for carsharing demand.

Usage imbalance degree is defined as the ratio of the difference between pick-up and drop-off to the sum of pick-up and drop-off in specific time sections of the spatial unit. Half an hour is used as the statistic time interval; then the difference between pick-up and drop-off of each statistic interval is aggregated by corresponding T time section and further divided with the sum of pick-up and drop-off in time section T.

Let

$$P_T = \begin{bmatrix} p_{11} & \cdots & p_{1n_T} \\ \vdots & \ddots & \vdots \\ p_{m_T 1} & \cdots & p_{m_T n_T} \end{bmatrix},$$

$$D_T = \begin{bmatrix} d_{11} & \cdots & d_{1n_T} \\ \vdots & \ddots & \vdots \\ d_{m_T 1} & \cdots & d_{m_T n_T} \end{bmatrix}, \tag{5}$$

where P_T and D_T are the pick-up matrix and drop-off matrix in time section T with half-hour statistic interval.

$$\mathrm{IM}_T = \frac{\sum_{i=1}^{m_T} \sum_{j=1}^{n_T} \left| p_{ij} - d_{ij} \right|}{\sum_{i=1}^{m_T} \sum_{j=1}^{n_T} p_{ij} + d_{ij}}, \tag{6}$$

where IM_T is the imbalance degree of the spatial unit, m_T is the amount of day in time section T, n_T is the statistic unit in time section T, and p_{ij} and d_{ij} are the pick-up and drop-off quantity in statistic unit of ith day and jth day, respectively.

Given that several types of POIs coexist in the same area with varying degrees, features have multiple collinearity problems that cannot be ignored. Meanwhile, there are many features for samples. Therefore, a linear regression model with $L1$ and $L2$ prior as a regularizer, called adaptive elastic net (AEN) regression [30], was developed to predict the usage intensity and the degree of imbalance. The method can be viewed as a combination of elastic net [31] and the adaptive least absolute shrinkage and selection operator (LASSO) [32], which overcome the lack of adaptive LASSO (instability for high-dimensional data) and lack of the oracle property for the elastic net. The AEN is defined as follows:

$$\hat{\boldsymbol{\beta}}(\mathrm{AEN}) = \left(1 + \frac{\lambda_2}{n}\right) \cdot \left\{ \underset{\boldsymbol{\beta}}{\operatorname{argmin}} \|y - \mathbf{X}\boldsymbol{\beta}\|_2^2 + \lambda_2 \|\boldsymbol{\beta}\|_2^2 + \lambda_1^* \sum_{j=1}^{p} \widehat{w}_j \left|\beta_j\right| \right\}, \tag{7}$$

where

$$\widehat{w}_j = \left(\left|\hat{\boldsymbol{\beta}}(\mathrm{EN})\right|\right)^{-\gamma}, \quad j = 1, 2, \ldots, p$$
$$\hat{\boldsymbol{\beta}}(\mathrm{EN}) = \left(1 + \frac{\lambda_2}{n}\right) \cdot \left\{ \underset{\boldsymbol{\beta}}{\operatorname{argmin}} \|\mathbf{y} - \mathbf{X}\boldsymbol{\beta}\|_2^2 + \lambda_2 \|\boldsymbol{\beta}\|_2^2 + \lambda_1 \|\boldsymbol{\beta}\|_1 \right\}, \tag{8}$$

where $\hat{\boldsymbol{\beta}}(\mathrm{EN})$ is an elastic net algorithm, p denotes the number of features, $\|\boldsymbol{\beta}\|_1 = \sum_{j=1}^{p} |\beta|_1$ is the $l1$-norm, and $\|\boldsymbol{\beta}\|_2^2$ is $l2$-norm. λ_1, λ_2, and λ_1^* are weights for the $l1$-norm, $l2$-norm, and optimal λ_1, respectively. $\{\widehat{w}_j\}_{j=1}^{p}$ are the adaptive data-driven weights.

4. Results and Discussion

A total of 1500 ETs are used to estimate the feature importance so that the ranks and scores of features' importance are stable. We drop the features with importance score not greater than 0.015. The information of features' importance indicating that the contribution of each feature to prediction is provided in Figures 4 and 5. The features that are not filtered are used for building prediction models. The results of models are shown in Tables 3 and 4.

4.1. Usage Intensity Model. The result shows that R^2 of intensity models for eight time sections are between 0.404 and 0.583. The goodness of fit is better than other researches on this issue [15, 23, 24]. Some features show the same direction of influence on usage intensity. However, other factors play a positive or negative effect on the demand depending on the time period. Meanwhile, all factors have different weights for demand across different time periods. This causes the usage intensity to be different across the whole day as shown in Figure 2.

4.1.1. Station-Related Factors. The operational attributes of the spatial unit play an important role in usage intensity across all periods. The longer the first station operates in spatial unit, the greater the intensity is because the service and location of a station are getting familiarized by users gradually. Limited-access parking space has constantly provided negative effect on usage intensity. Exclusive parking space represents the supply level partly. It positively affects usage intensity. However, these factors are endogenous items, and the carsharing operator can improve these factors as possible as it can. The more important factors are exogenous variables, such as the built environment and transportation in the spatial unit.

4.1.2. Built Environment Factors. Built environment factors show the diverse effect on usage intensity in different time periods. College and university constantly has positive influence. In contrast, the industrial area shows negative effect. Residential, culture, public authority, and medical hygiene area impose negative influence on usage intensity in special time period. Other factors have opposite effect depending on the time period. Recreation area has a negative influence on usage intensity in workday's early peak and positive influence during nonworking time sections. More POI-related shopping is within the spatial unit, and usage intensity shows more increase in evening peak and night.

POI-mixed entropy plays a positive role in the usage intensity during working day at around 16:00–2:00. The average trip distance is a special factor. It has positive effect on the evening peak of working days and night of nonworking days, which implies that users might tend to use carsharing for further distance trip during these periods. Additionally, it has negative effect during 6:00–10:00 in nonworking days, which indicates that short distance trip of carsharing tends to be at 6:00–10:00 in nonworking days.

Areas with higher intersection density, mainly during nonworking time sections, improve usage intensity because of good accessibility. It is unexpected that the length of the local street and one-way road generally has negative effect and two-way road has a positive effect on usage intensity. Given that the local street and one-way street are more walk-friendly, the result is opposite to some research. This finding may be attributed to our use of intensity which includes pick-up and drop-off instead of pick-up only to count as an indicator of demand. Many local streets and one-way streets would make it more difficult for users to find parking space. In contrast, more two-way roads but less one-way and local streets means simpler road network.

4.1.3. Transportation Factors. Considering the transportation factors, unexpectedly, traditional car rental station has a positive effect on station usage intensity at 6:00–16:00 on working days and 10:00–20:00 on nonworking days. Despite being shown in literature that car rental and carsharing have competitive relationship in medium distance trip [1], this result indicates a more complicated relationship between carsharing and car rental. Moreover, intercity coach station has key role and positive relation during 6:00–16:00 on nonworking days, which has not been reported in any current literature. Intercity coach stations are generally far from the center of the city and passengers have no personal car while taking some packages, which could be the main reason for the demand in using carsharing to connect with intercity coach.

For public transportation, the existence of metro station negatively affects carsharing station usage intensity during morning peak and evening peak on working days. This finding could be attributed to the belief that the metro is more reliable for commuting compared to ground traffic, and commute to work is time-limited. Meanwhile, the metro imposes positive effect during nonworkday, which implies

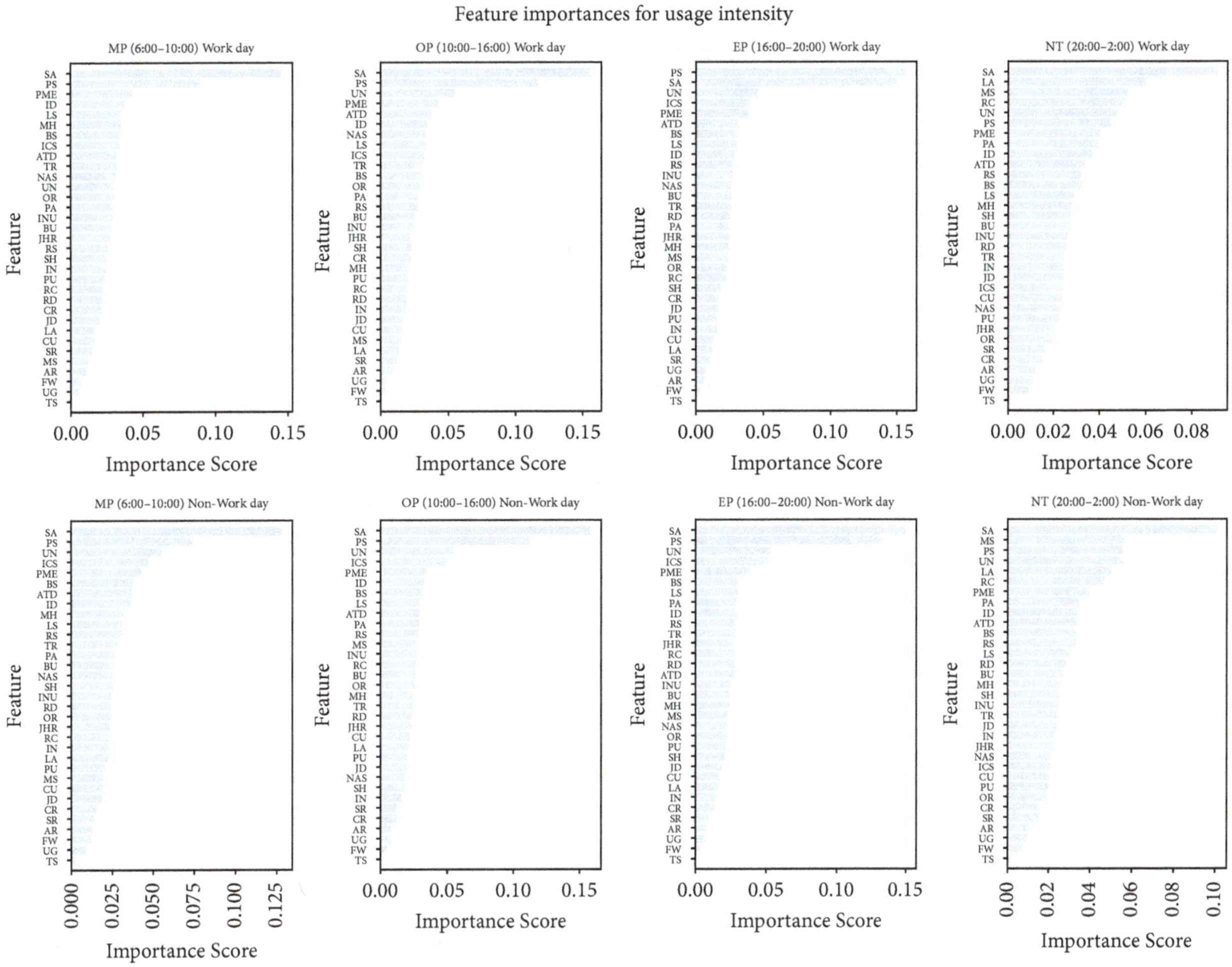

FIGURE 4: Feature importance for usage intensity.

that carsharing users are using the metro to connect carsharing during nonworkday. The metro competes with carsharing in rush hours but they cooperate with each other during nonworkdays. This relationship shows a policy potential for the government to promote diversified mobility without deteriorating ground traffic condition.

The bus stop has a positive effect during 6:00–20:00 in both workday and nonworkday, which is similar to literature [20]. This could be attributed to the good accessibility of the area near bus stops. Therefore, the exposed rate of carsharing will be high if the station is placed in nearby bus stop. This explains the nonsignificance during 10:00–20:00 in nonworkdays because the main purpose of nonworkdays is leisure, which requires higher sensitivity to comfort and lower sensitivity to price.

Public parking space has a significantly negative impact on usage intensity. The result implies that more public parking spaces result in more private vehicle trips rather than carsharing. Since private vehicle is very inefficient in using parking space, if a part of the public parking space is replaced with carsharing exclusive parking space gradually, it can (1) save huge area of high-value land in the center of the city and (2) reduce private vehicle usage.

4.2. Usage Imbalance Model. R^2 of usage imbalance models for eight time sections are between 0.217 and 0.514, which are worse than the usage intensity models. The worst imbalance model is that of the morning peak in a nonworkday. The less usage in the early morning of the nonworkday results in a few factors showing significant effect.

4.2.1. Station-Related Factors. With increasing operation age, the degree of imbalance decreases for all time periods. A spatial unit with more limited-access parking space means that it only serves lower proportion of users, and the demand diversity (purpose, departure time, and arriving time) within the spatial unit is lower. The same reason results in the similar appearance effect of underground-garage parking spaces. Therefore, the operator should locate less parking space on limited access and underground garage.

4.2.2. Built Environment Factors. At the built environment factor, residential, public authority, business, and industrial area continually play a negative role to increase the degree of imbalance in special time period. Recreation, medical hygiene, university, and shopping area have different effect in different time periods. Among them, the university has a

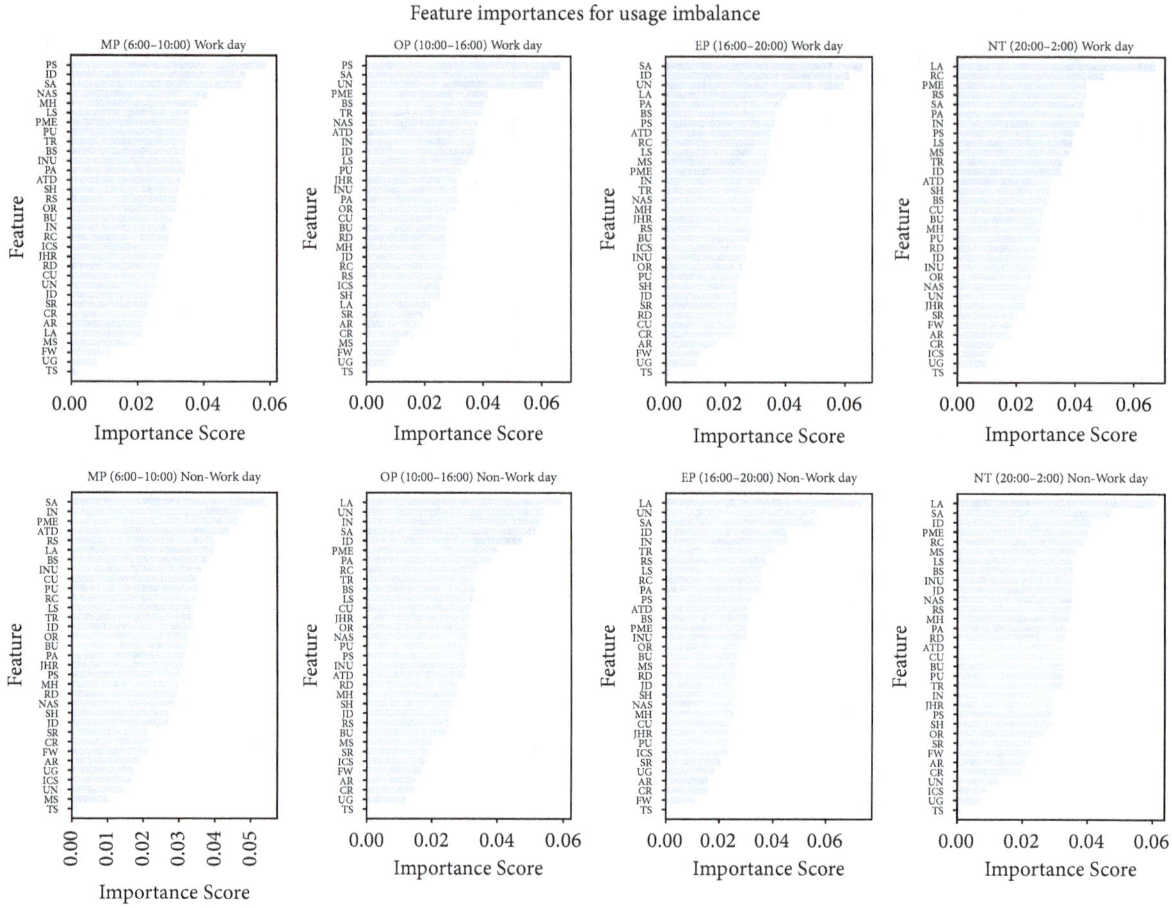

FIGURE 5: Feature importance for usage imbalance.

positive effect on station balance in most of the time sections, given that many adult students live in this area. These people have less time constraint, flexible travel time, and diverse travel purpose. However, it appears as a negative effect during 6:00–10:00 on nonworkday, which could mean that people living in these areas tend to go out of campus during this time section.

Intersection density represents accessibility partially. It can influence people who are unfamiliar with station location to access the station. More people using the station can generate and attract diversified and compensative usage of pick-up and drop-off. Our new finding is that a longer arterial road and secondary road lead to higher degree of usage imbalance in the spatial unit. By contrast, the local street results in more balanced carsharing spatial unit. Analogously, more two-way roads show higher degree of imbalance, and more one-way streets result in lower imbalance. This could be attributed to the increased possibility of imbalance in a specified area because of higher usage intensity.

POI-mixed entropy reduces the degree of imbalance during morning peak and night of workdays and non-workday nights. The diversified demand can reduce the usage imbalance. However, it increases the degree of station imbalance during 10:00–16:00 on working days, which could be attributed to the low usage intensity of the high diversity areas in this time section indicative of scattered pick-up and drop-off. It causes the imbalance of the demand for drop-off and pick-up during the statistical interval (half hour).

4.2.3. Transportation Variables. For transportation variables, metro stations play a significantly positive role in station usage balance because of the huge crowd nearby metro, and the trip purpose and time are diverse. On the other hand, it is implied that carsharing has a closer relationship with metro, and there might be a demand for connection between carsharing and metro. Therefore, it is implied that those two modes can compensate each other. By contrast, the bus stop shows opposite effect (negative) on usage imbalance. Given that the main difference between the bus and the metro is that the former runs on the ground where the uncertainty of trip duration is larger, the significant finding implies that carsharing attracts a part of the bus passengers unidirectionally, even in early peak and evening peak on workday. Therefore, from the government viewpoint, carsharing station should not be located near a bus stop, which results in a transit trip transferring to a car trip. Besides, car rental station appears to have positive effect on station usage balance in partial time section. It is implied that there is a demand of using carsharing to connect with car rental.

The results of imbalance model are shown in Table 4.

TABLE 3: The result of monthly usage intensity model.

Features	Workday				Nonworkday			
	06–09 h	10–15 h	16–19 h	20–01 h	06–09 h	10–15 h	16–19 h	20–01 h
Area attributes								
OA	0.790	0.664	0.628	0.679	0.748	0.652	0.598	0.597
LA	−0.073	−0.040	−0.123	−0.387	−0.178	−0.183	−0.163	−0.323
EPS	0.506	0.645	0.560	0.348	0.293	0.427	0.435	0.265
Built environment								
Density								
RS	−0.081				−0.052		−0.030	
PA	−0.016		−0.024	−0.448	−0.171	−0.067	−0.164	−0.283
RC	−0.055			0.318		0.017	0.151	0.019
CU	−0.149		−0.032	−0.153	−0.053			−0.026
MH				−0.059			−0.093	−0.071
BU	−0.047		0.012		−0.193	−0.017	−0.054	
UN	0.154	0.209	0.175	0.220	0.228	0.213	0.209	0.233
IN	−0.211	−0.104	−0.150	−0.192	−0.185	−0.148	−0.155	−0.061
SH	−0.087	−0.025	−0.017	0.077	−0.040		0.033	0.038
Design								
ID		0.103	0.173	0.484	0.307	0.301	0.356	0.413
LS	−0.050			0.071	−0.056		−0.045	
OR	−0.004			−0.099	−0.045	−0.063	−0.087	−0.076
TR				0.013				0.026
Diversity								
PME		−0.020		0.336	0.059		0.042	0.300
Destination accessibility								
ATD		0.186	0.236		−0.179			0.131
Transportation								
MS	−0.163		−0.192	0.188	0.094	0.136	0.107	0.187
BS	0.112	0.162	0.210		0.239	0.136	0.101	
CR	0.164	0.171	0.064			0.012	0.029	
ICS	0.186	0.211	0.189	0.315	0.327	0.288	0.306	0.270
PP	−0.218		−0.184	−0.166			−0.229	−0.175
R2	0.422	0.404	0.412	0.583	0.511	0.488	0.504	0.539

Combining these two models, the significant features can be arranged as shown in Figure 6. The features within the range of the dotted lines and located on x-axis or y-axis only have significant impact on single dependent variables. Meanwhile, the others in the outer side beyond the dotted range are significant on both dependent variables.

Given an average value of area attributes as shown in Table 2, we get some appropriate location of carsharing station based on the result of usage intensity model and usage imbalance model, respectively. We divide location of Shanghai into three levels in proportion as 25%, 50%, and 25%, firstly. Then combining results of two models, the location can be divided into five levels, prior, recommended, medium, not recommended, avoid, as Figure 7 shows. We find that central area takes a relatively large proportion of prior level area to locating. Given that a lot of central areas of city are appropriate to locating carsharing station and carsharing is more efficient in using parking space, we suggest that more carsharing exclusive parking space can be used to replace public parking space to decrease usage of private vehicle and save parking space. Moreover, many suburb locations are evaluated as prior or recommended level by models. This means that the usage scenarios of carsharing are wider than central area. If these suburb areas can be developed adequately, the usage scenarios related to outskirts will take on more trips.

5. Conclusions

This study focused on the largest station-based OWC program in Shanghai, China. There are many approaches to estimate carsharing demand according to research objects.

Table 4: The results of imbalance model.

Features	Workday				Nonworkday			
	06–09 h	10–15 h	16–19 h	20–01 h	06–09 h	10–15 h	16–19 h	20–01 h
Area attributes								
OA	−0.189	−0.146	−0.189	−0.003	−0.044	−0.102	−0.117	−0.012
LA	0.011		0.079	0.075		0.080	0.100	0.002
UG				0.147			0.020	
Built environment								
Density								
RS	−0.120		−0.015		−0.210			
PA	0.028		0.083			0.021	0.025	
RC				0.125		−0.051	−0.067	0.173
MH	−0.028		0.136					
BU			0.104			0.116	0.077	
UN	−0.064	−0.101	−0.101	−0.052	0.041	−0.076	−0.092	
IN	0.044	0.095	0.082			0.024	0.014	
SH	−0.010	0.022	0.047			−0.011	0.045	
Design								
ID	−0.204	−0.088	−0.174			−0.097	−0.119	−0.099
AR	0.113		0.103			0.018	0.026	
SR	0.351		0.265			0.107		
LS	−0.071		−0.054	−0.056		0.014	−0.049	
OR			−0.058			−0.001	−0.050	
TR		0.034	0.085				0.061	
Diversity								
PME	−0.048	0.058	0.024	−0.139		0.052	0.004	−0.059
Destination accessibility								
ATD	−0.071	−0.133	−0.107			−0.046	−0.125	
Transportation								
MS	−0.181		−0.046	−0.060		−0.046	−0.060	−0.096
BS	0.119	0.049	0.151		0.014	0.013		
CR	−0.030	−0.051	−0.166		0.016			
PP	−0.158		−0.127					
R2	0.424	0.394	0.514	0.329	0.217	0.419	0.447	0.315

However, the station-based one-way system is rarely investigated. Meanwhile, many research investigations focus on the usage rate, vehicle hour traveled (VHT), and many others, but the station usage imbalance has not yet been investigated. This study addressed this gap.

In this study, multiple linear regression models and beta regression model are developed to analyze how different factors affect station usage intensity and degree of station imbalance across different periods. The conclusions are summarized as follows:

(1) The attributes of spatial unit constantly appear to have significant effect on the demand characteristics. However, many built environment and transportation factors have a different effect on the demand in different time periods. This is the main reason why carsharing demand appears to be dynamic across time periods.

(2) For usage intensity, the university, high POI-mixed entropy, high intersection density area, and area including a metro station, bus stop, car rental station, and intercity coach have positive influence on usage intensity. However, industrial, residential, culture, public authority, and medical hygiene areas show negative effect in different time periods, in which layout should be avoided by carsharing stations.

(3) For the degree of usage imbalance, it will decrease along with the increase in operation age. Limited-access parking space enhances usage imbalance. Residential, public authority, business, and industrial areas continually play a negative role to increase the degree of imbalance in special time period. The area with the university, high intersection density, high POI-mixed entropy, and more local streets and one-way roads lead to more balanced operational area.

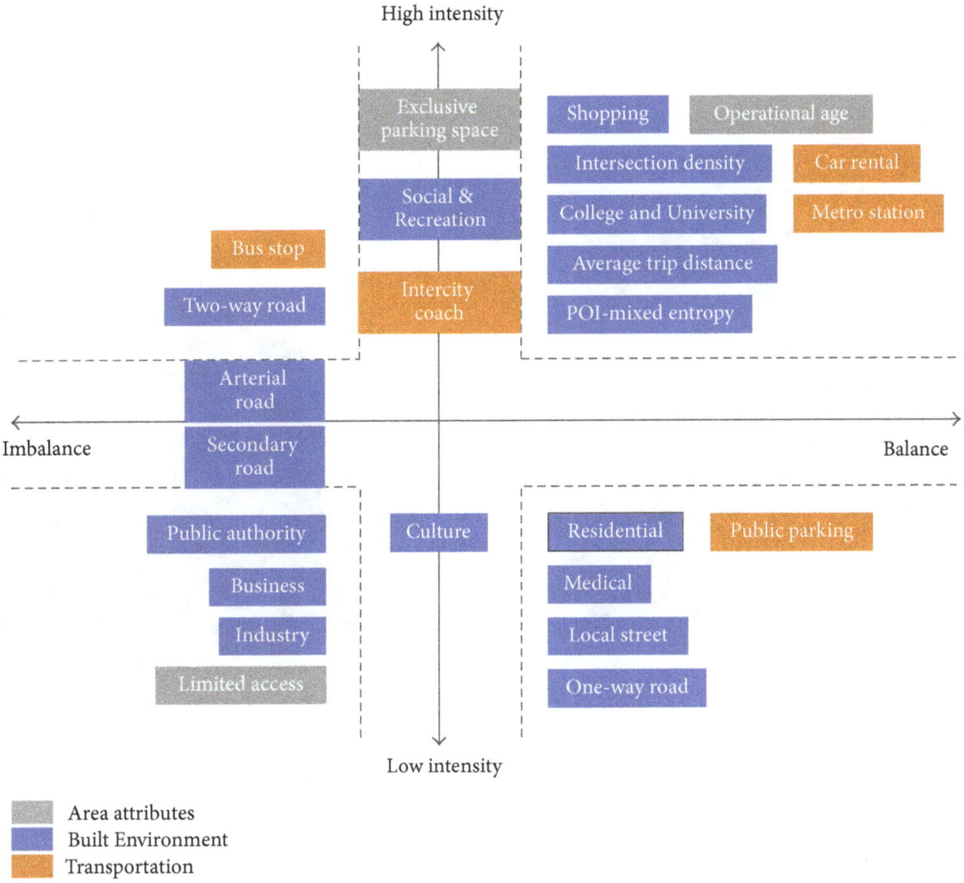

FIGURE 6: Influence diagram of statistically significant independent variables.

(4) Areas with adequate public parking space will attract more personal vehicle use rather than carsharing trip. Given that carsharing is more efficient in using parking space, we suggest that public parking spaces should be gradually converted to carsharing exclusive parking space. This will increase the usage efficiency of the limited number of parking spaces and reduce personal vehicle usage while having a flexible car trip still available.

(5) For public transportation, the metro and bus are significantly different for carsharing. The metro has a strong advantage over carsharing in the morning and evening peak on workdays because of its certainty of trip duration. Thus, carsharing cannot attract passengers from the metro in rush hour. Meanwhile, they appear to connect with each other in another time period, which is a complementary relationship. However, the bus is similar to carsharing, which runs on the ground but lacks the comfort and personality of carsharing. Thus, carsharing has a related advantage over the bus, which results in some bus passengers transferring to carsharing unidirectionally. Therefore, we suggest that the government should encourage carsharing station layout near a metro station but not a bus stop.

Usage intensity is related to profits, and the degree of station imbalance is related to dispatching cost. From the carsharing operator viewpoint, the purpose of the carsharing station is to minimize the cost to obtain the maximum benefit. Thus, the results shown in Figure 6 can be viewed as a guidance of carsharing station layout for maximizing benefit. The features in the first quadrant lead to higher usage intensity and lower imbalance degree; meanwhile, features in the third quadrant result in lower usage intensity and higher imbalance degree. Therefore, carsharing station should be given priority to locating at area with features in the first quadrant and setting up stations in areas with features in the third quadrant should be avoided. Other factors can be selected as secondary, such as stations nearby metro stations, which only decrease station usage intensity during peak time section on workdays. However, it might be a good choice to select the station near other stations so that the imbalance level can be dramatically decreased during most of the time sections.

The method of modeling for different time sections reveals, to a certain extent, the temporal dynamics patterns of the demand, which can provide guidance for vehicle relocation. In college and university areas, the imbalance level is high at 6:00–9:00 on nonworking days, which shows that extra dispatch is needed during this time section. However, the research conclusion is built upon long-term measurement

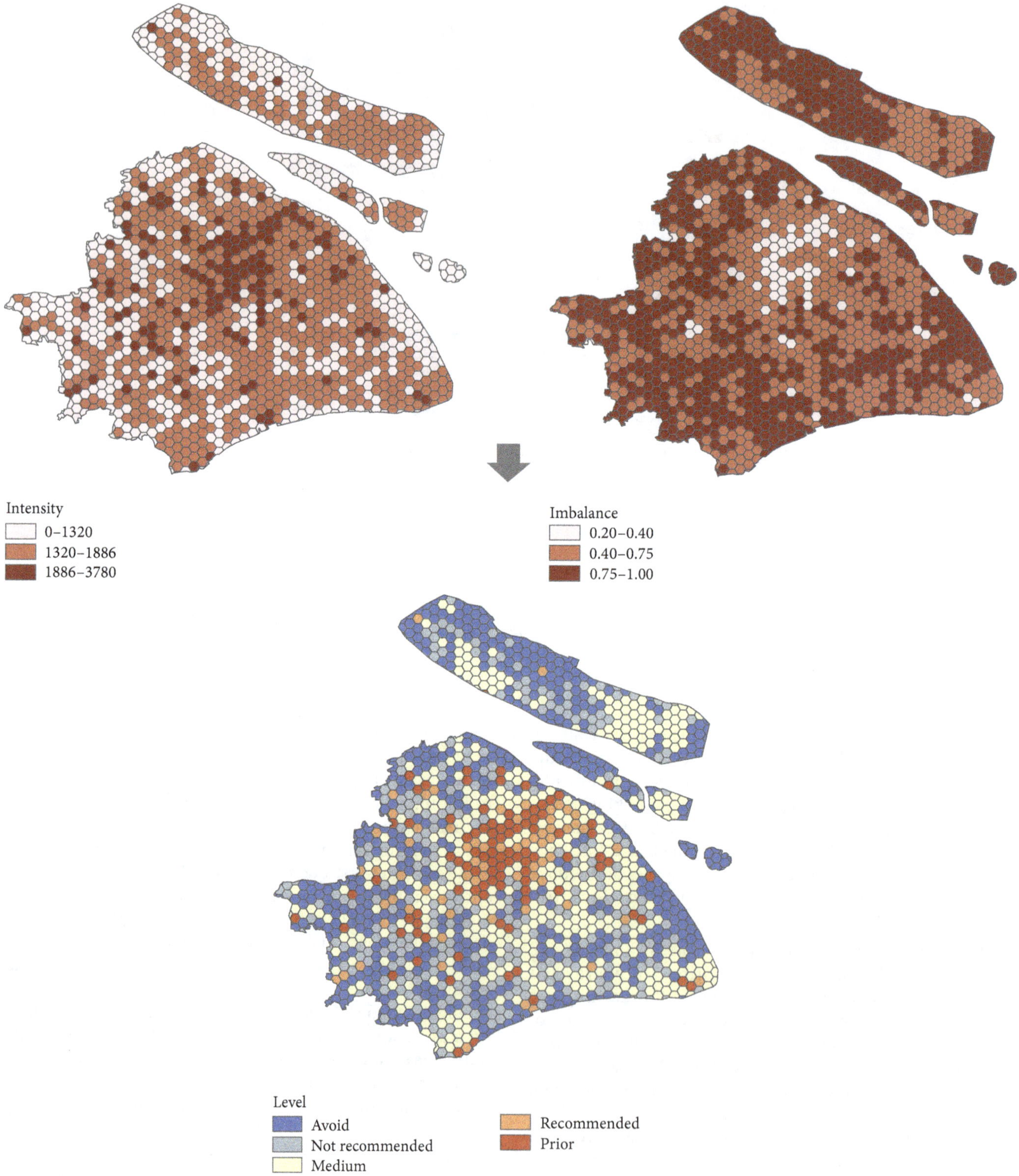

FIGURE 7: Combining usage intensity model and imbalance model to locating carsharing station in Shanghai.

(three months). Thus, it can provide a noninstant dispatch strategy. We believe that it is strategically advantageous to arrange vehicle in advance based on demand dynamics pattern concluded by this research. Then, an instant dispatch method is used for adjustment accordingly.

There are three main limitations in this research:

(1) The statistics radium station is 800 m, and it only refers to the value in the research of public transit. Although the range of 800 m is widely used in carsharing areas [24], the service range of carsharing stations in different zones and different traffic conditions can vary.

(2) The categorization of time section is only based on the time distribution feature of bookings, but more reasonable time categorization shall be an improvement direction.

(3) In the calculation of station imbalance level, statistic time interval is very important. Too small interval might cause high imbalance level, while too big interval may cause low level of imbalance. We infer that statistic time interval should depend on different usage intensities in each spatial unit, but this limitation will be improved in future research.

Conflicts of Interest

The authors declare that they have no conflicts of interest regarding the publication of this paper.

Acknowledgments

The authors would like to acknowledge the Shanghai International Automobile City Co., Ltd. and Global Carsharing & Rental Co., Ltd. for providing the precious data of EVCARD in this research. This study is supported by the National Natural Science Foundation of China (71734004), China National Key Technology R&D Program (2015BAG11B01), and Open Research Funding of "Gaofeng" Discipline (2016J012307).

References

[1] A. Millard-Ball, "Where and how it succeeds," *Transportation Research Board*, 2005.

[2] E. Martin, S. Shaheen, and J. Lidicker, "Impact of carsharing on household vehicle holdings: Results from North American shared-use vehicle survey," *Transportation Research Record: Journal of the Transportation Research Board*, vol. 2143, pp. 150–158, 2010.

[3] J. T. Schure, F. Napolitan, and R. Hutchinson, "Cumulative impacts of carsharing and unbundled parking on vehicle ownership and mode choice," *Transportation Research Record*, no. 2319, pp. 96–104, 2012.

[4] S. A. Shaheen, C. Rodier, and G. Murray, *Carsharing and Public Parking Policies: Assessing Benefits, Costs, and Best Practices in North America*, 2010.

[5] E. W. Martin and S. A. Shaheen, "Greenhouse gas emission impacts of carsharing in North America," *IEEE Transactions on Intelligent Transportation Systems*, vol. 12, no. 4, pp. 1074–1086, 2011.

[6] H. Nijland, J. Van Meerkerk, and A. Hoen, *Impact of Car Sharing on Mobility and CO_2 Emissions*, PBL Note, 2015.

[7] A. Bieszczat and J. Schwieterman, *Are Taxes on Carsharing Too High? A Review of the Public Benefits and Tax Burden of an Expanding Transportation Sector*, Chaddick Institute for Metropolitan Development, DePaul University, 2011.

[8] J. Firnkorn and M. Müller, "Free-floating electric carsharing-fleets in smart cities: The dawning of a post-private car era in urban environments?" *Environmental Science & Policy*, vol. 45, pp. 30–40, 2015.

[9] G. D. Kim, J. Park, and J. D. Woo, *Investigating the Characteristics of Carsharing Usage Pattern for Public Rental Housing Complexes: A Case Study in South Korea*, 2017.

[10] F. Ferrero, G. Perboli, and A. Vesco, *Car-Sharing Services—Part a Taxonomy and Annotated Review*, Montréal, Canada, 2015.

[11] R. Katzev, "Car Sharing: A New Approach to Urban Transportation Problems," *Analyses of Social Issues and Public Policy*, vol. 3, no. 1, pp. 65–86, 2003.

[12] C. Costain, C. Ardron, and K. N. Habib, "Synopsis of users' behaviour of a carsharing program: A case study in Toronto," *Transportation Research Part A: Policy and Practice*, vol. 46, no. 3, pp. 421–434, 2012.

[13] K. M. N. Habib, C. Morency, M. T. Islam, and V. Grasset, "Modelling users' behaviour of a carsharing program: Application of a joint hazard and zero inflated dynamic ordered probability model," *Transportation Research Part A: Policy and Practice*, vol. 46, no. 2, pp. 241–254, 2012.

[14] A. De Lorimier and A. M. El-Geneidy, "Understanding the factors affecting vehicle usage and availability in carsharing networks: a case study of communauto carsharing system from Montréal, Canada," *International Journal of Sustainable Transportation*, vol. 7, no. 1, pp. 35–51, 2012.

[15] K. Kim, "Can carsharing meet the mobility needs for the low-income neighborhoods? Lessons from carsharing usage patterns in New York City," *Transportation Research Part A: Policy and Practice*, vol. 77, pp. 249–260, 2015.

[16] J. Kang, K. Hwang, and S. Park, "Finding factors that influence carsharing usage: Case study in seoul," *Sustainability*, vol. 8, no. 8, p. 709, 2016.

[17] R. Seign and K. Bogenberger, "Model-Based Design of Free-Floating Carsharing Systems," in *Proceedings of the Transportation Research Board 94th Annual Meeting*, 2015.

[18] M. Khan and R. Machemehl, *The Impact of Land-Use Variables on Free-Floating Carsharing Vehicle Rental Choice and Parking Duration*, Seeing Cities Through Big Data, Springer International Publishing, 2017.

[19] S. Schmöller and K. Bogenberger, "Analyzing External Factors on the Spatial and Temporal Demand of Car Sharing Systems," *Procedia - Social and Behavioral Sciences*, vol. 111, pp. 8–17, 2014.

[20] S. Wagner, T. Brandt, and D. Neumann, "In free float: Developing Business Analytics support for carsharing providers," *OMEGA - The International Journal of Management Science*, vol. 59, pp. 4–14, 2016.

[21] K. Klemmer, S. Wagner, C. Willing, and T. Brandt, *Explaining Spatio-Temporal Dynamics in Carsharing: A Case Study of Amsterdam*, 2016.

[22] S. Schmöller, S. Weikl, J. Müller, and K. Bogenberger, "Empirical analysis of free-floating carsharing usage: The munich and berlin case," *Transportation Research Part C: Emerging Technologies*, vol. 56, pp. 34–51, 2015.

[23] T. Stillwater, P. L. Mokhtarian, and S. A. Shaheen, "Carsharing and the built environment: Geographic information system based study of one U.S. operator," *Transportation Research Record: Journal of the Transportation Research Board*, vol. 2110, pp. 27–34, 2009.

[24] C. Celsor and A. Millard-Ball, "Where does carsharing work? Using geographic information systems to assess market potential," *Transportation Research Record: Journal of the Transportation Research Board*, vol. 1992, pp. 61–69, 2007.

[25] Y. Jiang, P. Gu, F. Chen et al., *Measuring Transit-Oriented Development in Quantity and Quality: A Case of 24 Cities with Urban Rail Systems in China*, 2017.

[26] R. Cervero and K. Kockelman, "Travel demand and the 3Ds: density, diversity, and design," *Transportation Research Part D: Transport and Environment*, vol. 2, no. 3, pp. 199–219, 1997.

[27] R. Ewing and R. Cervero, "Travel and the built environment: a meta-analysis," *Journal of the American Planning Association*, vol. 76, no. 3, pp. 265–294, 2010.

[28] L. Breiman, "Random forests," *Machine Learning*, vol. 45, no. 1, pp. 5–32, 2001.

[29] P. Geurts, D. Ernst, and L. Wehenkel, "Extremely randomized trees," *Machine Learning*, vol. 63, no. 1, pp. 3–42, 2006.

[30] H. Zou and H. H. Zhang, "On the adaptive elastic-net with a diverging number of parameters," *Annals of Statistics*, vol. 37, no. 4, pp. 1733–1751, 2009.

[31] H. Zou and T. Hastie, "Regularization and variable selection via the elastic net," *Journal of the Royal Statistical Society*, vol. 67, no. 2, pp. 768-768, 2005.

[32] H. Zou, "The Adaptive Lasso and Its Oracle Properties," *Publications of the American Statistical Association*, vol. 101, no. 476, pp. 1418–1429, 2006.

3

Dynamic Eco-Driving Speed Guidance at Signalized Intersections: Multivehicle Driving Simulator based Experimental Study

Peng Chen[1,2] **Cong Yan,**[1] **Jian Sun**[3] **Yunpeng Wang,**[1,2]
Shenyang Chen,[4] **and Keping Li**[3]

[1] *Beijing Key Laboratory for Cooperative Vehicle Infrastructure Systems and Safety Control, School of Transportation Science and Engineering, Beihang University, Xue Yuan Road No. 37, Hai Dian District, Beijing 100191, China*
[2] *Beijing Advanced Innovation Center for Big Data and Brain Computing, Beihang University, Xue Yuan Road No. 37, Hai Dian District, Beijing 100191, China*
[3] *Department of Traffic Engineering and Key Laboratory of Road and Traffic Engineering, Ministry of Education, Tongji University, Shanghai 201804, China*
[4] *Department of Civil and Environmental Engineering, University of California, Davis, CA 95616, USA*

Correspondence should be addressed to Jian Sun; sunjian@tongji.edu.cn

Academic Editor: Antonino Vitetta

Variations in vehicle fuel consumption and gas emissions are usually associated with changes in cruise speed and the aggressiveness of drivers' acceleration/deceleration, especially at traffic signals. In an attempt to enhance vehicle fuel efficiency on arterials, this study developed a dynamic eco-driving speed guidance strategy (DESGS) using real-time signal timing and vehicle positioning information in a connected vehicle (CV) environment. DESGS mainly aims to optimize the fuel/emission speed profiles for vehicles approaching signalized intersections. An optimization-based rolling horizon and a dynamic programming approach were proposed to track the optimal guided velocity for individual vehicles along the travel segment. In addition, a vehicle specific power (VSP) based approach was integrated into DESGS to estimate the fuel consumption and CO_2 emissions. To evaluate the effectiveness of the overall strategy, 15 experienced drivers were recruited to participate in interactive speed guidance experiments using multivehicle driving simulators. It was found that compared to vehicles without speed guidance, those with DESGS had a significantly reduced number of stops and approximately 25% less fuel consumption and CO_2 emissions.

1. Introduction

The transportation sector not only accounts for a vast quantity of fuel consumption but also serves as the largest atmospheric carbon emitter in the world. Concerns over energy security, environmental protection, and economic benefits have prompted renewed public and private interest in an ecological energy-efficient driving style that can reduce fuel consumption and greenhouse gas emissions. Eco-driving serves as one effective and cost-effective strategy to enhance the fuel efficiency of the transportation section [1–3]. The essential idea of eco-driving is to provide real-time driving advice to individual vehicles so that the drivers can adjust their driving behavior or take driving actions to reduce fuel consumption and emission levels [4].

The Intelligent Speed Adaptation (ISA) initiative in the UK aimed to develop driver assistive devices that recommend desired speeds to avoid hard braking [5]. Even though the primary objective of an ISA device is safer driving, its users benefitted from reduced fuel consumption and emissions due to smoother speed variations [6]. From an energy and emissions standpoint, this initiative laid the foundation for eco-driving, which refers to driving in an ecological and economical way.

The effects of driving behavior on fuel consumption and emissions have been widely studied [7–9]. Yang et al. [10]

summarized that frequent acceleration associated with stop-and-go waves, excessive speeds, slow movement on congested roads, and extra idling times are major causes of increased fuel consumption and emissions. It is well recognized that maintaining a constant velocity and avoiding unnecessary acceleration and deceleration are the key principles of eco-driving [11].

To develop eco-driving strategies on freeways appears straightforward considering that the traffic flow is continuous and vehicles are seldom interrupted by traffic control devices [12]. The eco-driving strategy can simply compute advisory speeds or speed limits for drivers. Unlike freeways, the traffic flows on arterials are essentially interrupted because of complex interactions between volatile traffic patterns and signal control as well as the interdependent relationships of vehicle queues at neighboring intersections under different levels of congestion [13]. The dependency of speed variations on external factors, such as neighboring vehicles, signal status, and other infrastructural limitations, results in yet limited use of eco-driving on arterials.

Numerical empirical studies show a positive relationship between vehicle emissions and fuel consumption with delays at traffic signals [14, 15]. According to the state of practice, signal coordination ranks as one of the most effective control strategies for solving energy and fuel consumption problems on arterials. The core concept of this strategy is to coordinate a number of traffic signals by adjusting the offset between neighboring intersections to optimize travel speeds, reduce delays, and minimize stops. However, constraints in field operation, such as oversized spacing between adjacent intersections, complicated traffic composition, and the diversity of driving behavior, may lead to decreased reliability and stability of the coordinated signal control system, which is characterized by a discrete form of traffic or a large difference between running speed and preset speed [16, 17]. In such cases, signal coordination fails to reach its design goals and may result in increased fuel consumption, trip time, and engine and brake wear.

Recently, the emerging connected vehicle (CV) technology initiatively led by the U.S. Department of Transportation has encouraged increased planning and implementation [18]. Vehicle-to-Vehicle (V2V) and Vehicle-to-Infrastructure (V2I) communications offer the potential to obtain real-time vehicle positions and traffic signal timing at downstream locations. Speed can be adjusted for timely arrival at green lights, which enables an eco-friendly driving style. Thus, improving the traffic environment and decreasing emissions with arterial eco-driving strategies have become an achievable goal [19]. There has been increasing efforts to develop eco-driving strategies for assisting drivers to travel along signalized arterials.

Mandava et al. [20] developed an arterial velocity planning algorithm that provides dynamic speed advice to the driver to maximize the probability of having a green light when approaching signalized intersections. The algorithm aimed to minimize the acceleration/deceleration rates within the speed limit while avoiding complete stops. Last, it was tested by using a stochastic simulation technique along a 10-intersection signalized corridor. However, it was argued that lowering acceleration/deceleration levels does not necessarily indicate reducing fuel consumption. Asadi and Vahidi [21] used upcoming traffic signal information within the vehicle's adaptive cruise control system to reduce idle time at stop lights and fuel consumption. The control objectives were set as timely arrival at green light with minimal use of braking, maintaining safe distance between vehicles, and cruising at or near set speed. Barth et al. [22] developed a dynamic eco-driving velocity planning algorithm to minimize acceleration/deceleration rates, the total tractive power demand, and the idling time for probe vehicles, so that fuel consumption and emissions can be reduced. Rakha and Kamalanathsharma [23] presented seven examples with different speeds and throttles through a sample scenario with a single signalized intersection. After assuming or calculating the other parameters, they determined which case had the most fuel-optimal speed profile. Recently, Yang et al. [10] developed an Eco-Cooperative Adaptive Cruise Control (Eco-CACC) algorithm to compute the fuel-optimum vehicle trajectory through a signalized intersection by ensuring that the vehicle arrives at the intersection stop bar just as the last queued vehicle is discharged. The key of the algorithm is to predict the queue length ahead of the probe vehicles using the LWR and the release time of the queue using the signal phasing and timing (SPaT) data. Last, the algorithm was tested using the INTEGRATION microscopic traffic simulation software. It was found that the queue prediction enhanced the Eco-CACC algorithm to produce the smoothest vehicle trajectories and to result in the highest fuel savings, approximately 11.4%.

In the aforementioned studies, the goal of reducing fuel consumption was usually transformed into simple functions of acceleration/deceleration rates [20, 22] or the time of arrival at the intersection [10, 21, 23] (Sun and Henry, 2015; Wu et al, 2015a, 2015b). None of these models used an explicit objective function to minimize the total fuel consumption. In addition, most of these systems do not provide predictive eco-driving control strategies that optimize the vehicle controls in advance. Moreover, their findings were obtained through theoretical calculations or microscopic simulation instead of experimental validation or field testing. Last but not least, even for the experimental studies, they failed to consider driver responses and the interactions between human drivers and traffic flow.

Thus, this study developed a dynamic eco-driving speed guidance strategy (DESGS) to help optimize the fuel/emission speed profiles for vehicles approaching signalized intersections. Based on the signal timing and vehicle position information, an optimization algorithm was proposed for drivers to achieve minimum fuel consumption and CO_2 emissions. In addition, a vehicle specific power (VSP) based approach was integrated into DESGS to estimate the fuel consumption and CO_2 emissions. Furthermore, the proposed strategy was investigated on a typical arterial in Shanghai, China, using multivehicle driving simulators, which would enable us to integrate the real driver responses. As an initial step, the developed DESGS helps highlight the importance of retaining microscopic fuel consumption models in the optimization function and in the predictive

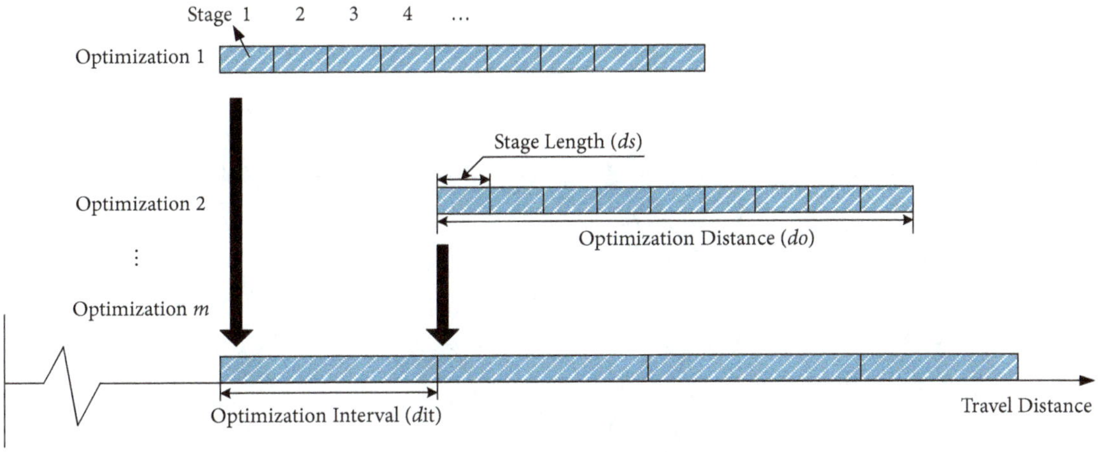

FIGURE 1: Rolling horizon optimization logic schematic.

model to minimize fuel consumption and CO_2 emissions for a given road topography.

The paper is organized as follows. Section 2 presents the methodology of DESGS for eco-driving speed calculations. Section 3 presents a case study based on multivehicle driving simulators for a signalized arterial. Section 4 analyzes the benefits and efficiency of the proposed DESGS. Section 5 offers concluding remarks and directions for future work.

2. Methodology

The DESGS consists of two building blocks, that is, an optimization module and an eco-index calculation module. The guided velocity is dynamically adjusted based on the vehicle's spatial-temporal trajectories using an optimization-based rolling horizon and a dynamic programming (DP) approach. Given the topological information and driving characteristics of the test vehicles, the eco-index calculation module can estimate the fuel consumption rates based on the vehicle operating conditions. The optimization module is then used to find an optimal control plan that minimizes the total fuel consumption and CO_2 emissions, while satisfying the preset minimum and maximum vehicle speeds.

2.1. Optimization Module

2.1.1. Rolling Horizon Optimization Logic. The optimization module is the most important component that defines the structure of the system. A rolling horizon approach is used for optimization based on the most recent predictions and observations. As illustrated in Figure 1, this module consists of three parameters: (i) the stage length (d_s) is the unit of discretization; (ii) the optimization distance (d_o), which is also referred to as the look-ahead distance, is the unit of optimization; and (iii) the optimization interval (d_{it}) represents the optimization frequency.

In this study, a guided speed decision is made for the optimization distance (d_o) at the position where a vehicle starts to receive speed advice. Here, d_o indicates the distance from the current position to the estimated position where the vehicle will pass through the intersection. To calculate the stage length (d_s) at the beginning of stage i, this study assumes that a vehicle begins to accelerate or decelerate at a feasible rate from an initial velocity to a guided velocity and then cruises at this velocity over this distance. Considering the minimum and maximum values of the highway speed limit, the engine force, and the braking torque ability, d_s is typically 100–200 m. In the optimization interval (d_{it}), a vehicle cruises at the guided velocity over stage i ($i = 2, 3, 4, \ldots$) after traveling through d_s of stage 1 until reaching the end of d_{it}. Since the guided velocity may not agree with the vehicle's current velocity, d_{it} should be sufficiently long to avoid the driver frequently changing vehicle speed or causing driver boredom. On the other hand, when d_{it} is too long, the chance of the vehicle hitting a red light increases, corresponding to a required speed change and a guidance failure. Thus, d_{it} was assigned a value of 400 m in this study. A new decision is made with updated parameters at the end of d_{it}.

2.1.2. Dynamic Programming (DP) implementation. There are three basic approaches for addressing optimal control problems, that is, DP, direct methods [24, 25] and indirect methods [26, 27]. DP adopts the principle of *optimality of subarcs* to recursively compute feedback control. In the continuous case, this method leads to the Hamilton-Jacobi-Bellman equation. DP can easily handle nonconvex control problems and inequality constraints [28]. Thus, this study uses DP to solve the minimum fuel vehicle driving problem.

At the beginning of each optimization interval (d_{it}), the optimization module finds an optimal control plan using Dijkstra's shortest path algorithm, which was developed specifically for this application in light of the cost function matrix.

As shown in Figure 2, the variables in the solid circle represent all the possible guided velocities (v_g) that may be displayed by each guided position. The value of v_g is given by

$$v_g = 5 \cdot n, \quad n = 1, 2, 3, \ldots \qquad (1)$$
$$v_{\min} \le v_g \le v_{\max},$$

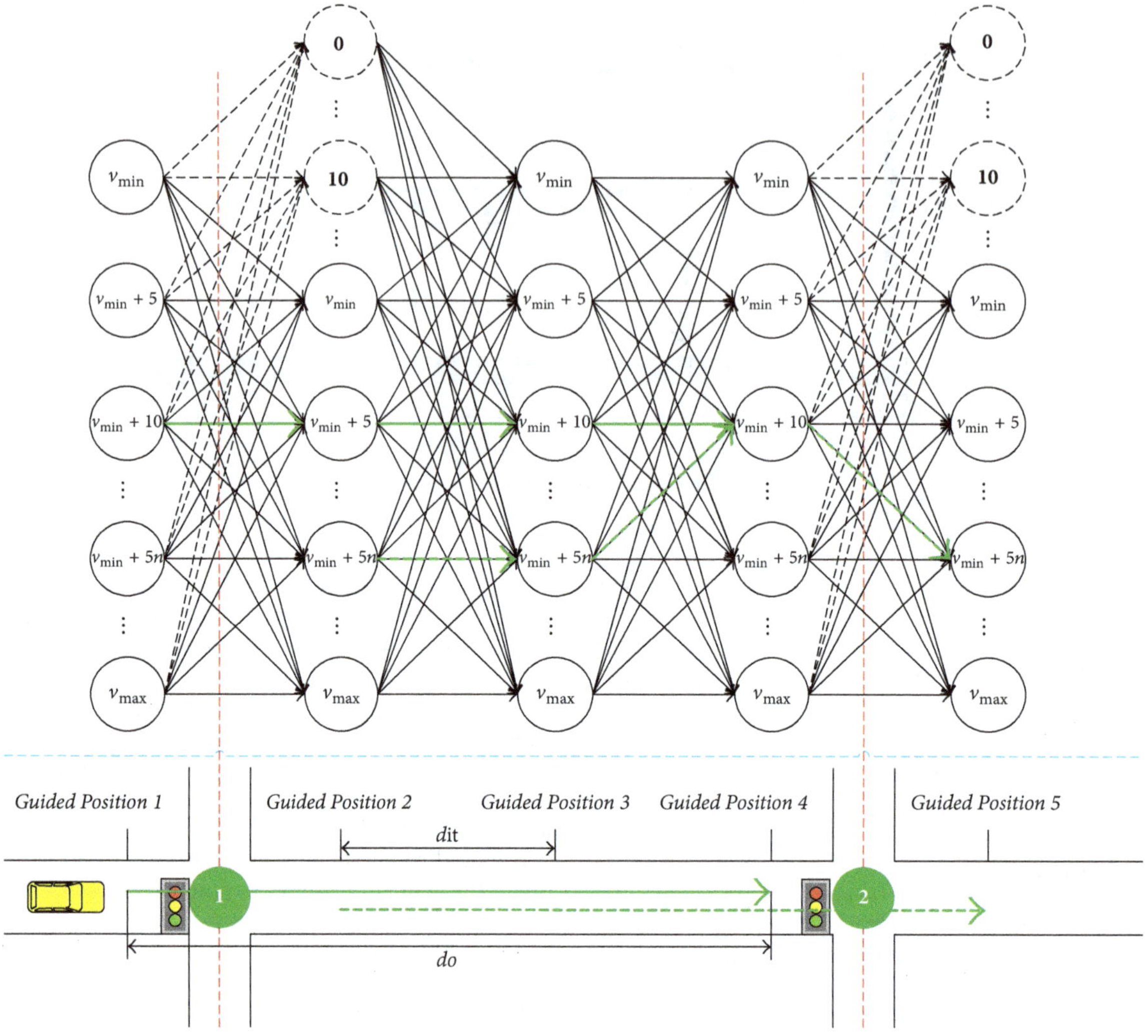

FIGURE 2: The DESGS calculation using DP.

where v_g is set as multiples of 5 km/h to facilitate the DESGS calculation process; v_{min} is the minimum speed limit of the study link, which is set to protect the speed guidance strategy from impairing traffic mobility or safety; v_{max} is the maximum speed limit, which is taken as the smaller value between the maximum speed limit of the link and the highest vehicle speed based on power performance. The numbers in the dotted circle represent the possible actual velocities when the vehicle crosses the intersection stop line because of the obstruction of a red light. Note that v_k' is the velocity of the vehicle when it arrives at the intersection stop line. The vehicle travels at a specific v_g from the last guided position to the downstream intersection stop line. In this process, when the signal light is always green, $v_k' = v_g$; when the driver observes that the light is red and starts to decelerate but the light turns green as the vehicle reaches the stop line, $0 < v_k' \leq v_g$; when the signal is always red, $v_k' = 0$.

The distance between the adjacent guided positions is defined as one optimization interval (d_{it}). For example, as a vehicle reaches guided position 1, the optimization module will calculate the fuel consumption along the travel segment considering all the possible combinations of guided velocity over one optimization distance (d_o). This distance is based on the signal timing parameters at the next downstream intersection, the queue length at the approach, and/or the leading vehicles. In the computation process, the influence of the signal lights between adjacent guided positions, which may lead to the vehicle traveling on the segment, does not consider the recommended velocity at former positions. For example, the guided velocity displayed at position 1 is 70 km/h, but the vehicle traveling at this speed to the downstream intersection will meet a red light. In this case, the vehicle will immediately brake and decelerate to the stop line. After the vehicle clears the intersection, its task is to accelerate back to the initial velocity (70 km/h), assuming that drivers completely conform to the guided velocity. This process will increase redundant fuel consumption and emissions due to the accelerations and decelerations required at the signals. Thus, we can easily infer that this guidance effect is not ideal. When a vehicle travels the same distance, for example, from guided position 1 to position 2 and from guided position 2 to position 3, with the same speed change value, for example, from 70 km/h to 60 km/h, the fuel consumption may be different because the vehicle may meet a red light in one of the segments, resulting in excess fuel consumption.

After comparing the total fuel consumption of each combination, the "guided velocity ribbon" will be determined for the smallest fuel consumption, as illustrated by the solid green line in Figure 2. That is, the actual guided velocity issued by guided position 1 is ($v_{\min}$ + 10) km/h. Similarly, as the vehicle reaches guided position 2, the green dotted line is obtained using the same algorithm. The speed advice provided by the guided position is in real-time based on the vehicle's current velocity, the signal phase and timing information, and the distance to the next downstream intersection. If the driver does not follow the guidance, he/she will receive new speed advice at the next guided position. However, if the driver fully complies with the "guided velocity ribbon," the actual guided velocity at the overlap between two adjacent optimization distance (d_o) should be consistent.

2.2. Objective Function and Constraints. When a vehicle receives speed advice, its initial speed may differ from the guided speed. Additionally, the current state of the signal could be either red or green. In either case, the vehicle must adjust its speed according to the system state at the beginning of each stage 1.

In this regard, the DESGS aims to provide an explicit objective function to minimize the total fuel consumption when vehicles pass several intersections. The control performance index at each step for the kth vehicle is evaluated in terms of a cost function L to calculate the fuel consumption rates.

$$L = w_1 * \mathrm{EF}_{(v_k, v_k'')} + w_2 * \left| v_k - v_g \right| * \mathrm{EF}_{(v_g)} + w_3 * F_{\mathrm{brake}}^{k} * \left(\mathrm{EF}_{(v_k, v_k')} + \mathrm{EF}_{(v_k', v_k)} \right), \tag{2}$$

where w_1 is the weight factor for the fuel consumption, w_2 is the weight factor for the deviation from the target speed, w_3 is the weight factor for the braking effort, and they are all set as one-third in this study for simplicity; v_k is the initial velocity of the kth vehicle, v_k'' is its final velocity, v_g is the guided velocity, and v_k' is the velocity when the kth vehicle crosses the intersection stop line; $\mathrm{EF}_{(v_g)}$ is the fuel consumption at v_g over one optimization interval (d_{it}), $\mathrm{EF}_{(v_k, v_k'')}$ is the fuel consumption from v_k to v_k'' over one d_{it}, and F_{brake}^{k} is the braking force of the kth vehicle. The indexes w_1 and w_2 penalize the deviations of vehicle speed v_k from the guided velocity v_g, and w_3 indicates a reduced use of brake force.

To achieve DESGS, we need to add certain constraints for controlling the movement of vehicles. In this study, the speed limit, acceleration and deceleration limits, and the minimum safe following distance are imposed as pointwise-in-time inequality constraints [21]. These constraints are effective over the future prediction horizon.

(1) Speed Limit Constraint. The speed limit constraint is as follows:

$$v_{\min} \le v_k \le v_{\max}. \tag{3}$$

As mentioned above, to reduce the fuel consumption and emissions, sharp accelerations and decelerations and idling time should be avoided as much as possible. Note that t_t is the total time taken for the vehicle to reach the next signalized intersection after receiving speed advice, and t_s is the time until then next signal state changed where $s \in \{\text{green}, \text{red}\}$. For example, t_{green} denotes the time when the signal turns green from the instant when the vehicle receives speed advice. t_{green} and t_{red} have number values. Suppose that $t_{\text{green}}(i)$ is the start of the ith green signal state and that $t_{\text{red}}(i)$ is the start of the ith red signal state. In this case, v_g can also be determined by

(i) $(v_k t_a + (1/2) a t_a^2) + v_g (t_t - t_a) = d_k$

(ii) $t_a \le t_t$

(iii) $t_{\text{green}}(i) \le t_t < t_{\text{red}}(i)$

(iv) $v_{\min} \le v_g = v_k + a t_a \le v_{\max}$.

In constraint (i), t_a is the portion of time spent accelerating or decelerating at acceleration or deceleration rate a, and d_k is the distance between the kth vehicle and the next downstream intersection. Constraints (i) through (iii) ensure that the vehicle will receive a green signal when arriving at the intersection. Constraint (iv) limits the guided velocity to the roadway speed limit.

(2) Acceleration and Deceleration Limit Constraints. The acceleration and deceleration are manipulated to track the target speed as closely as possible while maintaining a safe distance from the front vehicle. The bounds on the acceleration and deceleration are represented by

$$\begin{aligned} 0 &\le a_{\text{acceleration}}^{k} \le a_{\text{acceleration}}^{\max} \\ 0 &\le a_{\text{deceleration}}^{k} \le a_{\text{deceleration}}^{\max}, \end{aligned} \tag{4}$$

where $a_{\text{acceleration}}^{\max}$ and $a_{\text{deceleration}}^{\max}$ depend on the vehicle type.

(3) The Minimum Safe following Distance Constraint. The minimum safe distance between a vehicle and the preceding vehicle is a function of the vehicle speed and is chosen as [29]

$$x_p - x_k \ge \alpha \cdot v_k + \beta, \tag{5}$$

where β is a "static gap" parameter that determines the minimum distance needed for the vehicle to stop and α is a "dynamic gap" parameter that indicates the extra time gap given increased speed. When the vehicle is approaching a red light, the light is considered similar to a stopped vehicle, and the position x_p is fixed at the light. This ensures that the vehicle will come to a stop at a distance of β from the light ($x_p \ge x_k + \beta$).

2.3. Eco-Index Calculation Module. In the eco-index calculation module, the vehicle fuel consumption and CO_2 emissions models consider the real-time speed, acceleration, and deceleration. In macroscopic models, the average speed and speed correction factors (SCFs) are used as the basic input parameters [30]. However, the SCFs were developed using data collected with fixed driving cycles. Thus, this approach may not adequately capture the effects of driving and vehicle dynamics on fuel consumption and emissions, especially when assessing traffic management strategies [31]. CMEM [32] was developed at a microscopic level to predict second-by-second fuel consumption and emissions using operational activities of a vehicle. Implementing the same basic physical principles as CMEM, the PERE model [33] was designed to generate fuel consumption rates for MOVES, which is the new generation emission model from the United States Environmental Protection Agency, based on the vehicle's on-road load. However, both CMEM and PERE require interfacing with various traffic models. The complexity mainly arises from the following requirements: (1) detailed parameters for vehicle classification, engine, and transmission and (2) application programming interface (API) combined with the speed optimization model for real-time calculation of fuel consumption and emissions. Thus, these models do not apply to fuel consumption impact assessments based on the state of traffic flow.

Studies have shown a corresponding relationship between the fuel consumption per unit distance and vehicle speed and acceleration [11, 31, 34]; however, speed or acceleration alone is inadequate for researching fuel consumption. Thus, there is a need is to discover another parameter associated with both the speed and acceleration to discuss the regularity of fuel consumption.

2.3.1. Vehicle Specific Power (VSP). The normalization parameter used in this study is based on characterizing the VSP of individual vehicles. The VSP has been shown to be a strong descriptor of vehicle fuel consumption and emissions [35] and is defined as the instantaneous tractive power per unit mass of the vehicle. For a typical light-duty vehicle, VSP (in kW/metric ton) can be calculated as

$$\mathrm{VSP} = v \cdot (1.1 \cdot a + 0.132) + 0.000302 \cdot v^3, \quad (6)$$

where v is the vehicle speed (m/s) and a is the acceleration (m/s^2). Using (6), we can conclude that when the speed is zero, VSP must be zero and that when acceleration is negative, VSP may be a negative value. Thus, VSP is still a vector although it is being used as a descriptor of the motor vehicle power demand. When a vehicle slows down, the load force acting on the vehicle and traveling direction is inverted, corresponding to a negative value of VSP.

VSP-based approaches have recently been accepted in fuel consumption and emission modeling for two main reasons. First, VSP considers the power needed to overcome aerodynamic drag and rolling resistance and that needed to increase the kinetic and potential energy of a vehicle. Thus, VSP has a direct physical interpretation and strong statistical correlations with fuel consumption and emissions. Second, as shown in (6), by using typical parameters for a certain vehicle type, VSP can be calculated based on vehicle speed and acceleration.

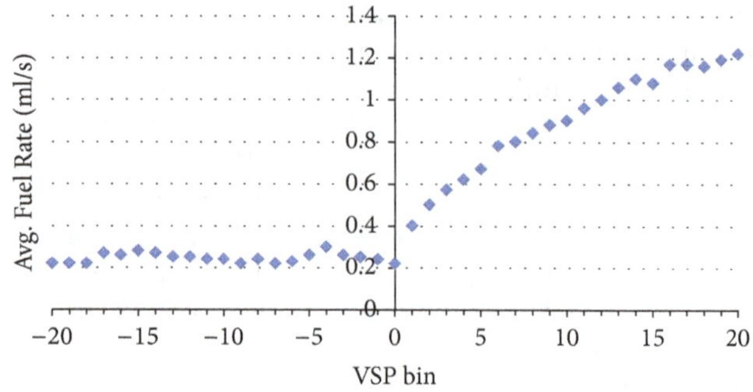

Figure 3: Average fuel rates (mL/s) in each VSP bin for a 2007 model Volkswagen.

2.3.2. Statistical Modeling of Fuel Consumption and VSP. Considerable research has been conducted on modeling the relationship between fuel consumption and VSP [36–38]. This study selected a 2007 Volkswagen model as an example to investigate the relationship between the fuel consumption and VSP distribution of light-duty vehicles.

To analyze the relationship between the VSP and fuel consumption per second more accurately, a binning approach is used. The VSP data are binned into 1 kW/t categories, and the average fuel consumption rate and the 95% confidence interval for each VSP bin are calculated [38]. The results are shown in Figure 3 for the 2007 Volkswagen model. The average fuel rates increase monotonically with VSP for positive bins, while they tend to be relatively lower and roughly constant for negative bins.

(1) Piecewise Modeling of Instantaneous Fuel Consumption and VSP. Statistical analysis found that, in urban transportation, negative VSP bins accounts for approximately 15% of the fuel consumption. Of the total fuel consumption, 68% occurs when VSP is positive, and 17% occurs when the value of VSP is zero. Based on the above analysis, a piecewise model was adopted. The piecewise function of the instantaneous fuel rate (FR, mL/s) versus VSP is shown as follows:

$$\mathrm{FR} = \begin{cases} 0.2508, & \text{for VSP} < -0.5 \\ 0.22, & \text{for } -0.5 \le \text{VSP} < 0.5 \\ 0.0411 \cdot \text{VSP} + 0.4629, & \text{for VSP} \ge 0.5. \end{cases} \quad (7)$$

In the regression equation, the standard deviation is 0.13 when VSP < -0.5, and $R^2 = 0.9618$ when VSP ≥ 0.5. Thus, the regression results are significant. In addition, the above piecewise function is not continuous. The values of FR in the interval $-0.5 \le$ VSP < 0.5 are lower than those in other ranges, which is caused by the idle speed control of the vehicle.

(2) Fuel Consumption per Unit Distance. Considering that FR is the instantaneous fuel consumption in each time unit (second), we take the integrals of FR and instantaneous velocity in

this time period to obtain the total fuel consumption and total distance traveled, respectively. The fuel consumption per unit distance (FU) of this light-duty vehicle can then be calculated as follows:

$$\mathrm{FU} = \frac{\int \mathrm{FR}\, dt}{\left(\int v\, dt/d\right)}, \qquad (8)$$

where FU is the fuel consumption per kilometer, mL/km; v is the instantaneous velocity, m/s; and d is the unit distance, 1,000 m.

The speed data obtained from the actual test usually have a unit of meters per second, or even per 0.1 second, and are not continuous variables. Thus, we can use (9) instead of (8).

$$\mathrm{FU} = \frac{\sum_i \mathrm{FR}_i}{\left(\sum_i v_i/d\right)}. \qquad (9)$$

From the above analysis, it can be concluded that if continuous speed data are acquired, we can calculate the fuel consumption per unit distance of a specific vehicle. If a vehicle is driving at a constant speed v ($v_{\min} \le v \le v_{\max}$), (9) can be expressed as

$$\mathrm{FU}' = d \cdot \frac{\left(0.0411 \cdot \left(0.132 \cdot v + 0.000302 \cdot v^3\right) + 0.4629\right)}{v}. \qquad (10)$$

Assuming that $d = 1{,}000$ m, we can obtain the minimum value of FU$'$. The calculation results indicate that the most fuel-efficient speed for this light-duty vehicle is 95.51 km/h and that the minimum fuel consumption (per kilometer) is 31.61 mL/km. The theoretical calculations of this study and the values based on actual tests match well, demonstrating that the model can represent the actual situation well.

2.3.3. CO_2 Emissions Model. Studies have found that, in urban driving, the trends of CO_2 emissions and fuel consumption are highly consistent, with correlations of up to 0.99 [38]. According to the mass conservation law, after fuel combustion in the engine, the total carbon mass in the exhaust gas is equal to that in the fuel. Carbon atoms exist in the forms of carbon dioxide (CO_2), carbon monoxide (CO), and hydrocarbons (HC) in the exhaust gas. The instantaneous fuel consumption is calculated from emissions of CO_2, CO, and HC using the carbon balance method, as shown in

$$\mathrm{FR} \cdot \mathrm{SG} = \left(\mathrm{ER}_{\mathrm{CO}_2} \cdot F_{\mathrm{CO}_2} + \mathrm{ER}_{\mathrm{CO}} \cdot F_{\mathrm{CO}} + \mathrm{ER}_{\mathrm{HC}} \cdot F_{\mathrm{HC}}\right) \cdot \frac{1}{g_C}, \qquad (11)$$

where FR is the instantaneous fuel rate, mL/s; SG is the fuel density, 0.749 g/mL at 15°C; $\mathrm{ER}_{\mathrm{CO}_2}$, $\mathrm{ER}_{\mathrm{CO}}$, and $\mathrm{ER}_{\mathrm{HC}}$ are the emission rates of CO_2, CO, and HC, respectively, g/s; F_{CO_2}, F_{CO}, and F_{HC} are the weight fractions of carbon in CO_2, CO, and HC emissions and are 0.273, 0.429, and 0.866 for gasoline vehicles, respectively; and g_C is the carbon percentage of gasoline by weight, which has a typical value of 86.4%.

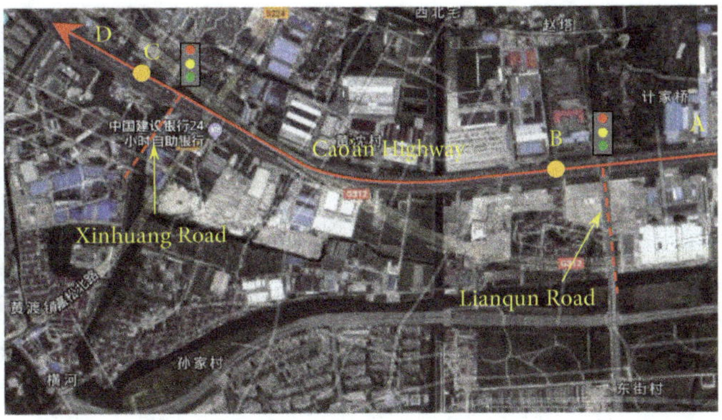

FIGURE 4: Test bed snapshot.

The proportions of CO and HC in all the exhaust gas are less than 2% and are especially low in strict emission control conditions; thus, they have almost no effect on the balance equation. Thus, the influences of CO and HC were neglected in this study when analyzing the relationship between CO_2 emissions and fuel consumption. Thus, (11) can be simplified as

$$\mathrm{ER}_{\mathrm{CO}_2} = \mathrm{FR} \cdot 2.370. \qquad (12)$$

3. Case Study

3.1. Test Bed. A section of Cao'an Highway near the Jiading Campus of Tongji University, Shanghai, China, was selected as the test bed for analysis. The experimental section is 1,700 m long, with eight lanes in two directions. As illustrated in Figure 4, two adjacent signalized intersections are 1,200 m apart and a curved segment is included between them.

In the experiment, drivers traveled from point A to point D. To eliminate additional errors, such as starting and stopping and unskilled operation, only the data collected between point B (50 m from the downstream Lianqun Road) and point C (100 m from the downstream Xianghuang Road) were taken as valid. In addition, during the experiment the vehicle speed and acceleration were collected every 0.1 seconds.

The other parameters, for example, traffic composition and link capacity, were collected via field surveys. In all the experiments, the vehicle mass is assumed to be $m = 1365$ kg. $v_{\max}$ is set to 80 km/h according to the road speed limit, and $v_{\min}$ is set to 40 km/h considering that a driver is likely to deviate from the desired speed if the guided velocity is lower than 40 km/h. The maximum acceleration and maximum deceleration are assumed to be $a^{\max}_{\text{acceleration}} = 3.5\,\mathrm{m/s^2}$ and $a^{\max}_{\text{deceleration}} = -7.0\,\mathrm{m/s^2}$, respectively. In most cases, the vehicles travel at a constant feasible deceleration, called desired deceleration, which is assumed to be $a^{\text{desired}}_{\text{deceleration}} = -3.0\,\mathrm{m/s^2}$.

3.2. Experiments. The dynamic eco-driving speed guidance system consists of the drivers, vehicles, driving environment,

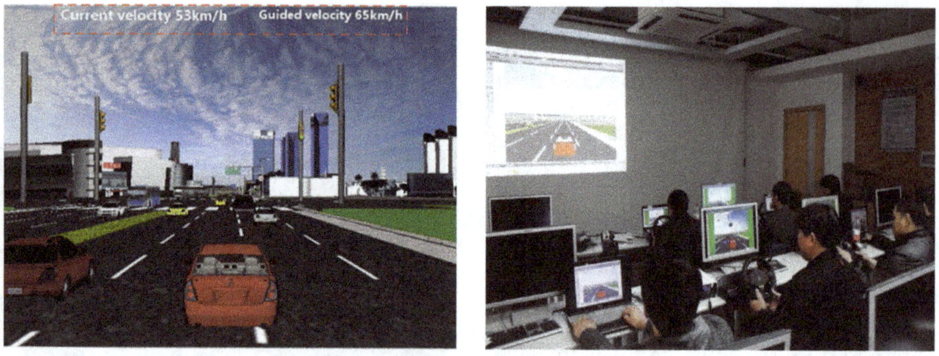

(a) Scene screenshot (b) Multivehicle driving experiment

FIGURE 5: Multivehicle driving simulator based experiments.

and speed guidance strategy. Experiments were conducted using driving simulators because we investigated a nonexistent situation and the experiment could create potentially dangerous situations due to unknown effects. The driving simulator experiment can also describe the influence of driver's psychology and behavior stimulated by the dynamic speed information. In this study, the program-oriented multidriving simulator was chosen as the experimental platform. Each computer was equipped with a special steering wheel, gear, pedals, and other hardware devices. 3DS Max and VIRTOOLS modeling software were used to build the virtual driving scene, as shown in Figure 5(a). The DESGS was programmed using the script language provided by VIRTOOLS. Using an intranet environment, the vehicles could be driven and interact in the same experimental setup, as shown in Figure 5(b).

Fifteen typical drivers from Tongji University and the nearby community were recruited to participate in the experiments. The ages of the participants were between 21 and 50 and followed a normal distribution. All the participants had held their driving licenses for more than five years and drove more than 8,000 kilometers annually.

The participants completed the driving experiments in two scenarios: vehicles with and without eco-speed guidance. Each scenario was simulated 24 times. The data, including time, vehicle position coordinates (x, y), current velocity, and guided velocity, were collected.

4. Results of Analysis

The 3rd vehicle of the group was placed in one of two scenarios: vehicles without speed guidance and vehicles with eco-speed guidance. An example of an eco-driving vehicle velocity profile compared to a non-eco-driving vehicle velocity profile is shown in Figure 6. The eco-driving vehicle had a much smoother velocity trajectory, yielding lower fuel consumption, as shown in Figure 7. The velocity profile shows that the vehicle never stopped with hard braking at red lights. With predictive use of signal information, the eco-driving vehicle scheduled its velocity based on the guided velocity, thus passing through intersections without coming to a stop.

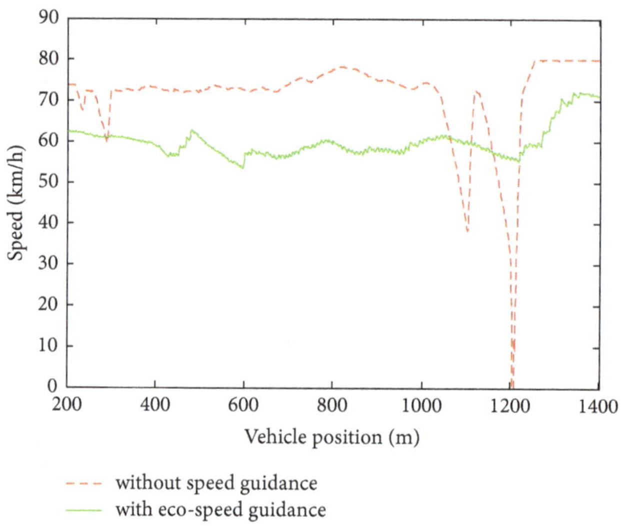

FIGURE 6: Velocity trajectory comparisons with and without speed guidance.

Figure 7 illustrates the instantaneous fuel rate comparison of two vehicles in different scenarios. Without speed guidance, the vehicles on the segment were driven randomly but usually had to stop at the downstream intersection. Frequent acceleration or deceleration required much fuel. The fuel rate curve was observed to be flat with speed guidance, increasing only when the speed advice was updated.

For comparison, the travel time, number of stops, fuel consumption, and CO_2 emissions were selected as the evaluation indexes. Figure 8 compares the results for the vehicles with and without eco-speed guidance. The results for both cases are given in terms of the average value of all the velocity profiles in the scenarios. According to Figure 8, for the vehicles with eco-speed guidance, the number of stops was significantly reduced; the total fuel consumption decreased from 53.01 mL to 39.88 mL (approximately 25% lower); and the CO_2 emissions decreased from 125.71 g to 94.56 g (approximately 25% lower). The average travel time was approximately 4% shorter for the vehicles with eco-speed

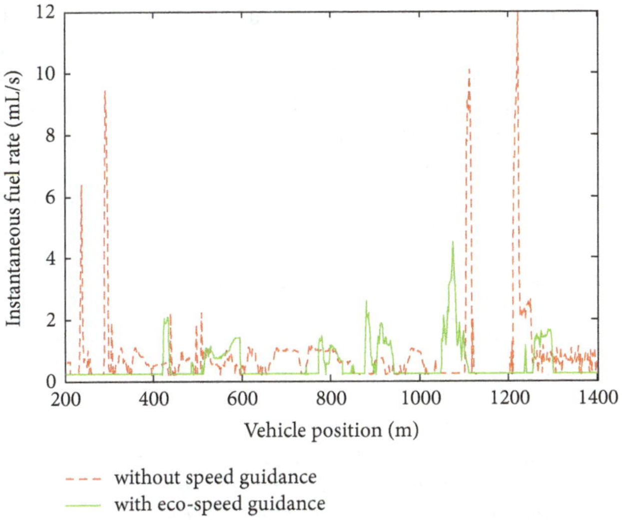

FIGURE 7: Instantaneous fuel rate comparison for vehicles with and without speed guidance.

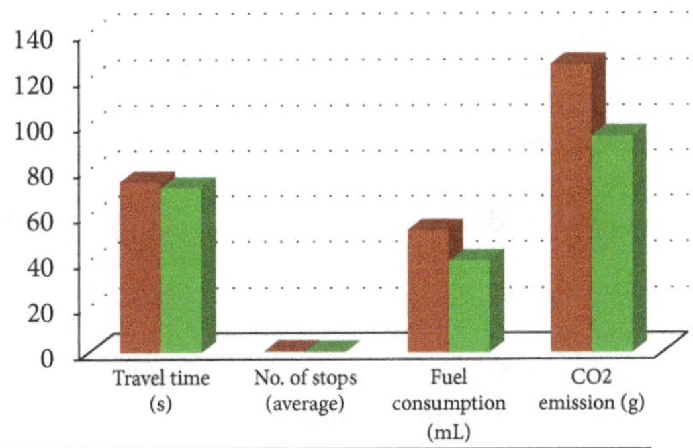

	Travel time (s)	No. of stops (average)	Fuel consumption (mL)	CO2 emission (g)
without speed guidance	75	0.60	53.01	125.71
with eco-speed guidance	72	0.07	39.88	94.56

FIGURE 8: Comparison of evaluation indexes for the vehicles with and without speed guidance.

guidance compared to the vehicles without speed guidance, but this difference is not significant. The energy savings and decrease in emissions are remarkable.

5. Conclusions and Future Work

To reduce fuel consumption and CO_2 emissions for traveling along signalized arterials, a dynamic eco-driving speed guidance strategy (DESGS) was established in this study using real-time signal timing and vehicle positioning information in a CV environment. The developed algorithms help optimize the fuel/emission speed profiles for vehicles approaching signalized intersections. To evaluate the effectiveness of the overall strategy, 15 typical drivers were recruited to participate in interactive experiments using multivehicle driving simulators. The main conclusions obtained in this study are summarized as follows:

(i) The optimization-based rolling horizon and DP approach can dynamically adjust the speed advice according to the vehicle's spatial-temporal trajectories. The eco-driving vehicles had much smoother velocity variations than the unguided vehicles.

(ii) The results given by the VSP-based fuel consumption model demonstrated good agreement with the field consumption curve. Thus, the VSP model was integrated into the optimal eco-driving speed calculation model.

(iii) The DESGS showed remarkable energy savings and decreased emissions. Both fuel consumption and CO_2 emissions were reduced by approximately 25% compared with the unguided scenarios.

(iv) The DESGS also showed a significant decrease in the number of stops and a small decrease in the travel time compared to the unguided scenarios.

There are still some ongoing efforts for the study. First, for simplicity, this study only investigated vehicles' fuel/emission speed profiles in a complete CV environment. That is, all the vehicles are CVs which are able to exchange traffic information and communicate with traffic signal controllers to obtain signal information. While in a partial CV environment, only limited vehicles can receive V2V and V2I information and a key impact factor on DESGS would be market penetration rate ranging from 0 to 100%. It is supposed that different market penetration rates of CVs would result in different savings of fuel consumption and CO_2 emissions through implementing DESGS. Besides, the robustness of the strategy to errors in wireless communication and driver responses should also be analyzed. Next, other impact factors, for example, the length of controlled segment, traffic demands (corresponding to undersaturated or oversaturated conditions), and signal phasing and timing, are supposed to be included in a more comprehensive analysis of the strategy performance. Furthermore, the developed DESGS can be extended to multiple signalized intersections along the arterial. Note that, in the case of multilane approaches, the potential impact of lane changing behavior in practice needs to be carefully considered. These will be our future work.

Conflicts of Interest

The authors declare that there are no conflicts of interest regarding the publication of this paper.

Acknowledgments

The authors appreciate the National Natural Science Foundation of China (nos. U1564212, 51508014, and 51422812), the Shanghai Science and Technology Project of International Cooperation (16510711400), and the Fundamental Research Funds for the Central Universities for their support of this research. Thanks also are due to Mr. Dening Niu and Mr. Shenyang Chen for comments and suggestions on an earlier version of this paper submitted to the 92nd Annual Meeting of the Transportation Research Board for presentation only.

References

[1] International Energy Agency, "Energy efficiency policy recommendations," in *Support of the G8 Plan of Action*, IEA, Paris, France, 2008.

[2] M. Barth and K. Boriboonsomsin, "Energy and emissions impacts of a freeway-based dynamic eco-driving system," *Transportation Research Part D: Transport and Environment*, vol. 14, no. 6, pp. 400–410, 2009.

[3] N. Jollands, P. Waide, M. Ellis et al., "The 25 IEA energy efficiency policy recommendations to the G8 Gleneagles Plan of Action," *Energy Policy*, vol. 38, no. 11, pp. 6409–6418, 2010.

[4] J. N. Barkenbus, "Eco-driving: an overlooked climate change initiative," *Energy Policy*, vol. 38, no. 2, pp. 762–769, 2010.

[5] K. Brookhuis and D. De Waard, "Limiting speed, towards an intelligent speed adapter (ISA)," *Transportation Research Part F: Traffic Psychology and Behaviour*, vol. 2, no. 2, pp. 81–90, 1999.

[6] K. Boriboonsomsin, O. Servin, and M. Barth, *Selection of Control Speeds in Dynamic Intelligent Speed Adaptation System: A Preliminary Analysis*, EScholarship, University of California, Calif, USA, 2008.

[7] H. Johansson, P. Gustafsson, M. Henke, and M. Rosengren, "Impact of eco driving on emissions. transport and air pollution," in *Proceedings of the 12th Symposium*, Avignon, France, 2003.

[8] J. Barbé and G. Boy, "On-board system design to optimize energy management," in *Proceedings of the European Annual Conference on Human Decision-Making and Manual Control (EAM'06)*, Valenciennes, France, 2006.

[9] O. H. Koskinen, "Improving vehicle fuel economy and reducing emissions by driving technique," in *Proceedings of the 15th ITS World Congress*, New York, NY, USA, 2008.

[10] H. Yang, H. Rakha, and M. V. Ala, "Eco-Cooperative Adaptive Cruise Control at Signalized Intersections Considering Queue Effects," *IEEE Transactions on Intelligent Transportation Systems*, vol. 18, no. 6, pp. 1575–1585, 2017.

[11] K. Ahn, H. Rakha, A. Trani, and M. van Aerde, "Estimating vehicle fuel consumption and emissions based on instantaneous speed and acceleration levels," *Journal of Transportation Engineering*, vol. 128, no. 2, pp. 182–190, 2002.

[12] Z. Zhang, Y. Wang, P. Chen, Z. He, and G. Yu, "Probe data-driven travel time forecasting for urban expressways by matching similar spatiotemporal traffic patterns," *Transportation Research Part C: Emerging Technologies*, vol. 85, pp. 476–493, 2017.

[13] M. Chen, G. Yu, P. Chen, and Y. Wang, "A copula-based approach for estimating the travel time reliability of urban arterial," *Transportation Research Part C: Emerging Technologies*, vol. 82, pp. 1–23, 2017.

[14] G. S. Pierre and J. Ehrlich, "Impact of intelligent speed adaptation systems on fuel consumption and driver behavior," in *Proceedings of the 15th ITS World Congress*, New York, NY, USA, 2008.

[15] M. Li, K. Boriboonsomsin, G. Wu, W. B. Zhang, and M. Barth, "Traffic energy and emission reductions at signalized intersections: a study of the benefits of advanced driver information," *International Journal of Intelligent Transportation Systems Research*, vol. 7, no. 1, pp. 49–58, 2009.

[16] P. Chen, K. Yin, and J. Sun, "Application of finite mixture of regression model with varying mixing probabilities to estimation of urban arterial travel times," *Transportation Research Record*, vol. 2442, pp. 96–105, 2014.

[17] P. Chen, G. Yu, X. Wu, Y. Ren, and Y. Li, "Estimation of red-light running frequency using high-resolution traffic and signal data," *Accident Analysis & Prevention*, vol. 102, pp. 235–247, 2017.

[18] Research and Innovative Technology Administration. Connected Vehicle Research, http://www.its.dot.gov/connected_vehicle/connected_vehicle.htm, 2012.

[19] Z. Papp, "Situational awareness in intelligent vehicles," in *Handbook of Intelligent Vehicles*, A. Eskandarian, Ed., pp. 62–78, Springer, 2012.

[20] S. Mandava, K. Boriboonsomsin, and M. Barth, "Arterial velocity planning based on traffic signal information under light traffic conditions," in *Proceedings of the 12th International IEEE Conference on Intelligent Transportation Systems*, St. Louis, Mo, USA, 2009.

[21] B. Asadi and A. Vahidi, "Predictive cruise control: utilizing upcoming traffic signal information for improving fuel economy and reducing trip time," *IEEE Transactions on Control Systems Technology*, vol. 19, no. 3, pp. 707–714, 2011.

[22] M. Barth, S. Mandava, K. Boriboonsomsin, and H. Xia, "Dynamic ECO-driving for arterial corridors," in *Proceedings of the IEEE Forum on Integrated and Sustainable Transportation System (FISTS '11)*, pp. 182–188, IEEE, Vienna, Austria, July 2011.

[23] H. Rakha and R. K. Kamalanathsharma, "Eco-driving at signalized intersections using V2I communication," in *Proceedings of the 14th IEEE International Intelligent Transportation Systems Conference (ITSC '11)*, pp. 341–346, IEEE, Washington, DC, USA, October 2011.

[24] H. Bock and K. Plitt, "A multiple shooting algorithm for direct solution of optimal control problems," in *Proceedings of the 9th IFAC World Congress*, Pergamon Press, Budapest, Romania, 1984.

[25] D. Kraft, "On converting optimal control problems into nonlinear programming problems," in *Computational Mathematical Programming*, K. Schittkowski, Ed., vol. 15, pp. 261–280, NATO ASI, 1985.

[26] A. E. Bryson and Y. C. Ho, *Applied Optimal Control*, Taylor & Francis, New York, NY, USA, 1975.

[27] A. Fröberg and L. Nielsen, "Optimal fuel and gear ratio control for heavy trucks with piece wise affine engine characteristics," in *Proceedings of the 5th IFAC Symposium on Advances in Automotive Control*, Monterey Coast, Calif, USA, 2007.

[28] B. Saerens, H. Rakha, and E. V. D. Bulck, "Assessment of eco-cruise control calculation methods," in *Proceedings of the 91st Annual Meeting of the Transportation Research Board*, Washington, DC, USA, 2012.

[29] A. Vahidi and A. Eskandarian, "Research advances in intelligent collision avoidance and adaptive cruise control," *IEEE Transactions on Intelligent Transportation Systems*, vol. 4, no. 3, pp. 132–153, 2003.

[30] C. Brzezinski and P. Enns, "Final facility specific speed correction factors," Printed on Recycled Paper EPA-420-R-01-060, EPA, Washington, DC, USA, 2001.

[31] A. Cappiello, I. Chabini, E. Nam, A. Lue, and M. Abou Zeid, "A statistical model of vehicle emissions and fuel consumption," in *Proceedings of the IEEE 5th International Conference on Intelligent Transportation Systems.*, pp. 801–809, Singapore.

[32] G. Scora and M. Barth, *User's Guide: Comprehensive Modal Emissions Mode (CMEM)*, University of California, Riverside, Calif, USA, 3.01 edition, 2006.

[33] E. K. Nam, "Fuel consumption modeling of conventional and advanced technology vehicles in the physical emission

rate estimator (PERE)," U.S. Environmental Protection Agency EPA420-P-05-001, EPA.

[34] R. Joumard, A. J. Hickman, and J. Nemerlin, "Modelling of emissions and consumption in urban areas (MODEM)," INRETS, LEN 9213, Bron, France, 1992.

[35] J. L. Jiménez-Palacios, *Understanding and Quantifying Motor Vehicle Emissions with Vehicle Specific Power and TILDAS Remote Sensing [Doctoral, thesis]*, Massachusetts Institute of Technology, Cambridge, Mass, USA, 1999.

[36] H. Wang, L. Fu, Y. Zhou, and H. Li, "Modelling of the fuel consumption for passenger cars regarding driving characteristics," *Transportation Research Part D: Transport and Environment*, vol. 13, no. 7, pp. 479–482, 2008.

[37] G. Song and L. Yu, "Estimation of fuel efficiency of road traffic by a characterization of VSP and speed based on FCD," in *Proceedings of the 88th Annual Meeting of the Transportation Research Board*, Washington, DC, USA, 2009.

[38] G. Song and L. Yu, "A practical model approach for evaluation of fuel efficiency for road traffic," in *Proceedings of the 87th Annual Meeting of the Transportation Research Board*, Washington, DC, USA, 2008.

4

Model-Based Optimization of Velocity Strategy for Lightweight Electric Racing Cars

Mirosław Targosz, Wojciech Skarka, and Piotr Przystałka

Silesian University of Technology, Institute of Fundamentals of Machinery Design, 18A Konarskiego Street, 44-100 Gliwice, Poland

Correspondence should be addressed to Wojciech Skarka; Wojciech.Skarka@polsl.pl

Academic Editor: Sara Moridpour

The article presents a method for optimizing driving strategies aimed at minimizing energy consumption while driving. The method was developed for the needs of an electric powered racing vehicle built for the purposes of the Shell Eco-marathon (SEM), the most famous and largest race of energy efficient vehicles. Model-based optimization was used to determine the driving strategy. The numerical model was elaborated in *Simulink* environment, which includes both the electric vehicle model and the environment, i.e., the race track as well as the vehicle environment and the atmospheric conditions. The vehicle model itself includes vehicle dynamic model, numerical model describing issues concerning resistance of rolling tire, resistance of the propulsion system, aerodynamic phenomena, model of the electric motor, and control system. For the purpose of identifying design and functional features of individual subassemblies and components, numerical and stand tests were carried out. The model itself was tested on the research tracks to tune the model and determine the calculation parameters. The evolutionary algorithms, which are available in the *MATLAB Global Optimization Toolbox*, were used for optimization. In the race conditions, the model was verified during SEM races in Rotterdam where the race vehicle scored the result consistent with the results of simulation calculations. In the following years, the experience gathered by the team gave us the vice Championship in the SEM 2016 in London.

1. Introduction

The issue of energy saving in road transport is getting more and more important. It is especially significant in the context of electric drives in vehicles becoming widespread. Reduction of energy consumption together with the development of new energy sources of greater capacity is seen as a method of overcoming the main limitation of electric vehicles which is range. *Shell Eco-marathon* car race is the testing ground for new solutions in designing energy saving vehicles. A team of scientists and students from the Silesian University of Technology has been taking part in it since 2012, in energy saving vehicles with electric drives, which are designed and built by them in the following categories: Prototype, UrbanConcept with Battery Electric, and UrbanConcept with Hydrogen Fuel Cell Stack. Significant reduction in energy consumption is achieved by the usage of two methods [1]. The first one includes design changes which result in lower energy consumption whereas the latter one introduces new strategies of driving and drive steering, which allows minimalizing the energy consumption on a given route. By means of this method, it is possible to reduce considerably the energy consumption in vehicles and in particular in electric vehicles. Currently, electric vehicles (city cars: segment B) can reach the result of about 5km/kWh of energy, whereas respective vehicles of UrbanConcept Battery Electric category reach up to 200 km/kWh in simulated conditions of driving in a city. However, prototype battery electric vehicles which are different type as far as their structure is concerned reach the result of 1000 km/kWh. It shows the potential of the design solutions and the strategy of driving. In both categories numerical simulation vehicle models are used, which allow determining the direction of designing changes as well as planning the proper strategy of driving. These models are also used for optimization based on Model-Based Optimization methodology (MBO). This proprietary methodology used for vehicles development and planning the driving strategy has been described in this paper. Vehicles modelling and in particular electric vehicles modelling are widely known and become of an increased interest to research, in particular with hybrid drive or power source. The necessity to model phenomena and objects is commonly known and creates substantial branch of science and engineering activities. The research and analysis of mechanical systems, technological processes, and other phenomena in real world are possible thanks to the use of proper mathematical apparatus which is based on previously assumed mathematical models of these processes [2]. According to this, by means of mathematical model, one should understand mathematical description of an object, process, or phenomenon which it represents.

Compared to cars with internal combustion engines, electric cars have a relatively simpler drive system; however, in order to achieve a significant reduction in energy consumption in the electric car's drive system, additional systems such as the energy recovery system, additional mechatronic subsystems, and advanced control systems are integrated, resulting in a continuous increase in the complexity of the electric drive systems currently used. The complexity of the technical means influences the complexity of the mathematical model of the vehicle. Among the models describing the dynamics of vehicles, which are used in the analysis and optimization of energy consumption, two approaches are generally observed due to the type of model used and the following models are distinguished:

(i) Analytical ones in which the model response to the signal is recorded as a motion equation

(ii) Models in the black box concept where the model is based on experimental data

The created vehicle simulation model can be a combination of two model types. To some extent, the mathematical description of a given phenomenon is based on known dependencies, while another fragment of the model is created in the concept of a black box. Taking into account the direction of information processing in the model, the following models [3] are distinguished:

(i) Simple: where the calculations made by the model start with the engine, energy is transferred to the wheels of the vehicle, and vehicle behaviour can be analyzed

(ii) Reverse: in which the behaviour of the body of the vehicle is modelled, e.g., the speed, and on this basis the required torque and the speed of rotation of the motor shaft are determined.

The choice of model type is mainly determined by the need for which the model is to be used. For modelling of dynamics of electric vehicles, many computer programs are used; among them noteworthy is the *ADVISOR* (*Advanced Vehicle Simulator*) platform [4–6] written as a program in the *MATLAB-Simulink* package. The program in its libraries gives you the opportunity to analyze many of the solutions of the power system, drive system, or vehicle body. Another platform running in the *MATLAB-Simulink* environment is *PSAT* (*The Powertrain System Analysis Toolkit*), which simulates many predefined solutions for conventional, electric, and other vehicles [7]. The evolution of *PSAT* is *Autonomy* [8] available on *LMS Imagine.Lab*. In addition to these simulation environments, where models are abstract and which take into account the mechatronic nature of the electric vehicle's propulsion system, there are many specialized dynamic analysis programs utilizing the multibody formalism, such as the *LS-Dyna*, *Adams*, or *MotionSolve* software. Simulation models designed to analyze and optimize energy consumption should allow for less time consuming calculations. This is especially important in optimization tasks, where simulation is performed many times (often over several tens of thousands of times). You can say that the faster the model is, the more abstract it is. In paper [9], this is demonstrated by the synchronous motor, written in *MATLAB* and *Simplorer*. In order to reduce the computational time, the model should be written in a low-level language, e.g., C, as given in [10] and the simulation time can be reduced by up to 20 times compared to a program written in *MATLAB's* own language.

2. Research Methodology

In order to optimize the control strategy for a light, electric racing vehicle, the general methodology described below is presented, which explains the basic ideas and above all the definitions of the key concepts from energy consumption domain. The general functional structure of the drive system with particular attention to drive control and the optimization scheme are also presented below.

Energy Consumption E of the Movement. Movement of the vehicle is a consequence of the longitudinal force, which overcoming the force of inertia and resistance against the movement performs work on a particular road section [12]. The energy that is associated with this work is called energy intensity and, in the case of wheeled vehicles, it can be expressed as

$$E = \int_0^{D_c} F_N(s)\,ds \tag{1}$$

where F_N *is* driving force and D_C is travelled distance.

Energy consumption of the movement determines the amount of energy supplied to the drive wheels and **does not depend on the nature of the drive unit and the transmission**.

Total Energy Consumption E_c. It determines total energy expenditure. In the case of an internal combustion engine unit expressed as the product of the fuel and its calorific value, or in the case of an electrical unit, the energy is taken from an electric source. *Total energy consumption E_c* is the sum of the energy consumption E and the power losses in the motor ΔE_m and in the transmission system ΔE_{TS} [12]:

$$E_C = E + \Delta E_m + \Delta E_{TS} \tag{2}$$

The losses from (2) are usually taken into account in *efficiency* η:

$$\eta = \frac{E}{E_C} \tag{3}$$

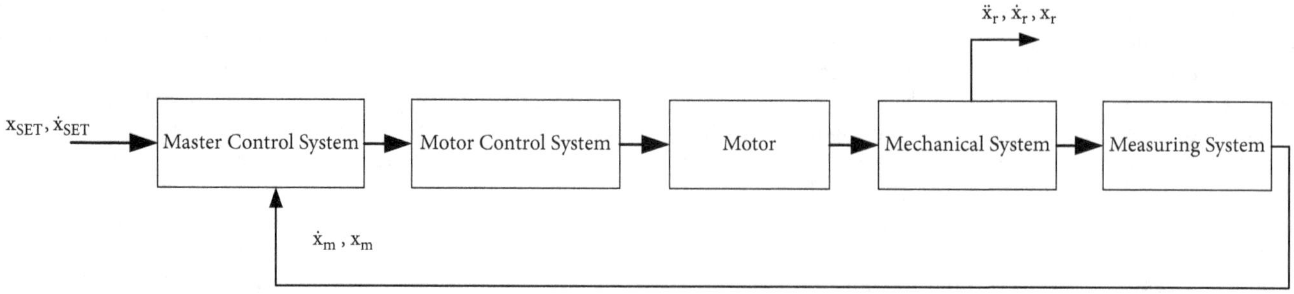

FIGURE 1: Functional diagram of the electric vehicle drive system.

The overall efficiency of the drive system is the product of the motor efficiency η_m and the efficiency of the transmission η_{TS}:

$$\eta = \eta_m \eta_{TS} \tag{4}$$

The total energy consumption E_c, of the process, which is defined as driving of the given distance D_C, is influenced by many factors, the vehicle's structural parameters, the energy losses generated when generating and transmitting the driving torque, and the control method. A function describing the speed of a vehicle along a certain route is the so-called speed profile [12, 13]. The speed profile is usually determined by the driver of the vehicle, by means of appropriate accelerator pedal control and shift gear selection or by electronic control. A cruise control system controls the power delivered to the drive wheels in order to maintain a certain speed.

If the vehicle is equipped with one drive system, it can be presented in the functional diagram as shown in Figure 1.

Specific blocks describe the performance of the individual vehicle systems:

(i) master control system: a speed profile generator, usually the driver, by adjusting the accelerator pedal, deciding on the profile, and depending on the road conditions;

(ii) motor control system: generator of appropriate motor control values (current, voltage);

(iii) motor: a machine that transforms the energy of a source into mechanical energy;

(iv) mechanical system of the vehicle: torque transmission from the motor shaft to the wheels of the vehicle (clutches, gears, etc.);

(v) measurement system: acting as feedback to provide information on the current position and speed of the vehicle.

The optimization task can be reduced to the following:

(i) Optimization of design parameters p_k and settings p_n, e.g., power of the drive, TS ratio, i.e., optimizing the characteristics of the vehicle describing the *mechanical system* and *current controller*. The design of the system and its mathematical description in the form of the model are decisive for the optimization of these parameters.

(ii) Optimizing the control strategy, i.e., searching for the optimum speed V_{SET} profile, or directing the control signal to the current controller U_{SET}.

(iii) Simultaneous optimization of control parameters and strategy.

You can use the Model-Based Optimization method to solve the optimization task. The scheme of the optimization is shown in Figure 2.

In this method, decision variables, whose value is determined by the optimization algorithm, are input variables for the simulation model. Numerical simulation of driving and obtained results during the simulation allows determining the result of the objective function. Repeating the simulation results for the subsequent driving parameters controlled by the optimization algorithm allows the calculation of successive results of the objective function. Frequently repeated results are aimed at finding the minimum of the target function and thus the optimal solution. Stopping the optimization algorithm loop depends on the type of the algorithm used. This usually occurs after a predetermined number of loops, after a lack of change in the value of the criterion function or other constraints.

3. Building the Model of the Electric Vehicle

To study the total energy consumption E_c of an electric vehicle, it is necessary to build a model to analyze the transfer of energy from the source of energy to the drive wheels of the vehicle. Figure 3 shows the functional diagram of the electric vehicle drive system. This diagram is the starting point for creating a simulation model.

According to the definition, the total energy consumption depends on the energy consumption of the movement E, that is, on the work generated by the driving force and the energy losses generated by the motor ΔE_m as well as on the energy loss of the transmission ΔE_{TS}. Accordingly, a mathematical model of the vehicle should be created describing the phenomenon of resistance and loss of energy in transmission system. The mathematical model of the electric vehicle is shown in this section. The model is based on equations of motion of a wheeled vehicle, biaxial with front axle steering, and rear wheel drive. The focus is on the representation of the driving forces and the resistance of movement acting in the direction of the longitudinal axis of

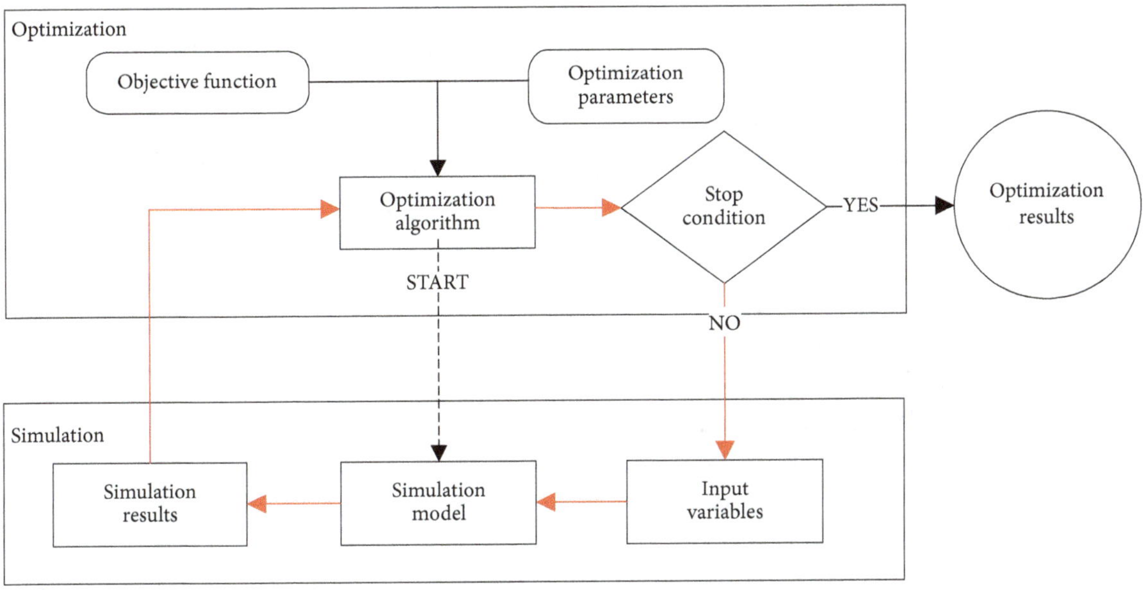

FIGURE 2: Schematic diagram of the optimization using model [11].

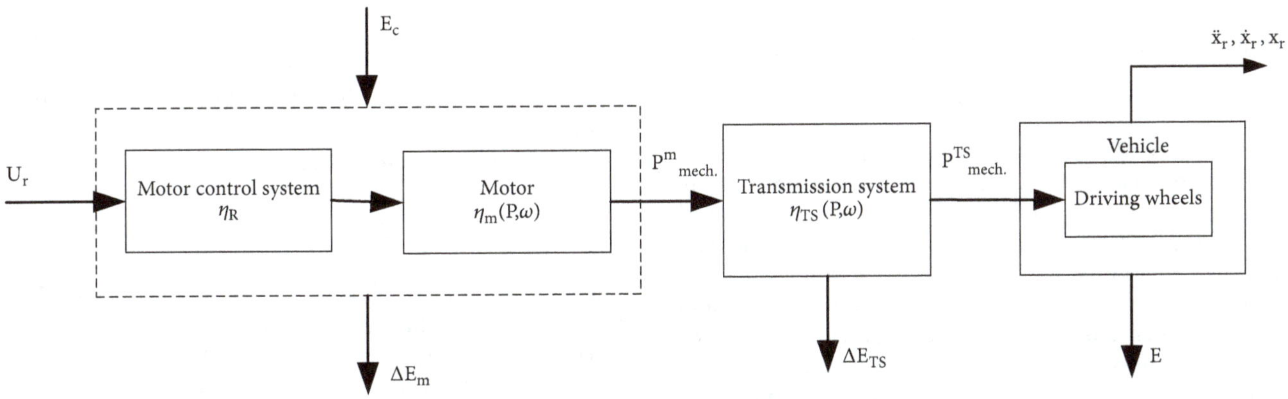

FIGURE 3: Functional diagram of an electric vehicle.

the vehicle. The model described here is sufficient to simulate and optimize the parameters of an object that affect the total energy consumption E_c of an object. There was no need to determine the values of forces and displacements affecting individual vehicle components, as in the case of, for example, strength analyses. These issues are not the subject of this study, so the forces acting in the directions of the other axes of the vehicle are not dealt with, except for the forces which, as a result of their impact, affect the total energy consumption.

3.1. Mathematical Model of the Vehicle. The theory of vehicle motion is widely published and has been presented many times in the scientific works [14, 15]. The following section provides formulas and symbols used to build the vehicle model. A detailed description of the presented issues should be sought, among others, in the above-mentioned bibliography. Figure 4 shows the forces acting on the vehicle moving along the path of the inclination angle α.

The following designations were used: F_N driving force, which is the quotient of the driving torque and the dynamic radius of the wheel; F_{T1} and F_{T2} of the rolling resistance of the wheels of front and rear axles, respectively; Q_m vehicle gravity force; F_P force of air resistance; $F_W = Q_m \sin(\alpha)$ uphill resistance; F_B inertia force; Z_1 and Z_2 ground reactions to the front and rear axle wheels; $m_z = m\delta_z$ reduced mass.

The equation of motion of the vehicle in the driving direction can be written as the sum of the driving forces and drag forces parallel to the road surface as

$$m_Z\ddot{x} = F_N - F_T - F_W - F_P \quad (5)$$

The following simplifications were adopted. The drive torque applied to the rear wheel does not affect the pressure distribution Z_1 and Z_2. In fact, in vehicles with rear wheel drive, the rear is downforced whereas the front axle is relieved. The reaction values Z_1 and Z_2 were statistically determined. The coefficients of rolling resistance of all wheels are identical.

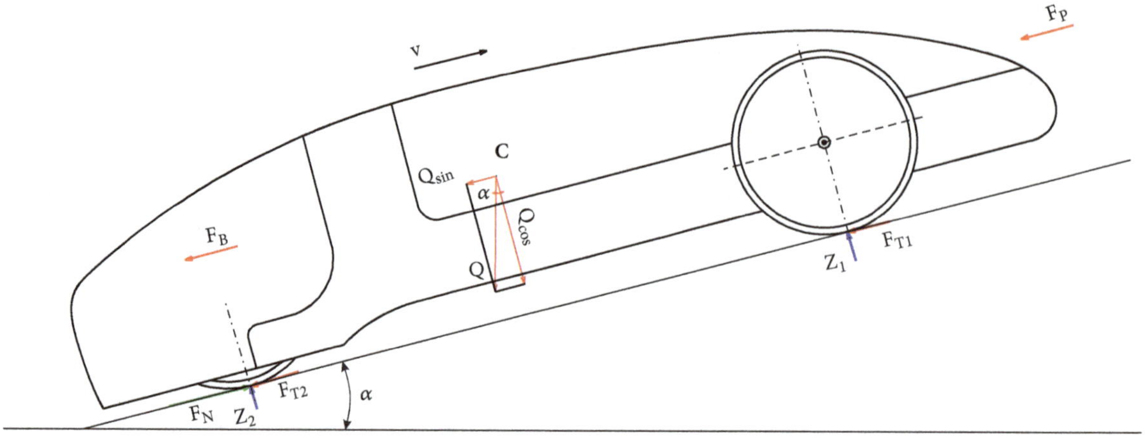

FIGURE 4: The system of forces acting on the vehicle.

Resistance forces depend on many factors and parameters; many of them are highly nonlinear. Modelling of wheel rolling resistance is the subject of separate research and the model used is very complex so in practice, a simplified model is often used.

The rolling resistance of a single wheel is as follows:

$$F_{T_i} = c_{rr} Z_i \tag{6}$$

c_{rr} is the coefficient of rolling resistance that depends on the type of road surface, the type of tire, or the pressure in the tire. This factor also depends on the speed of travel and increases with its growth; this is related to the increase in tire deformation. The value of the rolling resistance coefficient is usually given by the tire manufacturer, and exemplary values and empirical models on the dependence of this coefficient on the speed function can be found in the literature [16]. For low travel speeds (<40km/h), the value of the rolling resistance coefficient is assumed to be constant, independent of speed. In linear motion, the total rolling resistance is the sum of the rolling resistance of all vehicle wheels.

If an identical rolling resistance coefficient for each wheel is assumed, which is often justified, given the similar construction of the front and rear axle wheels and identical tires and the same pressure operation, the sum of the rolling resistance shall be written as

$$F_T = c_{rr} Q_m \cos\alpha \tag{7}$$

The total wheel resistance should take into account the tire rolling resistance, bearing losses, and so forth.

The essence of resistance in bearings is identical to the tire rolling resistance described previously. The bearing torque is proportional to the bearing load, friction coefficient, and bearing radius. Bearing friction is generally neglected, as the bearing resistance value is small compared to the tire rolling resistance [16].

Similarly to the value of rolling resistance, the turning resistance can be expressed as the product of the weight of the car and the coefficient of turning, although, according to literature, at low speeds it plays negligibly smaller role than the rolling resistance [16].

$$F_s = \mathrm{mg} f_z \tag{8}$$

f_s is the coefficient of turning and depends on the pressure force on the axles, tires, and so-called *intensity of turning*, in other words, ratio of centrifugal and vertical acceleration v^2/R_g.

The value of air resistance F_P can be determined from

$$F_P = \gamma \frac{v_{sum}^{\ 2}}{2} A c_x \tag{9}$$

where γ is the density of air. v_{sum} [m/s] is the total air velocity resulting from the linear velocity of the vehicle v_p and wind speed v_w and is the vector sum, A is the vehicle front area [m^2], and c_x is the aerodynamic drag coefficient.

In this study, a TS based on a constant gear ratio was used. PM BLDC (Permanent Magnet Brushless Direct Current) motors are very often used in electric power systems [17]. The mathematical description of all phenomena related to magnetism, electrical, thermal, or mechanical phenomena can lead to a considerable complexity of the model, so simplified models are often used. The electrical part of motor's model can be found in work [9]. The electromagnetic torque generated by the motor is proportional to the flowing current and the so-called *the motor torque constant* K_m and is given by [17]

$$M_e = 2K_m i_z \tag{10}$$

In steady state, the electromagnetic torque M_e is equal to the torque on the motor shaft, which is the sum of the load torque and the torque of the losses. The efficiency of an electric motor is defined as

$$\eta_m = \frac{P_{mech_m}}{P_{elect}} \tag{11}$$

P_{mech_m} is the mechanical output power on the motor shaft and P_{elect} is the electrical power input. Relations describing

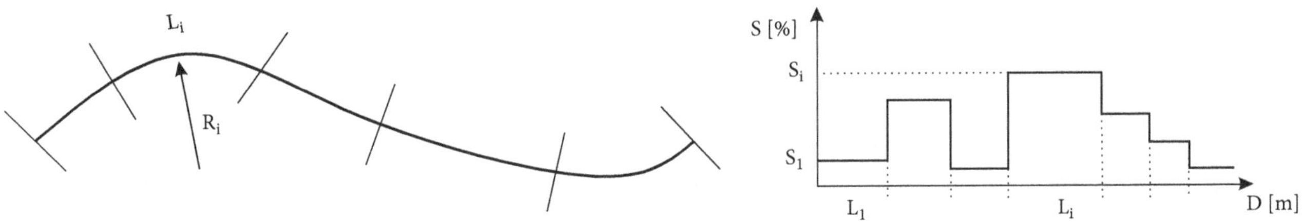

FIGURE 5: Driving path and its inclination profile.

the phenomenon of loss can be found in professional literature [10].

The motor torque is transmitted to the drive wheel via the TS transmission. The mechanical power on the drive wheel depends on the efficiency of the TS:

$$P_{mech_p} = \eta_{TS} P_{mech_m} = \eta_{TS} \eta_m P_{elect} \quad (12)$$

The efficiency η_{TS} is a function of many factors; i.e., it depends on the powertrain design, the transmitted power, and quality of the system. For TS based on tension transmission, efficiency can be approx. 90%. The efficiency of the tie link transmission is a function of inter alia transmission ratio i_{TS}, interaxial distance, belt angle, tension force, rotational speed, or torque [18–20]. In the case of lightweight vehicles in the TS design, it is often not necessary to use variable gear transmissions. Speed and torque control is achieved by the control system. On the one hand, it is an advantage because the TS is smaller in weight and less complicated, but on the other hand it forces the electric drive to operate in a wide range of torque and rotational speed, which affects the motor energy efficiency.

3.2. Building the Model of the Driving Route. The parameters of the path that the vehicle travels on have a significant effect on the resistance values shown in Section 3.1. In the paper, the path of ride is determined by the exact path of the vehicle on a designated route. It is assumed that the best path is known and it is determined based on the team's experience (it means that the quickest best path along the track is not calculated by the optimization method). The difference between the path and the track is as follows. A track is a system of physically existing roads between points A and B, and when the path does not exist physically, it is the trace that leaves the selected point of the vehicle (e.g., center of gravity) on the route surface. With this in mind, the journey along the route can take place on a number of paths. The route is only a limitation.

Figure 5 shows a schematic drawing of one path and its elevation profile. The description of the path is formed by its discretization, i.e., the division into finite numbers of elementary sections of $\mathbf{O}_{t_i}$, $i \in \{1, 2, ..., n\}$. Each elementary segment is described by a parameter vector.

$\mathbf{O}_{t_i} = [L_{t_i}, R_{t_i}, S_{t_i}, E_{t_i}]$, where $L_{t_i}[m]$ is the length of the path, $R_{t_i}[m]$ is the radius of curvature, $S_{t_i}[\%]$ is the inclination of the path, and E_{t_i} represents the type and condition of the path. The length of the path is the sum of the lengths of the L_{t_i} sections.

These basic data are essential for the correct determination of motion resistance on each segment. It is assumed that any change in the inclination or radius of curvature makes it necessary to define further vector $\mathbf{O}_{t_i}$.

Bearing in mind that vehicles usually move in the open air, a suitable model should be created describing the external conditions. For this purpose, the $\mathbf{O}_{t_i}$ vector describing the nature of the path should include additional columns such as velocity and direction of the wind v_{w_i} and v_{k_i}, atmospheric pressure p_{atm_i}, and air temperature t_{atm_i}. In this situation, the vector describing the path and external conditions takes the form: $\mathbf{O}_{t_i} = [L_{t_i}, R_{t_i}, S_{t_i}, E_{t_i}, v_{w_i}, v_{k_i}, p_{atm_i}, t_{atm_i}]$. The speed and direction of the wind have a significant effect on the value of air resistance. Determining the wind speed is relatively difficult and not precise. Temporary winds may have different directions and a much higher amplitude than the average. Preliminary investigations have been carried out considering the uncertainty of weather conditions. During optimization, where optimal control strategies were sought, the wind speed was randomly changed. The concept of optimizing the control strategy from taking into account the changing weather conditions was to design a family of strategies. It would be possible to select one optimal strategy or part of it in the real world because of the existing and changing weather conditions.

3.3. Control System Design. Two modes of the control system are taken into consideration in this study. They reflect the physical implementations of the control system used in the built-in electric race vehicles of Prototype class [21] and UrbanConcept class [22]. Figures 6 and 7 represent block diagrams of the drive system of the vehicle for different control strategies. The control scheme A (Figure 6) is used in a case when a driver is able to control an acceleration of the vehicle by means of a binary signal. In this case, the goal of the optimization task is to find the control strategy taking into account two criteria, i.e., the determined travel time as well as the minimal power consumption of the vehicle. The control strategy is used to specify in which parts of the route the driver should press or release the acceleration treadle. When the setpoint signal $U_{SET} \in \{0, 1\}$ is assumed, then the optimization task is to search for logic values of the control signal as the function corresponding to the position of the vehicle $x(t)$, $U_{SET} = f(x(t))$. The supervisory role of the driver is included in the functional scheme as a decision module. In this study, the behaviour of the driver is not modelled in the simulation. It means that he/she realises the strategy in

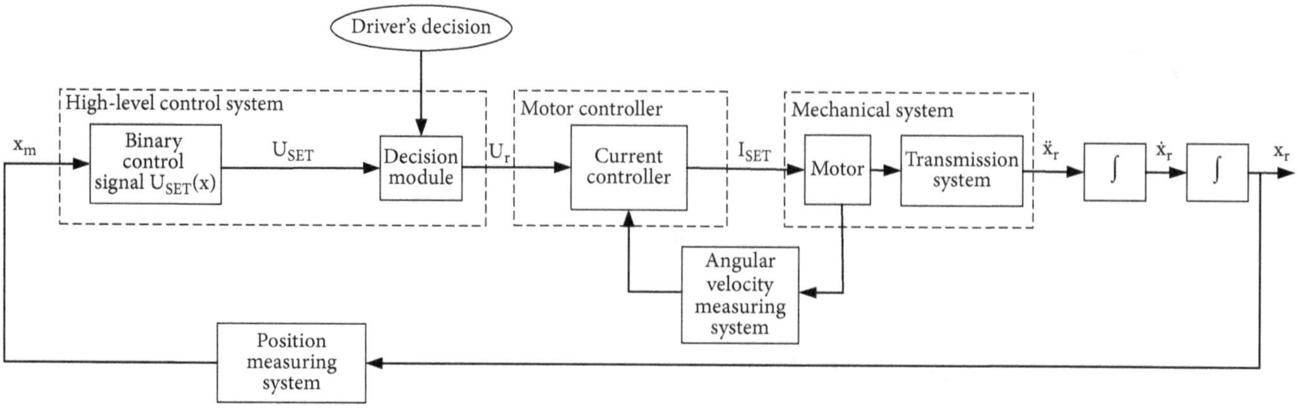

FIGURE 6: Drive system diagram: control scheme A.

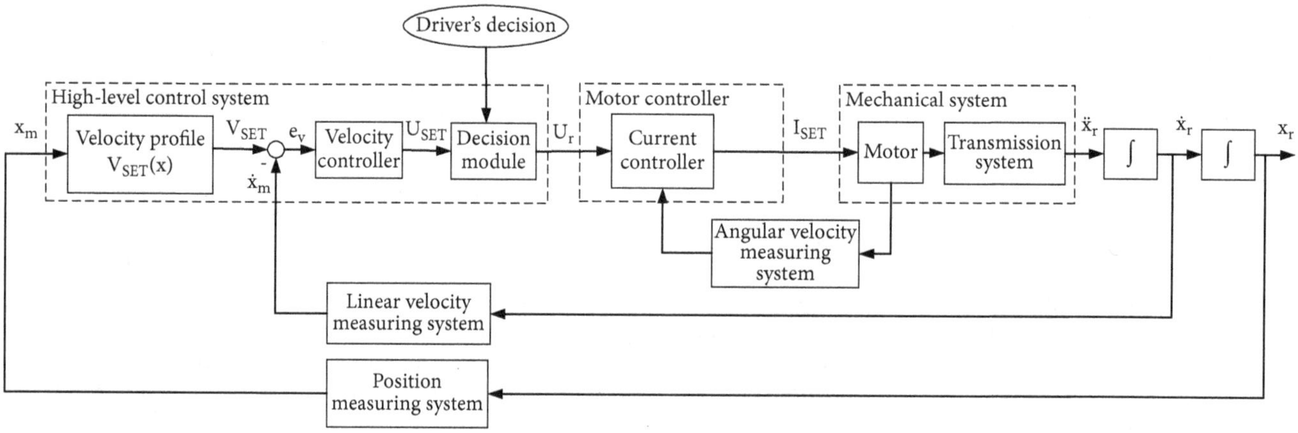

FIGURE 7: Drive system diagram: control scheme B.

the exact way. The input control signal U_r is always different from the signal U_{SET} in real conditions. It means that the signal U_{SET} is mainly employed by the driver to know when the acceleration treadle should be used.

In the case of the control scheme B (Figure 7), the general goal of the optimization task is the same as in the previous case. However, the formal definition of the problem is quite different because in this case the optimal form of the velocity signal is searched according to the position of the vehicle on the route $V_{SET} = f(x(t))$. The velocity controller compares the setpoint signal V_{SET} with the measured velocity of the vehicle and for certain controller setting as well as for control error e_v it generates the control signal U_{SET}. As in the previous case, the supervisory role of the driver is included in the decision module. However, the main difference is that in the optimization process the setpoint signal V_{SET} can be limited in the direct way.

According to control schemes A and B, the input control signal U_r is used by the motor control unit where the emended current regulator normally outputs the corresponding signals to the motor terminals. The driving torque of the motor shaft is transmitted through the transmission system to the wheels of the vehicle.

4. Evolutionary Optimization of the Velocity Strategy

4.1. Formulation of the Optimization Problem. One of the ways to define and to solve the dynamic optimization task of the velocity strategy of the vehicle is to reduce such a problem to the static optimization task through the so-called discretization of the independent variable. If one assumes that the independent variable is the position of the vehicle (a travelled distance) $x(t)$, then the velocity strategy **st** can be represented by a vector of instantaneous values U_{SET} or V_{SET} for discrete parts of the route. Figure 8 illustrates the idea of dividing the route into segments.

The path is divided into n segments O_i each of the length L_i, where the sum of the lengths of the segments is equal to the length of the total path. For each segment, it is necessary to determine the U_{SET_i} or V_{SET_i} setpoint. The task is to find the optimal velocity strategy for control scheme A in the following form:

$$\mathbf{st}_{opt_U} = \left[U_{SET_1} \;\; U_{SET_2} \;\; \cdots \;\; U_{SET_n}\right]^T, \quad U_{SET_i} \in \{0, 1\} \qquad (13)$$

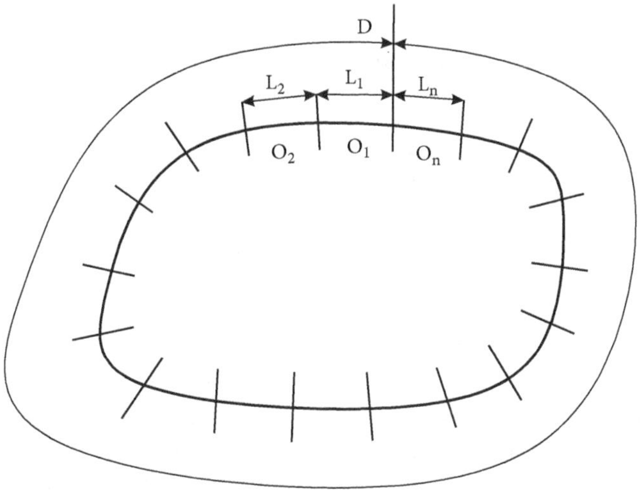

FIGURE 8: Concept of discretization of the route.

whereas in the second case for control scheme B

$$\mathbf{st}_{opt_V} = \begin{bmatrix} V_{SET_1} & V_{SET_2} & \cdots & V_{SET_n} \end{bmatrix}^T, \quad V_{SET_i} \in \{V_{min}, V_{max}\} \tag{14}$$

where $V_{\min}$ and $V_{\max}$ can be determined in different ways, e.g., based on the minimum/maximum vehicle speed or based on the simple calculation corresponding to guidelines of the race competition.

Evolutionary algorithms can be used to solve this problem. In the case of the search for strategy $\mathbf{st}_{opt_U}$ it is preferable to use a classical genetic algorithm. In this instance, the genes in a chromosome of the individual can be encoded directly as the binary values of the control signal U_{SET_i}. Therefore, the length of the chromosome N_{ch} is equal to the number of sections of the route.

In accordance with the control scheme B, the velocity profile is determined as the vector including setpoint values of the velocity $\mathbf{st}_{opt_V}$. The components of the $\mathbf{st}_{opt_V}$ vector are floating point values; hence it is possible to use an evolutionary strategy proposed in [23]. Each of the genes in the chromosome of the individual corresponds to the setpoint value of the velocity during a part of the route.

4.2. Chromosome Codding for Different Velocity Strategies. For a vehicle that is equipped with a binary control system of the motor, the chromosome can be represented as the optimal vector of the control signal in the following form $\mathbf{st}_{opt_U} = [U_{SET_1} \ U_{SET_2} \ \cdots \ U_{SET_n}]^T$, $U_{SET_i} \in \{0, 1\}$ for each segment O_i of the route (for control scheme A) or the optimal vector of the control signal in the second form $\mathbf{st}_{opt_V} = [V_{SET_1} \ V_{SET_2} \ \cdots \ V_{SET_n}]^T$, $V_{SET_i} \in [V_{\min}, V_{\max}]$ for each segment O_i of the route (for control scheme B).

In this optimization task, the total distance of the road D_C is composed of SEM race laps $N_{to} \in \{1, 2, ..., N\}$. The division of the whole route on a finite number of segments O_i is not recommended, because it can lead to unnecessary increase in the length of the vector **st**. If the route consists of a minimum of four laps, it is possible to limit the search of vector **st** to only three types of laps:

(i) the initial lap marked by O_{i_s}

(ii) the middle laps marked by O_{i_m}

(iii) the last lap marked as O_{i_f}

Such assumption can be proven for at least three reasons:

(1) In the practical implementation of velocity strategies, the driver should not implement different control rules for each lap. The master control unit informs the driver about the current velocity strategy, but despite all, the driver should remember this strategy to some extent.

(2) The limitation of the velocity strategy only to the three types of laps reduces the length of the vector **st**. In this way, the number of elements of the vector **st** corresponds to the number of decision variables of the optimization task.

(3) Such division is reasonable by the nature of the laps. In the initial lap, it is necessary to accelerate the vehicle. Within the middle laps, the velocity profile of the vehicle should be the same. During the last lap, the previously acquired kinetic energy E_k is utilized.

The task of finding the optimal vector $\mathbf{st}_{opt}$ can be solved using two types of evolutionary algorithms:

(i) Classical genetic algorithm for control without a velocity regulator (according to control scheme A presented in Figure 6)

(ii) Evolutionary strategy for the control system of the vehicle with a velocity regulator (according to control scheme B shown in Figure 7)

The usage of two algorithms forces the appropriate coding of the velocity strategy, which is described in the next two paragraphs.

4.2.1. Chromosome Coding For Control Scheme A. The phenotype of each individual determines the control vector $\mathbf{st}_u$ that means each gene of the individual of the population is the value of the control signal U_{SET_i} for each section of the segment O_i of the route. Assuming that we are looking for a velocity strategy of only three types of laps, the phenotype can be written as (for short form $\mathrm{U} \equiv U_{\mathrm{SET}}$)

$$\mathbf{st}_U = \begin{bmatrix} U_{1s} & U_{2s} & \cdots & U_{ns} & U_{1m} & U_{2m} & \cdots & U_{ns} & U_{1f} & U_{2f} & \cdots & U_{nf} \end{bmatrix}^T, \tag{15}$$

where U_{i_s}, U_{i_m}, U_{i_f} are the control signals for the initial lap, middle laps, and the last lap.

4.2.2. Chromosome Coding for Control Scheme B. In this case, the elements in the vector V_{SET_i} change over a wide range and can assign real number values. In the evolutionary strategy, the phenotype of each individual takes the following form:

$$\mathbf{st}_V = [\,V_{1s}\ V_{2s}\ \cdots\ V_{ns}\ V_{1m}\ V_{2m}\ \cdots\ V_{ns}\ V_{1f}\ V_{2f}\ \cdots\ V_{nf}\,]^T, \quad (16)$$

$V_{i_s}, V_{i_m}, V_{i_f}$ are the setting values of velocity for the initial lap, middle laps, and the last lap.

In the control mode, according to scheme B, the control signal is generated by the velocity controller based on the control error:

$$e_V = V_{SET} - V_{sim}, \quad (17)$$

where

$$U_{SET} = \begin{cases} 1 & for\ e_V \geq 1 \\ 0 & otherwise \end{cases} \quad (18)$$

The coding for control scheme B in comparison with the coding for control scheme A has undoubtedly one advantage, namely, the possibility of introducing a velocity limit during the simulation of the passage of a given section of the route. A velocity limitation is of great practical importance. In many situations, it is necessary to reduce the velocity of the vehicle during certain parts of the route, for example, during cornering. Accordingly, in case of control scheme B, two velocity vectors are designated as follows:

(i) $\mathbf{st}_{V15}$ for $V_{SET_i} = [0,\ 15]$ m/s: although it is formally limited to 15m/s, it can never be achieved because the maximum vehicle speed (based on maximum engine speed) is approximately 15m/s;

(ii) $\mathbf{st}_{V10}$ for $V_{SET_i} = [0,\ 10]$ m/s.

4.3. The Fitness Functions. In the case of evolutionary optimization, the fitness function f_{val} has to be designed. This section describes the formulas used to calculate the fitness function of an individual that is needed for solving a task of evolutionary optimization. Each chromosome is a potential solution. Each control strategy **st** is used during numerical simulation. The value of the fitness function f_{val} can be calculated based on the outcomes of the simulation such as

(i) the total energy consumption r_{sim} [km/kWh]

(ii) the simulation time t_{sim}

(iii) the distance x_{sim}

and limitations such as

(i) the maximum travel time T_{max}

(ii) the total distance of the vehicle D_C

The value of the function is also used to evaluate the velocity strategy of the vehicle. In the preliminary studies [24], the following fitness function f_{val} was used.

Fitness Function 1

$$f_{val_1} = w_1 f_1 + w_2 f_2 + w_3 f_3, \quad (19)$$

where

$$f_1 = \left[1 + r_{sim}^{\lambda_1}\right]^{\lambda_1}, \quad (20)$$

$$f_2 = \left[H\left(D_c - x_{sim}\right)\frac{|D_c - x_{sim}|}{D_c}\right]^{\lambda_2}, \quad (21)$$

$$f_3 = \left[H\left(t_{sim} - T_{\max}\right)\frac{|T_{\max} - t_{sim}|}{T_{\max}}\right]^{\lambda_2}, \quad (22)$$

where the weights $w_1 = 0.50E05$, $w_2 = w_3 = 0.25$ and coefficients $\lambda_1 = -2$, $\lambda_2 = 1$ were chosen experimentally. The penalty functions f_2 and f_3 were introduced. The Heaviside step function $H(.)$ is used to penalize individuals representing unacceptable solutions [25, 26], e.g., those that exceed the maximum time of the simulation $H(t_{sim} - T_{max})$ or do not reach the specified distance $H(D_c - x_{sim})$.

Fitness Function 2. In this study, the authors have proposed the fitness function which would lead to the convergence of the evolutionary algorithm without searching for optimal values of weights. Such a function is shown and described below:

$$f_{val_2} = \frac{1 + f_4 + f_5}{f_1 f_2 f_3}, \quad (23)$$

$$f_1 = r_{sim}, \quad (24)$$

$$f_2 = \cos\left(\frac{|D_c - d_{sim}|}{D_c}\right), \quad (25)$$

$$f_3 = \cos\left(\frac{|T_{\max} - t_{sim}|}{T_{\max}}\right), \quad (26)$$

$$f_4 = \chi \cdot H\left(D_c - d_{sim}\right), \quad (27)$$

$$f_5 = \chi \cdot H\left(T_{\max} - t_{sim}\right), \quad (28)$$

where f_1 is a measure related to the simulated energy consumption of the vehicle, f_2 and f_3 correspond to the limitations such as the maximum travel time and the total distance of the vehicle, f_4 and f_5 also correspond to the limitations but these play strong penalty role, and χ is arbitrarily large value, e.g., significantly higher than maximal values f_2=f_3 =1. In this variant of the fitness function *death penalty* [26] is applied for unacceptable individuals in the form of factors f_4 and f_5. Individuals who do not reach a specified distance or exceed the time limit are eliminated. These two criteria act as a constraint and correspond to the limitations formulated in the guidelines of the race competition. On the other hand, the criteria f_2 and f_3 have continuous nature and play very important role in the fitness function. They are very needed to evaluate continuously the individuals that are very close to and very far from the extremum.

The argument of cos(.) function is not related to an angle. However, the usage of the cos(.) function is needed to

TABLE 1: Research object main parameters.

Type of the motor	PM BLDC
Transmission unit	Toothed synchronous belt
The minimum turning radius	8 [m]
The battery type and capacity	Lithium-ion polymer 2100[mAh]
Mass of the vehicle	42 [kg]
Mass with the driver	min. 92 [kg]
Diameter of wheels	0.478 [m]
Moment of inertia of the front wheel I_{k_P}	0.023 $[kgm^2]$
Moment of inertia of the rear wheel I_{k_T}	0.023$[kgm^2]$
Drag coefficient c_x (see Table 2)	0,22 – 0,23
Frontal area A $[m^2]$ (see Table 2)	0.275 - 0.297
The total rolling coefficient c_{rr} (see Table 4)	0,0035 – 0,0058,

eliminate the possibility of dividing by zero. In addition, the form of the cos(.) function is preferred. At the initial stage of optimization, when the simulation time t_{sim} and the distance x_{sim} differ significantly from the maximum time T_{max} and the specified distance D_c, a small change will have a greater effect on the value of the fitness function. At the final stage of optimization, the function f_{val_2} is more dependent on the simulated energy consumption of the vehicle related to the factorf_1, and thanks to the cos(.) function, it is less dependent to the factors f_2 and f_3. The proposed function does not require the introduction of weights and therefore it is easier to use this function in other applications, e.g., to find optimal velocity strategies for different routes.

It should be noted that fitness functions f_{val} have been written in such a way as to ensure that the optimization algorithm will look for a minimum of these functions.

5. Case Study

As described in Sections 3 and 4, process of modelling and optimization was applied and verified using a prototype electric vehicle on street race circuit in Rotterdam. The methodology was also carried out on the experimental rides track of the Fiat Chrysler Automobiles Poland factory, located in Tychy. Based on (5)-(12), the simulation model was created in the MATLAB/Simulink® environment. Optimization was also carried out in MATLAB using the evolutionary algorithms available in the *Global Optimization Toolbox*. The library of functions for simulating, optimizing, analyzing, and presenting results has been developed.

5.1. Research Object. Light electric vehicle was a research object. The vehicle shown in Figure 9 is a three-wheeled self-supporting structure. The vehicle fuselage is made of composites based on epoxy resin and woven of carbon and aramid fibre [21].

The dimensions of the vehicle and basic data are shown in Figure 10 and in Table 1.

Identification of parameters for the mathematical model of a research object was carried out by numerical simulations, tests on specially designed test bench [27]

FIGURE 9: Research object: prototype electric vehicle "MuSHELLka" (photo by M. Wyleżoł, 2012).

and during conducted experiments in real environment [28].

The vehicle can be driven by two types of drive unit:

(i) DT1: drive unit based on PM BLDC motors and belt transmission. Drive wheel is equipped with a one-way clutch.

(ii) DT2: a high-torque PM BLDC motor mounted inside the drive wheel. The torque generated by the motor is directly transmitted to the rear drive wheel.

The catalogue data can be used for creating the mathematical model of the DT1 motor, but for identifying the exact parameters, it was necessary to carry out experimental tests. The manufacturer of the DT2 motor does not provide much information, so it was necessary to test the drive unit as well. DT1 drivetrain efficiency was estimated using a BG75x50 motor with toothed synchronous belt transmission (z1= 28 teeth, z2 = 200 teeth, i_{TS} = 7.14). The map of the efficiency was shown on Figure 11(a). The same test was carried out for DT2 drive unit with a motor directly mounted on the drive wheel (Figure 11(b)). The obtained efficiency maps were used to build the simulation model of the tested drivetrain systems.

5.2. Determination of Air Drag Coefficient. The air drag coefficient c_x was estimated as a result of CFD analysis using ANSYS software [24]; and it was also verified by the research in wind tunnel at the Institute of Aviation in Warsaw. Frontal area of vehicle A was determined from a three-dimensional model in CAD environment, Catia V5. The values of drag coefficient c_x and frontal area A are shown in Table 2.

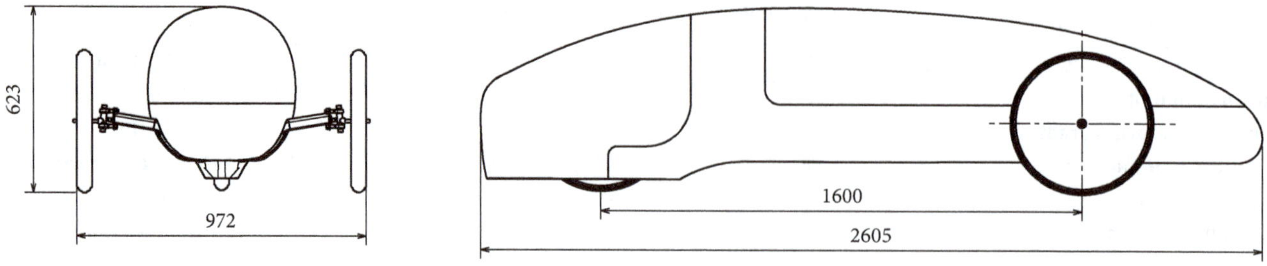

FIGURE 10: Dimensions of the vehicle.

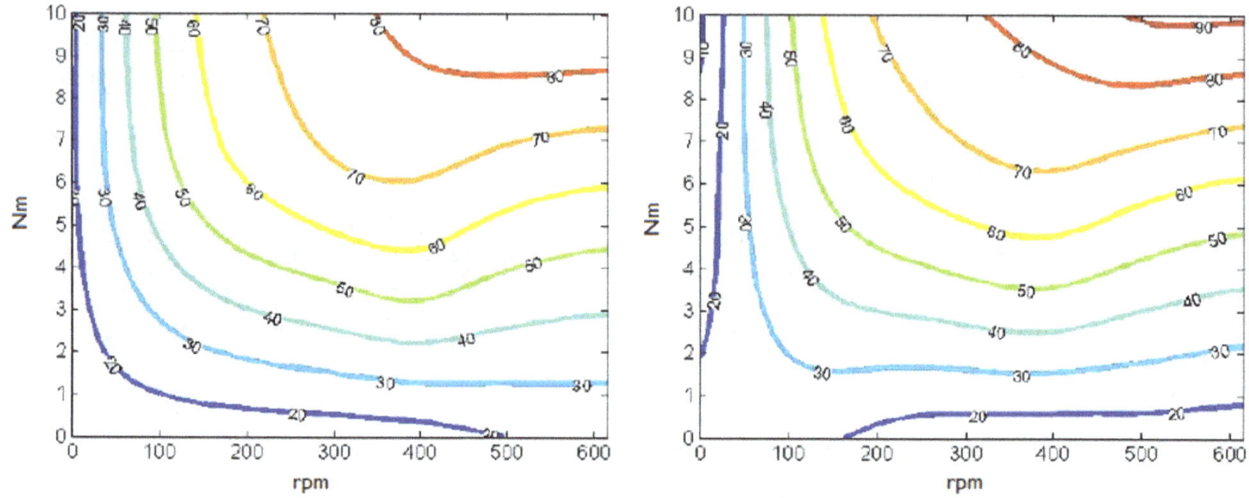

(a) DT_1 (b) DT_2

FIGURE 11: Maps of the efficiency of tested drivetrain systems.

TABLE 2: The air drag coefficient and frontal area.

c_x: numerical simulation	c_x: experimental research	Frontal area A [m^2]
0.23	0.22	{0.275, 0.297}

5.3. Determination of Rolling Resistance. According to formula (7), the rolling resistance depends on the weight of the vehicle (Table 1), the tires rolling resistance coefficient, and the resistance of the bearings. Depending on the drive unit (DT1, DT2) used in the vehicle, additional resistance has to be added. In the DT1 system, there is a one-way clutch which also generates some resistance torque. The DT2 system cannot be mechanically disconnected and therefore has a higher rolling resistance. As a result of the coast down test, the total rolling resistance of the vehicle has been identified for the four types of tires. The test was performed only with the DT1 system, i.e., the system with the clutch. Table 3 shows the basic parameters of the tires used in the experiments.

The prototype vehicle can be equipped with specially manufactured *Michelin* tires with a low rolling resistant coefficient of $c_{rr} = 0.0014$.

In addition to the special *Michelin* tires, three other types of tires were tested. The identification of rolling resistance coefficients was also estimated using a coast down test. The measurements were made as follows:

(i) The driver accelerated the vehicle to a fixed speed.

(ii) At the moment of crossing the marked gate, the driver pressed the marker button to save the flag in the acquired data.

(iii) The run continued until free braking, after which the driver switched off the measuring system.

(iv) The measurements were made at both directions of the road, to minimize influence of the wind.

(v) Before the attempts, wind velocity, temperature, and atmospheric pressure were measured.

(vi) The distance of the road was measured using a measurement system mounted on the vehicle and it was also verified using distance measuring wheel.

On the basis of the experimental data, the total rolling resistance coefficient of the vehicle was determined. Table 4 shows the coefficients of rolling resistance of the vehicle depending on the tire used during the test. Total resistance includes the tire rolling coefficient, the bearings resistance, and the resistance of the clutch. Verification of the data derived from the simulation and the experiment was conducted using

TABLE 3: The types of tires used in the prototype vehicle.

Tire's marker	Name of the tire	Tire's symbol	Shape	Type of tread	Working pressure
W_1	Michelin Shell Eco-marathon	45/75R16	Rectangular	Slick	max.500 [kPa]
W_2	Schwalbe Ul-Termo	23-406	Round	Slick	600-1100 [kPa]
W_3	Continental Sport Contact	28-406	Round	Slick	600 [kPa]
W_4	Schwalbe Marathon	40-406	Round	Symmetrical	400-700 [kPa]

TABLE 4: The total rolling coefficient depending on the tires.

Tire's marker	Name of the tire	The total rolling coefficient c_{rr}	*mMAPE* [%]
W_1	*Michelin ShellEco-marathon*	0.0035	7.84
W_2	*Schwalbe Ultermo*	0.0043	11.03
W_3	*Continental SportContact*	0.0058	12.52
W_4	*Schwalbe Marathon*	0.0053	4.45

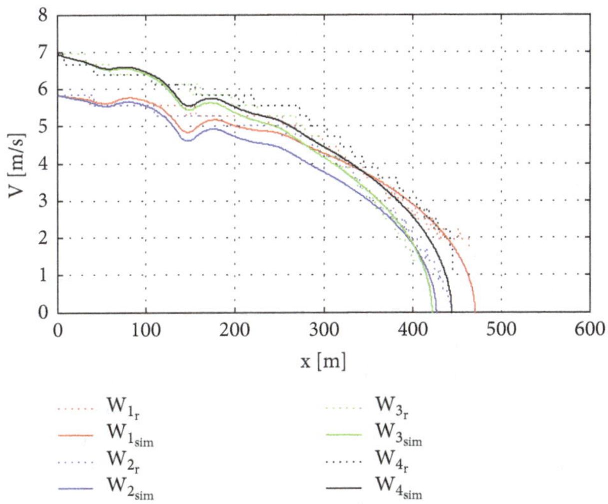

FIGURE 12: Velocity of the vehicle during coast down test.

mMAPE (*modifiable Mean Absolute Percentage Error*). For velocity simulation *mMAPE* can be determined by

$$mMAPE = 100\% \frac{1}{n} \sum_{i=1}^{n} \left| \frac{V_r - V_{sim}}{V_{max} - V_{min}} \right|. \quad (29)$$

Figure 12 shows the experimental data obtained in a coast down test compared with the results obtained during the computer simulation.

5.4. Model of the Race Circuit. In accordance with the methodology presented in Section 3.2, three routes have been described:

(i) Route 1: a street race circuit at the *Shell Eco-marathon* competition in Rotterdam.

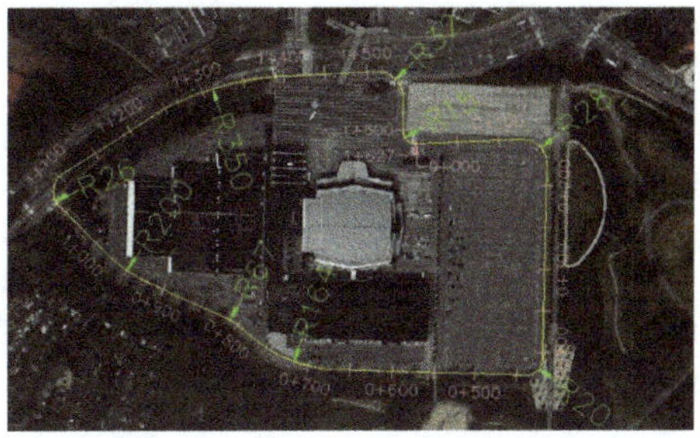

FIGURE 13: Plan view of the path of the drive on route 1 in Rotterdam.

(ii) Routes 2 and 3: on the experimental rides track of the Fiat Chrysler Automobiles Poland factory, located in Tychy, in the south of Poland.

The *Shell Eco-marathon* competition is held on a street circuit in Rotterdam. One lap is 1630 meters long. Vehicles have to cover ten laps in no longer than 39 minutes so such requirements impose an average minimum speed of 25 km/h.

The trajectory of the vehicle path was developed using satellite images. The vehicle trajectory was determined using the maximum curvature possible that could be obtained on the existing roads.

Movement of the vehicle in a curved motion causes the centrifugal force. Centrifugal force creates additional resistance. This resistance is inversely proportional to the radius of the path (8), so it is justified to select the largest radius. In addition, if the centrifugal force exceeds the limit value, it could provide to the side slip of the vehicle. According to drawings in *AutoCAD Civil 3D*, the lengths of straight sections and the lengths and the radius of arcs were obtained. Figure 13 shows a plan view of the path of the drive.

Driving the prototype vehicle on public roads is prohibited and can be dangerous. Vehicle testing was also carried

TABLE 5: Travel times and total lengths of the routes.

Route	Length of the lap D [m]	The number of laps	The total length of the route D_c [m]	Travel time T_{max} [s]
Route 1 Rotterdam	1630	10	16 300	2340
Route 2 Tychy	640	5	3200	460
Route 3 Tychy	1000	5	5000	720

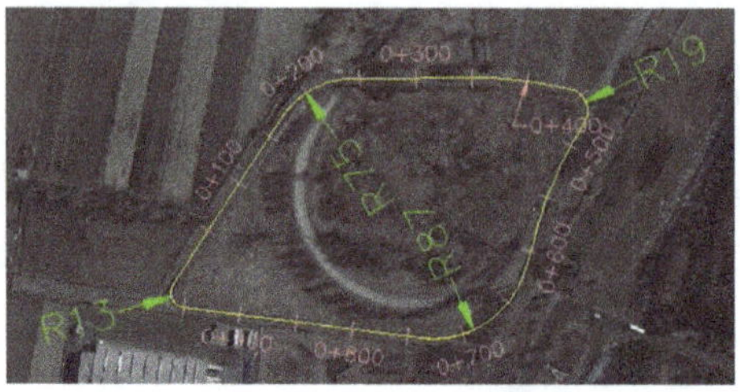

FIGURE 14: Plan view of the path of the drive on route 3 in Tychy.

out on the experimental rides track of the Fiat Chrysler Automobiles Poland factory.

For the track in Tychy geodetic measurements were taken using the *DGPS* (*Differential Global Positioning System*) technology. The measuring points were collected along the axis of the road, with the distance between points amounting to 10 meters. If on the measured route, direction or elevation were changed, the frequency of measurement was increased. Geodetic coordinates were also imported into *AutoCAD Civil 3D* and were used to prepare data concerning the route that was imported into the simulation environment *MATLAB-Simulink*. Figures 14 and 15 show the path and elevation profile, respectively.

Selected routes 2 and 3 have significant differences compared to route 1, especially in the height profile. This set of routes was selected to verify the effect of the route profile on the achieved results and to verify the proposed optimization methods for various road conditions. The routes in Tychy are located in an open area; there are no large buildings, which guarantees stability of wind conditions. The technical condition of the surface on routes 2 and 3 is much worse than on route 1, which has a significant influence on the rolling resistance. Rolling resistance was also identified and introduced into the simulation model. For route 2, the c_{rr} coefficient was 0.0056 and for route 3 was 0.0068.

5.5. Verification of the Simulation Model. In order to verify the compatibility of the identified simulation model, a series of numerical simulations and experimental tests were performed. The first experiment consisted of running 300 [m] with a control strategy of 50 [m] with the working motor, and after that the vehicle was moving freely for the next 50 [m]. The motor is turned on three times. Figures 16(a) and 16(b) show the velocity of the vehicle and the power of the motor in function of the distance determined during the experiment (red curves) and during the numerical simulations (green curves).

Verification of the data derived from the simulation and the experiment was conducted using *mMAPE* (29). For motor power simulation, *mMAPE* can be determined by

$$mMAPE_P = 100\%\frac{1}{n}\sum_{i=1}^{n}\left|\frac{P_r - P_{sim}}{P_{max} - P_{min}}\right|, \qquad (30)$$

where P_r is the motor power measured during experiment and P_{sim} is the motor power conducted in simulation.

As a result of the model validation tests, estimation error of the velocity was on level between 2.35% and 5.34%. However, the motor power error level was greater. It was between 8.02% and 9.98%.

5.6. Determination of the Optimal Control Strategy. The length of route 1 and the number of laps are determined by the organizers of the Shell Eco-marathon competition. The length of routes 2 and 3 is derived from the existing road plan, where it was possible to carry out trials in real conditions. The number of laps for routes 2 and 3 was assumed arbitrarily.

The travel time for route 1 is based on the rules of the race, which determine the minimum average speed of the vehicle. Based on the average speed and the length of routes 2 and 3, the travel times for these routes were determined. The maximum travel times and the total length of the route, as well as the number and length of one lap *D*, are summarized in Table 5.

Evolutionary algorithms which are available in the *MATLAB Global Optimization Toolbox* were used. From the MATLAB workspace, the algorithm ran a simulation model in a *Simulink* environment. During optimization while using evolutionary algorithms, the maximum number of fitness function evaluations $N_{f_{val}}$ is the product of the population size N_{pop} and number of generations $N_{gen.}$

It was assumed that the total number of fitness function evaluations was equal to 10^4; therefore, should this value be exceeded, the calculation would be cancelled. Therefore, the total number of generations $N_{gen.}$ is dependent on the value of *population size* N_{pop}.

As part of this work phase, the vectors $\mathbf{st}_U$, $\mathbf{st}_{V\ 15}$, and $\mathbf{st}_{V\ 10}$ of the optimum control strategy for an electric vehicle for routes 1, 2, and 3 were determined. During evolutionary optimization, the heuristic method of crossing was used, and the probability of crossing was established on the basis of previous studies at the level of 0.8 [24, 29]. At the

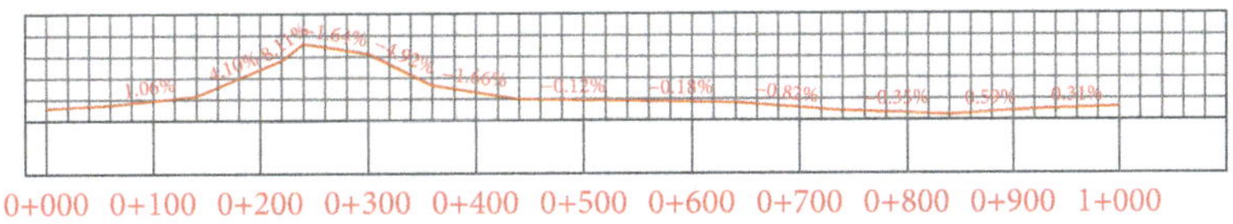

FIGURE 15: Profile of the path on route 3 in Tychy.

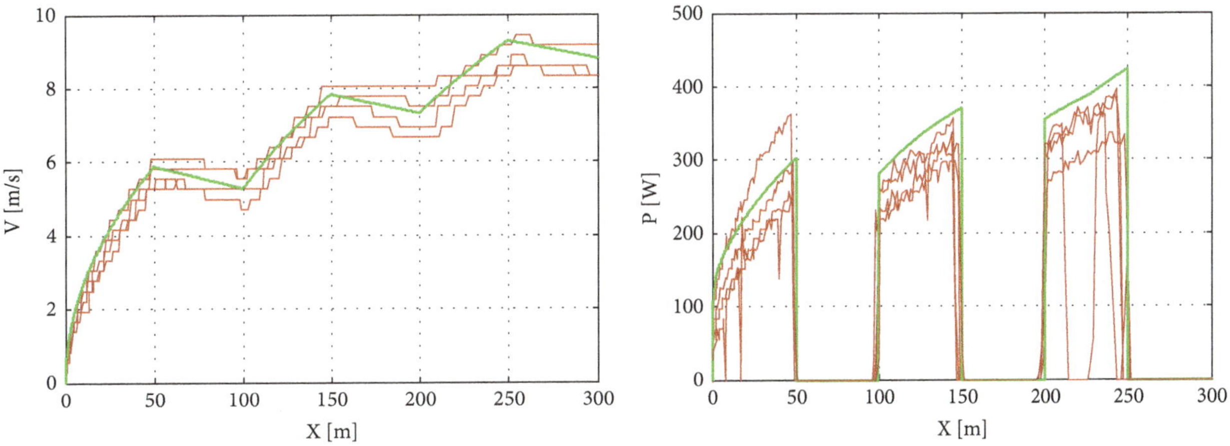

(a) DT_1 (b) DT_1

FIGURE 16: The velocity and the power determined with used DT1 drive unit.

stage of reproduction, the elitist succession method was applied where the best adapted two individuals in the current population passed to the next population. An adaptive method of mutation was used. The remaining parameters of the algorithm remain the default for the MATLAB *Global Optimization Toolbox*.

The first case was the search $\mathbf{st}_U$ vector for route 1. Figures 17(a)–17(c) show the vehicle velocity V_{sim} obtained by computer simulation. The graph of the $\mathbf{st}_U$ control signal is graphically displayed. Figure 17 shows the results where the optimal control strategy $\mathbf{st}_{U_{opt}}$ was used, and population was N_{pop}=100 and N_{gen}=100. The total energy consumption was estimated as r_{sim}=527,633 [km / kWh]. Figure 17(a) shows the vehicle speed and control signal $\mathbf{st}_U$ during the first lap while Figure 17(b) shows center laps and Figure 17(c) shows the last lap. Figure 17(d) shows the curves of the mean values of the fitness functions f_{val}, depending on the number of calculations of these functions, during the optimization of the control strategy. There are eight curves in the graph (Figure 17(d)) for each of the N_{pop} parameters used in the algorithm.

When analyzing the vehicle velocity chart on the last lap of route 1 (Figure 17(c)), it can be seen that, according to the determined control strategy, the propulsion system of the vehicle should no longer be used. The vehicle should use the stored kinetic energy.

The analysis of vehicle velocity V_{sim} during the last lap indicates that at the moment of reaching the given distance the speed was still about 2 [m/s]. This demonstrates that the simulated vehicle has kinetic energy that would have to be dispersed by the braking system or recovered by the electric motor. Having such a speed does not mean that there is still possibility for improvement. Lower speed at the end of the route has to cause higher speed of the earlier parts of the route. This speed is the optimum speed due to the chosen criterion. From the analysis of the shape of the fitness function (Figure 17(d)), it can be seen that, for route 1, it is preferable to use a smaller population of N_{pop}=25 and increase the number of N_{gen}=400. The curve shape for this case is steep and it falls to the minimum value of the fitness function, which may indicate that the algorithm has reached a convergence.

Results of the optimization where B type of coding was used were analyzed. The results obtained as the mean of the r_{sim} result together with the standard deviation from all the tests were summarized in Table 6.

By analyzing the results in Table 6, it can be seen that for route 1 better results were obtained (reduced simulated total energy consumption) using A coding compared to B, especially for $\mathbf{st}_{V\,15}$, whose results are characterized by the highest standard deviation. For routes 2 and 3, the same as for route 1, the procedure was followed. Optimization strategies have been optimized for all variants of control vector $\mathbf{st}_U$, $\mathbf{st}_{V\,15}$, $\mathbf{st}_{10}$.

The results using the control strategy $\mathbf{st}_{V\,10}$ are shown in Figures 18(a)–18(c) (N_{pop}=100 where r_{sim}=351,795 [km / kWh]). Figure 18(d) illustrates the graph of the value of the fitness function.

The results obtained for route 2 are presented in Table 7.

For route 2 compared to route 1, the best results were achieved using the $\mathbf{st}_{\mathbf{V10}}$ vector. When using vectors $\mathbf{st}_U$ and $\mathbf{st}_{\mathbf{V15}}$ the results were slightly worse. Having analyzed the size

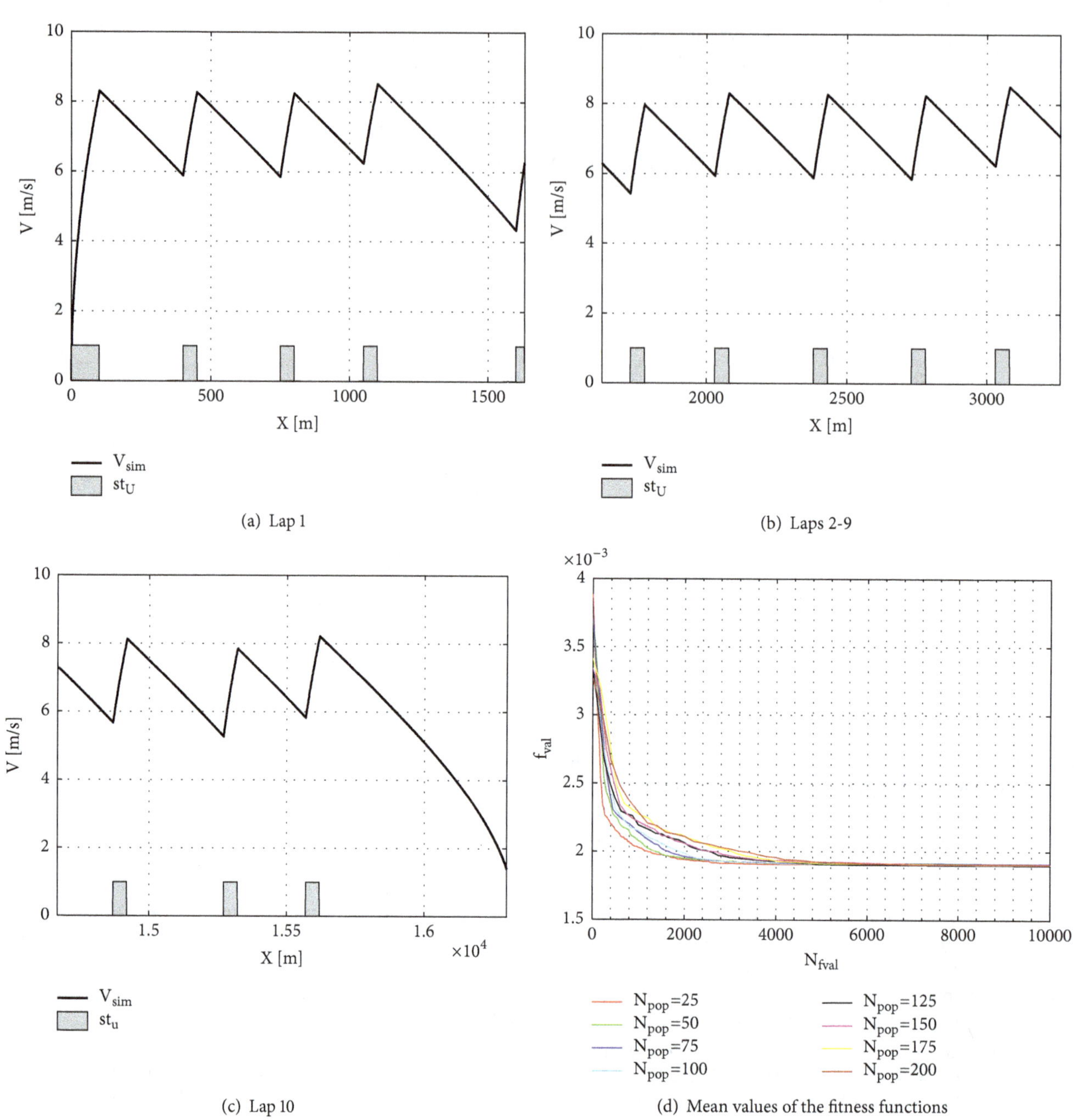

FIGURE 17: Vehicle speed and control signal $\mathbf{st}_U$ for route 1.

TABLE 6: The results for route 1.

	The type of coding, *A*		The type of coding, *B*			
	$\mathbf{st}_U$		$\mathbf{st}_{V15}$		$\mathbf{st}_{V10}$	
N_{pop}	r_{sim} [km/kWh]	*std*	r_{sim} [km/kWh]	*std*	r_{sim} [km/kWh]	*std*
25	524,65	1,459	428,45	17,41	515,69	5,4775
50	525,79	1,0053	444,31	22,486	514,92	3,4094
75	524,97	3,9913	438,42	28,173	512,56	8,0682
100	525,89	1,8128	422,32	17,97	511,66	4,9379
125	526,25	1,3862	427,33	39,616	513,41	7,0765
150	524,3	2,979	443,5	47,701	514,86	2,7086
175	525,13	2,511	414,28	27,299	518,38	1,9358
200	525,79	1,9666	423,37	25,776	512,04	7,8357

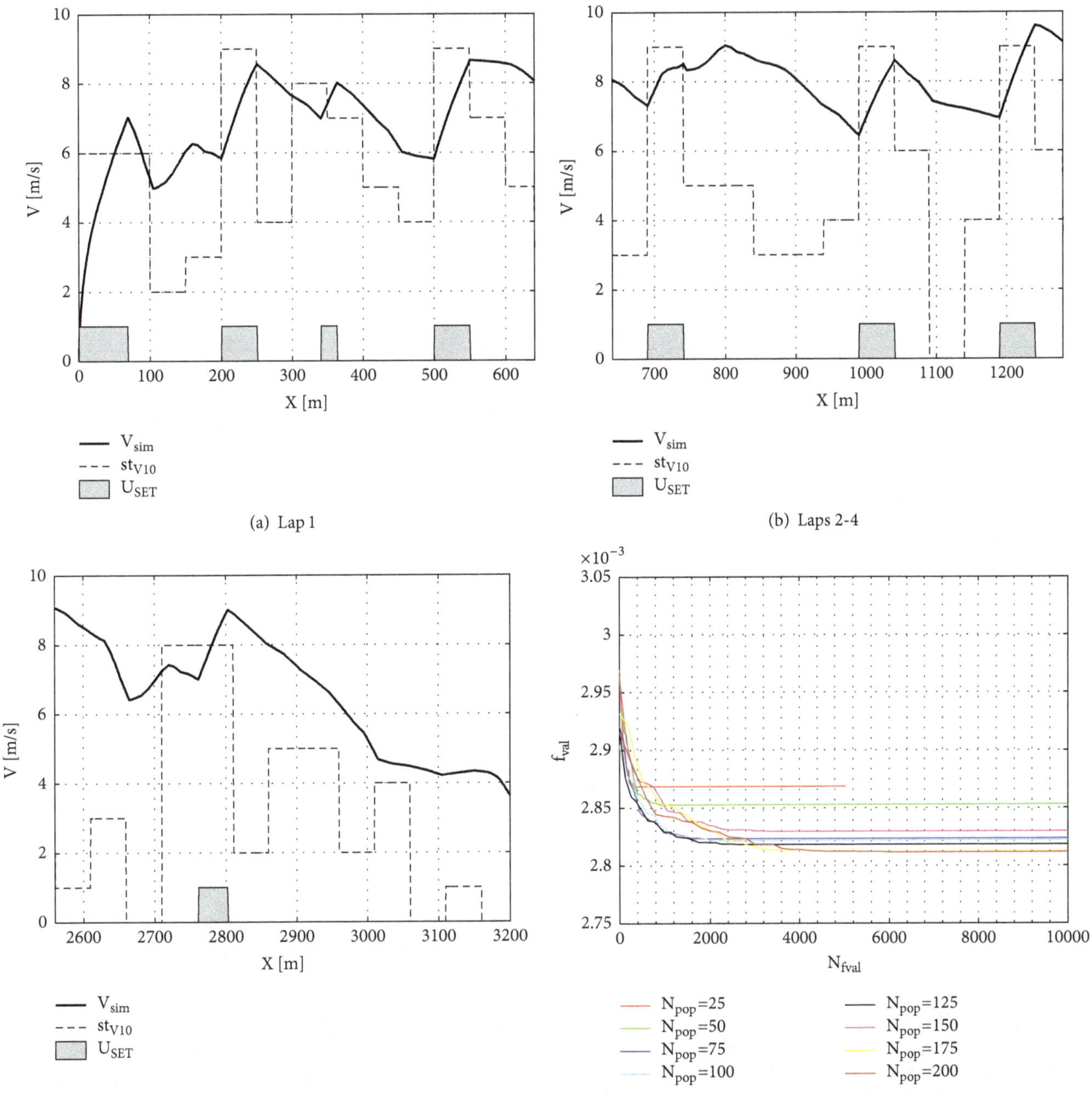

(a) Lap 1

(b) Laps 2-4

(c) Lap 5

(d) Mean values of the fitness functions

FIGURE 18: Vehicle speed and control signal $\mathbf{st}_{V10}$ for route 2.

of the population used in the optimization process, it should be noted that for route 2 it is preferable to use higher value of N_{pop}. For all the analyzed cases, the best average results for N_{pop}=200 were obtained. Table 8 summarizes the results for route 3.

For route 3, the average results were reported for a larger population of N_{pop} with B coding. In the case of using vector with A type coding, no characteristic change was noted. As expected, comparing the standard deviation between algorithms using B coding, the standard deviation is less for the less variance range.

Evolutionary algorithms are not complete search algorithms; therefore, it is not possible to determine whether the solution found is best for the chosen criterion and the constraints assumed.

5.7. Applying the Optimal Control Strategy in Real Conditions. Electric vehicle energy consumption optimization was finally verified in real conditions at the European edition of the *Shell Eco-marathon* held on 16-18 May 2014 in Rotterdam.

Optimal control strategy $\mathbf{st}_U$ was displayed by a prototype visualization system. Visualization unit is installed in the

TABLE 7: The results for route 2.

	The type of coding, A				The type of coding, B	
	$\mathbf{st}_U$		$\mathbf{st}_{V\ 15}$		$\mathbf{st}_{V\ 10}$	
N_{pop}	r_{sim} [km/kWh]	std	r_{sim} [km/kWh]	std	r_{sim} [km/kWh]	std
25	341,38	6,6734	342,94	2,3839	348,73	5,3532
50	345,73	4,6359	350,03	4,62	350,6	3,8862
75	345,12	5,5134	349,61	3,3476	354,19	0,67858
100	348,36	4,5332	352,53	2,6117	354,4	1,9295
125	346,32	5,2704	350,17	4,9039	354,83	1,4756
150	345,52	6,3714	353,26	1,4885	353,43	0,71015
175	350,63	1,8728	352,74	4,7141	355,64	0,38022
200	350,73	1,4496	354,0	1,3389	355,69	1,4942

TABLE 8: The results for route 3.

	The type of coding, A				The type of coding, B	
	$\mathbf{st}_U$		$\mathbf{st}_{V\ 15}$		$\mathbf{st}_{V\ 10}$	
N_{pop}	r_{sim} [km/kWh]	std	r_{sim} [km/kWh]	std	r_{sim} [km/kWh]	std
25	280,71	10,173	286,43	7,323	309,94	2,776
50	293,46	6,142	295,81	7,894	308,74	4,399
75	297,11	5,002	292,17	7,367	311,41	1,856
100	298,26	3,231	295,15	7,022	309,57	4,049
125	297,81	2,980	298,66	12,977	311,68	3,445
150	294,53	3,346	302,17	9,623	312,88	1,249
175	292,34	2,317	306,21	6,816	314,81	1,261
200	296,86	2,337	305,31	11,737	310,89	4,880

TABLE 9: Comparison of simulation results with experimental tests.

Number of attempts	r_{SE-m}	r_r	r_{sim}	r_{opt}	e_m	e_i	e_{st}
Attempt 1	365,4	360,8	-	-	1,25	-	-
Attempt 2	481,3	508,2	514,8	526,5	5,58	1,29	3,6
Attempt 3	487,3	512,7	519,8	526,5	5,21	1,39	2,69
Attempt 4	Failed attempt, damaged tires during 1st lap						

steering wheel of the vehicle. The visualization system shows the basic data such as instantaneous speed, average speed, travel time, and distance. In addition, the visualization system provides information on which part of the track the driver should press the acceleration button to execute the optimal control strategy. Measurement of the position on the track is carried out by the odometrical measuring system. Due to the possibility of accumulating errors in the measurement of the road, the driver will reset the measuring system every time he/she exceeds the lap line.

Table 9 presents the results of all the measurements conducted during the *Shell Eco-marathon*. The columns in the Table 9 contain

(i) r_{SEm} [km/kWh]: the result recorded during the race by the measuring device provided by the organizer; the results are available on the Shell Eco-marathon official website;

(ii) r_r [kW/kWh]: the result based on the indications of a prototype measuring system permanently installed in the vehicle;

(iii) r_{sim}: the result estimated by computer simulation using recorded real control signal U_r;

(iv) r_{opt}: the result obtained by computer simulation using the optimal control strategy U_{opt}.

Table 9 also includes the following:

(i) Measurement error:

$$e_m = \frac{|r_{SEm} - r_r|}{r_{SEm}} 100\%. \tag{31}$$

(ii) Identification error:

$$e_i = \frac{|r_r - r_{sim}|}{r_r} 100\%. \tag{32}$$

(iii) Strategy implementation error:

$$e_{st} = \frac{|r_r - r_{opt}|}{r_r} 100\%. \tag{33}$$

There were four measurement trials available. During run 1, a much weaker result was obtained than estimated by simulation. The driver performed run 1 without a defined control strategy. As part of the experiment, the driver and the support team were asked to select the optimal tires for the vehicle and to implement the proposed control strategy. During run 1, only the travel time of individual laps was checked. According to the choice of the team in run 1, tires W2 were selected (Table 4). During this test the worst result was obtained about 24% and 25% worse than in the other trials. In subsequent runs 2 and 3, the vehicle was equipped with the best W1 tires (Table 4). The driver has implemented an algorithm-based optimal control strategy. Strategy information was provided by the visualization system. During run 3 the best result was recorded. This result was taken into account in the classification of competitions, where the team placed in 12 positions out of 28 classified teams. Trials 4, due to damage to the tires, were not successful.

Measurement error was from 1.25% to 5.58%, depending on the sample. The measurement system installed permanently in the vehicle is a prototype system. This system has been calibrated using available laboratory equipment. Measurement errors indicate that other measuring devices of the higher accuracy class should be used.

Based on the measurement data obtained during test drives, the model was tuned, thus reducing the identification error. The competition took place in stable weather; the divergence of results was small and was about 1.29% for sample 2 and 1.39% for sample 3.

The difference between the results determined during numerical simulation and that obtained in the real run was 3.6% for sample 2 and 2.69% for sample 3. The difference at this level was considered to be small. Searching for optimal control strategy has not taken into account possible "errors" which could be made by the driver and which result from the current situation on the route, for example, selecting the path and using the route by other participants in the race. Including driver behaviour in the simulation model should be the subject of further work.

Figure 19 shows the vehicle velocity together with the control signals recorded during run 2. The black curves represent the real vehicle velocity V_r and the control signal. Green curves are the optimum control strategy $\mathbf{st}_U$ and the optimal velocity V_{opt} determined by the simulation. The graph shows the entire distance travelled (10 laps). The analysis of the graph shows that the driver did not perform the exact planned strategy. Errors in the implementation of the strategy are related to the disruptions. During the competition, the driver is not alone on the track; in certain situations the driver cannot overtake other participants and cannot take the optimal path. In addition, weather condition such as wind has influence on the motion resistance. In such cases, the driver has to modify the optimal control strategy.

In Figure 20, the velocity recorded during the race V_r (black) and the velocity obtained by the computer simulation V_{sim} (red) were shown. V_{sim} was determined by computer simulation by using real control signal. The purpose of this simulation was to investigate the correctness of identification of the simulation model and to identify the identification error.

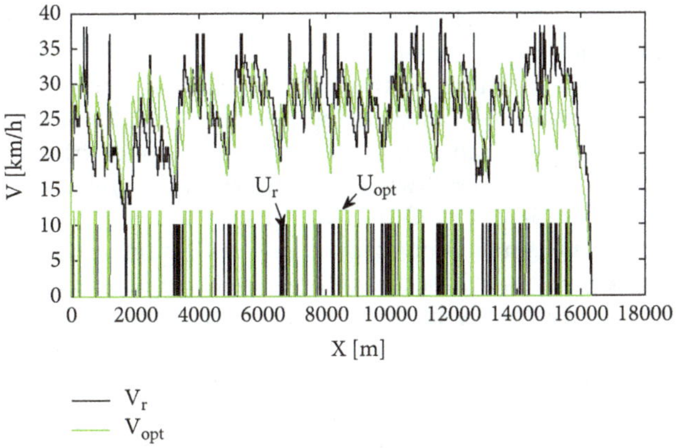

FIGURE 19: Comparison of speed and control signal during attempt 2.

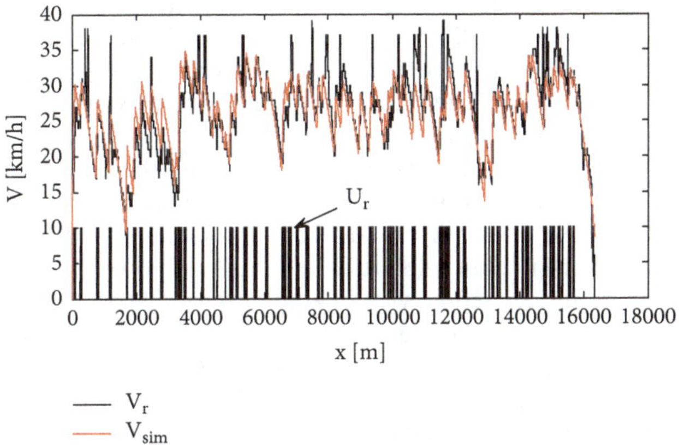

FIGURE 20: Comparison of speed registered in real conditions with the speed determined by computer simulation.

Compared to the *Shell Eco-marathon* competitions held in 2012 and 2013, our performance has improved. The results in these years were equal to 425 [km / kWh] and 454 [km / kWh], respectively. In 2012 and 2013, the control strategy was chosen on the basis of the simulation model, but the model was not adequately identified, and the control strategy had to be modified. In previous editions of the race, the implementation of optimal control strategy was not supported by the visualization system, which would inform the driver of the need to take specific actions (pressing or releasing the accelerator button).

6. Conclusion

The presented method was developed and tested for a prototype electric racing car that has been designed for the *Shell Eco-marathon* race. As it was shown the results of the numerical simulations were very similar to the outcomes of the real races in the context of the final result of energy consumption and the profile of velocity of the car during the race. The method has been developed and applied since 2012 and led in 2014 to achieving the optimal result for this type of the vehicle.

The optimization method was evaluated. The results achieved during the race attempts where the optimal control strategy was applied (481,3 and 487,3 km/kWh) were compared to the result where the strategy was proposed by the experts (365.4 km/kWh), who already had contact with the results of similar optimization models. Optimal attempts were better by over 30%, despite errors in the implementation of the strategy by the driver. In the case of experts who did not have such experiences, the differences were even greater.

Further improvement of the control strategy is impossible for this type of a vehicle and the best way to reduce energy consumption is by improving the mechanical design. It was the reason to develop new racing cars for UrbanConcept Battery Electric and Hydrogen competitions. The developed research methodology for vehicle optimization and control strategies has been developed and implemented in the Shell Eco-marathon in 2015, which has improved the score to 504 km / kWh. However, the research has shown that the potential for improved performance of the *MuSHELLka* vehicle has been exhausted. No major structural changes in the vehicle can significantly improve the result, so in the following years, the race was abandoned and the focus was on the development of the *Hydro*GENIUS, the electric vehicle of UrbanConcept class with fuel cell stack as power source, which eventually scored vice Championship in *SEM* 2016. In the further work, it is planned to improve the optimization results by using the presented methodology in the case of the models of newly constructed vehicles. The development of the optimization method should take into consideration the uncertainty resulting from the given velocity strategy considering the real conditions of the race. It is also possible to use this method for controlling vehicles in traffic. To do this, it is necessary to completely change the way of the route modelling and the driving conditions modelling, as well as to develop a new optimization method for obtaining results in real time.

Conflicts of Interest

The authors declare that there are no conflicts of interest regarding the publication of this paper.

References

[1] W. Skarka, "Reducing the Energy Consumption of Electric Vehicles," in *Proceedings of the 22nd ISPE-Inc International Conference on Concurrent Engineering Location*, R. Curran, N. Wognum, M. Borsato et al., Eds., Advances in Transdisciplinary Engineering, pp. 500–509, IOS Press, Delft, Netherlands, 2015.

[2] R. Andrzejewski and J. Awrejcewicz, *Nonlinear Dynamics of a Wheeled Vehicle*, vol. 10, Springer-Verlag, New York, NY, USA, 2005.

[3] D. W. Gao, C. Mi, and A. Emadi, "Modeling and simulation of electric and hybrid vehicles," *Proceedings of the IEEE*, vol. 95, no. 4, pp. 729–745, 2007.

[4] "Advisor documentation," http://adv-vehiclesim.sourceforge .net.

[5] T. Markel, A. Brooker, T. Hendricks et al., "ADVISOR: A systems analysis tool for advanced vehicle modeling," *Journal of Power Sources*, vol. 110, no. 1, pp. 255–266, 2002.

[6] K. B. Wipke, M. R. Cuddy, and S. D. Burch, "ADVISOR 2.1: a user-friendly advanced powertrain simulation using a combined backward/forward approach," *IEEE Transactions on Vehicular Technology*, vol. 48, no. 6, pp. 1751–1761, 1999.

[7] "Website," 2014, http://www.transportation.anl.gov/modeling.

[8] "Website," 2014, http://www.autonomie.net.

[9] K. Krykowski, J. Hetmańczyk, Z. Gałuszkiewicz, and R. Miksiewicz, "Computer analysis of high-speed PM BLDC motor properties," *COMPEL - The International Journal for Computation and Mathematics in Electrical and Electronic Engineering*, vol. 30, no. 3, pp. 941–956, 2011.

[10] M. Vaillant, M. Eckert, and F. Gauterin, "Energy management strategy to be used in design space exploration for electric powertrain optimization," in *Proceedings of the 2014 9th International Conference on Ecological Vehicles and Renewable Energies, EVER 2014*, mco, March 2014.

[11] A. T. Nguyen, S. Reiter, and P. Rigo, "A review on simulation-based optimization methods applied to building performance analysis," *Applied Energy*, vol. 113, pp. 1043–1058, 2014.

[12] W. Silka and W. Siłka, *Energy Intensity in the Car Traffic*, In Polish, WNT. Warszawa, 1997.

[13] T. J. Barlow, S. Latham, I. S. Mrcrae, and P. Boulter, *A reference book of driving cycles for use in the measurement of road vehicle emissions, TRL*, 2009, https://www.gov.uk/government/uploads/system/...data/.../.

[14] R. N. Jazar, *Vehicle Dynamics: Theory and Application*, Springer US, Boston, MA, 2008.

[15] M. Meywerk, "Vehicle Dynamics," in *West Sussex*, John Wiley & Sons, Lecture Notes to the MOOC Vehicle Dynamics of iversity, 1st edition, 2015.

[16] M. Mitschke and H. Wallentowitz, *Dynamik der Kraftfahrzeuge*, Springer, Berlin, Germany, 1972.

[17] T. Stefański and Ł. Zawarczyński, "Parametric identification of PM motor mathematical models," *Przegląd Elektrotechniczny*, vol. 88, no. 4 B, pp. 224–229, 2012.

[18] A. Cameron, "Measuring drive-train efficiency," *Human Power*, vol. 46.

[19] B. Rohloff and P. Greb, "Efficiency measurements of bicycle transmissions a neverending story," *Human Power*, vol. 55, p. 11, 2004.

[20] J. B. Spicer, C. J. K. Richardson, M. J. Ehrlich, J. R. Bernstein, M. Fukuda, and M. Terada, "Effects of frictional loss on bicycle chain drive efficiency," *Journal of Mechanical Design*, vol. 123, no. 4, pp. 598–605, 2001.

[21] K. Sternal, A. Cholewa, W. Skarka, and M. Targosz, "Electric vehicle for the students' shell eco-marathon competition. Design of the car and telemetry system," *Communications in Computer and Information Science*, vol. 329, pp. 26–33, 2012.

[22] K. Cichoński and W. Skarka, "Innovative control system for high efficiency electric urban vehicle," *Communications in Computer and Information Science*, vol. 531, pp. 121–130, 2015.

[23] M. J. Er and R. J. Oentaryo, "Computational Intelligence: Methods and Techniques (Rutkowski, L.; 2008) [Book Review]," *IEEE Computational Intelligence Magazine*, vol. 6, no. 4, pp. 76–78, 2011.

[24] M. Targosz, M. Szumowski, W. Skarka, and P. Przystałka, "Velocity Planning of an Electric Vehicle Using an Evolutionary

Algorithm," *Communications in Computer and Information Science*, vol. 395, pp. 171–177, 2013.

[25] M. S. Ismail, M. Moghavvemi, and T. M. I. Mahlia, "Genetic algorithm based optimization on modeling and design of hybrid renewable energy systems," *Energy Conversion and Management*, vol. 85, pp. 120–130, 2014.

[26] Z. Michalewicz, "Evolutionary algorithms for constrained parameter optimization problems," *Evolutionary Computation*, vol. 4, no. 1, pp. 1–32, 1996.

[27] M. Targosz, "Test bench for efficiency evaluation of belt and chain transmissions," in *Proceedings of the Conference Proceedings XII International Technical System Degradation Conferences, Liptovsky Mikulas*, 2013.

[28] M. Targosz, K. Cichonski, M. Glen, A. Szymon, K. Cichoński, and M. Gleń, "An experimental investigation of the energy consumption for lightweight electric vehicle," in *Proceedings of the 13th International Technical Systems Degradation Conference*, 2014.

[29] M. Targosz, W. Skarka, P. Przystalka, and P. Przystałka, "Simulation and optimization methodology of prototype electric vehicle," in *Proceedings of the 13th International Design Conference DESIGN 2014*, D. Marjanovic, M. Storga, N. Pavkovic, and N. Bojčetic, Eds., pp. 2014-1349, Croatia, 2014.

Investigation on Range Anxiety and Safety Buffer of Battery Electric Vehicle Drivers

Quan Yuan,[1] Wei Hao,[2,3] Haotian Su,[1] Guanwen Bing,[1] Xinyuan Gui,[1] and Abolfazl Safikhani[4]

[1] *State Key Laboratory of Automotive Safety and Energy, Tsinghua University, Beijing 100084, China*
[2] *Hunan Key Laboratory of Smart Roadway and Cooperative Vehicle-Infrastructure Systems, Changsha University of Science and Technology, Changsha, Hunan 410114, China*
[3] *Key Laboratory of Safety Design and Reliability Technology for Engineering Vehicle, Changsha University of Science and Technology, Changsha, Hunan 410114, China*
[4] *Department of Statistics, Columbia University in the City of New York, NY 10031, USA*

Correspondence should be addressed to Wei Hao; haowei@csust.edu.cn

Academic Editor: Yair Wiseman

Drivers tend to have more range anxiety compared with driving traditional fuel vehicles if they are driving battery electric vehicle (BEV) with a long trip. Range anxiety could potentially have negative effect on driver's emotions and behaviors. In order to understand this behavior and improve the related safety issues, this paper will focus on BEV drivers' study in China. A survey on BEV drivers' actual range anxiety as well as the effect of range anxiety on drivers' behaviors is conducted in this research. Levels of feelings and attitudes of the interviewees are quantized with Likert scales using mathematical tools of the relationship. Safety buffer is defined as a measurement of the period given range anxiety starting to significantly intervene in driver's operation. The research reveals the proportional quotative relationship between BEV drivers' safety buffer and the mileage of trip. Factors, including driving experience, satisfactory level of recharge accessibility, and resistibility to emotions, are found to be significant contributing factors to influence the perceived range anxiety level of BEV drivers. This research will provide implications to the future study on the interface design of BEV.

1. Introduction

Battery electric vehicle (BEV) drivers typically tend to have more range anxiety compared to traditional fuel vehicle drivers. High levels of range anxiety can potentially lead to negative effects on BEV drivers' emotion and even cause dangerous driving behaviors. Due to this potential safety issue, several research projects ([1, 2]) have been developed in order to understand range anxiety including appraisal models and influential factors, and further, several solutions are proposed in the literature to reduce the range anxiety.

Nadine Rauh et al. [1] examined the assumption that range anxiety disturbs only inexperienced BEV drivers, and therefore it vanishes as driver gains more driving experience. In the experiment conducted in [1], drivers with different levels of driving experiences were given a critical range situation, where the range remaining in the BEV was less than the trip length. By measuring the range appraisal and range stress with different scales, researchers discovered that driving experience imposed strong effects on all measured variables. With more experience, BEV drivers tend to have less negative range appraisal and hence range anxiety. Therefore, it is important to increase efficiency and effectiveness of the learning process for BEV drivers.

To enhance the understanding of factors which can relieve range anxiety, Thomas Franke et al. [2] designed a field study setting to examine several factors that possibly contribute to lower everyday range stress (ERS). The study revealed the fact that variables such as less encounter with critical range situations, higher practical experience, subjective range competence, tolerance of low range, and experienced trustworthiness of the range estimation system were related to lower ERS, and range stress is found to be related to range satisfaction and battery electric vehicle acceptance.

Also, there are few studies on checking the influence of in-vehicle information systems (IVISs) on range anxiety. For example, Matthias Eisel et al. [3] performed a battery electric vehicle field experiment under a real traffic situation. The crew measured the participants' psychometric range appraisal and psychophysiological feedback and came to the conclusion that individuals perceived the critical range situation as less challenging and threatening with the provided IVISs. However, despite the fact that the IVISs led to a reduction of mean value of stress throughout the driving task, researchers found that participants with the IVISs had higher levels of stress perception. This result indicates that the IVISs can lead to an increased awareness of the depletion of range resources over time.

To gain insight in individual differences in range stress when BEV drivers are faced with a critical range situation for the first time, Thomas Franke et al. [4] designed a field experiment where participants were given a trip of which the range was tailored close to the BEV's range capacity. The results were helpful to formulate strategies aiming at reducing early experienced range stress.

There are many solutions proposed in the literature to reduce BEV drivers' range anxiety. Vincent R. Tannahill et al. [5] explored the future of range anxiety solutions. A driver alerting algorithm is proposed, which was able to reduce range anxiety. The key improvement of the algorithm is the advanced accuracy of BEV range estimation. No complex computations are involved in the algorithm. Hence the algorithm can be implemented on low-cost microcontrollers and still can achieve accurate results.

Mahmoud Faraj and Otman Basir [6] proposed a path arrangement model based on battery capability and energy cost analysis. The model compares the battery capacity of the BEV with the least energy cost of a round trip to the charging station. If the battery capability is less than the energy cost to the charging station, the model recommends the driver to recharge, preventing the driver from being stranded. An accurate estimation on the available range can also be given by the model to the BEV driver. This model is able to reduce range anxiety and further promotes the widespread use of BEVs.

Kaveh Sarrafan et al. [7] proposed an algorithm which makes real-time recommendations on BEV charging strategy. The algorithm is based on the combined calculation of an accurate state of charge (SoC) estimation method and GPS. With necessary initial data input, the algorithm can calculate and predict the BEV's remaining range at the trip destination. The function is realized through a real-time indicator system run by the algorithm. The user experience could potentially be improved by reduced range anxiety during the trip.

Vincent R. Tannahill et al. [8] came up with a new method of range estimation. Based on a state of charge (SoC) estimation method already proposed, the new method accounts for a wide range of environmental and behavioral factors. Hence, the estimation of range turns out to be more accurate than the results drawn by conventional methods which consider merely the SoC and vehicle efficiency. This new range estimation method can be used to notify the driver of the range capability of the BEV and can propose suggestions on whether the BEV needs recharge before reaching the destination.

Battery swapping system is an alternative way of reducing range anxiety. Properly allocating spare batteries will benefit the efficiency of the battery swapping network. Michael Dreyfuss and Yahel Giat [9] revealed the importance of estimating customers' tolerable battery swapping waiting time, which was considered negligible in previous research. Taking the waiting time into consideration, this research found out it will reduce the time of battery swapping. Some features of the optimal spare battery allocation were also found, which will help the design of future battery swapping system.

With regard to the current state of range anxiety research, this paper is intended to verify and forward several previously implemented researches. More specifically, this paper tests the safety buffer model and attempts to fit it into different occasions of trip mileage. In addition, broadened scope of factors are checked in this paper with potential influence on range anxiety. Behavioral effects of range anxiety are another topic discussed in this paper on range anxiety research.

2. Data and Methods

2.1. Basic Assumptions. Range anxiety is described as the fear of being left in the middle of a trip due to a BEV's empty battery. Besides, there are other factors which contribute to the anxiety, such as the inadequate charging infrastructures and the trustworthiness of the vehicle's SoC display.

It is supposed that there is a minimum gap between vehicles' remaining mileage and trips' range which can protect a driver against range anxiety. In Thomas Franke's research [10], the gap was referred to as the safety buffer. In Franke's research, the safety buffer was measured from two aspects. First aspect is described as the difference between the values of vehicles' remaining mileage and the trips' range, and second aspect is depicted by the proportion of the trip's range. The analysis results indicated that the proportion was roughly around 80%.

The analysis results were derived from the study conducted in Germany, of which the authors were using conditions of BEVs which are greatly different from what it is utilized in China. The scenarios mentioned in [11] are fixed, where drivers are given a 168 km capable vehicle to drive from its full comfort range or given a 60 km-trip to select a specific range which makes them feel comfortable. Also, it is believed that range anxiety affects different modes in trips with different lengths [12]. Consequently, three trip scenarios with particular length are set up. The 10 km-trip represents short trip situations; the 30 km-trip represents middle trip situations; and the 60 km-trip represents long trip situations. Influential factors for each trip length is expected to be studied, which is helpful to formulate range anxiety reduction strategies in different circumstances. This paper provides the results based on Chinese BEV working condition, covering short-range, middle-range, and long-range trip conditions.

The factors that possibly have influences on range anxiety are also dug into the research. Information is collected from the interviewees. They include basic attributes: gender, age, defined figures, and degree of range anxiety. The defined figures cover two aspects including driver's satisfaction level towards vehicle and driving style. The two figures all consist of more than one measurement. Driver's satisfaction level is defined from three aspects: the driver's satisfaction level of vehicle range, the driver's satisfaction level of recharge accessibility, and the driver's trust level of range information displayed by the vehicle. In few previous works [13–15], better satisfaction towards vehicle range capacity leads to expansion of range comfort zone. The driver's driving style is defined from two dimensions: the driver's decisiveness and the driver's susceptibility. Factors are expected to be dug out that related to range anxiety and make solutions which can help with particular population against range anxiety.

When facing range anxiety, drivers would behave with the intention to reduce the risk of being stranded in the middle of a trip. The approaches to the reduction of risk are different including prolonging the remaining range of the vehicle or reaching a recharge infrastructure as soon as possible. Different approaches are reflected on the drivers as changes of driving behaviors. The changes of driving behaviors can bring additional stress to the driver and even can lead to dangerous behaviors. To help the drivers optimize their driving behaviors, the most common behaviors and the preferred behaviors differ for different types of drivers which is something needed to be studied. The results reveal some insights to advance the BEV human-machine interface.

2.2. Questionnaire. The questionnaire includes two major parts. The first part (Appendix Part I) is mainly about the driver's basic information, including gender, age, and driving age. Based on the previous reference [16], the second part (Appendix Part II) is designed with several 5-point Likert scales and multiple-choice questions on the range anxiety's influence. The Likert scales on the driver's satisfaction level towards vehicle and driving style are set with the intention that the higher score always refers to a better situation. For example, a 5-point answer may represent a situation where the driver shows the greatest trust towards the vehicle's range display, while the Likert scales on the safety buffer are set that the higher the score is, the more buffer is needed, which can make the results more straightforward.

The participants are all currently BEV drivers. Some of the recognized BEV owners were selected from the *Autohome* forum (which is the most visited auto website in the world). After an in-depth conversation, the recognized owners were asked to help spread the questionnaire to other BEV owners. Through the conversation, it was observed that the recognized owners did spread the questionnaire to authentic BEV owners. Besides, some particular measurements in the questionnaire were checked to filter the invalid questionnaires. For drivers with fuel vehicle driving experience, the *previous fuel vehicle fuel consumption per 100 km* was selected as a filter measurement. Only questionnaires with the value above 2 L and under 40 L passed the filter, taking drivers with hybrid power vehicle or luxury vehicle driving experience into consideration. For all participants, the *type of BEV currently driven* was examined to make sure that the interviewees are BEV drivers. Similarly, the *range of BEV currently driven* was selected as a filter. Questionnaires with a value between 50 km and 400 km would pass the filter. After the filter process, 208 questionnaires were considered valid for further research.

2.3. Statistical Method. To check the reliability of the questionnaire, particularly the Likert scales proposed in our approach, Cronbach's Alpha (α) was calculated and verified. All the Likert scales yield good internal reliability with $\alpha > 0.60$ ($\alpha = 0.758$).

The factors related to range anxiety are expected to be investigated qualitatively. Chi-square tests are conducted to examine the correlations between influential factors and safety buffer. Chi-square tests are also carried out for verifying the correlations between influential factors and driving behaviors. By checking the P values from the Chi-square tests of each combination of data, several factors are marked which are correlated with safety buffer or certain behaviors. Usually, data in combination with p value < 0.05 can be considered significantly correlated and data in combination with p value<0.01 can be considered very significantly correlated.

3. Results

3.1. Descriptive Statistics. Table 1 displays the characteristics of our research sample. Among the 208 interviewees, 156 interviewees are male (75% of all) and 52 interviewees are female (25% of all). The mean value of the interviewees' age is 32.99 (SD = 7.686), ranging from 19 to 67. The mean value of the interviewees' driving experience is 6.55 years (SD = 5.121), ranging from 0 to 22. Company personnel is the most occurred profession among all interviewees, appearing 78 times (78% of all). 169 interviewees have had fuel vehicle driving experience, with mean fuel consumption at 9.12 L/100 km (SD = 5.529). Geely Emgrand BEV appeared 65 times in the sample, which is the most among all types of BEV. The mean maximum range of the BEVs owned by the interviewees is 222.31 km (SD = 80.523), varying from 50 km to 400 km. Household charging-pile is the most commonly used recharge facility. The average recharge frequency of the interviewees is 3.33 times per week (SD = 3.045).

The results of the Likert scales on the driver's satisfaction level of the BEV are displayed in Table 2. The average point of the driver's satisfaction level towards the range is 3.41 (SD = 1.185), with 112 interviewees (53.9% of all) above average. The average point of the driver's satisfaction level towards the accessibility of recharge is 3.54 (SD = 1.100), with 121 interviewees (58.2% of all) above average. The average point of the driver's trust level towards the range information displayed by the BEV is 3.64 (SD = 1.135), with 133 interviewees (64.0% of all) above average. And the mean value of all 3 scales mentioned is 3.53 points (SD = 1.014). There are more interviewees with satisfaction level above the average than those with satisfaction level below the average. Based on the statistical data, the following conclusion can be reached that

Table 1: Sample characteristics (N = 208).

Variable	Mean	SD	Range	N	%
Gender					
Male				156	75.0
Female				52	25.0
Age	32.99	7.686	19-67		
Driving experience (years)	6.55	5.121	0-22		
Profession					
Job hunter				5	2.4
Civil servant				21	10.1
Teaching or scientific research personnel				21	10.1
Individual laborer				17	8.2
Retired				5	2.4
Production or service personnel				42	20.2
Company personnel				78	37.5
Student				5	2.4
Others				14	6.7
Fuel vehicle driving experience					
Yes				169	81.3
Fuel consumption (L/100 km)	9.12	5.529	2-30		
Refueling frequency (times/month)	3.27	3.289	1-20		
No				39	18.8
Type of BEV currently driven					
BJEV EV160				31	14.9
BJEV EV200				24	11.5
BYD E6				28	13.5
Geely Emgrand EV				65	31.3
JAC iEV5				8	3.9
Tesla Model S				11	5.3
Tesla Model X				4	1.9
Venucia e30				5	2.4
Denza				5	2.4
Others				27	13.0
Max range of currently driven BEV (km)	222.31	80.523	50-400		
Most commonly used method of recharge					
Public AC charging-pile				9	4.3
Public DC charging-pile				37	17.8
House hold charging-pile				116	55.8
Portable charger (220V AC)				39	18.8
Others				7	3.4
Recharging frequency (times/week)	3.33	3.045	0.5-25		
Use of BEV					
Official				8	3.8
Private				180	86.5
Rental				11	5.3
Service				6	2.9
Others				3	1.4

the interviewees were generally feeling contented with the BEVs they owned. Cronbach's Alpha for this group of Likert scales is 0.868.

The results of the Likert scales on the driver's driving style are displayed in Table 2. The average point of the driver's decisiveness is 3.87 (SD = 0.913), with 154 interviewees (74.1% of all) above the average. The average point of the driver's resistibility against emotions is 3.55 (SD = 0.989), with 123 interviewees (59.1% of all) above the average. Two drivers' driving style factors mentioned above are both believed to

TABLE 2: Statistics of driver's satisfaction level and driving style (StD: strongly disagree, D: disagree, Neither: neither agree or disagree, A: agree, StA: strongly agree; the same to the following tables).

Statement	StD (1 pts)	D (2 pts)	Neither (3 pts)	A (4 pts)	StA (5 pts)	Mean	SD
Satisfaction level							
I am contented with my BEV's range.	18 (8.7%)	29 (13.9%)	49 (23.6%)	74 (35.6%)	38 (18.3%)	3.41	1.185
It is convenient to recharge my BEV.	10 (4.8%)	29 (13.9%)	48 (23.1%)	80 (38.5%)	41 (19.7%)	3.54	1.100
I trust my BEV's range remain display.	12 (5.8%)	25 (12.0%)	38 (18.3%)	84 (40.4%)	49 (23.6%)	3.64	1.135
General satisfaction level						3.53	1.014
Driving style							
I behave decisively when driving.	6 (2.9%)	9 (4.3%)	39 (18.8%)	106 (51.0%)	48 (23.1%)	3.87	0.913
Emotions hardly affect me when I am driving.	8 (3.8%)	22 (10.6%)	55 (26.4%)	93 (44.7%)	30 (14.4%)	3.55	0.989

TABLE 3: Statistics of safety buffer.

Statement	**10 km**	**12.5 km**	**15 km**	**17.5 km**	**20 km**	**Mean**	**SD**
Minimum comfort range remaining in a 10 km trip	24 (11.5%)	24 (11.5%)	53 (25.5%)	23 (11.1%)	84 (40.4%)	3.57	1.406
Statement	**30 km**	**37.5 km**	**45 km**	**52.5 km**	**60 km**	**Mean**	**SD**
Minimum comfort range remaining in a 30 km trip	30 (14.4%)	43 (20.7%)	69 (33.2%)	31 (14.9%)	35 (16.8%)	2.99	1.267
Statement	**60 km**	**67.5 km**	**75 km**	**82.5 km**	**90 km**	**Mean**	**SD**
Minimum comfort range remaining in a 60 km trip	32 (15.4%)	28 (13.5%)	68 (32.7%)	35 (16.8%)	45 (21.6%)	3.16	1.326
General safety buffer						3.24	1.151

be positively related to better road safety performances. Cronbach's Alpha for this group of Likert scales is 0.706.

The results of the Likert scales on the driver's safety buffer are displayed in Table 3. For the situation where there is 10 km left in a trip, the average point of the safety buffer is 3.57 (SD = 1.406), which equals 16.43 km of BEV range remaining, 164.3% of the trip mileage. For the situation where there is 30 km left in a trip, the average point of the safety buffer is 2.99 (SD = 1.267), which equals 44.93 km of BEV range remaining, 149.8% of the trip mileage. For the situation where there is 60 km left in a trip, the average point of the safety buffer is 3.16 (SD = 1.326), which equals 76.20 km of BEV range remaining, 127.0% of the trip mileage. The comparison of net safety buffer and trip mileage under different trip scenarios is displayed in Figure 1. The net safety buffer for a 10 km-trip is 6.43 km in value and 64.3% in portion. The net safety buffer for a 30 km-trip is 14.93 km in value and 49.8% in portion. The net safety buffer for a 60 km-trip is 16.20 km in value and 27.0% in portion. For trips with longer mileages like 30 km or 60 km, the net safety buffer is close to 19.2 km as shown in the previous research. In addition, there seems to be stability with the buffer value; the gap between the values of 10 km-trip and 30 km-trip is huge whereas the gap between 30 km-trip and 60 km-trip is small. Also, the portion of the net buffer is higher when the trip is short and lower when the trip becomes longer and the buffer stays unchanged. Under 10 km-trip and 60 km-trip, the portion of interviewees choosing the longest safety range (40.4% under 10 km-trip, 21.6% under 60 km-trip) is much higher than that of the second longest range (11.1% under 10 km-trip, 16.8% under 60 km-trip). This may be due to the uncertainty the drivers have when the BEV range remaining is low or the trip mileage remaining is high. Cronbach's Alpha for this group of Likert scales is 0.829. For

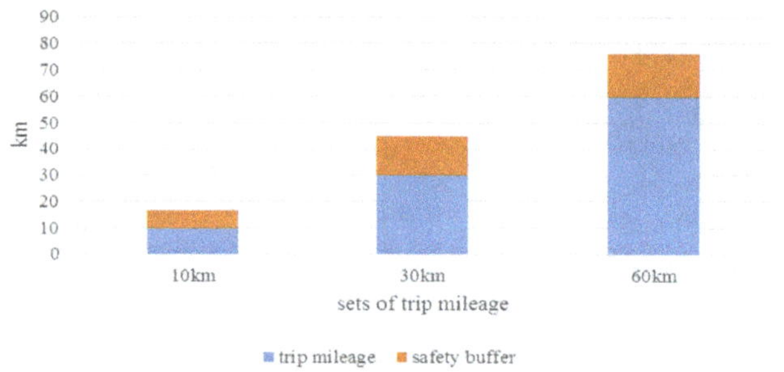

FIGURE 1: Trip mileage and safety buffer.

the convenience of further analysis, general safety buffer is defined as the mean value of the interviewee's safety buffer points under the three trip situations.

The statistics of the behaviors taken by drivers related to the range anxiety are displayed in Table 4. Among the 7 types of behaviors listed, *speed reduction* (97 interviewees, 46.6% of all), *change behaviors* (70 interviewees, 33.7% of all), and *seeking nearby charging-pile* (52 interviewees, 25.0% of all) are the three most frequently taken behaviors.

3.2. Correlation Analysis. Chi-square tests were conducted to find out factors related to driver's safety buffer and anxiety related behaviors. Results with P values less than 0.05 are considered as a sign that there is a significant difference between the observed frequency distribution and expected frequency distribution. The difference can be considered very significant in results with P values under 0.01. By checking the P value of each result, groups of figure combination can be discovered when the frequency distribution differs significantly from expected distribution. In groups where the involved figures

TABLE 4: Statistics of driver's behaviors taken related to range anxiety.

Behavior	Distracted by anxiety	Rearrange the route	Seeking nearby charging-pile	Change behaviors	Speed reduction	Get more disturbed	Grab lines
Number of interviewees taking the behaviors above.	26 (12.5%)	40 (19.2%)	52 (25.0%)	70 (33.7%)	97 (46.6%)	47 (22.6%)	26 (12.5%)

TABLE 5: Chi-square test results of factors that may be related to *general safety buffer* (factors with p value < 0.05 marked with ** and factors with p value < 0.01 marked with * * *, the same to the following tables).

Factor	Below average buffer	Above average buffer	p value
Driving experience**			0.049
≤6.55 years	72	57	
	33	46	
Satisfaction level of recharge accessibility**			0.018
Below average	45	60	
Above average	61	42	
Resistibility to emotions**			0.045
Below average	50	35	
Above average	55	68	

TABLE 6: Chi-square test results of factors that may be related to safety buffer (10 km-trip).

Factor	Below average buffer	Above average buffer	p value
10 km-trip			
Driving experience**			0.035
≤6.55 years	70	31	
>6.55	59	48	
Satisfaction level of max BEV range**			0.020
Below average	55	41	
Above average	46	66	
Decisiveness when driving**			0.014
Below average	34	20	
Above average	67	87	
60 km-trip			
Max range of BEV**			0.028
≤222.31 km	76	35	
>222.31	52	45	
Resistibility to emotions* * *			0.001
Below average	64	21	
Above average	64	59	

obey a frequency distribution different from the expected distribution, the two figures are closely related. Comparing the observed frequencies with the expected frequencies, the way in which one factor affects a driver's safety buffer or behaviors is found out.

To explore factors that may have influence on driver's safety buffer, Chi-square tests are conducted on different factors including driver characteristics, satisfaction level, and driving style. In order to dig out the difference of the time when each factor starts to work, the Chi-square P values of various factors under different trip mileage situations are examined. The results are shown in Tables 5, 6, and 7.

Among the drivers' characteristics on which Chi-square tests are conducted, *driving experience* (p value = 0.049) shows a significant relationship with *general safety buffer*. Comparing the results with the expected frequencies, the length of driving experience is found to have a positive relationship with the general safety index. This might be due to the difference between the driving mentalities of drivers with various driving experiences. Those with more driving experiences might be more cautious about leaving adequate safety buffer for a trip.

In the scales of driver's satisfaction level, the driver's *satisfaction level of recharge accessibility* (p value = 0.018)

TABLE 7: Chi-square test results of factors that may be related to grabbing lines behavior.

Factor	Grabbing lines	Not grabbing lines	p value
Gender**			0.029
Male	24	132	
Female	2	50	
Age**			0.040
≤32.99 years	19	94	
	7	88	
Fuel vehicle driving experience***			0.009
Yes	26	143	
No	0	39	
Max range of BEV**			0.040
≤222.31 km	9	102	
	17	80	
Decisiveness when driving**			0.023
Below average	2	52	
Above average	24	130	
Resistibility to emotions***			0.005
Below average	4	81	
Above average	22	101	

shows a significant relationship with *general safety buffer*. After comparing the results with the expectation, it turns out that with better recharge accessibility satisfaction comes less safety buffer. It is supposed that better recharge accessibility satisfaction reduces the driver's anticipation of the possibility for running out of energy in the middle of a trip. Therefore, less buffer is required by the driver to offset the anxiety.

In the scales of driver's driving style, the driver's *resistibility to emotions* (p value = 0.045) shows a significant relationship with *general safety buffer*. After comparing the results with the expectation, it is concluded that better *resistibility to emotions* results in more safety buffer. It is supposed that better *resistibility to emotions* yields from a series of driving habits and driver characteristics, which leads to the choice of bigger safety buffer. The *resistibility to emotions* and *general safety buffer* might be two related dependent variables to some other independent variables, which is left for further investigation in further research.

For the discussion of influential factors to different trip mileage situations, data are collected of driver's safety buffer for different trip mileage situations and Chi-square tests are conducted on the data, respectively. In the 10 km-trip situation, *driving experience* (p value = 0.035), *satisfaction level of max BEV range* (p value = 0.020), and *decisiveness when driving* (p value = 0.014) show a significant relationship with the safety buffer. Drivers with longer *driving experience* tend to have bigger safety buffer, which accords to the driving experience's influence on *general safety buffer*. Drivers with higher *satisfaction level of max BEV range* tend to have safety buffer of a smaller size. Drivers with greater *decisiveness when driving* tend to have smaller safety buffer. For the 30 km-trip situation, no factor is found to have a significant relationship with the safety buffer. In the 60 km-trip situation, *max range of BEV* (p value = 0.028) shows a significant relationship with the safety buffer and *resistibility to emotions* (p value = 0.001) shows a very significant relationship with the safety buffer. Drivers driving BEV with longer max range tend to have bigger safety buffer. Drivers with better *resistibility to emotions* tend to have bigger safety buffer.

To discover how potential factors can influence a driver's acts under range anxiety, Chi-square tests are conducted on each listed behavior. *Resistibility to emotions* (p value = 0.001) shows a very significant relationship with the behavior of *speed reduction*. Drivers with weaker *resistibility to emotions* tend to reduce driving speed under range anxiety. For the behavior of *getting more disturbed*, *general satisfaction level* (p value = 0.038) shows a significant relationship with it. Driver with lower *general satisfaction level* are more likely to get disturbed under range anxiety. For the behavior of *grabbing lines*, there are a series of factors that are linked to it. *Gender* (p value = 0.029) shows a significant relationship with the *grabbing lines*. Under range anxiety, male drivers are more likely to grab lines than female drivers. *Age* (p value = 0.040) shows a significant relationship with the *grabbing lines*. Younger drivers are more likely to grab lines under range anxiety. *Fuel vehicle driving experience* (p value = 0.009) shows a significant relationship with the *grabbing lines*. Drivers with *fuel vehicle driving experience* are more likely to grab lines. *Max BEV range* (p value = 0.040) shows a significant relationship with the *grabbing lines*. Drivers driving BEV with longer max range are more likely to grab lines. *Decisiveness when driving* (p value = 0.023) shows a significant relationship with the *grabbing lines*. Drivers with stronger decisiveness are more likely to grab lines. *Resistibility to emotions* (p value = 0.005) shows a significant relationship with the *grabbing lines*. Drivers with stronger resistibility are more likely to grab lines.

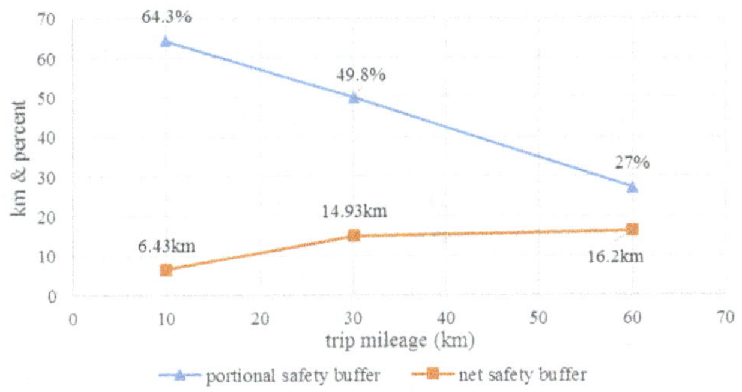

Figure 2: Net safety buffer and fractional safety buffer.

4. Discussions

4.1. Safety Buffer and Its Influential Factors. The safety buffer under different trip mileage circumstances is evaluated through the mean value of the Likert scale point. Data of net safety buffer and fractional safety buffer is shown in Figure 2. For 10 km-trip, the average net safety buffer is 6.43 km, which takes 64.3%. For 30 km-trip, the average net safety buffer is 14.93 km, which takes 49.8%. For 60 km-trip, the average net safety buffer is 16.20 km, which takes 27.0%. Research conducted by Thomas Franke et al. [9] finds that the average net safety buffer is 25.0% of the expected trip range. The data is collected from the field experiment in Germany. The safety buffer in this paper coincides well with the previous results in the 60 km-trip occasion, 27.0% comparing to 25.0%. However, the safety buffer portion in 30 km-trip is 49.8%, much greater than 25.0% assumed in [9]. In addition, the net values of safety buffer in 30 km-trip and 60 km-trip are close, 14.93 km and 16.20 km, respectively, which are very similar to each other. Regarding the portion of the net buffer in 10 km-trip, which is much bigger than that of 30 km-trip, 64.3% comparing 49.8%, the safety buffer portion is believed to increase as the actual trip range shortens, while in long-range trip, this paper confirms the reliability of the result from [9].

In Likert scales of 10 km-trip and 60 km-trip, the portion of interviewees choosing the option with the largest buffer takes up much more in total samples than interviewees choosing the second largest buffer (40.4% to 11.1% in 10 km-trip, 21.6% to 16.8% in 60 km-trip). The distribution of the chosen option is quite biased from the normal distribution. The uncertainty of upcoming traffic and energy consumption are supposed to play a role in the abnormal frequency distribution. For 10 km-trip, the uncertainty of the energy consumption might play a bigger role. The SoC information display is usually less accurate when the SoC is low whereas the traffic condition is clearer to the driver when the trip mileage remaining is little as 10 km. The uncertainty of the energy consumption weights is more than the uncertainty of traffic to the driver in that circumstance. For 60 km-trip, the uncertainty of the upcoming traffic generates more anxiety than the uncertainty of the remaining energy, forcing some drivers to require more safety buffer.

Among the driver's characteristics correlating with general safety buffer, *driving experience* is found to be significantly correlated with safety buffer. The results show that drivers with more years of driving experience are more likely to have bigger safety buffer. Experienced drivers are supposed to be more cautious about sparing enough buffer for a trip. The difference between the driving mentality of fresh drivers and skilled drivers might cause the difference in safety buffer choice. The effect of driving experience on range anxiety is discussed in [1, 2], specifically on the driving experience of BEV. The results show that drivers with more experience of driving BEV are likely to have reduced range anxiety. Driving experience of BEV is not included in our research. However, assuming that the driver's driving experience of BEV is positively connected with the total driving experience and combining the results of [1, 2] with ours, the size of safety buffer can be found to be negatively connected to the level of range anxiety. That is currently a hypothesis and the connection between safety buffer and range anxiety level requires further investigation.

Among the defined factors, driver's satisfaction level and driving style are significantly correlated with *general safety buffer*. The *satisfaction level of recharge accessibility* is negatively connected with *general safety buffer*. It is assumed that better recharge accessibility comes up with less anticipated risk of being stranded in the middle of a trip due to an unavailable recharge. This result accords with [12] as well, which indicates that better satisfaction towards vehicle capacity expands the range comfort zone and reduces the safety buffer. The driver's *resistibility to emotions* is positively connected with *general safety buffer*. This result supports the claim made by [4] that emotional stability is partly related to the lower range stress. Drivers with better *resistibility to emotions* require more safety buffer. The *resistibility to emotions* and *general safety buffer* are assumed to be dependent variables to some other independent variables connected with driving and range anxiety level. Following the hypothesis above, it is possible that better *resistibility to emotions* leads to lower level of range anxiety and the safety buffer grows consequently.

Factors that work in situations of trips with different mileage are not the same. In a short trip, *driving experience, satisfaction level of max BEV range,* and *decisiveness when driving* show a significant relationship with safety buffer. Drivers with more *driving experience* require more safety buffer, which is according to the previous analysis. Drivers with higher *satisfaction level of max BEV range* require less safety buffer. Drivers with greater *decisiveness when driving* require less safety buffer. In a middle-range trip, no factor is found to be of great correlation with safety buffer. In a long trip, *max range of BEV* is positively related to safety buffer. Drivers who are driving BEVs with more max range require more safety buffer. These drivers are assumed to be more frequently driving with adequate remaining range, since the max range of their vehicles is higher. The advantage in max range makes drivers get used to driving with more buffer. Hence, they are less comfortable with low buffer situations and require more buffer. *Resistibility to emotions* is also correlated with the safety buffer. Drivers with better *resistibility to emotions* require more safety buffer, which accords to previous analysis.

4.2. Driver Behavior and Its Influential Factors. Among the 7 types of behavior listed, *speed reduction*, *change behaviors*, and *seeking nearby charging-pile* are the three most frequently taken behaviors. The factor of *change behaviors* represents the instability of behavior when drivers are under stress. *Speed reduction* and *seeking nearby charging-pile* represent two methods to reduce the risk of being stranded in the middle of a trip. To save energy and extend the range, drivers choose to reduce the speed. To recharge energy and substantially extend the range capacity, drivers choose to seek charging-pile. Other listed behaviors are also related to the two strategies, respectively, but not as featured as the two most chosen behaviors.

There are three of our listed behaviors showing co-relationship with one or more factors and there are rather many correlated factors with *grabbing lines*. For the behavior of *speed reduction*, *resistibility to emotions* is of significant correlation. Drivers with weaker *resistibility to emotions* tend to reduce driving speed under range anxiety. For the behavior of *getting more disturbed*, *general satisfaction level* shows a significant correlation. Drivers with lower *general satisfaction level* are more likely to get more disturbed in mind under range anxiety. The dissatisfaction of the BEV clearly worsens the stress of BEV driver under range anxiety.

For the behavior of *grabbing lines*, a series of factors are found. *Grabbing lines* is believed to be a behavior with the intention to approach recharge points quicker but it is a violation of traffic morality. So drivers are taking risks of getting involved in accidents by grabbing lines. Male drivers are more likely to grab lines, possibly because male drivers are more adventurous mentally. Younger drivers are more likely to grab lines than elder drivers, possibly for the similar reason that the young drivers are more adventurous. Drivers with *fuel vehicle driving experience* are more likely to grab lines, which is very interesting. No one in the total of 39 drivers without fuel vehicle experience reported grabbing lines. Grabbing lines might be too energy-risky for drivers who have no experience of driving fuel vehicles with great accessibility to energy. Considering the negative influence of grabbing lines and the potential risk of the drivers taking more hazardous driving behavior, the driving ethic of BEV drivers under range anxiety deserves further research. *Max BEV range* shows a significant relationship with *grabbing lines*. Drivers driving BEV with higher max range are more likely to grab lines. These drivers are considered to be used to their BEV's excellent range capacity and show more acceptance to more risky behaviors like grabbing lines. Drivers' *decisiveness when driving* and *resistibility to emotions* both show a positive connection with the possibility of grabbing lines under range anxiety. Drivers with stronger decisiveness are more likely to make quick decisions and thus more likely to take risky behaviors. Drivers with better resistibility to emotions are less sensitive to stress and thus take the risky behaviors like grabbing lines.

4.3. Supposed Method of Helping with Range Anxiety. A more accurate and trustworthy in-vehicle information system is of great help to lower the total range anxiety level of drivers, according to [3, 5, 10]. Methods are proposed to estimate the battery SoC or vehicle range in [5, 8, 14]. However, the information system also reminds the driver of the depletion of the BEV's range, potentially raising the speed of the vehicle making drivers gaining anxiety. Hence, it is important to give the driver only the needed information and it must be on time, not earlier or later. Otherwise, the driver will suffer from unnecessary range anxiety gain. Necessary driving behavior guidance should also be given to drivers under stress. From the previous analysis, it can be concluded how range anxiety leads drivers to behave themselves. Changing existing driving habits could cause unfamiliarity of driving. Some of the behaviors are distractive and even dangerous, potentially leading to traffic accidents. So proper guidance is very important to driving safety.

If the BEV receives the destination of a trip, working with GPS, the BEV might be able to calculate the route and the mileage remaining in the current trip. When the detected BEV range remaining is close to the safety buffer range for the trip, the BEV should give guidance on solving the range anxiety. One solution has been proposed in [6] that helps the driver to follow the energy-optimized travelling route, which could relieve the range anxiety. By recording the personal attributes and behaviors, the BEV can generate a specific trip solution for each driver. The changeable part of the solution could include the safety buffer range when the guidance intervenes, the preferred strategy of the driver, and the content of the driver's guidance. Even charging behaviors can make a difference on driver's range anxiety, if one combines the charging behavior model in [13] with the charging strategy recommendation algorithm in [7] together. With the development of the Internet of vehicles (IOV), the analysis can be completed by big data technology. Moreover, when driving different BEVs connected to the IOV, drivers can download the individual setting of human-machine interface that precisely cuts down range anxiety according to the drivers' preference. An exclusive range anxiety solution for each driver of BEV will benefit from cutting range anxiety to the minimum and prevent unsafe behaviors from taking place.

4.4. Study Limitations and Further Research. In the research of safety buffer for different trip mileage, three mileage scenarios are given, roughly representing three different types of trips. But in reality, range anxiety can take place at any point of a trip. The limitation of the questionnaire research is that not all situations are covered precisely. The net safety buffer increases with mileage but starts to fluctuate around 15 km at some point between 10 km and 30 km. The exact mileage has not been found when the buffer stops to consistently grow. Further research might analyze the safety buffer with data collected from field experiments, where the trip mileage could vary more randomly.

The driving style measurements, including *decisiveness when driving* and *resistibility to emotions*, might not

TABLE 8

	Statement	Strongly disagree	Disagree	Neither agree or disagree	Agree	Strongly agree
1.	I am content with my BEV's range.	□	□	□	□	□
2.	It is convenient to recharge my BEV.	□	□	□	□	□
3.	I trust my BEV's range remain display.	□	□	□	□	□
4.	I behave decisively when driving.	□	□	□	□	□
5.	Emotions hardly affect me when I am driving.	□	□	□	□	□

be adequate to thoroughly describe a driver's driving style. Therefore, further research can expand the scope of drivers' factors that possibly have influence on the driving style.

During our conversation with a BEV driver, the interviewee mentioned the issue about the rapid energy consumption in winter when the heater is on. He never experienced range anxiety, except the range anxiety caused by the rapid drop of available energy due to the heater. But extensive contents were not added immediately to the questionnaire. The anxiety caused by heater in the winter should be a common and crucial issue to the BEV drivers. Future research might focus on the particular range anxiety under the working conditions in winter with the heater on.

The interviewees of our research are all BEV drivers. However, range anxiety might also be keeping some people from owning a BEV. If future research is conducted on the range anxiety of non-BEV-owners, some proposals might be developed for BEV manufacturers to relieve the stress of non-BEV-owners. Some optimizations could be made on the strategies for product development and marketing which will be beneficial for the industry.

5. Conclusions

The average net safety buffer for 10 km-trip is 6.43 km, 64.3% in portion. The average net safety buffer for 30 km-trip is 14.93 km, 49.8% in portion, while the average net safety buffer for 60 km-trip is 16.20 km, 27.0% in portion. The net safety for middle and long mileage trips are roughly around 15 km. To the general safety buffer without considering the trip mileage, *driving experience*, *satisfaction level of recharge accessibility*, and *resistibility to emotions* are significantly correlated. The influential factors are different for trips with different lengths.

Drivers' most chosen behaviors under range anxiety are *speed reduction*, *change behaviors*, and *seeking nearby charging-pile*. Three behaviors have correlations with the drivers' attributes. For the behavior of *speed reduction*, *resistibility to emotions* is significantly correlated with each other. For the behavior of *getting more disturbed*, *general satisfaction level* is of significant correlation. For the behavior of *grabbing lines*, six factors are correlated, including *gender*, *age*, *fuel vehicle driving experience*, *max BEV range*, *decisiveness when driving*, and *resistibility to emotions*.

Our research provides references for the solution of range anxiety. Through analyzing the attributes of a BEV driver, the driver's safety buffer for different trips and preferred behaviors under range anxiety can be anticipated. With this result, a solution to the range anxiety can be generated for a particular BEV driver. This will reduce the range anxiety level without causing extra stress to the driver and the risk of traffic accident can also be cut down.

Appendix

Questionnaire of Battery Electric Vehicle (BEV) Driver's Range Anxiety

This research is carried out with scientific research intention. Your privacy is privileged.

Range anxiety is defines as the anxiety or stress of BEV drivers when they concern about being stranded in the middle of a trip for running out energy.

DRIVER CHARACTERISTICS

GENDER——AGE——DRIVEING EXPERIENCE——YEARS

PROFESSION:

Job hunter/Civil servant/Teaching or scientific research personnel/Individual laborer/Retired/Production or service personnel/Company personnel/Student/Others

FUEL VEHICLE DRIVING EXPERIENCE? Yes/ No

If yes, the fuel consumption of previous fuel vehicle——L/100km

Refueling frequency——times/month

TYPE OF BEV CURRENTLY DRIVEN

BJEV EV160/BJEV EV200/BYD E6/Geely Emgrand EV/JAC iEV5/Tesla Model S/Tesla Model X/ Venucia e30/Denza/Others———(write down the type)

MAX RANGE OF CURRENTLY DRIVEN BEV——km

MOST COMMONLY USED METHOD OF RECHARGE

Public AC charging-pile/Public DC charging-pile/House hold charging-pile/Portable charger (220V AC)/ Others

RECHARGING FREQUENCY——TIMES/WEEK

USE OF BEV

Official/Private/Rental/Service/Others

PART I

See Table 8

PART II

See Table 9

PART II

See Table 10

Thank you for your precious time spent on the questionnaire!

Table 9

	Statement	10 km	12.5 km	15 km	17.5 km	20 km
6.	Minimum comfort range remaining in a 10 km trip	□	□	□	□	□
	Statement	**30 km**	**37.5 km**	**45 km**	**52.5 km**	**60 km**
7.	Minimum comfort range remaining in a 30 km trip	□	□	□	□	□
	Statement	**60 km**	**67.5 km**	**75 km**	**82.5 km**	**90 km**
8.	Minimum comfort range remaining in a 60 km trip	□	□	□	□	□

Table 10

	Statement	Distracted by anxiety	Rearrange the route	Seeking nearby charging-pile	Change behaviors	Speed reduction	Get more disturbed	Grab lines
9.	Under range anxiety	□	□	□	□	□	□	□

Conflicts of Interest

The authors declare that they have no conflicts of interest.

Acknowledgments

This work is supported by the National Natural Science Fund of China (71701023), Hunan Key Laboratory of Smart Roadway and Cooperative Vehicle-Infrastructure Systems (no. 2017TP1016), and Student Research Training Project of Tsinghua University (1622T0092).

References

[1] N. Rauh, T. Franke, and J. F. Krems, "Understanding the impact of electric vehicle driving experience on range anxiety," *Human Factors*, vol. 57, pp. 177–187, 2015.

[2] T. Franke, N. Rauh, M. Günther, M. Trantow, and J. F. Krems, "Which Factors Can Protect Against Range Stress in Everyday Usage of Battery Electric Vehicles? Toward Enhancing Sustainability of Electric Mobility Systems," *Human Factors: The Journal of the Human Factors and Ergonomics Society*, vol. 58, no. 1, pp. 13–26, 2016.

[3] M. Eisel, I. Nastjuk, and L. M. Kolbe, "Understanding the influence of in-vehicle information systems on range stress – Insights from an electric vehicle field experiment," *Transportation Research Part F: Traffic Psychology and Behaviour*, vol. 43, pp. 199–211, 2016.

[4] T. Franke, N. Rauh, and J. F. Krems, "Individual differences in BEV drivers' range stress during first encounter of a critical range situation," *Applied Ergonomics*, vol. 57, pp. 28–35, 2016.

[5] V. R. Tannahill, D. Sutanto, K. M. Muttaqi, and A. Masrur, "Future vision for reduction of range anxiety by using an improved state of charge estimation algorithm for electric vehicle batteries implemented with low-cost microcontrollers," *IET Electrical Systems in Transportation*, vol. 5, no. 1, pp. 24–32, 2015.

[6] M. Faraj and O. Basir, "Range anxiety reduction in battery-powered vehicles," in *Proceedings of the IEEE Transportation Electrification Conference and Expo, ITEC '16*, USA, June 2016.

[7] K. Sarrafan, K. M. Muttaqi, D. Sutanto, and G. Town, "An intelligent driver alerting system for real-time range indicator embedded in electric vehicles," in *Proceedings of the 52nd Annual Meeting on IEEE Industry Application Society, IAS '16*, USA, October 2016.

[8] V. R. Tannahill, K. M. Muttaqi, and D. Sutanto, "Driver alerting system using range estimation of electric vehicles in real time under dynamically varying environmental conditions," *IET Electrical Systems in Transportation*, vol. 6, no. 2, pp. 107–116, 2016.

[9] M. Dreyfuss and Y. Giat, "Optimizing Spare Battery Allocation in an Electric Vehicle Battery Swapping System," in *Proceedings of the 6th International Conference on Operations Research and Enterprise Systems*, pp. 38–46, Porto, Portugal, Feburary 2017.

[10] T. Franke, I. Neumann, F. Bühler, P. Cocron, and J. F. Krems, "Experiencing range in an electric vehicle: understanding psychological barriers," *Applied Psychology*, vol. 61, no. 3, pp. 368–391, 2012.

[11] M. F. Jung, D. Sirkin, T. M. Gür, and M. Steinert, "Displayed uncertainty improves driving experience and behavior: The case of range anxiety in an electric car," in *Proceedings of the 33rd Annual CHI Conference on Human Factors in Computing Systems, CHI '15*, pp. 2201–2210, Seoul, Korea, April 2015.

[12] N. S. Pearre, W. Kempton, R. L. Guensler, and V. V. Elango, "Electric vehicles: How much range is required for a day's driving?" *Transportation Research Part C: Emerging Technologies*, vol. 19, no. 6, pp. 1171–1184, 2011.

[13] T. Franke and J. F. Krems, "Interacting with limited mobility resources: Psychological range levels in electric vehicle use," *Transportation Research Part A: Policy and Practice*, vol. 48, pp. 109–122, 2013.

[14] T. Franke and J. F. Krems, "Understanding charging behaviour of electric vehicle users," *Transportation Research Part F: Traffic Psychology and Behaviour*, vol. 21, pp. 75–89, 2013.

[15] G. Liu, M. Ouyang, L. Lu, J. Li, and X. Han, "Online estimation of lithium-ion battery remaining discharge capacity through differential voltage analysis," *Journal of Power Sources*, vol. 274, pp. 971–989, 2015.

[16] W. Zhang, J. Gu, K. Wang, Y. Du, and Z. Feng, "Drivers' willingness to drive in old age: a case study on young and mid-aged drivers in China," *Cognition, Technology & Work*, vol. 18, no. 4, pp. 733–743, 2016.

6

An Eco-Driving Advisory System for Continuous Signalized Intersections by Vehicular Ad Hoc Network

Wei-Hsun Lee and Jiang-Yi Li

Department of Transportation and Communication Management Science, National Cheng Kung University, No. 1, University Road, Tainan City 701, Taiwan

Correspondence should be addressed to Wei-Hsun Lee; leews@mail.ncku.edu.tw

Academic Editor: Mohammad Hamdan

With the vehicular ad hoc network (VANET) technology which support vehicle-to-vehicle (V2V) and vehicle to road side unit (V2R/R2V) communications, vehicles can preview the intersection signal plan such as signal countdown message. In this paper, an ecodriving advisory system (EDAS) is proposed to reduce CO_2 emissions and energy consumption by letting the vehicle continuously pass through multiple intersections with the minimum possibilities of stops. We extend the isolated intersection model to multiple continuous intersections scenario. A hybrid method combining three strategies including maximized throughput model (MTM), smooth speed model (SSM), and minimized acceleration and deceleration (MinADM) is designed, and it is compared with related works maximized throughput model (MaxTM), open traffic light control model (OTLCM), and predictive cruise control (PCC) models. Some issues for the practical application including safe car following, queue clearing, and gliding mode are discussed and conquered. Simulation results show that the proposed model outperforms OTLCM 25.1%~81.2% in the isolated intersection scenario for the CO_2 emissions and 20.5%~84.3% in averaged travel time. It also performs better than the compared PCC model in CO_2 emissions (19.9%~31.2%) as well as travel time (24.5%~35.9%) in the multiple intersections scenario.

1. Introduction

Greenhouse gases (GHG) are recognized as the main cause of the global warming, with CO_2 being the primary GHG emitted through human activities. Studies by the IEA [1, 2] show that transport was responsible for 23% of world CO_2 emissions in 2014, as shown in Figure 1(a). Within this, the fastest growth in emissions has been from the road transport sector, as shown in Figure 1(b), which increased by 64% since 1990 and accounted for about three-quarters of transport-related emissions in 2014 [1]. In other words, encouraging environmentally friendly driving practices (eco-driving, hereafter) by eliminating unnecessary vehicle acceleration and braking would reduce fuel consumption and CO_2 emissions and contribute to slowing down global warming. The development of eco-driving systems is thus attracting strong interest from academia, vehicle industry, and governments.

A typical driving trip consists of idling, accelerating, cruising, and decelerating, and the related CO_2 emissions depend on changes in driver behavior, road geometry, or traffic congestion [3]. Barth and Boriboonsomsin show that when a vehicle is idling it consumes more fuel and emits more exhaust fumes than when it cruises at a steady speed [3]. Similarly, Frey et al. [4] show that accelerating and decelerating cause more emissions than idling. In other words, avoiding unnecessary engine idling periods and stops and optimizing driving speeds will reduce CO_2 emissions and fuel consumption. An eco-driving advisory system (EDAS), which provides smooth driving suggestion according to current traffic dynamics and traffic signal plan, can help drivers travel in an environmentally friendly manner.

With vehicular ad hoc network (VANET), also named as connected vehicle technology, vehicles equipped with an on-board unit (OBU) can communicate with road side units (RSU) by vehicle-to-roadside (V2R/R2V) to preview the traffic signal plans and obtain real-time information such as waiting queue length and signal countdown message. Forward collision warning (FCW), which can detect the distance relative to the vehicle ahead, facilitates vehicle safety driving assistance such as adaptive cruise control (ACC). Assuming

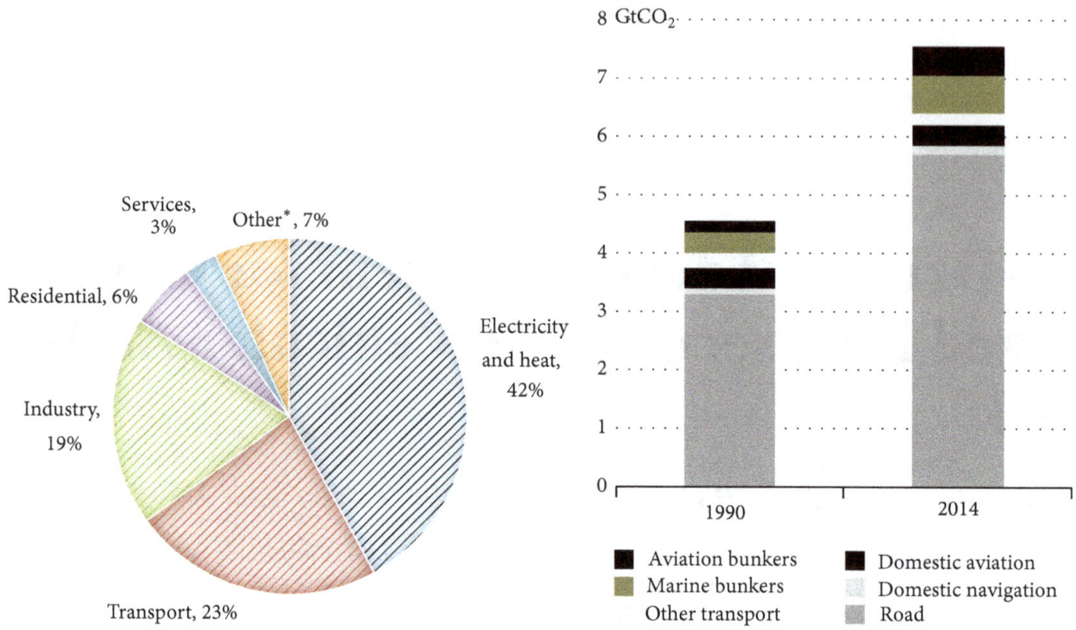

(a) World CO_2 emissions by sector

(b) CO_2 emissions from transport

FIGURE 1: Statistics on CO_2 emissions produced by the IEA [1].

that vehicles are equipped with VANET and FCW, OBU can obtain real-time information and calculate the optimal eco-speed to cross an intersection. Our previous work [5] proposed two decision tree based eco-driving suggestion models for an isolated signalized intersection, using OBU to calculate the best eco-driving speed based on real-time information including the traffic signal countdown messages and waiting queue length broadcasted by RSU. These two models are called MaxTM (maximize throughput model) and MinADM (minimize acceleration and deceleration model); simulation results show that MaxTM outperforms MinADM and Open Traffic Light Control Model (OTLCM) [6], being 5% to 102% better than MinADM and 13% to 209% better than OTLCM with regard to CO_2 emissions in the simulation cases, and 8% to 14% better than MinADM and 15% to 231% better than OTLCM in the real traffic cases [5].

MaxTM, MinADM [5], and OTLCM [6] are applied for the isolated intersection scenario; however, in practice vehicles may pass a number of intersections when traveling from origin to destination. In such cases, the isolated intersection eco-driving model may not always be appropriate for use with multiple intersections, since OBU does not acquire the necessary information from the next RSUs, and thus these models cannot provide an optimized speed for the sequential intersections. For example, in Figure 2(a) the vehicle may drive at an unnecessarily high speed if it applies the isolated EDAS model, and it has to stop at the second intersection as it does not have the traffic signal countdown message for the next intersection, as illustrated in Figure 2(b). With the real-time information of the traffic signal timing plans of the sequential intersections, it is possible to design a better eco-driving model to avoid these energy wasting cases, thus lowering fuel consumption and CO_2 emissions.

In this paper we extend our previous works [5, 7] to develop an EDAS that help drivers to avoid unnecessary acceleration, braking, engine idling periods, and stops and to optimize the driving speed in a continuous multiple intersections scenario. Several studies in the literature discuss the multiple intersection scenario. Alsabaan et al. proposed an EEFG model [8] by using decision tree to measure the optimal speed to pass multiple intersections. They then improved this in 2013 [9] by dividing the road region into two parts, one passable and the other unpassable. Asadi and Vahidi [10] use a predictive cruise control (PCC) algorithm in a mathematical programming model to calculate the most economic cruise speed without changing the velocity to pass through sequential intersections. Similar to the underlying concept of the MaxTM, the speed suggested by PCC [10] can be as high as the free flow speed in order to maximize the throughput in conditions when this is possible. The results of their work show that the proposed model can reduce fuel consumption by up to 47% compared to the baseline model. Katsaros et al. [11] proposed GLOSA (Green Light Optimized Speed Advisory), which combines GPS information with common vehicle sensors to obtain more accurate road slope estimates, which are then used to optimize the fuel consumption of the vehicle, thus achieving a 7% reduction in fuel consumption comparing to the baseline. In summary, while a number of works have proposed eco-driving advice systems for multiple intersections [8–11] they are all insufficient or inefficient for practical applications due to the following issues not fully considered.

(1) Safe Car Following. Previous studies either do not consider the issue of safely following cars [6, 8, 11] or following a car with a fixed time gap [5]. This approach may result

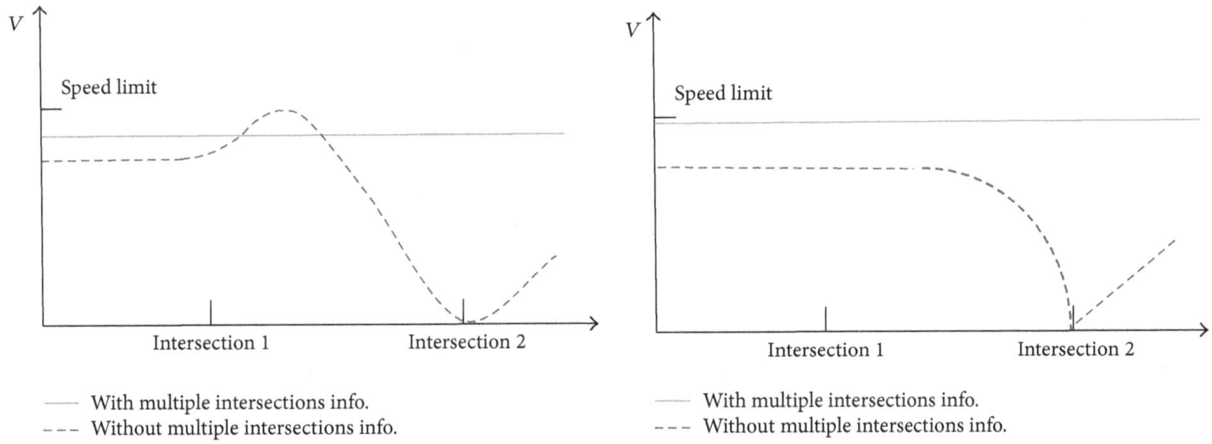

(a) Unnecessarily high speed

(b) Not enough speed

FIGURE 2: Comparison of the system with and without multiple intersection data.

in wasting road capacity at low speed conditions and may cause accidents at high speed ones. A dynamic time gap mechanism, based on the relative speed and distance of the preceding vehicle and the host vehicle, will increase the usage of road capacity.

(2) Queue Clearing. When vehicles are stopped by the traffic signals, the time needed for the waiting queue dissipating cannot be neglected, especially when we aim to enable a vehicle to pass smoothly through an intersection without stopping. In related works [6, 8–11], this issue is not considered so that the model would not be practical for realistic applications. Waiting queue dissipating time, which is related to the time needed from current time to the last vehicle in the waiting queue passing through the vehicle, must be considered. When the issue of queue clearing is considered, the resulting eco-driving model will be more accurate and of greater practical use.

(3) Gliding Mode. Only three vehicle moving control modes (acceleration, deacceleration, and maintaining current speed) are applied in the traditional models [6, 8–11]. However, vehicle gliding mode (free from gas pedal) leverages the engine brake to slow the vehicle without additional fuel consumption, thus causing less CO_2 emissions than these control modes. When the traffic signal plans are known in advance, vehicles can apply gliding to slow down to the suggestion speed.

In contrast to previous works, we take the above issues into consideration and propose more realistic and accurate suggestions with regard to the optimal eco-driving speed. Safety is the major concern so that the proposed model must follow the safe car following rule, speed limit, and traffic signal regulations. Two eco-driving strategies are proposed and compared in the continuous intersections scenario, namely, the maximized throughput model (MTM) and smooth speed model (SSM). These are illustrated in Figure 3, in which the green time window upper bounded by $S_H(i)$ and lower bounded by $S_L(i)$ (both in red dotted line) are

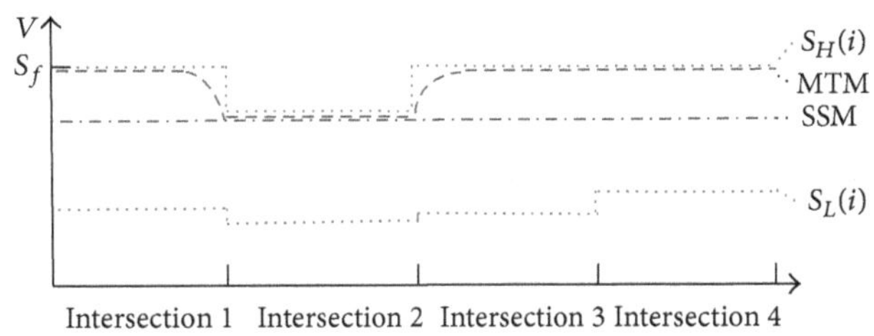

FIGURE 3: Comparison of two strategies: MTM and SSM.

first discovered in four continuous signalized intersections. The MTM strategy (blue dashed line) keeps the suggested speed as high as possible to avoid blocking the following vehicles, thus maximizing the throughput. On the other hand, the SSM strategy (green dashed line) adopts the smooth moving speed upper bound to avoid unnecessary acceleration and deceleration.

The rest of this paper is organized as follows. The assumptions, terminologies, and VANET protocol are presented in detail in Section 2. The algorithms used in MTM and SSM are proposed in Section 3. Two simulation experiments are then discussed in Section 4, including an isolated intersection scenario and multiple intersections scenario, and the proposed system models are compared with the PCC [10], OTLCM [6], and MaxTM [5] strategies. Finally, Section 5 concludes this paper and presents some directions for future works.

2. Eco-Driving System Protocol

Assume that each traffic signal controller is equipped with an RSU, and each vehicle is equipped with an OBU which integrates a location module (ex. GPS) and FCW module to detect the speed and distance of the front vehicle. With these capabilities, the main purpose of the OBU is to calculate the recommended speed (S_r) and provide driving suggestions (speed up/maintain speed/slow down/glide) to pass through as many intersections as possible by considering all the collected real-time information.

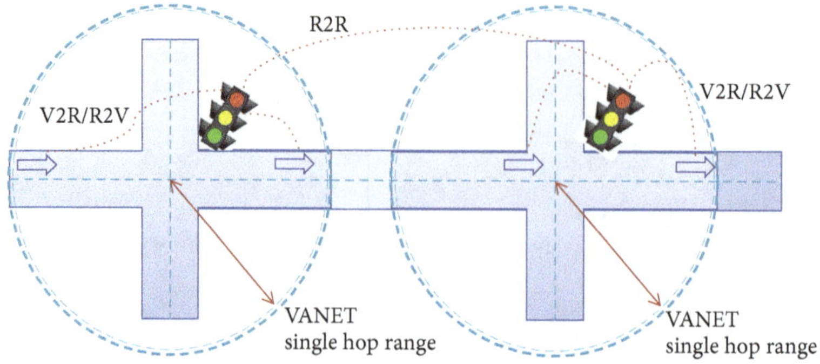

FIGURE 4: Eco-driving advisory system communication schematic.

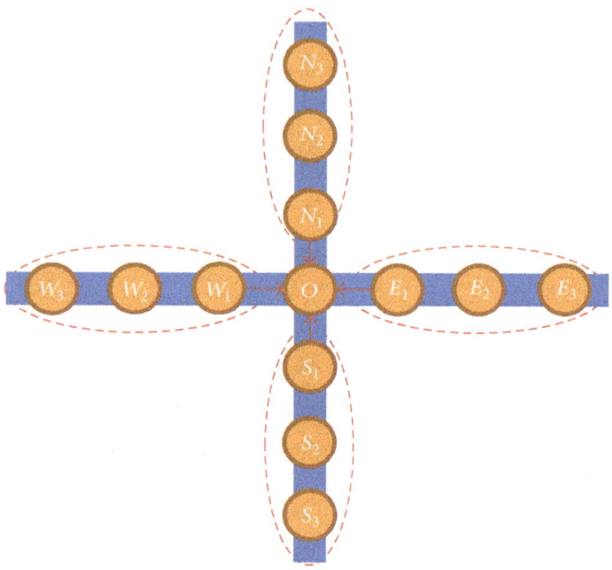

FIGURE 5: RSU-to-RSU information exchange schematic.

The EDAS model is designed on top of the VANET environment with V2R/R2V communication, as illustrated in Figure 4. OBU can collect RSU broadcasted information beyond the range of visual contact. RSU-to-RSU (R2R) communication among the neighborhood RSUs is also included in the system protocol to exchange the signal timing plan and waiting queue length information. As illustrated in Figure 5, every RSU (R_O) will exchange real-time information with three neighborhood RSUs (R_{Di}) in each direction, including data on the waiting queue, signal phase timing plan, road length, and traffic conditions.

The communication protocol of EDAS is designed as illustrated in Figure 6, which is composed of six steps of R2R, V2R, and R2V communication, as listed in Table 1 and explained as below.

Step 1 (R2R exchange basic configuration and signal plan). Each RSU, as the host RSU itself, will exchange its information with twelve neighborhood RSUs. The exchanged information, as shown in (1), is organized as two parts including basic configuration and signal plan. Basic configuration of an RSU (R_s), as defined in (1), includes RSU ID (R_i), location

TABLE 1

Parameters	Value
Road length (m)	200 m; 400 m; 600 m
Traffic volume (vph)	50; 100; 200; 400; 800
Traffic signal (sec) (green/red)	30/30; 45/45; 60/60
Spatial depth	Four intersections
Temporal depth	Four cycles
VANET range (m)	200
Cruise gliding acceleration	$-0.15\ \text{m/s}^2$
Idle gliding speed ($S_{\min}$)	2 m/s

(X, Y), road length to the next intersection in four directions ($L(n, s, e, w)$), and BSS (basic service set) identification in 802.11p (SSID). Traffic signal plan and current traffic information (R_p), as defined in (2), consisted of cycle time (T_p), main direction (north-south) green split (T_g), current phase (P_c, 0 indicates currently signal is green in north-south direction and red in east-west direction and 1 indicates the reverse case), and countdown remaining seconds in north-south and east-west direction ($R(n\text{-}s, e\text{-}w)$). With the above definitions, the R2R data exchange protocol is defined in (3), which includes RSU configuration (R_s), traffic signal plan (R_p), and timestamp (T_s).

$$R_s = \{R_i, X, Y, L(n, s, e, w), \text{SSID}\} \tag{1}$$

$$R_p = \{T_p, T_g, P_c, R(n\text{-}s, e\text{-}w)\} \tag{2}$$

$$\text{R2R}: \{R_s, R_p, T_s\}. \tag{3}$$

Step 2 (RSUs broadcast synchronized message to OBUs). Each RSU periodically broadcasts message to notify all the vehicles (OBUs) with the range of its transmissions. These messages are used to synchronize the OBUs and wake them up if they are not currently in working status. The message format is shown in (4), including broadcast RSU information (R_s) and timestamp for time synchronization. When an OBU receives one or more RSU broadcasted messages from different RSUs, it first decides which one is the host OBU based on the relative location and moving direction and computes the recommended speed using its current location,

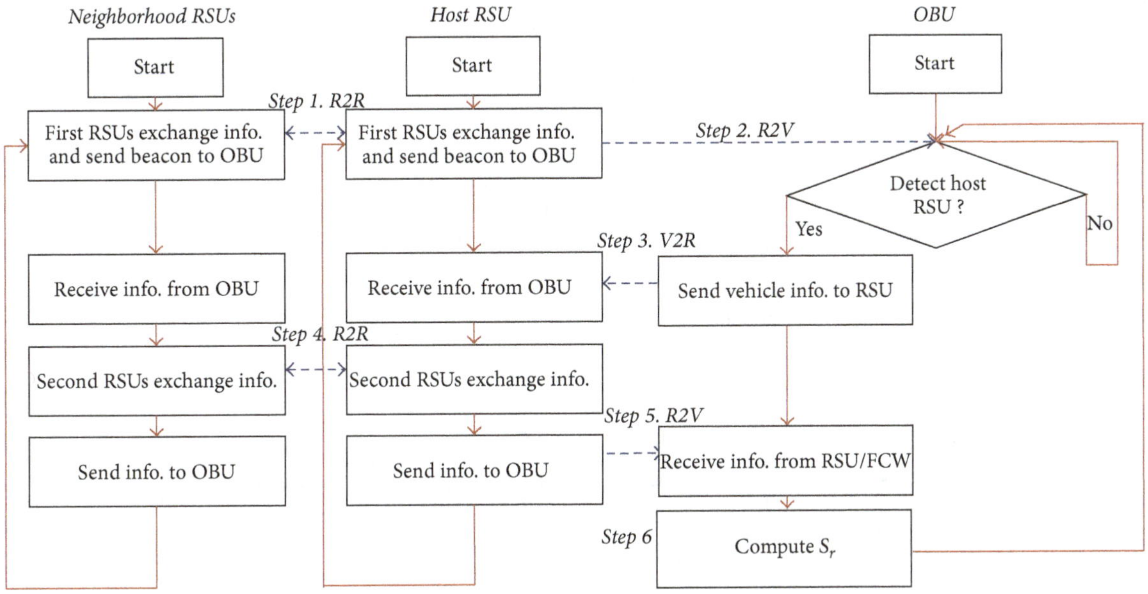

FIGURE 6: EDAS communication protocol.

speed, front vehicle speed and distance (from FCW), and the information collected from the host RSU, as explained in the next section.

$$\text{R2V} : \{R_s, T_s\}. \tag{4}$$

Step 3 (OBUs send information to RSUs). After receiving a synchronized message from the host RSU, an OBU will be in active mode and periodically broadcast information including OBU ID (O_i), current speed (S_c), moving direction (S_d), current acceleration (A_c), and position (X, Y), as shown in (5). The OBU also starts to detect the distance and speed of the front car via FCW at the same time.

$$\text{V2R} : \{O_i, S_c, S_d, A_c, X, Y\}. \tag{5}$$

Step 4 (RSUs exchange dynamic traffic information collected from OBUs). An RSU exchanges dynamic traffic information with the neighborhood RSUs in this step, calculating the waiting queue length in each directions by estimating the vehicles' positions using the collected OBU information. Each host RSU updates the neighborhood RSUs array each time an RSU exchange event occurs. The exchange message format is defined in (6), including RSU ID (R_i), current phase and signal plan (R_p), waiting queue length in each direction ($Q(n, s, e, w)$), and timestamp T_s.

$$\text{R2R} : \{R_i, R_p, Q(n, s, e, w), T_s\}. \tag{6}$$

Step 5 (RSUs broadcast dynamic signal message to OBUs). In this step, the summarized information of all the RSUs, including the RSU itself (R_o) and the twelve neighborhood RSUs ($R_{d,i}$), organized as an array as shown in (7), is periodically broadcasted to OBU. The information includes the RSU basic information (R_s), traffic signal timing plan (R_p) and waiting queue (Q) of host as well as neighborhood RSUs, and timestamp T_s.

$$\begin{aligned} \text{R2V} : \{&R_o [R_s, R_p, Q(n, s, e, w)], \\ &R_{d,i} [R_s, R_p, Q(n, s, e, w)], T_S\}. \end{aligned} \tag{7}$$

Step 6 (OBU calculates recommended speed). To calculate the suggested eco-speed, the OBU integrates the collected RSUs array information broadcasted in Step 5, GPS status (location, speed, and moving direction), and front vehicle information, including S_p (speed of front vehicle), and D_p (distance to the front vehicle) from FCW. The OBU itself repeats this step several times based on dynamic GPS status and FCW information until the new R2V packet arrives.

3. Eco-Driving Advisory System

With the EDAS protocol, OBU can collect the up-to-date information about upcoming traffic signals including RSU locations, road segment length, and traffic signal timing plan, and real-time traffic information, such as the waiting queue length in each direction for each subsequent RSU, and vehicle flow information collected from the host RSU. The goal of the EDAS is to calculate the optimal speed and suggest driving actions to maximize the possibility of moving through the subsequent intersections without stopped or waiting behind the signal. By comparing the current moving speed, the suggested speed will be converted into suggested driving actions such as speed up/brake/maintain speed/glide and can be presented in colored LED bar-liked graphical user interface.

Two critical decisions need to be made before calculating the recommended speed (S_r) based on the collected real-time information, and then the three eco-driving strategies are applied, as shown in Figure 7. The suggested driving

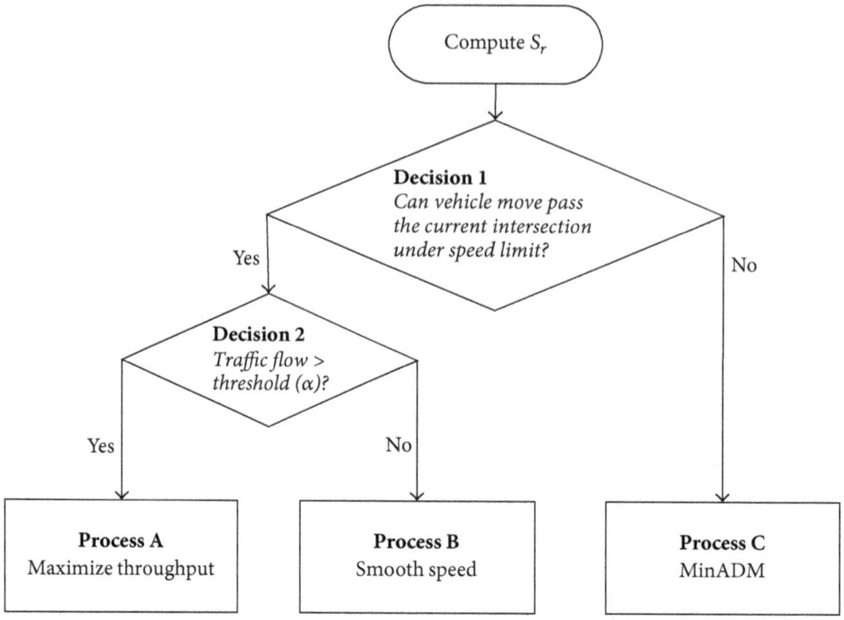

FIGURE 7: Three strategies are applied in computing S_r.

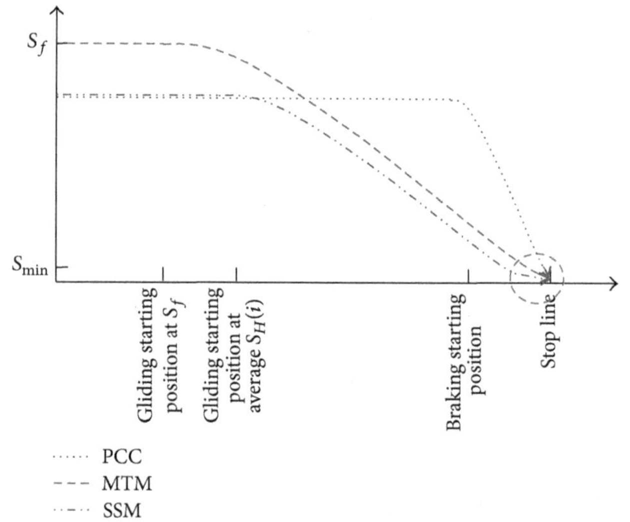

FIGURE 8: Eco-driving strategies comparison of the three models.

behaviors of the proposed eco-driving strategies MTM and SSM as well as the comparative eco-driving model PCC [10] are illustrated in Figure 8. The major difference between the MTM, SSM, and PCC [10] is that the proposed models apply vehicle gliding concept to leverage the engine brake. MTM adopts free flow speed (S_f) and thus applies gliding earlier than the SSM, which adopts smooth speed at upper bound of the speed range ($S_H(i)$). On the other hand, PCC [10] has to brake the vehicle in front of the stop line, where the averaged acceleration is much larger than gliding mode. The three modes, two decision criteria, and some issues about eco-driving are discussed in detail below.

Decision 1: Isolated or Multi-Intersections Mode. The top decision node determines if a vehicle should apply multi-intersection mode or isolated intersection mode by calculating whether a vehicle can pass through the current intersection under its current conditions with the appropriate speed limit. The decision goes to the left subtree if the answer to Decision-1 is "Yes," and otherwise it goes to the right subtree, where a modified isolated intersection model, the simplified MinADM, will be applied.

To find out whether a vehicle can pass the intersection under current traffic signal phase P_c (green (0), red (1)), phase remain time ($R(n\text{-}s, e\text{-}w)$), we can check if the speed needed for the vehicle to move the distance D_i by time period G_t falls into the speed limit range $[S_f, S_{\min}]$, as shown in (10). G_t is defined as the remaining green time if the current signal is green, and as the remaining red time plus the queue clearing time (T_Q) if the current signal is red, as illustrated in (9). The waiting queue clearance time (T_Q), as defined in (8), is obtained from the literature [12].

$$T_Q = \sum_{k=1}^{K} \left(\left| D_k - D_h \right| \right) \tag{8}$$

$$G_t = \begin{cases} R_i(ns), & \text{if } P_c = 0 \\ R_i(ew) + T_Q, & \text{if } P_c = 1 \end{cases} \tag{9}$$

$$\text{Decision 1} = \begin{cases} \text{Y}, & \text{if } \left[\dfrac{D_i}{G_t} \right] \in \left[S_f, S_{\min} \right] \\ \text{N}, & \text{if } \left[\dfrac{D_i}{G_t} \right] \notin \left[S_f, S_{\min} \right]. \end{cases} \tag{10}$$

Decision 2: Apply the MTM or SSM Strategy. Once Decision-1 is "Y," which means the vehicle can pass through current intersection without stop, the multi-intersections mode should be applied. In Decision-2, the EDAS will decide whether MTM or SSM should be adopted. As shown in Figure 3, the MTM strategy tries to maximize the traffic throughput so that the suggested speed will be as high as possible while remaining under the speed limit to reduce the possibility of blocking upstream vehicles. The SSM strategy, on the other hand, tries to minimize the acceleration and deceleration to keep the vehicle moving as smoothly as possible. The choice between these two strategies will be a compromise solution. On peak hours, adopting SSM may result in more serious traffic congestion, as it may block the upstream vehicles so that the number of vehicles stopped by the signal will rise, thus making fuel consumption and CO_2 emissions worse. On the other hand, adopting the MTM strategy on off-peak hour may increase the fuel consumption due to unnecessary acceleration.

For this reason we set a compromise factor α as the threshold to alternate between the throughput oriented and smooth speed models. The value of α is determined by the sensitivity test, and once α is determined OBU can decide to apply MTM or SSM based on the vehicle volume information collected from the host RSU.

$$\text{EDAS} = \begin{cases} \text{MTM}, & \text{if } \alpha \geq \text{threshold} \\ \text{SSM}, & \text{if } \alpha < \text{threshold}. \end{cases} \tag{11}$$

Process A: Maximized Throughput Model. In the MTM and SSM cases, the target of EDAS is to provide the optimal driving speed suggestion so that vehicles can move through multiple sequential intersections without being stopped by a signal. In order to maximize throughput and reduce the possibility of blocking the upstream vehicles, MTM suggests driving at as high a speed as possible while remaining safe and under constraints such as the speed limit, safely following the car in front, and so on. OBU suggests the highest speed under the calculated speed range to pass the next intersection when the vehicle passes the current intersection. By using highest speed $S_H(i,j)$ and lowest speed $S_L(i,j)$ from (12) which are the speeds needed to pass the ith intersection at the beginning and ending time of the green signal at the jth cycle is calculated, where TP_i and TG_i are the cycle time and green time of the ith intersection, respectively. For example, $S_H(2,1)$ indicates the speed needed to drive through the second intersection at the first green second of the next signal cycle ($j = 1$). $S_r(3)$ indicates the suggested speed to pass the third intersection, which is computed by calculating all the possible speeds in all possible cycles ($j = 0$ to 4) and intersecting this with the speed limitation $[S_f, S_{\min}]$. Based on this, the highest legal speed in the lowest cycle will be the MTM eco-driving suggested speed (S_R).

$$S_H(i,j) = \begin{cases} \dfrac{D_i}{R_i(ns) + j * TG_i}, & \text{if } P_c = 0 \\ \dfrac{D_i}{R_i(ew) + T_Q + j * TP_i}, & \text{if } P_c = 1 \end{cases}$$
$$S_L(i,j) = \begin{cases} \dfrac{D_i}{R_i(ns) + j * TP_i}, & \text{if } P_c = 0 \\ \dfrac{D_i}{R_i(ew) + TG_i + j * TP_i}, & \text{if } P_c = 1. \end{cases} \tag{12}$$

As illustrated in Algorithm 1, MTM first calculates the speed needed to pass the ith intersection at the jth cycle in the first green period ($S_H(i,j)$) or the last green period ($S_L(i,j)$); that is, it finds the green band for the vehicle to sequentially drive through the intersections without any signal stops. By using the MTM strategy, OBU will choose the highest speed in the legal speed range which is available by intersecting the calculated speed range ($[S_H(i,j), S_L(i,j)]$) with the legal speed range ($[S_f, S_{\min}]$). The $MaxSR[i]$ array in Algorithm 1 is applied to keep the calculated highest speed in each cycle. It is initialized as zero and is assigned the calculated speed during the iterations and will remain at zero if there is no legal speed that can be assigned because the intersection of the two speed ranges is empty. The results of the MTM are presented in the suggested speed array $S_r[i]$. For example, the array $S_r = [60, 60, 50, 0]$ indicates that the vehicle can continuously drive through three intersections without stopping, and the suggested speeds for these intersections will be 60, 60, and 50 km/hr.

Process B: Smooth Speed Model. If there is a low volume traffic flow then the MTM strategy may not be energy efficient, because unnecessary acceleration would not help the vehicle to pass through more intersections than the SSM strategy. The SSM suggests a smooth driving speed that passes as many intersections as possible at a stable speed. This enables more economic driving as well as a more comfortable experience, since it avoids sudden acceleration or deceleration. Here we introduce a parameter β to let the advised speed range from $S_H(i,j)$ to $S_L(i,j)$, and the recommended speed S_r in SSM can be obtained from

$$S_r(i) = S_L(i,j) + \beta * (S_H(i,j) - S_L(i,j)). \tag{13}$$

Process C: Isolated Intersection Model (MinADM). In this process, the vehicle condition is that it has to stop by the signal at the intersection under the current signal phase (either green or red), and there is no need to consider the probability of blocking any vehicles that are behind it. We thus adopt the minimized acceleration and deceleration model (MinADM), which is simplified from a model proposed in our previous work [5]. In this model, OBU is only concerned with maintaining speed and deciding when to begin slowing down by gliding [5]. The recommended speed S_r in this model is obtained from (14) and (15).

(1) When the current phase is green

$$S_r = \begin{cases} S_c + A_g \cdot t, & \text{if } \dfrac{S_c^2}{-2 * A_g} \leq d_I \leq \dfrac{S_c^2}{-2 * A_g} \\ S_c, & \text{otherwise.} \end{cases} \tag{14}$$

(2) When the current phase is red

$$S_r = \begin{cases} S_c + A_g \cdot t, & \text{if } \dfrac{S_c^2}{-2 * A_g} \leq d_I - Q \leq \dfrac{S_c^2}{-2 * A_g} + D_h \\ S_c, & \text{otherwise.} \end{cases} \tag{15}$$

```
Input: N,         // number of intersections
       M,         // number of cycles
       RSU[i];    // RSU array (D_i, R_i(ns), R_i(ew), TG_i, TP_i)
Output: S_r[i],   // suggested speed for passing intersections: 1···N
// finding out min, max speed for passing through ith intersection at jth cycle
   for (i = 1; i <= N; i++) {          // iterate N intersections
    for (j = 0; j < M; j++) {          // iterate M cycles, j = 0 means current cycle
      if (P_c = 0) {      // current signal is green
        S_H(i, j) = D_i/(RSU[i] · R_ns + j * RSU[i] · TG); //
        S_L(i, j) = D_i/(RSU[i] · R_ns + j * RSU[i] · TP);
      } else {        // current signal is red
        S_H(i, j) = D_i/(RSU[i] · R_ew + RSU[i] · T_Q + j * RSU[i] · TP);
        S_L(i, j) = D_i/(RSU[i] · R_ew + RSU[i] · TG + j * RSU[i] · TP);
        }
      }
   }
// calculate the suggested speed for each intersection i
  for (i = 1; i <= N; i++) {   // iterate N intersections
    MaxSR[i] = 0;                    // initialize suggested speed
    for (j = 0; j < M; j++) {     // iterate M cycles to get max legal speed
      if ((S_f < S_L(i, j)) continue;                // no solution
      if (S_H(i, j) ≤ S_f) S_r[i] = S_H(i, j) else S_r[i] = S_f;
      if (MaxSR[i] < S_r[i]) MaxSR[i] = S_r[i];   // choose the fastest
    }
    S_r[i] = MaxSR[i]
  }
  Output S_r[i]
```

ALGORITHM 1: MTM eco-driving suggested speed (S_r).

The Issue of Safely Following Cars. In EDAS the top priority is the issue of safely car following, whatever the suggested speed is, OBU must obey the car following rule to prevent the vehicle from being closer than the safe distance. In such cases, OBU will dynamically calculate the suggested speed to maintain a safe distance from the front vehicle.

If the FCW module detects a front vehicle, OBU will switch to car following mode due to safety concerns. In this study, we apply ISO 15623 [13] to keep a safe distance between the focal vehicle and the one in front. Aa shown in (16), the safe distance (D_s) can be calculated by the vehicle's current speed (S_c), front car current speed (S_p), average maximum deceleration of both the vehicles (A_d), OBU refresh time t, and additional safe headway space (D_h, and in this study we set this as 1 meter). The suggested safe speed S_s for maintaining the safe distance D_s in the car following mode is defined in (17), which can be inferred by combining Newton's three law of motions with (16).

$$D_s = S_c \cdot t + \left(\frac{{S_c}^2 - {S_p}^2}{2A_d} \right) + D_h \tag{16}$$

$$S_s = 2S_p - S_c + A_p \cdot t - \frac{2}{t} \left(D_s - D_p \right) \tag{17}$$

Driving Action Suggestion. With the suggested moving speed S_r (without a front vehicle) or S_s (with a front vehicle), OBU can suggest appropriate eco-driving actions, including speeding up, maintaining current speed, gliding (no pressure applied on the gas pedal), maintaining the safe speed, or braking by comparing the suggested speed (S_s or S_r) and the vehicle's current speed (S_c).

4. Simulation Study and Discussions

Three experiments are designed to evaluate the performance of the proposed MTM and SSM by comparing then with three related models. In the isolated intersection scenario, MTM is compared with MaxTM [5] and OTLCM [6]. In the multiple intersection scenario, MTM and SSM are compared with MaxTM [5] and PCC [10]. The third experiment tests 10 real cases collected from local city traffic bureau in isolated scenario. The performance of all the models is evaluated and simulated using Arena [14], a general simulation software which is effective in modeling and analyzing processes or flows.

To simplify the proposed model and without loss of generality, the traffic signal in the yellow (Y) period is combined with that in the green (G) period, and all the red (AR) periods are combined into the red (R) period. The assumptions used in all the simulation experiments are as follows: (1) all the vehicles are with traditional combustion engine, hybrid electric vehicle (HEV), and plugin hybrid electric vehicle (PHEV) are not considered in this work; (2) each traffic signal is equipped with an RSU, and the OBU penetration rate is 100%; (3) there are no lane changes or overtaking; (4) drivers always follow the recommend speed;

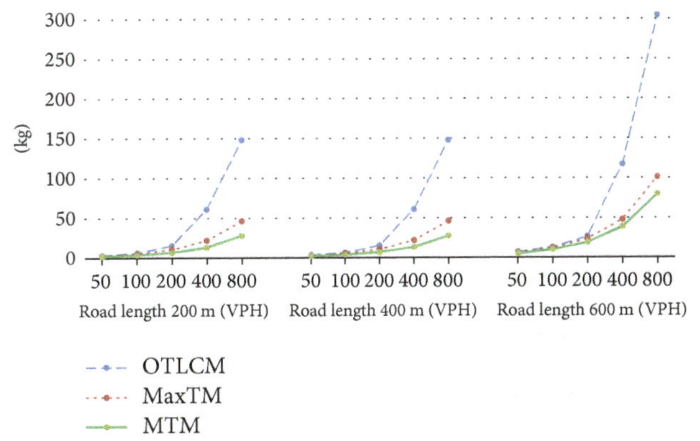

FIGURE 9: CO2 emission comparison in isolated intersection case.

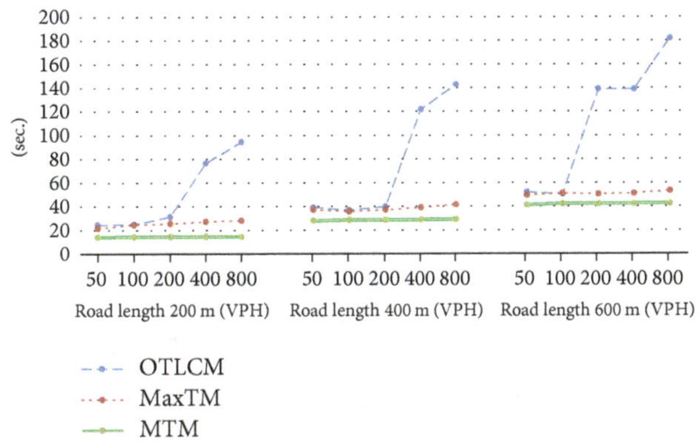

FIGURE 10: Averaged travel time comparison in isolated intersection case.

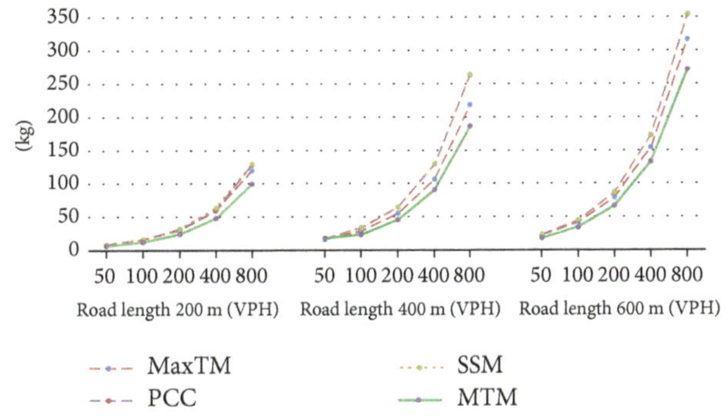

FIGURE 11: CO2 emission comparison in multiple intersections case.

(5) there is only one type of vehicle; (6) all the signal cycle times and green splits (the fraction of a cycle when the signal is green) are the same in one simulation; and (7) both RSU and OBU adopt the one-hop message broadcasting mechanism. The simulation parameters are summarized in Table 1. The VANET one-hop communication range is 200 m, the road length parameters are 200~600 m, the traffic volume (in each direction) parameters range from 50 to 800 vph, and the traffic signal green-red period parameters are 30/30, 45/45, and 60/60. For each OBU, four sequential intersections are simultaneously considered in the spatial dimension, and four cycles are considered for each RSU in the temporal dimension.

To evaluate and compare the CO_2 emissions we adopt the Models for Projecting Energy Consumption and Air Pollutants Emissions (MPECAPE) [15], which considers microscopic CO_2 emissions and was developed by the Institute of Transportation, Ministry Of Transportation and Communications (MOTC) in Taiwan. It is established in 2007 by MOTC using emissions data collected from various types of vehicles considering multiple vehicle driving scenarios, including on highways and in townships and urban areas.

Experiment 1 (isolated intersection scenario). In an isolated intersection located in the center of four road segments in four directions, each road segment has two lanes connected in two directions, one for each direction. The proposed MTM is compared with MaxTM [5] and OTLCM [6] with regard to CO_2 emissions and average travel time, as shown in Figures 9 and 10, respectively. The results show that in the case of road lengths 200 m, 400 m, and 600 m, the CO_2 emission of MTM is less than MaxTM [5] and OTLCM [6] models on the cases from sparse traffic condition (50 vph) to congested condition (800 vph). The travel time simulation also shows the similar result; MTM has the best performance averaged travel time and the travel time variability is very stable in all the road length and traffic volume combination cases.

Experiment 2 (multiple intersections scenario). A cross road network topology consisting of thirteen intersections, as shown in Figure 5, is applied in the experiment. The assumptions and parameters are the same as those in Experiment 1. The two proposed strategies, MTM and SSM, are compared with MaxTM [5] and PCC [10] with regard to CO_2 emissions and average travel time, as shown in Figures 11 to 12, respectively. In multiple intersections simulation, all the performances of these four models are quite well; MTM outperforms the other three models SSM, PCC [10], and MaxTM [5] in both CO_2 emissions, average travel time. Two interest points are revealed in this experiment. First, the performances of PCC [10] and SSM are almost the same due to the similar strategy and idea behind these models (except the gliding). Second, although MaxTM [5] is an isolated model which without the real-time knowledge of multiple intersections, it performs better than the PCC [10] and SSM. This may due to when SSM tries to smoothly pass through the multiple intersections, the green band (ranged from $S_H(i, j)$ to $S_L(i, j)$) will be narrower when the number of previewed intersections increased. This will result in the suggestion speed (S_r) advised by the SSM being much less than the MaxTM [5], which blocks the upstream vehicles and increases the possibilities of stops by signal for them.

Experiment 3 (real data collected from Tainan traffic bureau). In addition to these simulation cases, 10 real traffic cases including AM and PM peak hours as well as normal hours (as listed in Table 2) collected from the Transportation Bureau of Tainan City Government are also simulated. The collected parameters including road length, traffic flow volume, and traffic signal plan (green and red splits) are stated in Table 3. The proposed MTM is compared with MaxTM [5] and

TABLE 2: Case definition of the real traffic cases.

Case name	Duration	Classification
C1, C6	7:00 to 12:00 am	Hybrid
C2, C7	12:00 to 17:00 am	Normal
C3, C8	7:00 to 9:00 am	AM peak
C4, C8	17:00 to 19:00 pm	PM peak
C5, C10	19:00 to 24:00 pm	Night time

TABLE 3: Sample parameters of the real traffic cases simulation.

Case name	Time plan offset	VD ID	Road length	Traffic flow	Timing plan T_G	Timing plan T_R
C1	Initial	V009600	180 m	279.72	80 s	40 s
C2	Initial	V009600	180 m	383.80	80 s	40 s
C3	Initial	V009600	180 m	360.85	80 s	40 s
C4	Initial	V009600	180 m	526.06	90 s	40 s
C4	105 min	-	-	-	80 s	40 s
C5	Initial	V009600	180 m	419.02	80 s	40 s
C5	120 min	-	-	-	75 s	45 s
C5	195 min	-	-	-	55 s	35 s
C6	Initial	V052370	170 m	312.5	75 s	45 s
C7	Initial	V052370	170 m	443.35	75 s	45 s
C8	Initial	V052370	170 m	398.59	75 s	45 s
C9	Initial	V052370	170 m	507.54	95 s	35 s
C9	105 min	-	-	-	75 s	45 s
C10	Initial	V052370	170 m	401.51	75 s	45 s
C10	195 min	-	-	-	63 s	27 s

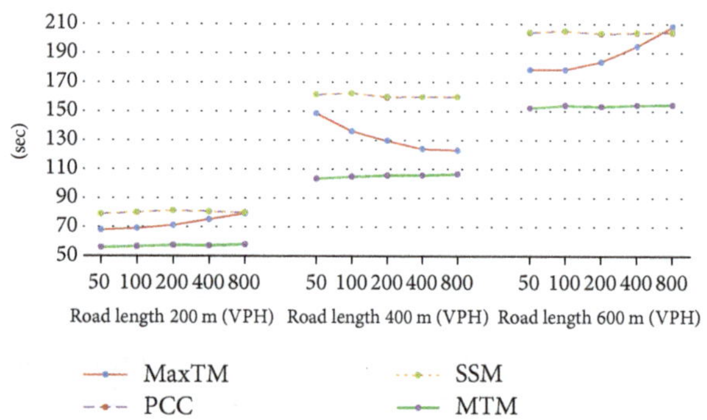

FIGURE 12: Averaged travel time comparison in multiple intersections case.

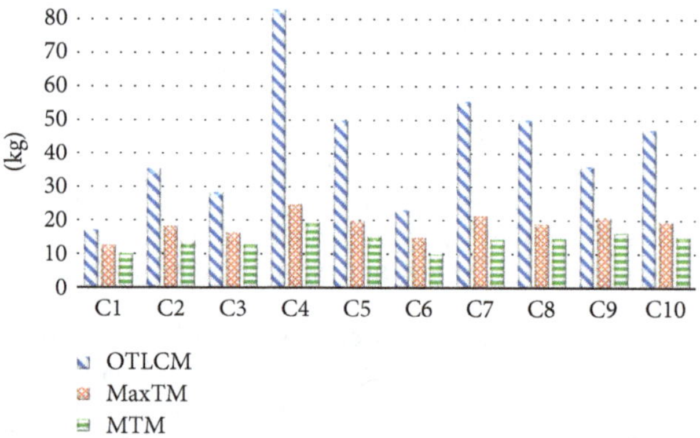

FIGURE 13: CO2 emission comparison in 10 real traffic cases.

OTLCM [6] with regard to CO_2 emissions and average travel time, as shown in Figures 13 and 14, respectively. The results show that the MTM outperforms MaxTM [5] and OTLCM [6] models in CO_2 emissions and average travel time on all the 10 cases.

5. Conclusions

With VANET, OBU on the vehicle can collect the information from RSUs and preview the intersection signal plans so that it can decide the optimal eco-driving speed to pass through multiple intersections with the minimum possibilities of stops. This work proposes a hybrid model that combines three strategies including MTM, SSM, and MinADM to provide eco-driving suggestions. With the proposed EDAS, the suggested driving actions can help vehicles avoid unnecessary acceleration and braking and unnecessary stops. Moreover, several practical issues with regard to applying EDAS in the real world are taken into consideration including a safe and efficient car following mode, signal queue clearing time, cruise gliding, and idle gliding.

Traditionally, three indexes are applied to measure the performance of an intersection in traffic management science, including minimizing waiting queue length, minimizing vehicle waiting time, and maximizing the average cycle

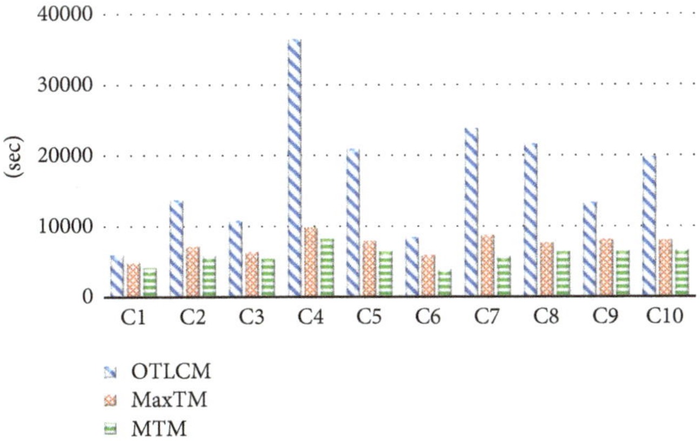

FIGURE 14: Travel time comparison in 10 real traffic cases.

time throughput. These indexes are generally considered to be positively correlated to each other, and thus if one index improves so will the two others. We adopt a strategy of maximizing the average cycle time throughput, and the simulation testing yielded promising results. Originally the MTM is designed for multiple intersections cases; however, the results show that it is 18.6%~40.1% better than MaxTM [5] and 25.1%~81.2% better than OTLCM [6] in CO_2 emissions, 16.6%~47.7% better than MaxTM [5], and 20.5%~84.3% better than OTLCM [6] in averaged travel time when MTM is applied to the isolated intersection scenario. For the multiple intersections scenario, the proposed MTM and SSM outperformed the PCC [10] model as well as the MaxTM [5]. Comparing to PCC [10], MTM is 19.9%~31.2% better in CO_2 emissions and 24.5%~35.9% better in travel time.

In real applications eco-driving recommendations depend heavily on traffic conditions [5]. While the maximize throughput strategy has good performance in congested as well as normal traffic conditions, it may not be suited for the case of very low traffic flow in off-peak hours, since the chance of blocking the rear vehicles is very low. In such conditions the SSM approach should be applied, as it may have better performance than in the MTM. In future work the traffic congestion threshold (α) will be studied in more detail and decided by sensitivity test experiments. Moreover, the smooth speed control parameter (β), which is applied to decide the suggested speed in SSM (as shown in (13)), can be further studied to acquire the optimal driving speed. Once these issues have been dealt with, the MTM and SSM can be combined into one eco-driving advisory system that is more practical for use with continuous signalized intersections.

Notations

R_i, R_s, R_p: RSU ID, configuration, and signal plan
R_o, $R_{d,i}$: Host RSU, ith neighborhood RSU in direction d
O_i: OBU ID
n, s, e, w: North, South, East, West
n-s, e-w: North and south direction, East and west direction
X, Y: Longitude and latitude
T_p: Cycle time period
$T_g(n\text{-}s)$: Green light time period in north-south direction
T_s: Time stamp
P_c: Current phase (0 = green, 1 = Red) in main direction (north-south)
$R(n\text{-}s, e\text{-}w)$: Countdown remaining period
L: Road length (L_n, L_s, L_e, L_w)
Q_d: Queue length in direction d
$T_Q(d)$: Queue dissipating time for direction d
S_f: Free flow speed (Or speed limit)
$S_{\min}$: Min speed limit
S_r: Recommend speed
S_s: Safety speed for car following
S_c: Current speed
S_p: Front car speed
S_d: Direction
$S_H(i, j)$: Pass speed upper limit for ith signal at jth cycle
$S_L(i, j)$: Pass speed bottom limit for ith signal at jth cycle
A_c: Current acceleration
A_p: Acceleration of the front car
A_g: Cruise gliding deceleration
D_P: Front car distance
D_h: Minimum discharge headway
D_k: Headway of the kth queued vehicle
SSID: RSU Wi-Fi SSID
α: Traffic congestion threshold
β: SSM control parameter.

Conflicts of Interest

The authors declare that there are no conflicts of interest regarding the publication of this paper.

Acknowledgments

This work is supported in part by Ministry of Science and Technology of the Republic of China under Grant no. MOST 106-2221-E-006-044-.

References

[1] IEA, *CO2 Emissions from Fuel Combustion 2016*: IEA, 2016.
[2] IEA. (2017). World Energy Outlook 2017, Nov. 2017.
[3] M. Barth and K. Boriboonsomsin, "Traffic congestion and greenhouse gases," *Access Magazine*, vol. 2-9, 2009.
[4] H. C. Frey, A. Unal, N. M. Rouphail, and J. D. Colyar, "On-road measurement of vehicle tailpipe emissions using a portable instrument," *Journal of the Air & Waste Management Association*, vol. 53, no. 8, pp. 992–1002, 2003.
[5] W.-H. Lee, Y.-C. Lai, and P.-Y. Chen, "A study on energy saving and CO2 emission reduction on signal countdown extension by vehicular Ad Hoc networks," *IEEE Transactions on Vehicular Technology*, vol. 64, no. 3, pp. 890–900, 2015.
[6] C. Li and S. Shimamoto, "An open traffic light control model for reducing vehicles' CO_2 emissions based on ETC vehicles," *IEEE*

Transactions on Vehicular Technology, vol. 61, no. 1, pp. 97–110, 2012.

[7] J.-Y. Lee, *A VANET Based ECO-Driving Advisory System for Reducing Fuel Consumption and CO2 Emissions on Signalized Multi-intersections [Master, thesis]*, National Cheng Kung University, 2014.

[8] M. Alsabaan, K. Naik, T. Khalifa, and A. Nayak, "Vehicular networks for reduction of fuel consumption and CO2 emission," in *Proceedings of the 8th IEEE International Conference on Industrial Informatics, INDIN 2010*, pp. 671–676, jpn, July 2010.

[9] M. Alsabaan, K. Naik, and T. Khalifa, "Optimization of fuel cost and emissions using V2V communications," *IEEE Transactions on Intelligent Transportation Systems*, vol. 14, no. 3, pp. 1449–1461, 2013.

[10] B. Asadi and A. Vahidi, "Predictive cruise control: utilizing upcoming traffic signal information for improving fuel economy and reducing trip time," *IEEE Transactions on Control Systems Technology*, vol. 19, no. 3, pp. 707–714, 2011.

[11] K. Katsaros, R. Kernchen, M. Dianati, and D. Rieck, "Performance study of a Green Light Optimized Speed Advisory (GLOSA) application using an integrated cooperative ITS simulation platform," in *Proceedings of the 7th International Wireless Communications and Mobile Computing Conference (IWCMC '11)*, pp. 918–923, Istanbul, Turkey, July 2011.

[12] J. A. Bonneson, "Change intervals and lost time at single-point urban interchanges," Highway Capacity and Traffic Flow (Transportation Research Record), 32, 1993.

[13] ISO. (2013). Intelligent transport systems – Forward vehicle collision warning systems -Performance requirements and test procedures *ISO15623:2013*.

[14] W. Kelton, R. Sadowski, and N. Zupick, *Simulation with Arena, 6/e*, McGraw-Hill, 6/e, 2015.

[15] K. Lin, *Integrating the Applications of Sustainable Transportation Planning Model and Models for Projecting Energy Consumption and Air Pollutants Emissions*, Institute of Transportation, Ministry of Transportation and Communications, 2010.

7

Vehicle Relocation Triggering Thresholds Determination in Electric Carsharing System under Stochastic Demand

Wanjing Ma[1] **Lei Wang**[1] **and Lin Li**[2]

[1]*The Key Laboratory of Road and Traffic Engineering of the Ministry of Education, Tongji University, Shanghai 201804, China*
[2]*Shanghai SH Intelligent Automotive Technology Co., Ltd., Shanghai 201805, China*

Correspondence should be addressed to Lin Li; lilin@siac-sh.com

Academic Editor: Jose E. Naranjo

Vehicle unbalance usually occurs in multistation electric carsharing systems. Threshold triggering method is one of the most practicable approaches for vehicle relocation, while determination of thresholds has not been sufficiently studied particularly for electric carsharing system. This paper presents an approach on determining the thresholds considering the stochastic demands and system states. Firstly, we establish a state transition model involving the stochastic variables to capture the dynamics of the number and battery status of vehicles as well as the traffic demands. Consequently, a dual-objective optimization model was developed to determine the proper values of thresholds. The solution algorithm employed the min–max robust optimization to tackle the uncertainty and the Pareto optimum to decide the solution under dual objectives. To test the distribution stochastic variables, we involve the orders data and the supplementary user survey. Comparison is conducted among three methods: the empirical rules, the deterministic method, and the stochastic method, where the results suggest that the stochastic method achieves better solutions on the dual objectives under stochastic demands.

1. Introduction

Carsharing mode has obtained extensive attention in either personal-vehicle-oriented or public-transit-oriented cities, including in Europe, North America [1], and some parts of Asia [2]. By 2015, there had been over 1.1 million members and 16,754 vehicles in the US since the first carsharing program launched in 1994 [3]. In China, from the emergence of carsharing services in 2009 to 2015, more than 330,000 people signed up for a carsharing membership, and about 5000 vehicles had been in operation, where half of them were electric vehicles (EV) and five of the eight organizations ran EVs and three of them were EV-only [4].

Traditional carsharing models require users to pick up and return vehicles at the same station and with strictly appointed time (as roundtrip carsharing). In order to increase the accessibility, provide better mobility, attract more users, and strengthen competitiveness, carsharing organizations (CSO) introduced flexible pick-up and drop-off location and time (namely, one-way carsharing) that allows users to pick up a vehicle on demand and return it to other stations without a preconcerted arrangement. It is suggested that one-way carsharing is experiencing a rapid worldwide expansion contemporarily [5], new one-way CSOs are emerging (e.g., Car2Go and DriveNow), and existing CSOs are gradually offering one-way capability in major cities (e.g., Zipcar).

However, accompanying the improvement of the flexibility, the distribution of fleets in one-way models potentially turns to be imbalanced with demands variation. The unbalance of demands and fleets makes it necessary to relocate vehicles for sustainable operation. On the other hand, the on-demand pick-up and drop-off rule also causes the preplanned relocation to be unfeasible, especially under stochastic demands, which requires the relocation mechanisms to take the randomness into account.

The threshold based relocation strategy has been developed to respond to the dynamic system changes [6, 7]. The thresholds are applied to keep the number of vehicles within a proper range in each station (as shown in Figure 1(a)). This mechanism was also concluded into two major steps (as

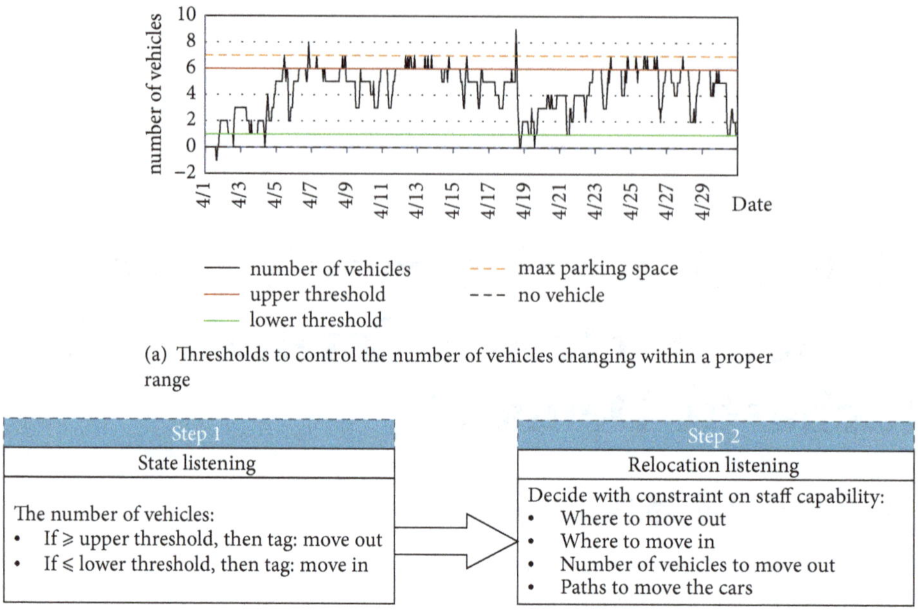

(a) Thresholds to control the number of vehicles changing within a proper range

(b) Two-step threshold based relocation method

FIGURE 1: The threshold based relocation approach.

shown in Figure 1(b)): (1) The system monitors the change of the number of vehicles (the state variable) in each station. If the number of vehicles of a station accumulated above an upper threshold, tag the station as which needs moving out cars (overflowing); if the number of vehicles of a station decreased below a lower threshold, tag the station as which needs moving in cars (lacking). (2) Relocation listening: The system monitors the relocation tag of each station and decides the optimum solutions of path to move vehicles from overflowing stations to lacking stations, with the constraints of relocation staff resources.

It should be noted that the thresholds setting is a fundamental work of the threshold based relocation mechanism. To the best of our knowledge, there had been some studies based on designated thresholds, but detailed study towards thresholds determination is still needed to be extended.

The main targets of this paper are as follows:

(i) To propose a quantitative method for the triggering thresholds determination

(ii) To capture the uncertainty of demands (the solutions of thresholds are decided through a "min-max" robust optimization to find the best solution in the worst condition)

(iii) To weigh both service quality and the relocation times, applying the Pareto front to find the nondominant sets for both the worst condition stochastic demand sets and the best objectives solution sets

(iv) To involve the battery and charging properties of EV as a major difference with conventional carsharing systems.

The rest of the paper is structured as follows: Section 2 reviews the literature concerning the carsharing vehicle relocation problem; Section 3 is the modeling of station vehicle changing and the robust optimization of thresholds under stochastic demand; Section 4 presents a solution algorithm; Section 5 presents the analysis and discussion; and the last section concludes the study.

2. Literature Review

2.1. Study on Vehicle Relocation Problems. Studies on carsharing mode generally fall into two categories: strategy planning level and operating level. Strategy level covers issues of environment and resources benefits [8, 9], market investigation [10, 11], users' behavior [12, 13], and network design [14, 15]. Operating level problems also received attention, but studies are still not sufficient to support the industrial development, especially for one-way model, which allows users to pick up and drop off vehicles at different stations. As one major difficulty in the system operation, the fleet imbalance problem makes the study on vehicle relocation even more urgent.

Vehicle relocation problem had been studied through many approaches, as what we have summarized in Table 1. For some preliminary one-way carsharing systems which require users to provide reservation information, mathematical programming approaches can be adopted without many obstacles. With demands known, the problem can be formulated as mixed integer programming and efficiently solved by feasible solvers [14, 16–18].

With the deeper penetration of on-demand/instant access one-way demand systems, which allow users to get access to

TABLE 1: Summary of studies on vehicle relocation of carsharing systems.

Types	Approaches	Representative studies
Reservation-based (with demands known)	Mathematical planning	Nourinejad (2014), Carlier (2015), Zakaria (2014), Boyacı (2015)
On-demand / instant access (with demands unknown)	Threshold based	Barth (1999), Kek (2009), Cepolina (2012, 2014)
	Demand prediction & mathematical programming	Bruglieri (2014), Repoux (2015)
	Rolling horizon & mathematical programming	Febbraro (2012), Santos (2015)
	Stochastic programming	Briest (2011), Nair (2011), Fan (2013)

the system without previous reservation, this policy makes the demands unknown. One approach is to predict demands then apply deterministic mathematical programming model [19, 20] or to collect short-term demands and then model with rolling horizon scheme [21, 22]. Besides optimization, a practical approach is to dynamically detect the number of available vehicles in stations and trigger relocation tactic by judgment of thresholds. This method has been regarded as an easy way to practice and enables fast implementation for carsharing operators [23].

2.2. Threshold Based Vehicle Relocation Problem. The concept of thresholds was firstly introduced by Barth (1999) [23], which involved a minimum threshold to judge a vehicle insufficiency condition and generate a request to relocate vehicle in. This threshold enables an intuitive vehicle relocation approach which was named as static relocation by that time. There had not been more detailed explanations on mechanism and setting of thresholds. Based on Barth's concept, Kek et al. (2006) [6] further discussed the role of thresholds, which are defined as the upper and lower limits on the number of vehicles in a station. When thresholds are reached, there should be an operation to remove or add vehicles at the station. Kek analyzed the effects of different range of thresholds from a logical perspective but did not quantitatively model the dynamics behind the system. Kek et al. (2009) provided more thinking about thresholds and explained more details of the function of different thresholds. However, thresholds were manual settings for triggering relocation requests and were not involved in optimization variables in that study.

Cepolina et al. (2012) [24] summarized existing studies intended to address the problem of vehicle relocation and categorized some of the approaches. Threshold based method was clearly organized. The authors deemed that the thresholds answer the question of where and when a relocation is required. The framework of thresholds was based on the proposal of Kek's paper. Cepolina et al. (2014) [25] applied thresholds to help planning a new urban carsharing system with fully automated personal vehicles, where they involved thresholds as a dependent variable in their model. The solution to this model is programmed as a parallel heuristic approach, which outputs the thresholds as a side product. Note that specific study on thresholds has not been conducted yet. Cepolina et al. (2015) [26] further developed three innovative strategies for urban carsharing systems where two of them were derived from thresholds method. They refer to article [25] when explaining the setting of thresholds. This confirms that the threshold method has potential for being developed for different application scenarios, and it would be valuable to provide further study on threshold determination.

Our previous work (2016) had presented a framework of threshold based relocation [27], which is based on the operating practice of EVCARD. We found that optimization methodology can hardly implement and handle operating in large scale networks and with stochastic demands, and threshold based relocation is fast practical and rather flexible to account for complex field circumstances. We model and determine the values of thresholds by a deterministic approach in the previous work. However, the randomness of demands and the property of electric vehicle are ignored at that time.

The purpose of this paper is to present a quantitative justification on thresholds determination. Moreover, the uncertainty of demands and trade-off between users' and operators' benefits are as well considered in this research.

3. Thresholds Determination Modeling

3.1. Problem Description. In each station, the number of vehicles is dynamically changing because vehicles would be picked up, dropped off, relocated in, and relocated out. For a single station, the changing range of vehicle amount depends on its own pick-ups and drop-offs, and its proper range is controlled by relocations.

As for the coherence with previous relocation-related works [7, 18], in this paper, time series is also discretized into time steps of 1 hour. During a time step, the vehicles dropped off, the vehicles relocated in, and the idle vehicles are all available for picking up and relocating out according to the matching between available state of charge (SOC) of these vehicles and the required travel distance.

If the number of vehicles at the station is below the lower threshold at one time step, the station requires vehicles to be relocated in. However, not all the required vehicles can be successfully moved into the station within this time step, and the exact arrivals of vehicles relocated in will endure a time-stepwise delay. For the number exceeding the upper threshold it is similar, and the exact departure of vehicles to be relocated out will also delay due to the availability of relocation staff.

Users want to select the vehicle whose battery life is the most suitable for their journey, and users will abandon the system at a proportion if no suitable vehicles are available. Vehicles will be charged immediately when arriving at station. Vehicles variation and depreciation are not concerned; battery charging is linear over time approximately; the battery status is denoted by the available travel distance, km.

By modeling the dynamics of the number of vehicles at a station, the proper upper and lower thresholds can be determined under the minimization of the number of relocations and the time ratio where either the vehicles or parking spaces are out of service.

3.2. State Transition

3.2.1. Transition of the Number of Vehicles. The number of vehicles at the station in time step t is formulated as

$$x_t = x_{t-1} + r_t + \delta_t^{(+)} - d_t^{(st)} - \delta_t^{(-)}, \quad \forall t = 1, 2, \cdots, T \tag{1}$$

3.2.2. Vehicles to Be Moved In. The number of vehicles required to be moved in at t is

$$\delta_t^{(n+)} = \max\left(\left(S_{lower} - x_{t-1}\right), 0\right) \tag{2}$$

For each vehicle $v_t = 1, 2, \cdots, \delta_t^{(n+)}$ that required to be moved in at t and actually moved in at $t + t_{v_t}^{(+)}$, the moving tags and the sum of the moving tags of all v_ts should be

$$vm_{t,t+t_{v_t}^{(+)}}^{(+),v_t} = 1 \tag{3}$$

$$\sum_{v_t=1}^{\delta_t^{(n+)}} vm_{t,t+t_{v_t}^{(+)}}^{(+),v_t} = \delta_t^{(n+)} \tag{4}$$

Each vehicle $v_t \in V_t^{(n+)}$ should be only moved once, from t to the maximum delay time step:

$$\sum_{v_t=1}^{\delta_t^{(n+)}} \sum_{td=0}^{\max(t_{v_t}^{(+)})} vm_{t,t+td}^{(+),v_t} = \delta_t^{(n+)} \tag{5}$$

The vehicles that actually moved in at t:

$$\sum_{k=\max(t-\max(t_{v_t}^{(+)}),1)}^{t} \sum_{v_k=1}^{\delta_t^{(n+)}} vm_{k,t}^{(+),v_k} = \delta_t^{(+)} \tag{6}$$

3.2.3. Vehicles to Be Moved Out. The number of vehicles required to be moved out at t is

$$\delta_t^{(n-)} = \max\left(\left(x_{t-1} - S_{upper}\right), 0\right) \tag{7}$$

For each vehicle $v_t = 1, 2, \cdots, \delta_t^{(n-)}$ that required to be moved out at t and actually moved out at $t + t_{v_t}^{(-)}$, the moving tags and the sum of the moving tags of all v_ts should equal $\delta_t^{(n-)}$:

$$vm_{t,t+t_{v_t}^{(-)}}^{(-),v_t} = 1 \tag{8}$$

$$\sum_{v_t=1}^{\delta_t^{(n-)}} vm_{t,t+t_{v_t}^{(-)}}^{(-),v_t} = \delta_t^{(n-)} \tag{9}$$

Each vehicle $v_t \in V_t^{(n-)}$ should be only moved once, from t to the maximum delay time step:

$$\sum_{v_t=1}^{\delta_t^{(n-)}} \sum_{td=0}^{\max(t_{v_t}^{(-)})} vm_{t,t+td}^{(-),v_t} = \delta_t^{(n-)} \tag{10}$$

The vehicles that should be actually moved out at t:

$$\sum_{k=\max(t-\max(t_{v_t}^{(-)}),1)}^{t} \sum_{v_k=1}^{\delta_t^{(n-)}} vm_{k,t}^{(-),v_k} = \delta_t^{*(-)} \tag{11}$$

3.2.4. Vehicles Matching according to Battery Status. Denote $B_t = B_{t-1}^{V(rm)} \cup B_t^{(+)} \cup B_t^{(R)}$ and $B_t^{(D)} = B_{t-1}^{D(rm)} \cup B_t^{(ge)}$. Supposing demands of pick-ups and requests to relocate vehicles out have equal opportunity to select vehicles, the matching processing should satisfy the following conditions.

For each available vehicle, there should be at most one demand matching it:

$$\sum_{b^{(D)} \in B_t^{(-)} \cup B_t^{(D)}} vs_t^{b,b^{(D)}} \leq 1, \quad \forall b \in B_t \tag{12}$$

For each demand, there should be at most one available vehicle matching it:

$$\sum_{b \in B_t} vs_t^{b,b^{(D)}} \leq 1, \quad \forall b^{(D)} \in B_t^{(-)} \cup B_t^{(D)} \tag{13}$$

For available vehicles and demands matching, we suppose a "best matching" principle that users and relocation staff will get the vehicles with the best utility, leading to a maximum of the total provided and demanded battery of matched vehicles:

$$\sum_{b \in B_t} \sum_{b^{(D)} \in B_t^{(-)} \cup B_t^{(D)}} b \times b^{(D)} \times vs_t^{b,b^{(D)}} - \sum_{b \in B_t} \sum_{b^{(D)} \in B_t^{(-)} \cup B_t^{(D)}} b \times b^{(D)} \times \overline{vs}_t^{b,b^{(D)}} \geq 0, \quad \forall \overline{vs}_t^{b,b^{(D)}} \in VS \tag{14}$$

where VS denotes a space containing all combinations of $vs_t^{b,b^{(D)}}$.

It should be noted that the matching between the demands and the available vehicles according to the desired and the provided battery status is actually even complicated, and further studies on this problem can be derived.

Hence, the actual number of vehicles to be relocated out at t should be

$$\delta_t^{(-)} = \sum_{b \in B_t} \sum_{b^{(D)} \in B_t^{(-)}} vs_t^{b,b^{(D)}} \tag{15}$$

The actual number of satisfied demands at t should be

$$d_t^{(st)} = \sum_{b \in B_t} \sum_{b^{(D)} \in B_t^{(D)}} vs_t^{b,b^{(D)}} \tag{16}$$

The battery status of the remaining vehicles at the end of time step t consists of a set:

$$B_t^{V(rm)} = \left\{ \min\left(b + l \cdot cv, B_{max}\right) \mid \forall b \in B_t, \sum_{b^{(D)} \in B_t^{(-)} \cup B_t^{(D)}} vs_t^{b,b^{(D)}} = 0 \right\} \tag{17}$$

The unsatisfied demand battery status set is

$$B_t^{D(us)} = \left\{ b \mid \forall b \in B_t, \sum_{b^{(D)} \in B_t^{(-)} \cup B_t^{(D)}} vs_t^{b,b^{(D)}} = 0 \right\} \tag{18}$$

The remaining demand battery status set is

$$B_t^{D(rm)} = \left\{ b \mid \forall b \in B_t^{D(us)}, \ p_t^b = 1 \right\} \tag{19}$$

3.3. Thresholds Determination. The optimization model of thresholds determination is formulated as follows.

Objective 1 (CSOs' perspective)

$$\min f_1\left(S_{upper}, S_{lower}\right) = -\left(RV \cdot \sum_{t=1}^{T} d_t^{(st)} - CR \cdot \sum_{t=1}^{T} \delta_t^{(-)} \right) \tag{20}$$

Objective 2 (users' perspective)

$$\min \quad f_2\left(S_{upper}, S_{lower}\right) = \frac{\left(\sum_{t=1}^{T} u_t^{(+)} + \sum_{t=1}^{T} u_t^{(-)}\right)}{T} \tag{21}$$

$$\text{s.t.} \quad S_{lower} \leq S_{upper} \tag{22}$$

$$S_{upper} \leq N \tag{23}$$

$$S_{lower} \geq 0 \tag{24}$$

$$S_{lower}, S_{upper} \text{ are integers,} \tag{25}$$

with state transition equations (1) to (19).

Objective 1 indicates that the CSOs' objective is to maximize the profits at the station. Formula (20) is to add up the revenues from all satisfied demands and subtract the costs of all the relocations. Formula (21) describes the ratio of the time that the system is unavailable for both pick-up and drop-off requirement to the total time, which indicates the service quality of the system. The decision variables S_{upper}, S_{lower} are hidden in both objective functions. Formulae (22) to (25) constrain the thresholds where the lower threshold should be equal to or lower than the upper threshold, the upper threshold should not exceed the number of parking space, and the lower threshold should not be below 0. The state transition equations are also constraints to guarantee the dynamics of the system.

4. Solution Algorithm

4.1. Solution Principles. The thresholds determination model under stochastic demands is facing three problems:

(a) The stochastic variables make the other parameters and variables uncertain during the state transition and let the definition of the minimization of the objectives be uncertain.

(b) The two objectives are conflicting with each other and hard to achieve at the same time, which is not explicit for decision, and compromises need to be made.

(c) The model is hard to be linearized due to the hidden property of decision variables, the dynamic change of system parameters, and the uncertainty of the stochastic variables and sets.

To solve this model, reformations are given as follows.

For problem (a), employ the min–max robust optimization to find out the best solutions under the worst conditions of the stochastic variables, and rewrite the objective as

$$\min_{S_{upper}, S_{lower}} \left\{ \max_{\substack{r_t, d_t^{(ge)}, b_{t,v_t}^{(R)}, b_{t,v_t}^{(ge)}, \\ b_{t,v_t}^{(+)}, b_{t,v_t}^{(-)}, t_{v_t}^{(+)}, t_{v_t}^{(-)}, p_t^b}} f_i\left(S_{upper}, S_{lower}\right), \ \forall t = 1, 2, \cdots, T, \ \forall i = 1, 2 \right\}, \quad \forall i = 1, 2 \tag{26}$$

In objective (26), the inner objective is to locate the maximum (worst) metrics and find out the values of the stochastic variables (auxiliary decision variables) under the worst conditions; the outer objective is to judge the minimum (best) metrics among worst conditions of all the S_{upper}s and S_{lower}s and decide the final solutions on S_{upper} and S_{lower}. This strengthens the robustness of the solutions, i.e., ensures the result thresholds to be capable during the worst condition under the stochastic demands.

Problem (b) can be addressed by introducing the Pareto optimum, which determines a set of nondominated solutions where improvement of one objective cannot be achieved without sacrifice of the other objective, and the objective metrics of these solutions lie on the Pareto front. To define the Pareto solution set, a decision vector $\vec{x}^*$ is Pareto optimal solution if there does not exist another $\vec{x}$ such that $f_i(\vec{x}) \leq f_i(\vec{x}^*)$, $\forall i = 1, 2$ and $f_j(\vec{x}) < f_j(\vec{x}^*)$ for at least one j. For both "min" and "max" steps the Pareto optimum creates rational solution sets and Pareto fronts to pick out compromised solutions under both objectives.

To construct the solution algorithm and cope with the problem (c), we involve the Monte Carlo approach to generate the stochastic seeds and determine the final solutions according to the previous rules. Detailed solution procedures are illustrated in Section 4.2.

4.2. Solution Procedure. In the study case of EVCARD, because the maximum N is 10 among all stations, the combinations of S_{upper} and S_{lower} can be enumerated and the random seeds can be generated through Monte Carlo approach, so the enumerative solution procedure is proposed as follows.

Step 1. For each S_{upper} and S_{lower} satisfying the constraints, create pairs of S_{upper} and S_{lower}; for each pair of thresholds we have the following.

Step 1.1. For stochastic variables $\{r_t, d_t^{(ge)}, b_{t,v_t}^{(R)}, b_{t,v_t}^{(ge)}, b_{t,v_t}^{(+)}, b_{t,v_t}^{(-)}, t_{v_t}^{(+)}, t_{v_t}^{(-)}, p_t^b\}$, which should follow their own distribution, generate R random seeds.

Step 1.2. For each random seed, calculate the two objectives through the state transition equations, denote these solutions as k_r, $\forall r = 1, \cdots, R$, with decision variables $S_{upper}^{k_r}$, $S_{lower}^{k_r}$ and objectives $f_1^{k_r}$, $f_2^{k_r}$; add k_r, $\forall r = 1, \cdots, R$ to a set K_a which includes all stochastic solutions.

Step 1.3. From the calculated objectives corresponding to the random seeds, find out the worst-case Pareto front of this pair of S_{upper} and S_{lower}; add these solutions to a set denoted as $K_{S_{upper},S_{lower}}^{w}$.

Step 1.4. If every pair of S_{upper} and S_{lower} has been processed, continue to Step 2.

Step 2. For all pairs of S_{upper} and S_{lower}, find out the best-case Pareto front in set $\bigcup_{S_{upper},S_{lower}} K_{S_{upper},S_{lower}}^{w}$, as well as the corresponding nondominated solutions; add these solutions to a set denoted as K.

Step 3. Pick out final decision. For solution $k \in K$ with decision variables S_{upper}^k, S_{lower}^k and objectives f_1^k, f_2^k, since $f_1^k \in [-\infty, +\infty)$ and $f_2^k \in [0, 1]$, to judge both objectives fairly, normalize the two objectives as

$$fn_1^k = \frac{\left(\left(f_1^k - \min_{m\in K_a} f_1^m\right) - \min_{m'\in K_a}\left(f_1^{m'} - \min_{m\in K_a} f_1^m\right)\right)}{\max_{m'\in K_a}\left(f_1^{m'} - \min_{m\in K_a} f_1^m\right)} \quad (27)$$

$$fn_2^k = \frac{\left(f_2^k - \min_{m\in K_a} f_2^m\right)}{\max_{m\in K_a} f_2^m}$$

Define the weighted objective as

$$f^k = \omega_1 fn_1^k + \omega_2 fn_1^k \quad (28)$$

where ω_1 and ω_2 are weights and $\omega_1 + \omega_2 = 1$, which represent the importance of each objective and could be set by the operators. For the operators that attach more importance to objective 1 over objective 2, ω_1 could be set higher than ω_2. Find out $k^* \in K$ where $f^{k^*} = \min_{k\in K} f^k$, and k^* is the final decision, where the thresholds are $S_{upper}^{k^*}$, $S_{lower}^{k^*}$, with the objectives of $f_1^{k^*}$, $f_2^{k^*}$.

5. Analysis and Discussion

5.1. Calibration of Stochastic Distribution. The data bed to calibrate the stochastic variables is based on the order record database of EVCARD during April 2015, which contained 2848 order records as well as the information of 56 stations, and the supplementary user survey conducted in March 2016, which retrieved 138 respondents' intention.

(1) Demand variables r_t and $d_t^{(ge)}$: One day is discretized into 24 time periods by the step width 1h, and each period over a month is independent and follows the same distribution. Poisson distributions are tested for them. Figure 2 displays the comparison between the real data histograms (displayed in blue bars) and the theoretical Poisson distributions (displayed in orange bars) at station No. 44 Jiading campus among all the time steps in one day. Their distribution hypothesis tests are conducted by χ^2-test under significance level 0.05, where h = 0 denotes not rejecting the null hypothesis of Poisson and h = 1 means reject.

(2) Continuous variables $b_{t,v_t}^{(R)}$, $b_{t,v_t}^{(ge)}$, $b_{t,v_t}^{(+)}$, $b_{t,v_t}^{(-)}$: The battery status follows continuous distribution. We investigated the battery status of returned vehicles and relocated vehicles through the orders and the users' desired battery status by the user survey. The histograms and the hypothesized distribution are illustrated in Figure 3. Under significance level 0.05, K-S tests were conducted to exam the fitness of continuous distribution and all of the four distributions ((a) uniform, (b) log-normal, (c) log-normal, and (d) normal) are not rejected.

(3) Discretized variables $t_{v_t}^{(+)}$, $t_{v_t}^{(-)}$, p_t^b: The relocation time delay is inferred from the order data and the demands remaining probability is inferred from the user survey. Figure 3 reveals the distribution of the discretized time delay and rejected demands remaining. Figure 3(e) (or Figure 3(f)) shows the probability of delay for 0 time step, 1 time step, 2 time steps, etc. Figure 3(g) refers to the abandon proportion (denoted by 0) and the remaining proportion (denoted by 1) when the demands are rejected due to the vehicle insufficiency as a 0-1 distribution, where more than a half users prefer to abandon using the system.

5.2. Numerical Example. The data of the example came from the station No. 44, Jiading Campus, of the EVCARD system during Apr. 2015. There are seven parking spaces ($N = 7$) at the station; the time step width $l = 1$h; for 30 days' data there are 720 time steps during the study period ($T = 720$). The general vehicle type can travel at most 100km and will be charged in 5h in AC charging mode ($cv = 20$, $B_{max} = 100$).

Figure 4 shows the generation of the random seeds by the scatters of the two objectives. In the figure, different colors

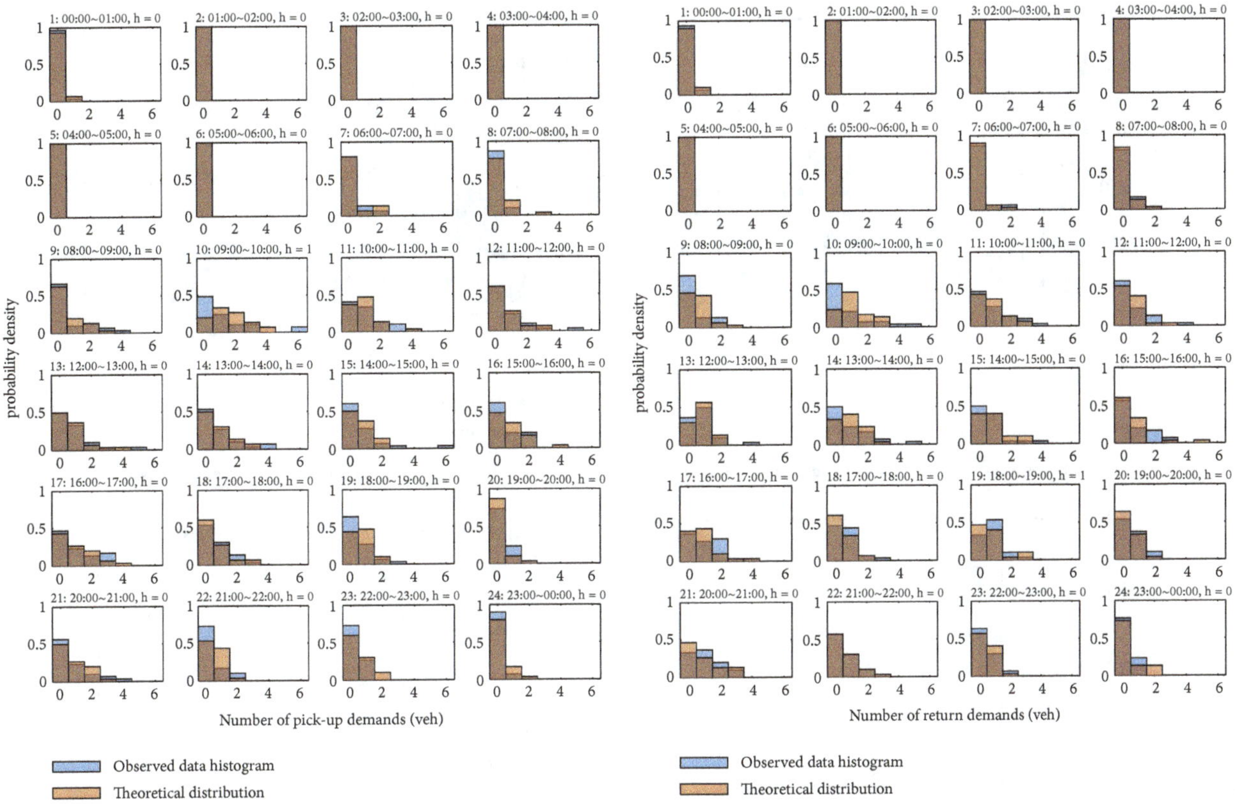

(a) Pick-up distribution

(b) Return distribution

FIGURE 2: Observed histogram and theoretical distribution of pick-up and return demands.

are used to distinguish different pairs of S_{upper} and S_{lower}; dots denote the objective values of all the random seeds, stars denote the worst-case Pareto front of each pair of S_{upper} and S_{lower}, and stars with circles denote the minimum Pareto front in all the worst conditions. Step 1 of the algorithm can be visualized in Figure 4(a) where the random seeds are generated for each pair of thresholds and the worst Pareto fronts of each pair of thresholds are highlighted in stars. Step 2 is visualized in Figure 4(b) which retained the worst-case Pareto fronts and the best-case Pareto front is approximated based on the worst-case Pareto fronts and marked by circles.

The nondominated solutions of thresholds and corresponding minimum Pareto front are listed in Table 1. Thresholds (3,4), (2,5), (3,5), (0,6), (1,6), and (2,6) are all reasonable to guarantee the worst condition caused by the random demands. To decide the final decision for CSOs according to Step 3 of the algorithm, we need to select values of weights. Generally, operators attach equal importance to both of the objectives, like the case of EVCARD in this study. We attached equal importance to both objectives in previous study [27] through deterministic approach (weight $\omega_1 = \omega_2$). For comparison, we also choose equal weights here. Under this metric, (2,6) will be selected as the final decision as the thresholds of the station.

5.3. Comparison among Determination Methods. Three means of thresholds determination methods are compared. (1) Empirical method: The upper thresholds are set by the number of parking space minus 1 to satisfy the drop-offs, and the lower thresholds are set as 1 to ensure a vehicle available. (2) Deterministic method: The random variables are substituted by the known order data, and optimal solutions of thresholds are calculated aiming to minimize the relocation times as well as the out-of-service rate. (3) Stochastic method is proposed by this paper.

Based on the same data bed, we calculated the thresholds of all the 56 stations using the three methods. Figure 5 illustrates a comparison among the results of the three methods. The results of the 56 stations using the three methods appear with both differences and indiscrimination, which are concluded in Table 2.

Notably, there are 27 in 56 stations where the three methods give all different solutions and six stations where the stochastic method gives different solutions from empirical method and deterministic method (as shown in Table 3), which means there are 33 stations (59%) where the stochastic method reaches other solutions. However, for 13 stations (23%) the stochastic method solves out the same solutions as the empirical rules do, and for 7 stations the three methods come to the same results; i.e., empirical method gives identical answer as the stochastic method does in 20 stations (36%). The deterministic method only gives identical results to the stochastic methods in 13 stations (23%).

Except the 7 stations where the three methods give the same solutions, we extract four representative stations under uncertain demands, respectively, from the condition that

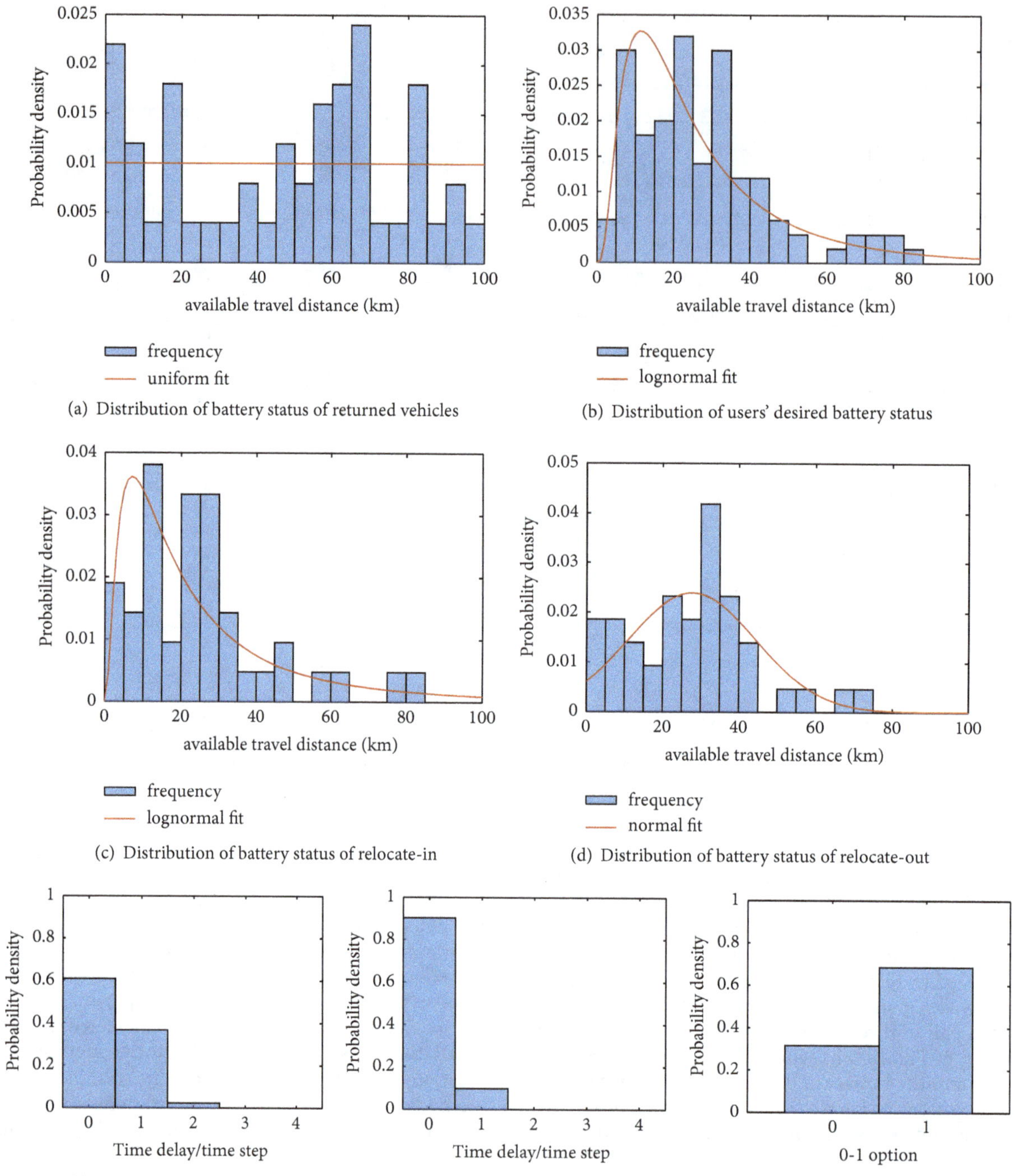

(a) Distribution of battery status of returned vehicles

(b) Distribution of users' desired battery status

(c) Distribution of battery status of relocate-in

(d) Distribution of battery status of relocate-out

(e) Relocate-in time delay

(f) Relocate-out time delay

(g) Demands remaining

FIGURE 3: Distribution of discretized variables.

the empirical method equals the deterministic method, the condition that the empirical method equals the stochastic method, the condition that the deterministic method equals the stochastic method, and the condition that they are all different, and the scatterplots and histograms of both objective metrics of four stations are shown in Figures 6(a), 6(b), 6(c), and 6(d), respectively. In Figure 6, random seeds are denoted by dots, the worst-case Pareto fronts are denoted by color circles and solid links, and the best-case Pareto fronts are denoted by black circles and dash links. For station No. 36 in Figure 6(a) the solution from stochastic method performs better than the solution from the empirical method and deterministic method in out-of-service rate with minor concession on profits. For station No. 34 in Figure 6(b), the stochastic method as well as empirical rules perform obviously better on out-of-service rate. For station No. 29 in Figure 6(c), the stochastic method and the deterministic method achieve absolutely lower out-of-service rate with

TABLE 2: The nondominated solutions set and the final decision under weighted method.

Lower threshold S_{lower}	Upper threshold S_{upper}	Profits of the operator f_1	out of service Rate f_2	Weighted objective f
3	4	-2407	0.001	0.500
2	5	-4343	0.004	0.260
3	5	-5023	0.010	0.187
0	6	-6230	0.193	0.471
0	6	-6339	0.210	0.497
1	6	-6113	0.054	0.155
2	**6**	**-5851**	**0.028**	**0.125**

TABLE 3: Differences and indiscrimination of results from the three methods.

	All the same	Emp eq. Det	Emp eq. Sto	Det eq. Sto	All different
The number of stations	7	6	13	3	27

Note: Emp denotes the empirical method; Det denotes the deterministic method; Sto denotes the stochastic method; eq. denotes equals.

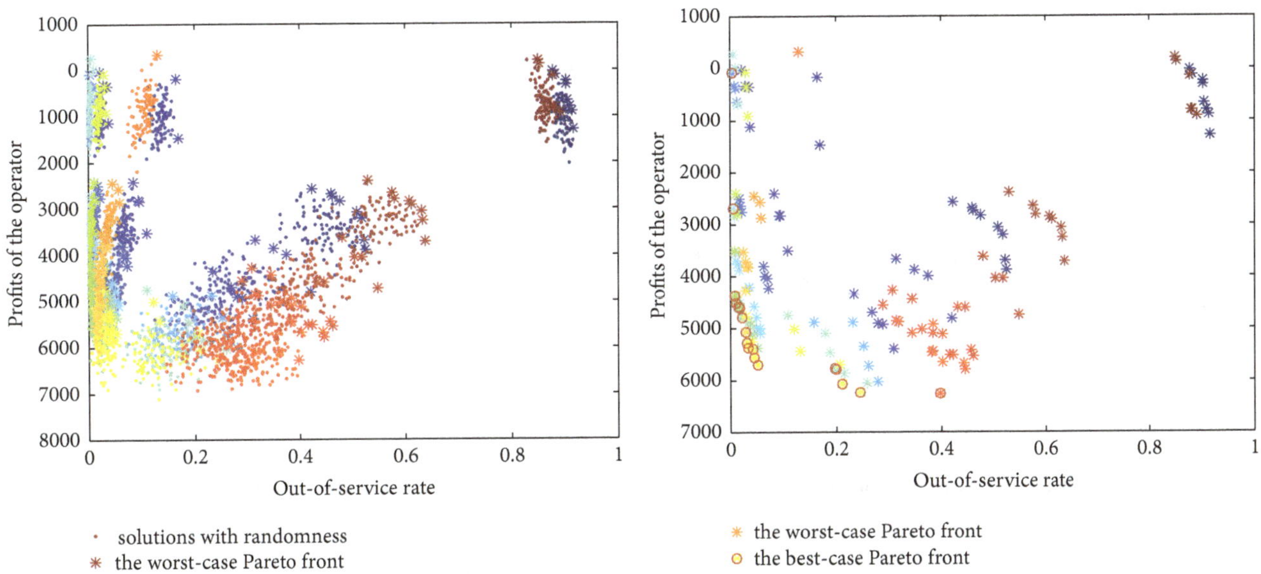

(a) Determination of the worst-case Pareto front

(b) Determination of the best-case Pareto front

FIGURE 4: The visualization of the algorithm of the example station.

approximate profits. For station No. 44 in Figure 6(d) the worst condition of stochastic method performs better than the worst condition of other methods, with little sacrificing on profits.

5.4. Variable Thresholds during a Day. The above model captures long term characteristics and determined a daily thresholds scheme, which supports day-to-day operation. However, for more specific situation, demands can appear with variation in different period of day. To account for variations depending on time-of-day, we specify the daily thresholds into variable thresholds during a day.

The decision variables should be extended to all time steps, and $\forall t = 1, 2, \cdots, T$, we have the following decision variables:

S^t_{upper}: upper threshold of the station at time step t

S^t_{lower}: lower threshold of the station at time step t.

And the formulae in state transition functions (2) and (7) should be modified as

$$\delta_t^{(n+)} = \max\left(\left(S^t_{lower} - x_{t-1}\right), 0\right) \tag{29}$$

$$\delta_t^{(n-)} = \max\left(\left(x_{t-1} - S^t_{upper}\right), 0\right) \tag{30}$$

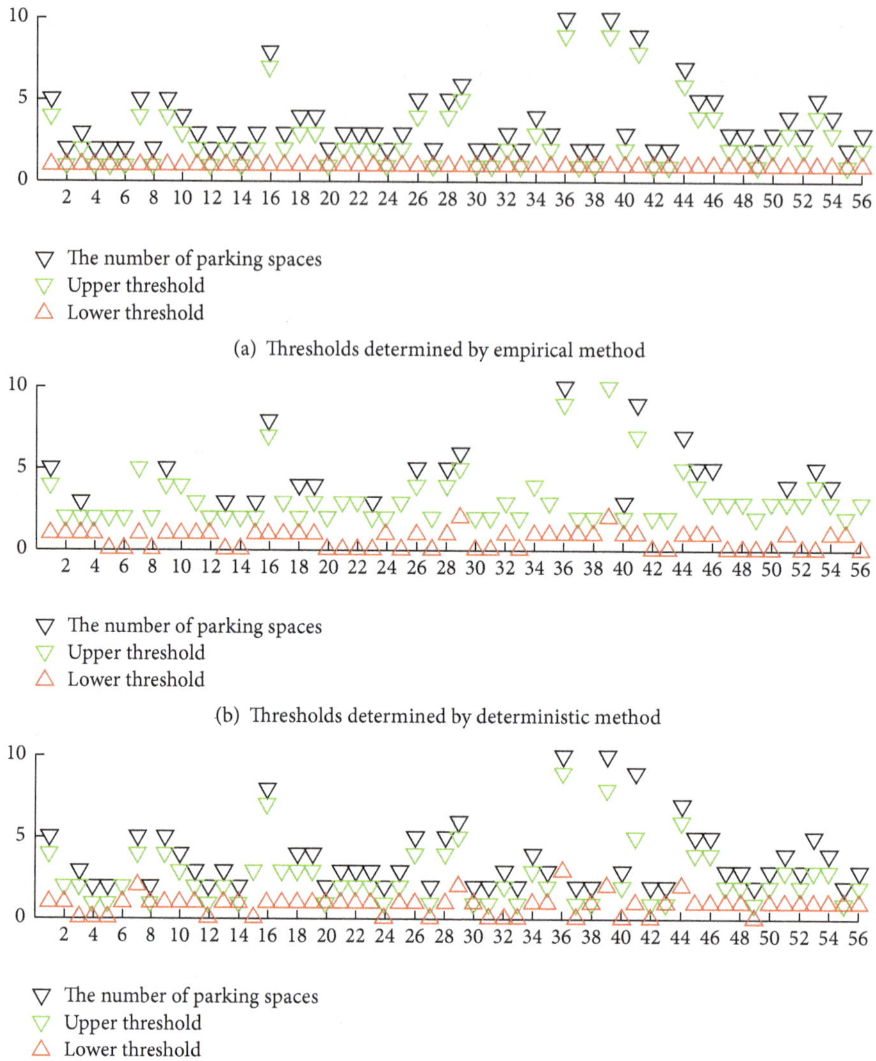

(a) Thresholds determined by empirical method

(b) Thresholds determined by deterministic method

(c) Thresholds determined by stochastic method

FIGURE 5: Result comparison among empirical method, deterministic method, and stochastic method.

The constraints regarding upper and lower thresholds should also be

$$
\begin{aligned}
& S_{lower}^{t} \le S_{upper}^{t}, \\
& S_{upper}^{t} \le N, \\
& S_{lower}^{t} \ge 0, \qquad (31) \\
& S_{lower}^{t}, S_{upper}^{t} \text{ are integers}, \\
& \forall t = 1, 2, \cdots, T
\end{aligned}
$$

The previous model can be fast solved due to the limited combination of two thresholds. In the extended model, more thresholds responding to different time steps significantly enlarge the space of the solutions. There are $N(N-1)/2$ combinations of two thresholds but $(N(N-1)/2)^t$ combinations of t time steps. Supposing there are 24 periods in a day as the research time range, the solution space turns out to be quite huge.

Considering the dynamics of the system happens by time step successively, a local optimum can be reached from a dynamic programming approach. We solve one pair of thresholds for one time step from $t = 1$ to T step by step. Firstly $(S_{lower}^{1}, S_{upper}^{1})$ can be solved. When solving $(S_{lower}^{t}, S_{upper}^{t})$, we deem the previous feasible solution of $(S_{lower}^{1}, S_{upper}^{1}), \ldots, (S_{lower}^{t-1}, S_{upper}^{t-1})$ as known variables solved in previous steps. In that case the algorithm in this paper is applicable to solve the variable thresholds problem.

Figure 7 shows the case to determine the thresholds of station No. 44. Figure 7(a) illustrates the demand pattern in different time steps, where boxplots indicate the randomness of demands. Figure 7(b) shows the result of $(S_{lower}^{t}, S_{upper}^{t})$, $\forall t = 1, 2, \cdots, 24$, during one day, where green triangles denote the upper thresholds and red triangles denote the

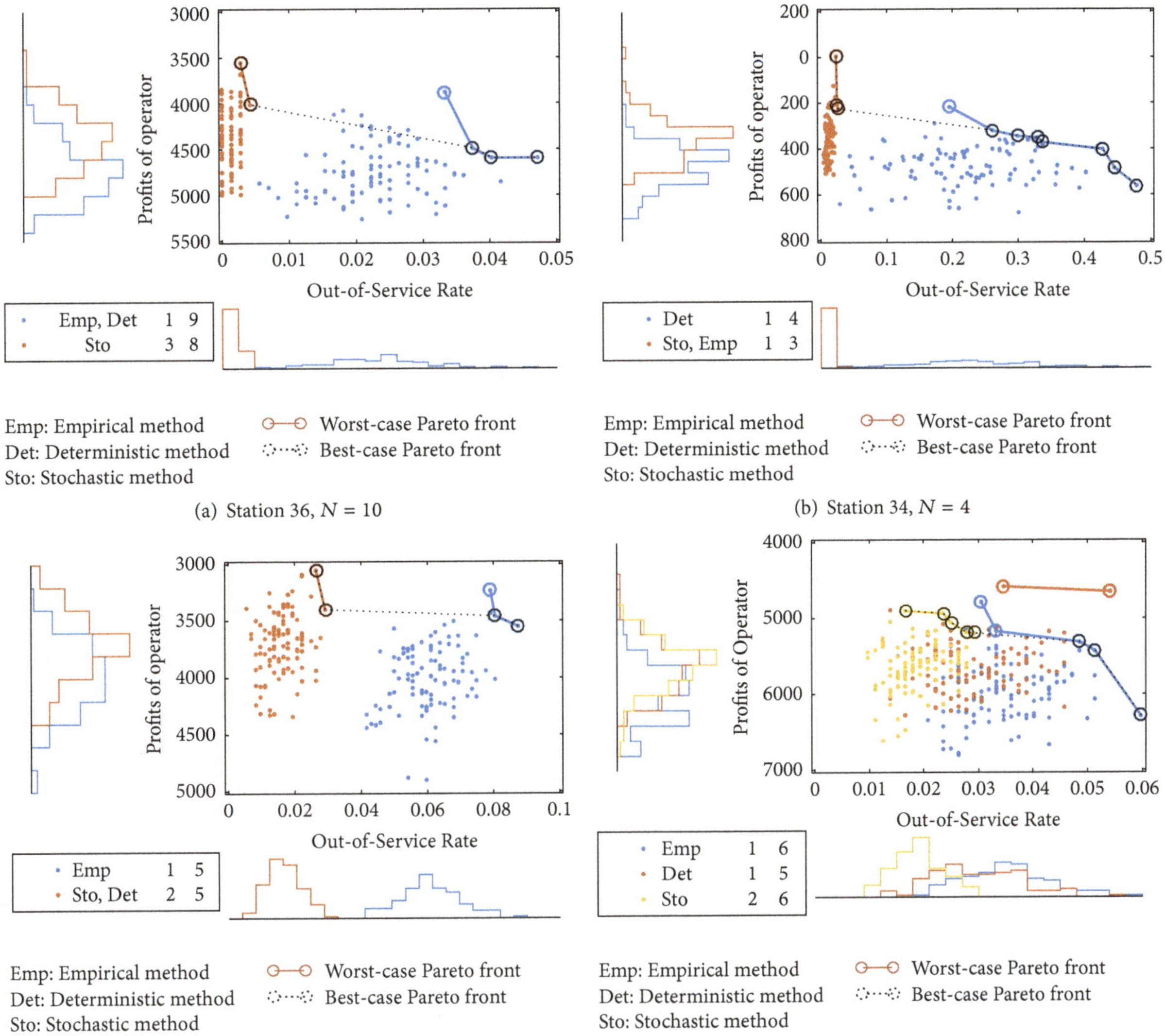

(a) Station 36, $N = 10$

(b) Station 34, $N = 4$

(c) Station 29, $N = 6$

(d) Station 44, $N = 7$

FIGURE 6: Comparison of the profit among the three methods.

lower thresholds. The solid line in this figure represents the average possible number of vehicles conserved in the station. The green and red dash line, respectively, indicate the possible maximum and minimum bound of vehicles conserved in the station due to the stochastic demands. The green and red dash lines confine a probable range of the numbers of vehicles at the station.

There are some findings from the result shown by the figure. During the wee hours (1:00 ~ 6:00) with low demands, the upper threshold is relatively low, which allows idle vehicles to be moved out, to support other stations which need vehicles to be moved in during the same period (if existing). When demands occur after 7:00, the lower threshold goes up for requiring vehicles to be moved in, accounting for the following peak hours. Until 16:00 demands slightly decrease, and the possible range of vehicle number at station during 16:00 seems to be relatively higher, so that the thresholds accordingly go down to let the vehicles to be moved out. During 18:00 the possible range of vehicle number seems to decrease and the thresholds allow vehicles to be moved in. After 20:00 the number of vehicles turns to be stable and the thresholds are wider to let the number of vehicles change naturally.

6. Conclusions

This paper presents an approach on determining the thresholds considering the stochastic demands and system states, which provides a quantitative justification for thresholds determination. Firstly, we establish a model to capture the dynamics of the system status involving stochasticity and an optimization tactic is employed to find out proper values of upper and lower thresholds, aiming at minimizing the relocation times and the out-of-service time rate. The solution algorithm is provided to determine the rational solutions for this model, employing the min–max robust optimization to

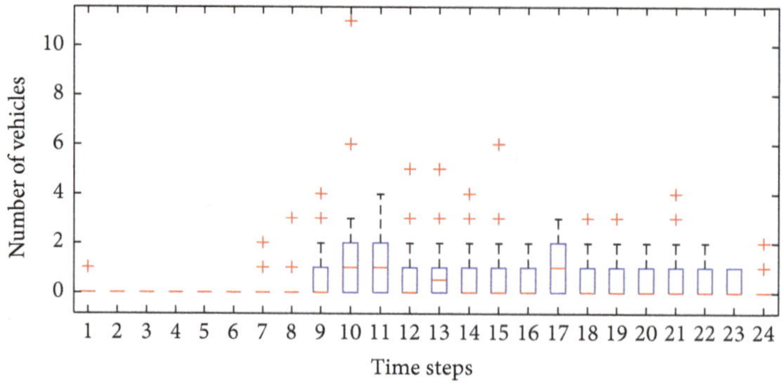

(a) Boxplot of stochastic demand pattern in different time steps

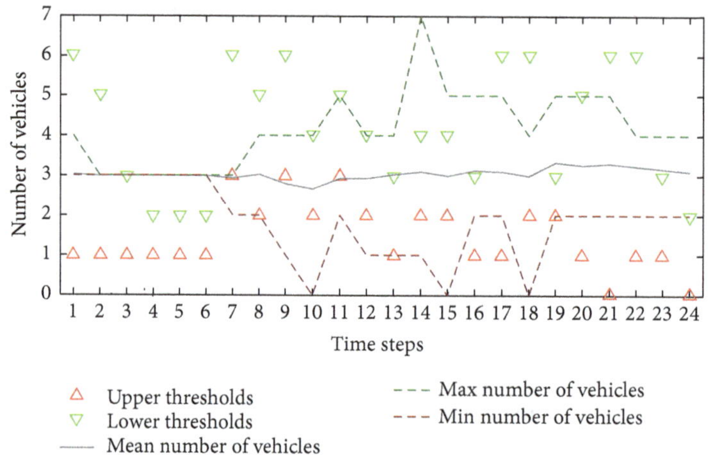

(b) Thresholds in different time steps

FIGURE 7: Solution of thresholds for different hours in a day.

tackle the uncertainty and the Pareto optimum to decide the solution under dual objectives.

To calibrate the stochastic variables, the continuous variables and the discretized variables are tested under the significance level of 0.05, which did not reject their hypothesis on their distribution, respectively. A numerical example was given to demonstrate the algorithm. A comparison was conducted among the empirical rules, the deterministic method, and the stochastic method, where the results suggest that the stochastic rules get the identical results with the empirical method at 23% of the stations and deterministic method at 5%. Among the stations where the stochastic method outputs different results from the empirical or deterministic method, the stochastic method achieves better solutions, which confirm that the stochastic method provides more reasonable justification towards uncertain demands.

Actually the threshold determination is a fundamental part of the threshold based relocation approach; the entire method considering stochastic demands and system status is not sufficiently studied. Although the threshold based method is an on-demand respond method, the efficiency and effectiveness of this method in practice have not been reported yet. In the future, dynamic control or operation approaches are promising to cope with the real time relocation problem, which will bring more profits to the carsharing system operation.

Notations

(a) Parameters

T: The number of time steps

t: Time step t, $t = 0, 1, 2, \cdots, T$

N: The number of parking spaces at the station

l: Time step width, h, in this paper $l = 1$

cv: Charging rate of the vehicle, km/h

CR: Cost of relocation for each trip

RV: Average revenue of each trip departing from the station

B_{max}: The maximum battery status, which is described as continuing travel distance, km.

(b) Decision Variables

(1) Primary Decision Variables

S_{upper}: Upper threshold of the station
S_{lower}: Lower threshold of the station.

(2) Auxiliary Decision Variables, $\forall t = 1, 2, \cdots, T$

$vm_{t,t'}^{(+),v_t}$: Binary variable (equals 1 when vehicle $v_t = 1, 2, \cdots, \delta_t^{(n+)}$ is required to be moved in at time step t and actually arrives at time step t', otherwise 0)
$vm_{t,t'}^{(-),v_t}$: Binary variable (equals 1 when vehicle $v_t = 1, 2, \cdots, \delta_t^{(n-)}$ is required to be moved out at time step t and actually departures at time step t', otherwise 0)
$vs_t^{b,b^{(D)}}$: Binary variable (equals 1 when vehicle with battery status b is matched with demand on battery status $b^{(D)}$, otherwise 0).

(3) Stochastic Variables, $\forall t = 1, 2, \cdots, T$

r_t: The number of returned vehicles in time step t
$d_t^{(ge)}$: The number of demand vehicles generated in time step t
$b_{t,v_t}^{(R)}$: The battery status of each returned vehicle in time step t, km (it follows a continuous distribution; $b_t^{(R)} \in B_t^{(R)}$, $v_t = 1, 2, \cdots, r_t$)
$b_{t,v_t}^{(ge)}$: The desired battery status of demands generated in time step t, km (it follows a continuous distribution; $b_t^{(ge)} \in B_t^{(ge)}$, $v_t = 1, 2, \cdots, d_t^{(ge)}$)
$b_{t,v_t}^{(+)}$: The battery status of each vehicle that is actually relocated into the station in time step t, km (it follows a continuous distribution; $b_t^{(+)} \in B_t^{(+)}$, $v_t = 1, 2, \cdots, \delta_t^{(+)}$)
$b_{t,v_t}^{(-)}$: The desired battery status of each vehicle that is relocated out of the station in time step t, km. (it follows a continuous distribution; $b_t^{(-)} \in B_t^{(-)}$, $v_t = 1, 2, \cdots, \delta_t^{*(-)}$)
$t_{v_t}^{(+)}$: Delay from the time step of the call of move-in to the time step of the arrival of v_t, due to the moving time consumption, following a discrete distribution ($v_t = 1, 2, \cdots, \delta_t^{(n+)}$)
$t_{v_t}^{(-)}$: Delay from the time step of the call of move-out to the time step of the departure of v_t, due to the availability of relocation staff, following a discrete distribution ($v_t = 1, 2, \cdots, \delta_t^{(n-)}$)
p_t^b: $\forall b \in B_t^{D(us)}$, equal to 1 if the rejected demands remain to the next time step and 0 if not, at the end of time step t.

(c) Dependent Variables, $\forall t = 1, 2, \cdots, T$

x_t: The number of remaining vehicles at the end of time step t ($x_0 = 0$ for initial condition)
d_t: The number of total demand vehicles in time step t
$d_t^{(st)}$: The number of demands that are satisfied in time step t
$\delta_t^{(n+)}$: The number of vehicles needed to be relocated into the station in time step t according to the threshold triggering
$\delta_t^{(n-)}$: The number of vehicles needed to be relocated out of the station in time step t according to the threshold triggering
$\delta_t^{(+)}$: The number of vehicles actually relocated into the station in time step t
$\delta_t^{(-)}$: The number of vehicles actually relocated out of the station in time step t due to the battery status
$\delta_t^{*(-)}$: The number of vehicles required to move out of the station in time step t due to the delay
$u_t^{(+)}$: The time mark that there is no parking space available in time step t ($u_t^{(+)} = 1, x_t \geq N; 0$, otherwise)
$u_t^{(-)}$: The time mark that there is no vehicle available in time step t ($u_t^{(-)} = 1, x_t \leq 0; 0$, otherwise).

(d) Sets, $\forall t = 1, 2, \cdots, T$

$B_t^{(R)}$: Battery status of returned vehicles in time step t
$B_t^{(ge)}$: Desired battery status of each generated demand vehicle in time step t
$B_t^{(+)}$: Battery status of vehicles to be relocated in in time step t
$B_t^{(-)}$: Battery status of vehicles to be relocated out in time step t
$B_t^{V(rm)}$: Battery status of the remaining vehicles at the end of time step t, $B_t^{V(rm)} = \emptyset$ for initial condition
$B_t^{D(us)}$: The unsatisfied demand battery status set at the end of time step t
$B_t^{D(rm)}$: The remaining demand battery status set at the end of time step t, $B_t^{D(rm)} = \emptyset$ for initial condition.

Conflicts of Interest

The authors declare that there are no conflicts of interest regarding the publication of this paper.

Acknowledgments

Special thanks are due to Shanghai International Automobile City Co., Ltd., for the collaborative research projects of "Research and Application of the Key Technology of Electric Car Rental Sharing System Based on Internet of Vehicles" and "The Threshold Method of Vehicle Allocation Technology". The research is supported by the National Natural Science Foundation of China (no. 51722809), the National Key Research and Development Program of China (no. 2016YFE0206800), the Fok Ying-Tong Education Foundation (no. 151076), Humanity and Social Science Youth Foundation of Ministry of Education of China (no. 16YJCZH070), and Ministry of House and Urban-Rural Development of China Project (2015K5006).

References

[1] S. Shaheen, D. Sperling, and C. Wagner, *Carsharing in Europe and North American: Past, Present, and Future*, University of California Transportation Center, 2001.

[2] M. Barth, S. A. Shaheen, T. Fukuda, and A. Fukuda, "Carsharing and station cars in Asia: Overview of Japan and Singapore," *Transportation Research Record*, no. 1986, pp. 106–115, 2006.

[3] S. Shaheen, N. Chan, A. Bansal, and A. Cohen, *Shared Mobility: A Sustainability Technologies Workshop: Definitions, Industry Developments, and Early Understanding'*, University of California, Berkeley, Transportation Sustainability Research Center; California Department of Transportation, 2015.

[4] A. Jung, "Fewer Cars, More Mobility: Can Carsharing Work in China?" *International Transportation*, vol. 67, no. 1, pp. 26–29, 2015.

[5] S. A. Shaheen, N. D. Chan, and H. Micheaux, "One-way carsharing's evolution and operator perspectives from the Americas," *Transportation*, vol. 42, no. 3, pp. 519–536, 2015.

[6] A. G. H. Kek, R. L. Cheu, and M. L. Chor, "Relocation simulation model for multiple-station shared-use vehicle systems," *Transportation Research Record*, no. 1986, pp. 81–88, 2006.

[7] A. G. H. Kek, R. L. Cheu, Q. Meng, and C. H. Fung, "A decision support system for vehicle relocation operations in carsharing systems," *Transportation Research Part E: Logistics and Transportation Review*, vol. 45, no. 1, pp. 149–158, 2009.

[8] J. Firnkorn and M. Müller, "What will be the environmental effects of new free-floating car-sharing systems? The case of car2go in Ulm," *Ecological Economics*, vol. 70, no. 8, pp. 1519–1528, 2011.

[9] M. Namazu and H. Dowlatabadi, "Characterizing the GHG emission impacts of carsharing: A case of Vancouver," *Environmental Research Letters*, vol. 10, no. 12, 2015.

[10] S. A. Shaheen, A. P. Cohen, and J. D. Roberts, "Carsharing in North America: Market growth, current developments, and future potential," *Transportation Research Record*, no. 1986, pp. 116–124, 2006.

[11] S. Le Vine, M. Lee-Gosselin, A. Sivakumar, and J. Polak, "A new approach to predict the market and impacts of round-trip and point-to-point carsharing systems: case study of London," *Transportation Research Part D: Transport and Environment*, vol. 32, pp. 218–229, 2014.

[12] K. M. N. Habib, C. Morency, M. T. Islam, and V. Grasset, "Modelling users' behaviour of a carsharing program: Application of a joint hazard and zero inflated dynamic ordered probability model," *Transportation Research Part A: Policy and Practice*, vol. 46, no. 2, pp. 241–254, 2012.

[13] C. Morency, K. M. N. Habib, V. Grasset, and M. T. Islam, "Understanding members' carsharing (activity) persistency by using econometric model," *Journal of Advanced Transportation*, vol. 46, no. 1, pp. 26–38, 2012.

[14] B. Boyaci, K. G. Zografos, and N. Geroliminis, "An optimization framework for the development of efficient one-way car-sharing systems," *European Journal of Operational Research*, vol. 240, no. 3, 2015.

[15] G. H. D. A. Correia and A. P. Antunes, "Optimization approach to depot location and trip selection in one-way carsharing systems," *Transportation Research Part E: Logistics and Transportation Review*, vol. 48, no. 1, pp. 233–247, 2012.

[16] M. Nourinejad and M. J. Roorda, "A dynamic carsharing decision support system," *Transportation Research Part E: Logistics and Transportation Review*, vol. 66, pp. 36–50, 2014.

[17] R. Zakaria, M. Dib, L. Moalic, and A. Caminada, "Car relocation for carsharing service: Comparison of CPLEX and greedy search," in *Proceedings of the 2014 IEEE Symposium on Computational Intelligence in Vehicles and Transportation Systems, CIVTS 2014*, pp. 51–58, USA, December 2014.

[18] A. Carlier, A. Munier-Kordon, and W. Klaudel, "Mathematical model for the study of relocation strategies in one-way carsharing systems," *Transportation Research Procedia*, vol. 10, pp. 374–383, 2015.

[19] M. Bruglieri, A. Colorni, and A. Luè, "The relocation problem for the one-way electric vehicle sharing: An Application to the Milan Case," in *Proceedings of the Social and Behavioral Sciences*, vol. 111, pp. 18–27, 2014.

[20] M. Repoux, B. Boyaci, and N. Geroliminis, "Simulation and Optimization of One-Way Car-Sharing Systems with Variant Relocation Policies," in *Transportation Research Board*, 2015.

[21] A. Febbraro, N. Sacco, and M. Saeednia, "One-Way Carsharing: Solving the Relocation Problem," *Transportation Research Record*, no. 2319, pp. 113–120, 2012.

[22] G. Santos and G. Correia, "A MIP model to optimize real time maintenance and relocation operations in one-way carsharing systems," *Transportation Research Procedia*, vol. 10, pp. 384–392, 2015.

[23] M. Barth and M. Todd, "Simulation model performance analysis of a multiple station shared vehicle system," *Transportation Research Part C: Emerging Technologies*, vol. 7, no. 4, pp. 237–259, 1999.

[24] E. M. Cepolina and A. Farina, "Urban car sharing:An overview of relocation strategies," in *Proceedings of the 18th International Conference on Urban Transport and the Environment, UT12*, pp. 419–431, Spain, May 2012.

[25] E. M. Cepolina and A. Farina, "A methodology for planning a new urban car sharing system with fully automated personal vehicles," *European Transport Research Review*, vol. 6, no. 2, pp. 191–204, 2014.

[26] E. M. Cepolina, A. Farina, C. Holloway, and N. Tyler, "Innovative strategies for urban car-sharing systems and a simulator to assess their performance," *Transportation Planning and Technology*, vol. 38, no. 4, pp. 375–391, 2015.

[27] G. Cao, L. Wang, Y. Jin et al., "Determination of the vehicle relocation triggering threshold in electric car-sharing system," *Lecture Notes in Electrical Engineering*, vol. 404, pp. 11–22, 2016.

8

An Ecolevel Estimation Method of Individual Driver Performance based on Driving Simulator Experiment

Yiping Wu, Xiaohua Zhao, Ying Yao, and Jian Rong

Beijing Key Laboratory of Traffic Engineering, Beijing University of Technology, 100 Ping Le Yuan, Beijing 100124, China

Correspondence should be addressed to Yiping Wu; wuyiping@bjut.edu.cn

Academic Editor: Richard S. Tay

Accurately acquiring the ecolevel of individual driver performance is the precondition for more targeted ecodriving behavior optimization. Because of obvious advantage in mining hidden relationship, machine learning was adopted to explore the complicated relationship between driver performance and vehicle fuel consumption and thus to predict the ecolevel of individual driver performance in this study. Based on driving simulator tests, data of driver performance and vehicle fuel consumption were collected. The ecolevel was indicated as the ecoscore corresponding to vehicle fuel consumption. The model input was designed as 10 feature indexes of driver performance (e.g., percentage number, mean value, standard deviation, and power of applying acceleration pedal). The output was treated as ecoscore. Taking a number of one hundred of data segments in vehicle starting process as training sample, the optimal structure, functions, and learning rate of a backpropagation neural network model with three layers were obtained, after repeated model simulation experiments. The validation test of 16 sample data items showed that the mean prediction accuracy of our developed model was 92.89%. In addition, comparative analysis displayed that the performance of backpropagation neural network based model was better than linear regression based model and random forest based model, from the aspects of elapsed time and prediction accuracy in estimating the ecolevel of driver performance. The study results provide an effective method to grasp the ecolevel of driver performance and further contribute to driving behavior optimization towards vehicle fuel consumption and emissions reduction.

1. Introduction

Previous researches have stated that driving behavior was a key factor influencing traffic accidents, congestion, and environment [1]. It is also indicated that driving behavior optimization has a big potential to improve traffic safety and efficiency and reduce vehicle fuel consumption and emissions [2, 3]. Specifically, ecodriving, the concept of changing driving behavior to improve fuel efficiency and reduce greenhouse (GHG) emissions of existing vehicles, has been widely reported to be effective in reducing fuel consumption and the reduction percent could even be 20 to 50% for excellent performers [4, 5]. Thus, comparing to vehicle manufacturing technology, new road materials, and other energy conservation and emission reduction measurements for motor vehicles, it seems to be more cost-effective and of higher efficiency to apply ecodriving behavior optimization for existing vehicle drivers.

As a primary condition for ecodriving behavior optimization, it is first necessary to know the ecolevel of individual driver's driving behavior exactly. Generally, driving behavior represents both driver performance and vehicle operating status. In the process of driving, vehicle operating status was mainly influenced by drivers' manipulation behavior of vehicle control units. Therefore, driver performance (e.g., pressing acceleration pedal, turning steering wheel, and shifting gear) seems to be more direct to represent drivers' ecodriving level. In the past years, a few studies tried to find methods about how to estimate individual driver performance's ecolevel reasonably. Most of the relative researches took driving experience or simple statistical analysis of hard accelerating or breaking times as the main reference for estimation criterion [6]. In addition, few quantitative evaluation approaches were based on the difference between the practical value and the desirable threshold of some specific driving performance parameters (e.g., average acceleration, deceleration, and speed) in given condition [7, 8]. But, almost all of these quantitative estimation methods stated here analyzed these feature parameters separately. As driver performance is random, continuous, and instantaneous during vehicle operation, the ecolevel evaluation based on single parameter (e.g., only considering acceleration or deceleration pedal) is not practical. Thus, according to these limited evaluation methods above, on one hand, the estimation accuracy was not satisfactory; on the other hand, the estimation results had no small gap for supplying targeted ecodriving behavior optimization.

In fact, the core of ecolevel estimation for driver performance is to precisely explore the relationship between drivers' manipulating behavior and vehicle fuel consumption and emissions during driving. In the process of driving, driver performance is a dynamic and comprehensive result from driver perception and decision-making about road condition, traffic control, environment stimulation, and their interactions [9]. Therefore, it appears to be much challenging to estimate the ecolevel of driver performance exactly because of its multidimension, complication, randomness, and uncertainty. Furthermore, the relationship between driver performance and vehicle fuel consumption or emissions might be chaotic and hidden but not totally linear or obvious. So, the statistical analysis based traditional estimation and prediction models are not sufficient for accurate ecolevel estimation of driver performance because of limitation in expressing complicated relationship [10].

Fortunately, machine learning with excellent data processing ability has been widely applied in deep data mining in traffic areas [9]. Moreover, as with obvious advantage of mining hidden features, machine learning has been treated as an effective approach for driver behavior modeling, evaluation, and prediction. At present, shallow machine learning models including decision tree [11], Hidden Markov Model [12], Network of Neuron [13], and Support Vector Machine [14, 15] have been widely used for individual driving habit modeling, traffic flow prediction, and unsafe driving behavior (e.g., driving drowsiness and drunk driving) identification. Moreover, the prediction and identification accuracy were acceptable.

Comparing these shallow machine learning models commonly used, it is not difficult to find that artificial neural network could realize the highly nonlinear reflection from input to output [16]. Hence, aiming at reliability prediction and evaluation, the method based on artificial neural network would be better. Among these artificial neural networks, backpropagation (BP) neural network is the most commonly used, especially in practical application [17, 18]. Moreover, BP network has apparent advantages in self-learning, self-organizing, good fault tolerance, and excellent nonlinear approximation ability [19]. So, system with complicated structure and perfect performance could be established by a large number of neurons with simple performance, and further the complex tasks could be resolved [20].

Summarizing above, to grasp the ecolevel of driver performance more accurately, BP neural network was introduced in this study. As with obvious advantage in driver performance perception and convenient controlling for test condition, interference factors, and driver safety, driving simulator technology was adopted in the current study for driver performance data collection in the process of driving. In general, vehicle fuel consumption reduction indicates individual driver's application of ecodriving behavior for a given vehicle in large degree, so vehicle fuel consumption is used to represent the ecolevel of driver performance. Based on our driving simulator experimental data with repeated tests, this paper aims to find the optimal structure, suitable network parameters and functions of the BP neural network based model for individual driver's ecolevel estimation. The study results established the relationship between driver performance and vehicle fuel consumption, which laid a foundation for more targeted ecodriving optimization and thus further contributed to reducing vehicle fuel consumption and emissions. The study results provide an effective method to grasp the ecolevel of driver performance and further contribute to driving behavior optimization towards vehicle fuel consumption and emissions reduction.

2. Data Collection

Driving simulator platform in Beijing University of Technology was used to collect driver performance data and its corresponding fuel consumption (see Figure 1). This driving simulator is a fix-based type, which consists of a renovated real car, computers, and video and audio equipment. The road scenarios are projected onto three big screens providing a total of 130 degrees of the driver's visual field. There were two side mirrors and one rear view mirror to show the traffic behind the simulated vehicle. This simulator records the actions of drivers and vehicles, and the parameters that describe vehicle operations are saved 30 times per second. The data of drivers' actions collected by this simulator mainly include drivers' maneuver for acceleration and deceleration pedal, clutch pedal, steering wheel, and gear. The data of vehicle operations recorded includes the engine revolution as well as speed, acceleration, deceleration, etc. Moreover, the validation of this simulator in driving behavior tests has been conducted in our previous studies, including both subjective evaluation by questionnaires [21] and comparative experimental tests in simulator and field [22].

The output mode of vehicle energy and emission in our driving simulator depended on filed test data in Beijing. The microscopic emissions model was based on vehicle specific power (VSP) distribution. The parameters in the VSP based emission model was calibrated by a large number of taxies operating in realistic driving conditions in Beijing [23]. On this basis, vehicle fuel consumption was calculated by the carbon balance method [24]. After comparative and validation analysis, it has been verified that the vehicle fuel consumption per 100 kilometers outputted by our driving simulator is consistent with the field test data collected by On-Board Diagnostics mounted on vehicles [25].

As most common rules or guidance of ecodriving behavior was based on various driving events (e.g., vehicle starting event, vehicle stopping event, and vehicle cruising event) [2, 3], our current experiment is designed to record subjects' driving performance in the process of vehicle starting event.

(a) Hardware equipment (b) Software system

Figure 1: Driving simulator in Beijing University of Technology.

Driver performance during vehicle starting process requires more comprehensive maneuvers (manual transmission car) and it is direct reflection of drivers' ecodriving level. The scenario used in this experiment was an urban road with two lanes in each direction. The data collection segments were designed at four signal intersections with digital display. The interval between each intersection and the beginning and ending of this scenario were designed as urban roads with the length of 1,000 meters. Between each starting event, no other events were designed and participants were asked to drive routinely and obey the traffic laws. In order to eliminate the interference of confounding factors on the experimental data, the scenarios used for each vehicle starting event in different locations are uniformed. The road used for data collection was straight and flat. No other vehicles were visible in the participant's driving direction and no vehicles existed in front of the target vehicle.

Before the main driving, every participant took 3-5 minutes for a practice drive to become familiar with the simulator's steering, braking dynamics, and other driving manipulation. For each driving at these four signal intersections, drivers started the vehicle when the traffic signal turns green. Subjects were asked to drive straightly through each intersection. In data processing and analyzing, the starting process was defined as the vehicle speed increased from zero to the location where 100 meters passed by.

A total of thirty drivers (eighteen male and twelve female drivers) participated in this study. All participants were licensed drivers with an average age of 30 (between the ages of 22 and 50 years, with a standard deviation of 9.5 years) and 3 to 20 years of driving experience (average=7.6 years and standard deviation =6.1 years). According to the experimental design, each participant experienced four times of vehicle starting process and thus a total of one hundred and twenty data segments of vehicle starting would be collected.

As the road used for vehicle starting is straight and flat, only driver performance of accelerator pedal, clutch pedal, and gear was used for individual ecolevel estimation. Figure 2 displays two types of driver performance with different fuel consumption. As shown in Figure 2, the higher fuel consumption during vehicle starting process is 23.87L/100km (see Figures 2(a), 2(c), and 2(e)) and the lower is 18.90L/100km (shown in Figures 2(b), 2(d), and 2(f)). It obviously indicates that the driver performance with high and low fuel consumption is significantly different.

3. Estimation Model Construction

Aiming at ecolevel estimation of driver performance, the input, output, and hidden layer of our BP neural network model should be driver performance parameters, ecolevel estimation results, and the relationship between driver performance and vehicle fuel consumption, respectively. In addition, as any continuous function in a closed interval could be approximated by BP neural network with single hidden layer, any reflection from n-dimension to m-dimension could be achieved through BP neural network with three layers [20]. Thus, this study focused on constructing an ecolevel estimation model of driver performance based on BP neural network with three layers.

According to the design of experimental data collected, a total of 120 data segments would be obtained. As one participant suffered from simulation sickness, data of twenty-nine successful candidates (seventeen male and twelve female drivers) were collected. Namely, the data of 116 segments was used in the following analysis. The age of these successful candidates was 29.5±9.4 years. The driving experience was 7.5±6.3 years. To construct ecolevel estimation model and analyze its applicability, one hundred data segments were randomly selected for model training and the remaining sixteen sample data were used for model accuracy test.

3.1. Input Design. Based on our driving simulator experimental data collected and combining with these influencing factors on vehicle fuel consumption found in our previous studies [6, 7, 25], the driver performance features in the process of vehicle starting was deeply analyzed. Finally, ten indexes were selected as the input parameters for BP neural network model, which mainly consists of drivers' maneuver for vehicle acceleration pedal, clutch pedal, and gear. As the data segment was limited to vehicle starting process and the road condition was straight and flat, drivers' maneuver for vehicle steering and deceleration pedal hardly occurred. Thus, the index of applying deceleration pedal was not included in the current research, although drivers' action behavior of deceleration pedal is usually an important

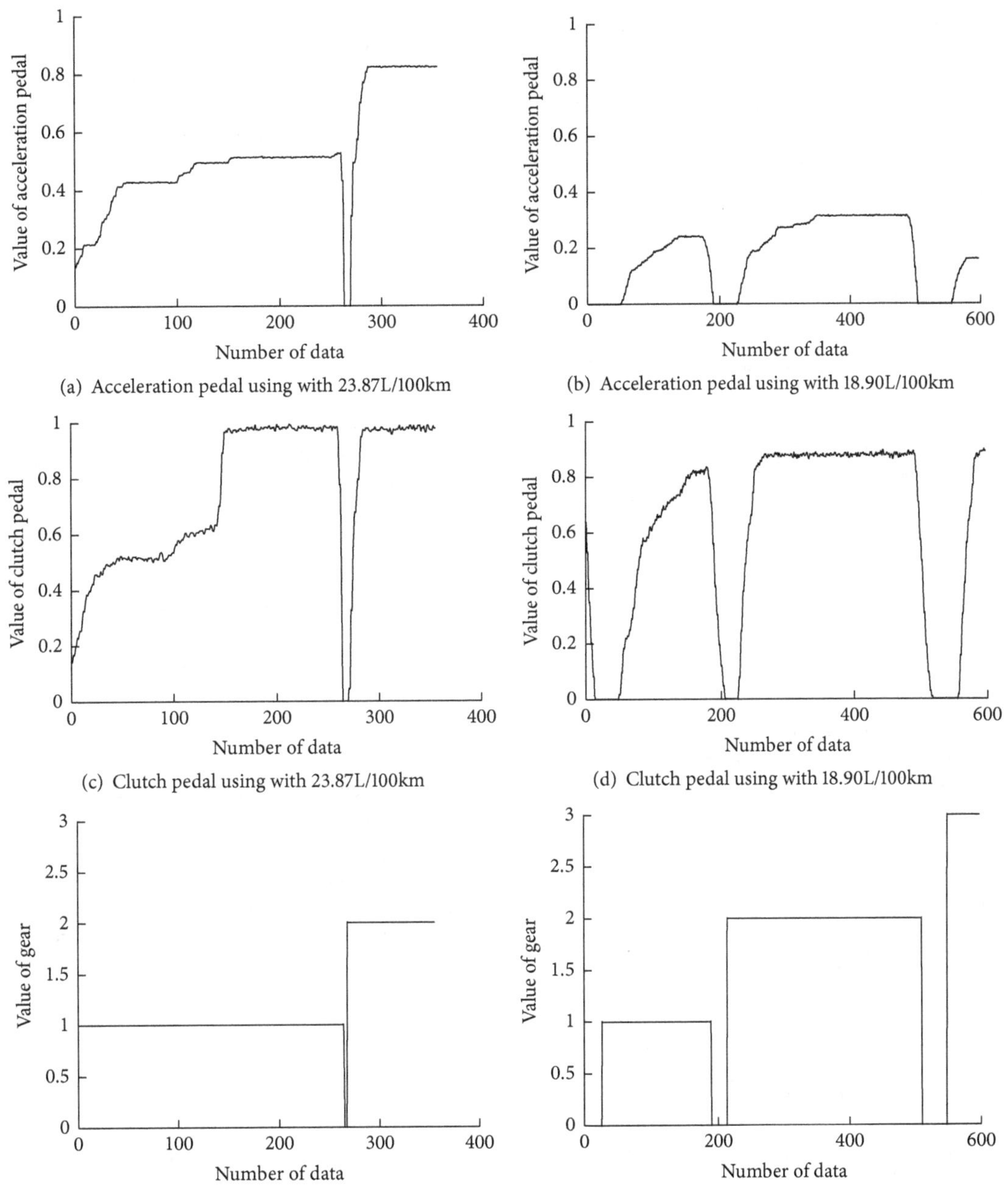

(a) Acceleration pedal using with 23.87L/100km

(b) Acceleration pedal using with 18.90L/100km

(c) Clutch pedal using with 23.87L/100km

(d) Clutch pedal using with 18.90L/100km

(e) Gear using with 23.87L/100km

(f) Gear using with 18.90L/100km

FIGURE 2: Driver performance in vehicle starting process with different fuel consumption.

representation of driver performance during other driving events. The ten indexes used for model input were described below.

(1) Percentage of number when applying acceleration pedal (*A_number*).

A_number is used to reflect drivers' operating frequency of acceleration pedal. *A_number* with higher value means more frequent maneuver of acceleration pedal, which might lead to higher fuel consumption. The calculation method of *A_number* is described as

$$A\_number = \frac{N_A}{N} \tag{1}$$

$$N_A = \sum_{i=1}^{N} n_i \tag{2}$$

$$n_i = \begin{cases} 1, & a_i \neq 0 \\ 0, & a_i = 0 \end{cases} \tag{3}$$

where

a_i is the value of acceleration pedal applying at time i;

n_i is the statistic value of acceleration pedal applying at time i. The value of n_i is 1 if the acceleration pedal is pressed at time i, otherwise n_i is zero;

N_A is sum of n_i during the given time or journey;

N is the total number of data recorded during the given time or journey.

(2) Mean value of applying acceleration pedal (*A_mean*)

A_mean is adopted to represent drivers' average pressing intensity of acceleration pedal, which is defined as (4). Vehicle fuel consumption would be increased when *A_mean* becomes larger.

$$A\_mean = \frac{\sum_{i=1}^{N} a_i \, (a_i \neq 0)}{N_A} \quad (4)$$

(3) Standard deviation of acceleration pedal applying value (*A_SD*)

In addition to *A_mean*, *A_SD* indicates the stability of drivers' maneuver on acceleration pedal. *A_SD* with larger value would have negative impact on reducing vehicle fuel consumption. The calculation method of *A_SD* is shown in

$$A\_SD = \sqrt{\frac{1}{N_A - 1} \sum_{i=1}^{N_A} (a_i - A\_mean)^2 \, (a_i \neq 0)} \quad (5)$$

(4) Percentage of number when acceleration pedal applying value is more than 2/3 of the whole pedal travel (*A_hard*)

A_hard represents the percentage of drivers' rapid acceleration behavior, which is defined as (6) to (8). Higher *A_hard* will increase vehicle fuel consumption significantly.

$$A\_hard = \frac{N_A'}{N_A} \quad (6)$$

$$N_A' = \sum_{i=1}^{N} m_i \quad (7)$$

$$m_i = \begin{cases} 1, & a_i \geq \frac{2}{3} \\ 0, & a_i < \frac{2}{3} \end{cases} \quad (8)$$

where m_i is the statistic value of applying acceleration pedal at time i. The value of m_i is 1 if the acceleration pedal applying is equal to or more than 2/3 of its whole pedal travel at time i, otherwise m_i is zero and N_A' is sum of m_i during the given time or journey.

(5) Power of pressing acceleration pedal (*A_power*)

The power of pressing acceleration pedal was defined by drivers' press intensity and duration, as shown in (9). The power of acceleration applying could be intuitively represented by the area surrounded by the x-axis (time), y-axis (value of acceleration pedal), and press intensity profile in a graph (e.g., Figures 2(a) and 2(b)). With the area increasing, it means that a driver presses on the acceleration pedal with more power.

$$A\_power = \int_{i=1}^{N} a_i \, di \quad (9)$$

(6) Percentage of number when applying clutch pedal (*C_number*)

In the same way with *A_number*, *C_number* is used to reflect drivers' operating frequency of clutch pedal. The calculation method of *C_number* is described as

$$C\_number = \frac{N_C}{N} \quad (10)$$

$$N_C = \sum_{i=1}^{N} z_i \quad (11)$$

$$z_i = \begin{cases} 1, & c_i \neq 0 \\ 0, & c_i = 0 \end{cases} \quad (12)$$

where

c_i is the value of clutch pedal pressing at time *i*;

z_i is the statistic value of clutch pedal applying at time *i*. The value of z_i is 1 if the clutch pedal is used at time *i*, otherwise z_i is zero;

N_C is sum of z_i during the given time or journey.

(7) Power of applying clutch pedal (*C_power*)

In the same way with *A_power*, *C_power* represents drivers' pressing intensity and duration of vehicle clutch. The definition of *C_power* is described as

$$C\_power = \int_{i=1}^{N} c_i \, di \quad (13)$$

(8) Percentage of number when gear is equal to 1 (*G_number*)

Vehicle fuel consumption would be significantly enlarged when drivers adopt a lower gear but higher driving speed. Shifting up the gear timely and accurately is one important rule guided in ecodriving behavior. Thus, *G_number* is adopted to indicate the percentage of using the first gear during vehicle starting process, which was illustrated in (14) to (15).

$$G\_number = \frac{N_G}{N} \quad (14)$$

$$N_G = \sum_{i=1}^{N} u_i \quad (15)$$

$$u_i = \begin{cases} 1, & g_i = 1 \\ 0, & g_i \neq 1 \end{cases} \quad (16)$$

where

g_i is the value of gear using at time *i*;

u_i is the statistic value of gear using at time *i*. The value of u_i is 1 if the gear locates in the first level at time *i*, otherwise u_i is zero;

N_G is sum of u_i during the given time or journey.

(9) Mode of gear (*G_mode*)

G_mode means the gear value with the highest appearance frequency, which would be highly related to vehicle fuel consumption during vehicle starting process. On one hand, it reflects drivers' preference of gear using; on the other hand, it indicates whether drivers shifted up the gear timely in the process of vehicle starting.

Table 1: Description Statistic results of ten feature indexes of model input.

	Maximum	Minimum	Average	Median	Standard deviation
A_number	0.99	0.55	0.86	0.87	0.09
A_mean	0.84	0.10	0.36	0.32	0.18
A_SD	0.32	0.05	0.13	0.11	0.07
A_hard	0.85	0.00	0.12	0.00	0.24
A_power	313.87	57.04	142.48	127.20	55.22
C_number	1.00	0.72	0.94	0.95	0.06
C_power	661.25	175.65	366.81	356.35	89.71
G_number	1.00	0.19	0.54	0.52	0.21
G_mode	2.00	1.00	1.35	1.00	0.48
ACG_percent	0.25	0.00	0.05	0.04	0.04

(10) Percentage of number when pressing acceleration pedal with no work (*ACG_percent*)

ACG_percent means that driver has pressed the acceleration pedal but no work was produced or exported. Higher *ACG_percent* would waste vehicle energy directly. As shown in (17) to (19), *ACG_percent* is defined as two aspects: (1) drivers pressed the acceleration pedal but the gear is zero or (2) acceleration pedal is pressed and the pedal travel of clutch is equal to or more than 2/3 at the same time.

$$ACG\_percent = \frac{N_{ACG}}{N} \tag{17}$$

$$N_{ACG} = \sum_{i=1}^{N} o_i \tag{18}$$

$$o_i = \begin{cases} 1, & a_i \neq 0 \& g_i = 0, \ or, \ a_i \neq 0 \& c_i \geq \frac{2}{3} \\ 0, & otherwise \end{cases} \tag{19}$$

where o_i is the statistic value of acceleration pedal, gear and clutch pedal at time i. The value of o_i is 1 if the acceleration pedal is pressed and the gear is zero or the pedal travel of clutch is equal to or more than 2/3 at time i, otherwise o_i is zero and N_{ACG} is sum of o_i during the given time or journey.

Summarizing above, the description statistic results of these ten indexes of all one hundred and sixteen data segments were calculated and displayed in Table 1.

3.2. Output Design. The output of BP neural network should be the ecolevel corresponding to various driver performance. The vehicle fuel consumption was used to represent the ecolevel. For perceptual intuition, the ecolevel should be designed as ecorank (e.g., excellent, good, fair, and poor) or ecoscore (e.g., 73.2 out of a perfect 100 points). In this study, ecoscore was selected to reflect the ecolevel of driver performance. Thus, the core issue of ecolevel estimation for individual driver performance was to predict drivers' ecoscore based on his or her driver performance.

In this study, we assumed that the ecoscore was from 40 (i.e., the highest fuel consumption) to 100 points (the lowest fuel consumption). Based on the data in the process

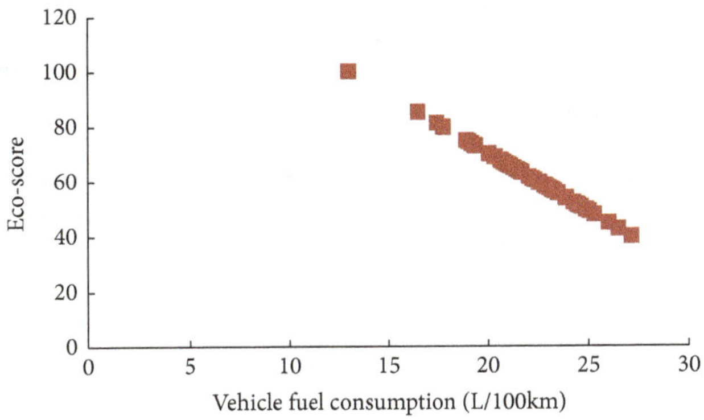

Figure 3: Relationship between fuel consumption and ecoscore during vehicle starting process.

of vehicle starting, the method of transforming vehicle fuel consumption to ecoscore is described as

$$Eco - score_i = 40 + \left(1 - \frac{FC_i - FC_{\min}}{FC_{\max} - FC_{\min}}\right) * 60 \tag{20}$$

where

$Eco - score_i$ is the ecoscore in vehicle starting process for driver *i*;

FC_i is the mean fuel consumption in vehicle starting process for driver *i*, L/100km;

FC_{max} is the maximum value of all FC_i;

FC_{min} is the minimum value of all FC_i.

According to (20) and the experimental data collected, the relationship between fuel consumption per 100 kilometers and ecoscore in the process of vehicle starting could be obtained, as shown in Figure 3.

3.3. Node Number of Hidden Layer Selection. The node number of hidden layer selection is a key issue for BP neutral network model construction. However, there is still a lack of certain methods about how to determine the suitable number of nodes in hidden layer for different tasks. According to our literature review, we found that BP neutral network model has been widely applied in traffic areas and there were many empirical equations to select the suitable number of hidden layer size. Generally, the range of possible neuron numbers

TABLE 2: Prediction error and training speed corresponding to different training functions.

Training function	Prediction error	Training speed (Second)
trainr ( )	0.089	8.133
trainbfg ( )	0.086	0.738
traingd ( )	0.082	0.545
traingdm ( )	0.093	0.544
trainlm ( )	0.072	0.286
traincgp ( )	0.081	0.963
traincgb ( )	0.074	1.070

would be obtained based on these empirical equations. To find the best number of hidden layer size, experimental test was conducted and the smaller prediction error was treated as control objective [19, 20, 26, 27]. So, previous study experience and experimental test were jointly used to find the optimal node number of hidden layer for the BP neutral network model towards ecolevel estimation of driver performance.

Firstly, based on the number of input and output indexes, the experiential equations (21) to (23) were used to calculate the range of node number [19, 20, 26, 27].

$$n_{node} = \sqrt{n_{input} + n_{output}} + \alpha \quad (21)$$

$$n_{node} = \sqrt{n_{input} \times n_{output}} \quad (22)$$

$$n_{node} = \log_2 n_{input} \quad (23)$$

where

n_{node} is the number of nodes in hidden layer;
n_{input} is the number of input indexes;
n_{ouput} is the number of input indexes;
α is constant, and the value is from 0 to 10.

As stated above, the number of input parameters is ten and the output number is one. Hence, the node number should be set as from 4 to 14 according to (21) to (23). Based on this, the node number range was designed as an arithmetic progression with a tolerance of 2 for optimal value test. To find the most suitable node number, the prediction error was treated as control objective. Namely, we should find the optimal value of node number from 4, 6, 8, 10, 12, and 14 under the restriction of the smallest prediction error.

In order to get a relatively stable prediction error corresponding to different node numbers, each BP neural network model with a given node number (i.e., 4, 6, 8, 10, 12, and 14) ran 10 times. The relationship between the average prediction error and the number of nodes in hidden layer was displayed in Figure 4. It revealed that, at the beginning, the prediction error decreased slightly with the node number increasing, while the prediction error increased when the node number further enlarged. Obviously, the mean prediction error is the smallest when the node number is 8. Therefore, the optimal number of the nodes in hidden layer of our established BP neural network should be 8.

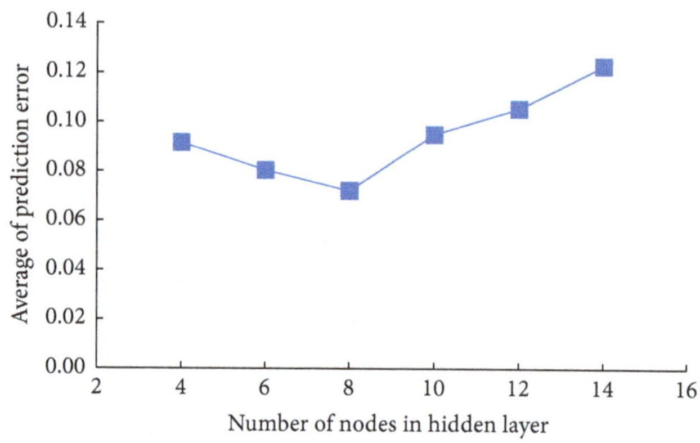

FIGURE 4: Relationship between neuron node number and average prediction error.

3.4. Function Selection. In the same way with other typical BP neural network models [26, 28], the transfer function from input to output layer was set as S-type tangent function "tansig" in this study. Similarly, the linear function "purelin" was set as the transfer function from hidden to output layer. Considering that different training functions have apparent effects on training time and might also influence prediction accuracy, experimental test was applied to find the most suitable training function from these commonly used. A total of 7 types of training functions were treated as candidates. The prediction accuracy and training speed were the evaluation indexes. To get steady prediction results, the BP neural network model with each training function was tested for ten times. The average prediction error and training speed corresponding to different training functions were calculated and illustrated in Table 2.

According to Table 2, the function of "trainlm" was finally selected as the optimal training function in our developed BP neural network model, because of both lower prediction error and more rapid training speed when comparing to other training functions.

3.5. Learning Rate Selection. Learning rate determines the weight variation for each loop iteration in the BP neural network. The learning rate with smaller value could make the prediction error the smallest at the end of iteration but the learning time might increase accordingly, and thus the model convergence rate would be slower. Generally, the range of learning rate was set as from 0.01 to 0.8 [28].

In this study, according to these typical applications of BP neural network model in traffic areas [19, 20, 26, 27], the

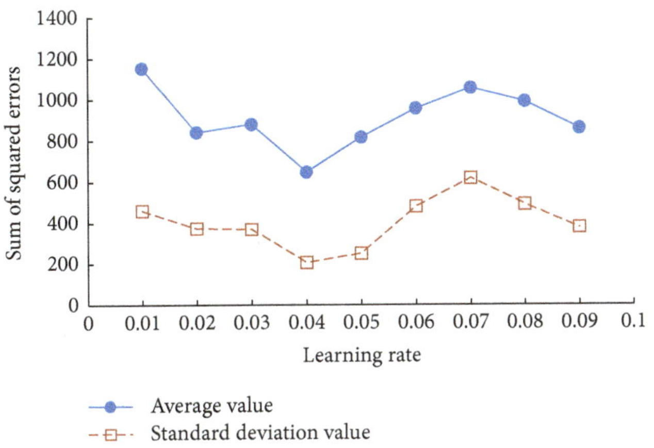

FIGURE 5: Relationship between learning rate and sum of squared errors.

range of learning rate was set as from 0.01 to 0.09. Then, the method of repeated trials was conducted, and the learning rate with the smallest sum of squared errors was treated as the optimal value. The learning rate range was designed as an arithmetic progression with a tolerance of 0.01 for candidate test. Each BP neural network model with a given learning rate ran 10 times. The relationships between learning rate and the average (AVG) and standard deviation (SD) of sum of squared errors were illustrated in Figure 5. It is obvious that both the AVG and SD are the smallest when the learning rate is 0.04. Therefore, the optimal value of learning rate of our established BP network should be 0.04.

3.6. Results of Model Construction. Summarizing the above analysis and discussion, the ecolevel estimation model of driver performance including input and output, number of hidden layers and its corresponding nodes, transfer and training functions, and learning rate could be obtained. Based on these test results, we also attempted to find the most suitable model format by changing the number of neurons in hidden layer, learning rate, and function at the same time. After repeated tests, the optimal structure of our BP neural network model could be described as Figure 6.

In this model, there are ten input indexes of driver performance and the output index is the ecoscore. This model has one hidden layer with eight neuron nodes. The transfer function from input to hidden layer and from hidden to output layer is "tansig" and "purelin," respectively. The suitable training function is "trainlm" and the optimal learning rate is 0.04. The accuracy rate calculated by the difference between prediction value based on training data and the train label is 96.81%.

4. Model Validation Test and Discussion

According to our experimental data collected and the data distribution for model training and validation test, model validation test results show that our established model has relatively high operating efficiency with the average elapsed time of 0.238 seconds. The prediction results were displayed in Figure 7.

As shown in Figure 7(a), it is clearly indicated that the fitting degree of our developed model is pretty good. The prediction value is around the original value for each vehicle starting process. Taking the absolute value of the difference between the original and predictive values divided by the original value as a forecasting performance measure, this average percentage error was 7.11%. Namely, the mean prediction accuracy of our developed model is 92.89%.

The prediction error for each driving process was shown in Figure 7(b). It revealed that most of the absolute prediction error was less than 5 points and the biggest absolute prediction error was not more than 10 points. Furthermore, to evaluate the performance of our developed model more completely, another five evaluating indexes about model prediction error were calculated and illustrated in Table 3. Summarizing these evaluating indicators, it demonstrated that the ecoscore prediction model based on BP neural network was effective. Thus, it is practicable to precisely estimate the ecolevel of driver performance based on BP network models.

Besides its own validation test, we also compared the estimation results based on our developed model (i.e., backpropagation neural network) with the linear regression model and random forest model. The comparing analysis was displayed below.

(1) Linear Regression Model. According to our experimental data collected and the data distribution for model training (the data base was the same with BP neural network model training), the linear regression model was obtained by using the stepwise method. Table 4 displays the model form, estimated coefficients, the probability of t-test for regression coefficients, the probability of F-test, and goodness-of-fit for the regression equations. The accuracy rate from the training data is 89.18%.

Using the remaining 16 data samples, the prediction accuracy of this estimation model was tested. Figure 8 displays the measured values and the predictive values calculated by regression model. Taking the absolute value of the difference between the measured and predictive values divided by the measured value as a forecasting performance measure, this average percentage error was 12.81%.

(2) Random Forest Model. Using the same data base with the BP neural network model training, the prediction model for ecoscore of driver performance based on random forest was obtained. Figure 9 shows the relationship between number of grown trees and out-of bag (OOB) regression error. The optimal number of grown trees was selected as 60, by analyzing the relationship between prediction accuracy and the number of grown trees and also the relationship between OOB error and the number of grown trees. After repeated tests, the elapsed time of this random forest based model is 0.521seconds. The accuracy rate from the training data is 94.56%.

Similarly, using the remaining 16 data samples, the prediction accuracy of this estimation model was also tested. Figure 10 displays the measured values and the predictive values calculated by random forest model. Taking the absolute value

TABLE 3: Statistical results of prediction error for ecoscore.

Mean absolute error (MAE)	Mean percentage error (MPE)	Mean absolute percentage error (MAPE)	Mean square error (MSE)	Standard deviation of error (SDE)
4.61	-4.47%	7.02%	35.59	5.97

TABLE 4: Linear regression model to predict ecoscore of individual driver performance.

Models	F-test	Goodness-of-fit (R^2)
*Eco_score=124.716 *** -0.188A_power *** -22.233G_number *** +13.406A_hard ** -5.122G_mode ** -19.636A_number **	$F_{(5,44)}$ =52.572 $p<0.001$	0.84

Significance codes: 0.001 '***'; 0.01 '**'; 0.05 '*'.

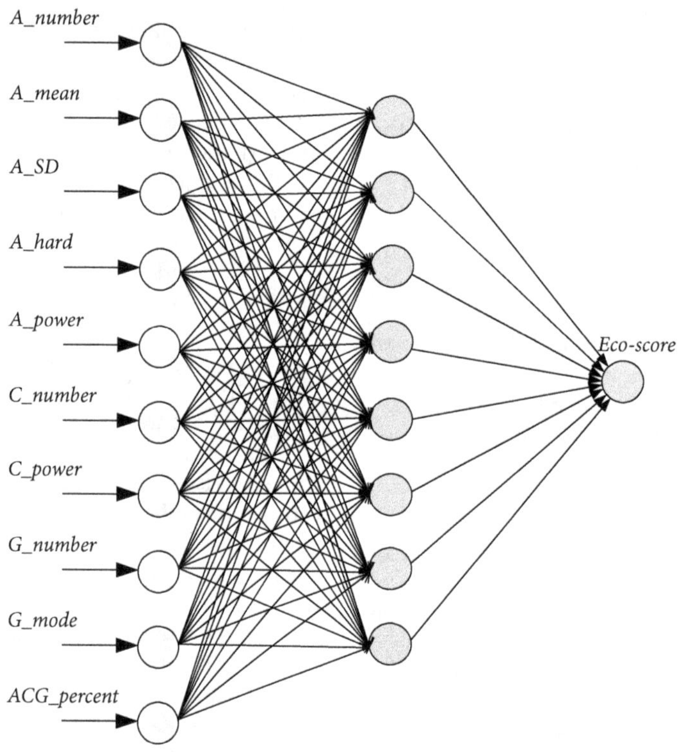

FIGURE 6: BP neural network structure for ecoscore estimation of driver performance.

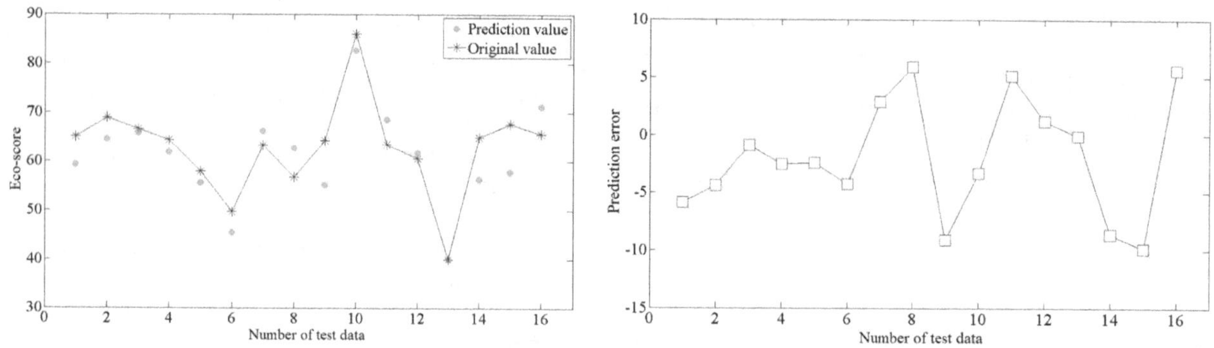

(a) Relationship between prediction and original value for each test data

(b) Prediction error for each test data

FIGURE 7: Accuracy test results of ecolevel estimation model.

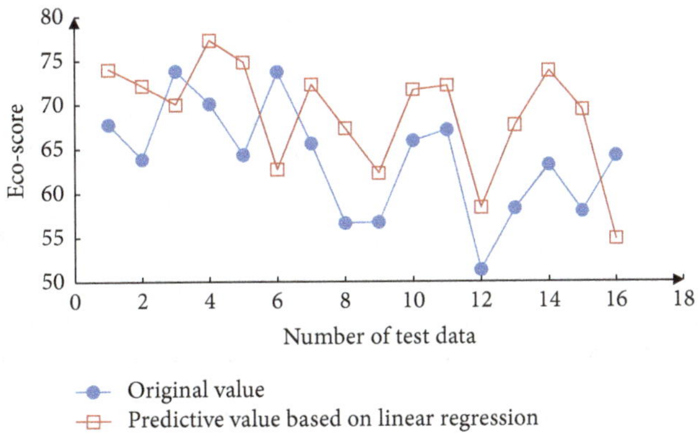

FIGURE 8: Relationship between original values and predictive values based on linear regression model.

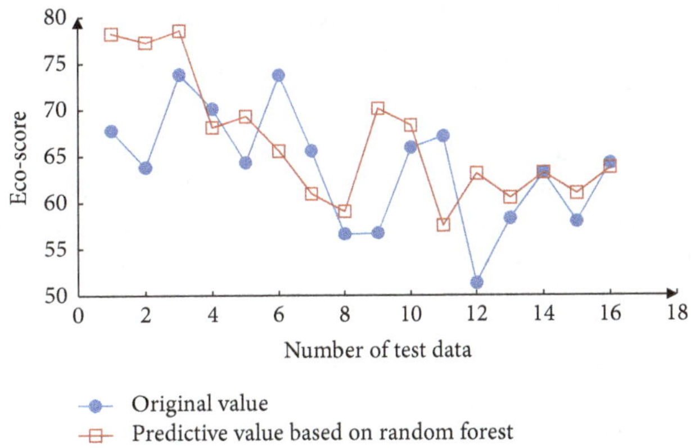

FIGURE 10: Relationship between original and predictive values based on random forest model.

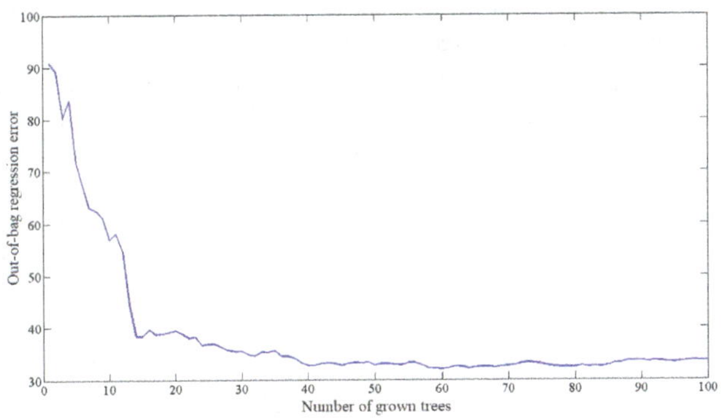

FIGURE 9: Relationship between number of grown trees and out-of-bag regression error.

of the difference between the measured and predictive values divided by the measured value as a forecasting performance measure, this average percentage of prediction error was 9.37%.

Summarizing above, the backpropagation neural network based model is better to estimate the ecolevel of driver performance, after comparing the elapsed time (except for linear regression model) and prediction accuracy of these models developed above. Thus, our study results indicated that BP neural network model in estimating drivers' ecolevel of driver performance was applicable and the model performance was good. The model structure, functions, and parameters used in our developed model were according to the repeated trial tests based on experimental data and thus increased the model reliability markedly.

The current developed model was focused on the ecolevel estimation of driver performance based on driving simulator experimental data. On one hand, the study results support simulator-based ecodriving training, such as dynamic ecodriving evaluation and real-time feedback system development. On the other hand, although this driving simulator has been approved to be effective in studying driving behaviors in many of our previous researches [21, 22], the current study results should be further validated through realistic driving data in the field. In fact, we have planned to establish the ecolevel estimation model based on vehicle operating status data on-road. With the rapid development of vehicle sensor data collection technology such as On-Board Diagnostics (OBD) + Global Position System (GPS), it is desirable and practicable to estimate the real-time ecolevel of driver behaviors in naturalistic driving conditions.

Besides, although the prediction accuracy of our developed model seems to be acceptable and improved more than 5% when comparing to linear regression model, only shallow learning model was used in this study. The deep learning model with stronger learning ability is expected to be more effective in driving behavior prediction and estimation. Meanwhile, we failed to introduce self-adaptive adjustment algorithm for learning rate selection in model training process. This more improved method for learning rate selection would be integrated into deep BP neural network model in future studies. In addition, our established model should be further compared with other more machine learning approaches (e.g., Support Vector Machine). Furthermore, a challenge that always existed for machine learning is the black-box effect of the model itself. Although the ecolevel of driver performance could be estimated accurately by BP neural network model, the difference of driver performance between drivers with perfect or poor ecoscore was still unknown. Much deeper researches to explore the hidden relationship between input parameters and output results is needed.

As limited to the high resource demands of driving simulator experiment, only one hundred and sixteen data segments were used for model construction and test (i.e., one hundred samples for model training and sixteen for model accuracy test). Meanwhile, only vehicle starting process was used to estimate the ecolevel of driver performance because starting a vehicle is more typical to reflect drivers' ecodriving level. In future studies, more driving events and bigger sample size are needed to further enhance the stability and applicability of our developed model.

5. Conclusions

Aiming at accurately estimating the ecolevel of individual driver performance, a BP neural network model with three layers was developed in our current study. Based on repeated experimental tests with the driver performance and vehicle fuel consumption data collected in a driving simulator, the optimal structure, functions, and learning rate of BP neural network models in the process of vehicle starting was obtained. The model accuracy was tested to be satisfactory. The study results provide an effective method to estimate the ecolevel of individual driver performance and further contribute to driving behavior optimization towards vehicle fuel consumption and emissions reduction.

Specifically, ten feature indexes of driver performance (e.g., percentage number, mean value, standard deviation, and power of applying acceleration pedal) were selected as model input parameters. The output is designed as the ecoscore to reflect individual's ecodriving level. This model has one hidden layer with eight neuron nodes. The transfer function from input to hidden layer and from hidden to output layer is "tansig" and "purelin," respectively. The suitable training function is "trainlm" and the optimal learning rate is 0.04. Taking 100 segments' data in starting process as training samples and 16 samples as model accuracy test, the mean prediction accuracy of our developed model is 92.89%. The performance of backpropagation neural network based model was better than linear regression based model (i.e., 87.19%) and random forest based model (i.e., 90.63%), from the aspects of elapsed time and prediction accuracy in estimating the ecolevel of driver performance.

Conflicts of Interest

The authors declare that there are no conflicts of interest regarding the publication of this paper.

Acknowledgments

The authors would like to show great appreciation for China Postdoctoral Science Foundation (no. 2017M620562) and the support from the National Natural Science Foundation of China (no. 61672067).

References

[1] Y. Bai, *Research on driving energy-saving technology. [Master, thesis]*, Changan University, Changan, 2011.

[2] R-Y. Lin, *The international situation for promoting eco-driving*, Industrial Technology Research Institute, Xinzhu, Taiwan, 2011.

[3] R. Luther and P. Basss, "Eco-driving scoping study," Tech. Rep., AA Research Foundation, Wellington: US, 2011.

[4] E. Martin, N. Chan, and S. Shaheen, "How public education on ecodriving can reduce both fuel use and greenhouse gas emissions," *Transportation Research Record*, no. 2287, pp. 163–173, 2012.

[5] C. Andrieu and G. S. Pierre, "Comparing Effects of Eco-driving Training and Simple Advices on Driving Behavior," *Procedia - Social and Behavioral Sciences*, vol. 54, pp. 211–220, 2012.

[6] X. Zhao, Y. Wu, J. Rong, and Y. Zhang, "Development of a driving simulator based eco-driving support system," *Transportation Research Part C: Emerging Technologies*, vol. 58, pp. 631–641, 2015.

[7] C. Chen, *Research on evaluation method of drivers eco-driving behavior on urban road. [Master, thesis]*, Beijing University of Technology, Beijing, China, 2016.

[8] C. Ma, *Study on the evaluation method of driving behavior habits based on OBD technology. [Master, thesis]*, Nanjing University, Nanjing, 2016.

[9] C. Miyajima, Y. Nishiwaki, K. Ozawa et al., "Driver modeling based on driving behavior and its evaluation in driver identification," *Proceedings of the IEEE*, vol. 95, no. 2, pp. 427–437, 2007.

[10] J-Y. Zhang and G-Y. Pan, "Comparison and application of multiple regression and BP neural network prediction model," *Journal of Kunming University of Science and Technology (Natural Science Edition)*, vol. 38, no. 6, pp. 61–67, 2013.

[11] X. Y. Wang and X. Y. Yang, "Study on decision mechanism of driving behavior based on decision tree," *Journal of System Simulation*, vol. 20, no. 2, pp. 415–420, 2008.

[12] S. S. Seyhan, F. N. Alpaslan, and M. Yavaş, "Simple and complex behavior learning using behavior hidden Markov model and CobART," *Neurocomputing*, vol. 103, pp. 121–131, 2013.

[13] Z-H. Yao, Y-S. Jiang, P. Han, X-L. Luo, and T. Xu, "Traffic Flow Prediction Model Based on Neural Network in Small Time Granularity," *Journal of Transportation Systems Engineering and Information Technology*, vol. 17, no. 1, pp. 67–73, 2017.

[14] C. Xu, S.-J. Pei, and X.-S. Wang, "Driver drowsiness detection based on non-intrusive metrics considering individual difference," *Zhongguo Gonglu Xuebao/China Journal of Highway and Transport*, vol. 29, no. 10, pp. 118–125, 2016.

[15] Z-L. Li, J-L. Han, X-H. Zhao, M-H. Zhu, and W-H. Dong, "Comparison of Drunk Driving Recognizing Methods Based on KNN and SVM," *Journal of Transportation Systems Engineering and Information Technology*, vol. 15, no. 5, pp. 246–251, 2015.

[16] G. Li and W.-Y. Chung, "Detection of driver drowsiness using wavelet analysis of heart rate variability and a support vector machine classifier," *Sensors*, vol. 13, no. 12, pp. 16494–16511, 2013.

[17] K. Aghabayk, M. Sarvi, N. Forouzideh, and W. Young, "Modelling heavy vehicle car-following behaviour in congested traffic conditions," *Journal of Advanced Transportation*, vol. 48, no. 8, pp. 1017–1029, 2014.

[18] M. Badhrudeen, J. Raj, and L. D. Vanajakshi, "Short-term prediction of traffic parameters - Performance comparison of a data-driven and less-data-required approaches," *Journal of Advanced Transportation*, vol. 50, no. 4, pp. 647–666, 2016.

[19] W. Wang, W. Zhang, H. Guo, H. Bubb, and K. Ikeuchi, "A safety-based approaching behavioural model with various driving characteristics," *Transportation Research Part C: Emerging Technologies*, vol. 19, no. 6, pp. 1202–1214, 2011.

[20] D. Kang, S-F. Ma, and S-Q. Zhong, "A microcosmic forecasting model for traffic safety based on BP neural network," *Journal of Transport Information*, vol. 29, no. 3, pp. 79–83, 2011.

[21] H. Ding, X. Zhao, J. Rong, and J. Ma, "Experimental research on the effectiveness of speed reduction markings based on driving simulation: a case study," *Accident Analysis & Prevention*, vol. 60, pp. 211–218, 2013.

[22] X. Zhao, Y. Wu, J. Rong, and J. Ma, "The effect of chevron alignment signs on driver performance on horizontal curves with different roadway geometries," *Accident Analysis & Prevention*, vol. 75, pp. 226–235, 2015.

[23] L. Xu, *Comparative analysis and optimization of car-following models for emission estimation. [Master, thesis]*, Beijing Jiaotong University, Beijing, 2012.

[24] X-M. Zhang, Y-S. Ge, Y. Zhang, X-K. Han, and L-P. Gao, "Research on fuel consumption measurement using carbon balance method," *Vehicle Engine*, vol. 3, pp. 56–58, 2005.

[25] Y.-P. Wu, X.-H. Zhao, J. Rong, Y. Liu, and L. Xu, "Potential of eco-driving in reducing fuel consumption and emissions based on a driving simulator," *Beijing Gongye Daxue Xuebao/Journal of Beijing University of Technology*, vol. 41, no. 8, pp. 1212–1218, 2015.

[26] S-H. Dong and Z-S. Shi, "Study on road traffic accident forecast based on BP neural network," *China Safety Science Journal*, vol. 20, no. 9, pp. 15–20, 2010.

[27] J. Li, X. Cheng, Y. Qin, Y. Zhang, and Z. Xing, "Reliability prediction of urban rail transit vehicle based on BP neural network," *Journal of Central South University (Science and Technology)*, vol. 44, no. 1, pp. 42–46, 2013.

[28] B. Liu and H-X. Guo, *MATLAB Neural Network Super Learning Manual*, Posts & Telecom Press, Beijing, China, 2014, MATLAB neural network super learning manual. Posts Telecom Press.

9

Fuel-Efficient Driving Strategies for Heavy-Duty Vehicles: A Platooning Approach based on Speed Profile Optimization

Sina Torabi and Mattias Wahde

Department of Mechanics and Maritime Sciences, Chalmers University of Technology, 412 96 Göteborg, Sweden

Correspondence should be addressed to Sina Torabi; sina.torabi@chalmers.se

Academic Editor: Dongjoo Park

A method for reducing the fuel consumption of a platoon of heavy-duty vehicles (HDVs) is described and evaluated in simulations for homogeneous and heterogeneous platoons. The method, which is based on speed profile optimization and is referred to as P-SPO, was applied to a set of road profiles of 10 km length, resulting in fuel reduction of 15.8% for a homogeneous platoon and between 16.8% and 17.4% for heterogeneous platoons of different mass configurations, relative to the combination of standard cruise control (for the lead vehicle) and adaptive cruise control (for the follower vehicle). In a direct comparison with MPC-based approaches, it was found that P-SPO outperforms the fuel savings of such methods by around 3 percentage points for the entire platoon, in similar settings. In P-SPO, unlike most common platooning approaches, each vehicle within the platoon receives its own optimized speed profile, thus eliminating the intervehicle distance control problem. Moreover, the P-SPO approach requires only a simple vehicle controller, rather than the two-layer control architecture used in MPC-based approaches.

1. Introduction

Heavy-duty vehicles (HDVs) account for about a quarter of road transport emissions and around 5% of total European CO_2 emissions [1]. Moreover, the demand for transportation of goods is expected to increase in the coming years. In order to deal with increasingly strict limitations on greenhouse gas emissions in the European Union (as well as all around the world), and also to cope with the likely increase in fossil fuel prices, HDV manufacturers and freight transportation sectors are under pressure to reduce fuel consumption. Heavy-duty vehicles driving at small intervehicle distances, a configuration known as *platooning*, can exhibit reduced fuel consumption through reduced air drag. For this reason, the research on platooning has gained interest from both industry and academia recently. Moreover, apart from the economical and environmental benefits of platooning, there is also the potential of reducing traffic congestion by better utilization of current transportation infrastructures.

In early attempts for controlling a vehicle platoon, the combination of cruise control (CC) and adaptive cruise control (ACC) was used. In this approach, the lead vehicle maintains a constant speed, using its cruise control, while the following vehicles in the platoon maintain a constant distance to the vehicle in front, using their adaptive cruise control [2–5]. Considering that the aerodynamic drag is responsible for as much as a quarter of the total fuel consumption, on flat roads the CC+ACC approach for platooning leads to fuel-optimal performance since it allows the vehicles to drive with short intervehicle distances. However, on a typical road, with *varying* topography, this approach could lead to excessive acceleration and braking during uphill and downhill segments of a road in order to maintain the set speed, or the intervehicle distance. Such accelerations increase the vehicle's fuel consumption significantly [6, 7] making the approach unsuitable for nonflat roads. Therefore, it is necessary to consider the impact of road topography in order to fully exploit the fuel saving potential of platooning. This can be achieved by allowing the speed of the vehicles to vary when driving on typical roads by generating fuel-efficient speed profiles, i.e., profiles representing the vehicle's speed as a function of its position along the road.

Platooning generally concerns optimization of the longitudinal dynamics of all the vehicles involved, in order to maximize fuel efficiency while guaranteeing safety. Most platooning work to date has been focused on generating a fuel-optimal platooning algorithm for the follower vehicles while the lead vehicle follows a fuel-efficient speed reference using, for instance, look-ahead control (LAC) [8–10]. These platooning controllers are, mostly, designed based on the model predictive control (MPC) framework that uses dynamic programming to compute an optimal speed reference over a short horizon (2-4 km) for the lead vehicle to follow, while the other vehicles follow the lead vehicle using a spacing policy, of which there are different kinds [11–13]. Since the speed reference is updated frequently (every few seconds [8, 10, 14, 15]), these methods require computationally heavy calculations to be performed iteratively during driving.

In recent MPC-based approaches for platooning in which the impact of road topography on fuel efficiency has been considered [9, 10], all vehicles follow the same speed profile, which is generated taking into account the characteristics and limitations of all vehicles. In [9], on a synthetic road of 4 km length (with one uphill segment followed by a downhill segment), the fuel consumption of the platoon was reduced by 14% on the downhill segment of the road and 0.7% during the uphill segment, relative to the combination of CC and ACC. In [10], fuel savings of around 13% (depending on the platoon configuration) were obtained, relative to each vehicle's fuel consumption when using CC, driving over a 45 km section of a typical Swedish highway.

An alternative approach, however, is to generate an optimal lead vehicle speed profile for a longer horizon. In our previous work [16], Caltagirone et al. showed that speed profile optimization (SPO) with respect to fuel consumption by using a simple stochastic optimization algorithm can reduce the fuel consumption by 15% (on average) for the entire platoon. The main advantage of using SPO, compared to MPC-based methods, is that SPO does not require iterative online computations: once the speed profiles have been generated, over the entire horizon, they can be used without any further calculations.

Much of our earlier work was, in fact, focused on the fuel consumption of a *single* vehicle. Using simulations, in [16], it was found that, with SPO, the lead vehicle achieved higher fuel saving than with MPC-based methods. The potential fuel saving of speed profile optimization, for a single truck, was further investigated in our paper [17], which involved both simulations and thorough testing with real vehicles. In that paper, a more direct comparison was made between SPO and MPC-based methods. In cases where larger speed variations were allowed, in real vehicles (and in simulations) the proposed SPO method obtained 11% savings, compared to typical savings of 3 to 7% (depending on the road) obtained with MPC-based methods. In addition, in the more difficult case of a narrower speed range, the proposed SPO method reduced fuel consumption by 6%, compared to the 3% achieved using an MPC-based controller, in the exact same road and with the same allowed speed range. Moreover, in the same paper, it was shown that, despite the simplicity of the implemented longitudinal model of the HDV, the fuel savings results obtained with *real* HDVs were similar to those obtained in the simulations.

This paper, however, is mainly focused on HDV *platooning* based on SPO, and will introduce a method referred to as Platooning SPO (P-SPO). As in [9, 10, 18], both homogeneous platoons and heterogeneous platoons will be considered. Here, a homogeneous platoon is defined as a platoon consisting of identical vehicles. In a heterogeneous platoon, the vehicles may differ with respect to their masses, engine types, etc. even though, here, only differences in mass will be considered. For heterogeneous platoons, in the approach proposed here (and contrary to the common leader-follower approaches in, for example, [9, 10]), an optimized speed profile is assigned to each vehicle. Each vehicle then follows its profile *independently* of the other vehicles. These profiles are optimized together, using a *genetic algorithm* [19] and considering primarily safety and then fuel efficiency, of the entire platoon. Moreover, in this work, it is assumed that, during the simulations, the platoon's motion is not affected by other traffic, an assumption that is common in studies involving platooning *methods* (see, e.g., [9, 10, 20]).

2. Road and Speed Profile Representation

In this paper, as in [17], both the road profiles and the speed profiles are represented by sequences of Bézier splines, which are referred to as composite Bézier curves; see, e.g., [21]. Here, two-dimensional cubic Bézier splines of the following general form are used to model the profiles:

$$\begin{aligned}\mathbf{x}(u) &\equiv (x(u), y(u))^T \\ &= \mathbf{P}_0(1-u)^3 + 3\mathbf{P}_1 u(1-u)^2 + 3\mathbf{P}_2 u^2(1-u) \\ &\quad + \mathbf{P}_3 u^3,\end{aligned} \tag{1}$$

where the vectors $\mathbf{P}_j$ are two-dimensional control points and u is a variable ranging from 0 to 1. In order to form a composite Bézier curve, a sequence of cubic Bézier splines are used in the following form:

$$\mathbf{x} \equiv (\mathbf{x}_i(u)), \quad i = 0, \ldots, n-1 \tag{2}$$

where $\mathbf{x}_i(u)$ is defined using (1) and n is the total number of splines in the composite curve. The number of splines for fitting a composite Bézier curve to a data set can be selected in various ways. For instance, it can be selected such that the composite curve goes through every data point in the data set. However, this approach is often not suitable due to the presence of noise in the collected data set. Moreover, reducing the number of splines also reduces the complexity of the problem as well as the size of the search space during optimization. Thus, in practice, the number of splines is typically much smaller than the number of data points; see Section 2.1 below.

2.1. Road Profiles. Since, in platooning, only the longitudinal dynamics of the vehicles is typically considered, the road profiles can be represented using the composite Bézier curves described in (2), where the two dimensions represent the longitudinal position along the road and the elevation, respectively. In other words, the road profiles can be modeled in the following form:

$$(s, z) \equiv (s_i(u), z_i(u)), \quad i = 0, \ldots, n-1 \tag{3}$$

where s is the longitudinal position and z is the elevation. With this representation, the road elevation can be written as $z = z(s)$ since, for any position s along the road, the corresponding spline index i and u–value can be obtained. Here, the number of splines used for representing the data set ranges from 18 to 22 for each road profile of 10 km. The accuracy of the fitted splines is discussed in Section 2.3 below.

2.2. Speed Profiles. Similar to the road profiles, the speed profiles were modeled using two-dimensional cubic composite Bézier curves in the following form:

$$(s, v) \equiv (s_i(u), v_i(u)), \quad i = 0, \ldots, n-1 \tag{4}$$

where v is the longitudinal speed of the HDV and s again is the longitudinal position along the road. As in the case of road profiles, one could find the corresponding spline index and u–value, for any given position s, so that the speed can be computed and can therefore be written as $v = v(s)$. Note that when the speed profiles are generated, same splines that were used in generating longitudinal position, $s_i(u)$, in generating road profiles, are used here.

2.3. Road Data. For the purpose of testing the performance of the platooning method (proposed below; see Section 3) in simulations, a set of road profiles was selected from a road between the two cities of Göteborg and Borås in Sweden. The road data were collected through the public site of Trafikverket (https://pmsv3.trafikverket.se) (Swedish Transport Administration). The collected data set contains, among other data, the elevation z, the covered distance along the road (from an absolute starting point), and the average slope (pitch) of the road (referred to as hilliness) for every 20 meters. The accuracy of the absolute elevation data is rather low (±4 meters). However, the accuracy of the slope, which is a more important measure for the cases considered here, is at 0.1%; that is, the error is at most 0.1 meters over 100 meters. For this paper, 10 road profiles of 10 km length were generated, 5 from the road from Göteborg to Borås and 5 in the opposite direction, by fitting composite Bézier curves (see Section 2.1) to each road section. The average absolute difference in slope *percentage*, between the fitted splines and the collected data set, is 0.026. An example of a fitted curve is shown in Figure 1 where the cubic Bézier splines are shown by gray lines separated by vertical line segments and the original data are shown by filled dots. As one can see from this figure, the cubic Bézier splines are of different lengths. Figure 2 shows the fitted spline for the 54.5 km road segment from which the road profiles were extracted. The altitude variation within each profile is at least 23.7 meters and at most 114.7 meters.

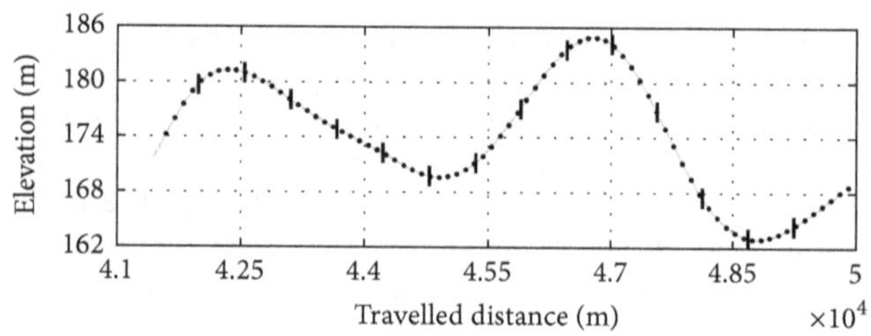

FIGURE 1: Comparison between the fitted Bézier splines and the original data, over a part of the road segment between Göteborg and Borås. The dots represent the original data and the gray curve represents the fitted splines. The vertical line segments separate splines.

3. Methods

In the approach used here, the fuel consumption minimization of the platoon relies on the generated optimized speed profiles. In this section, first, the HDV model will be briefly reviewed. Next, the baseline case against which the proposed platooning algorithm is compared will be introduced. Finally, the proposed approach for vehicle platooning based on speed profile optimization will be described.

3.1. HDV Model. The HDV model used here is similar to the standard models implemented in other platooning studies, e.g. [6, 8, 9, 14]. Here, as in [17], the parameters correspond to a 460 hp tractor-trailer combination. The HDV's mass varies in the range from 30 to 40 tonnes. The longitudinal dynamics of each vehicle is governed by

$$m(G)\dot{v} = F_{\mathrm{e}} - F_{\mathrm{b}} - F_{\mathrm{d}} - F_{\mathrm{r}} - F_{\mathrm{g}}, \tag{5}$$

where the terms on the right-hand side of the equation correspond to the forces experienced by the vehicle, namely, the engine force F_e, the braking force F_b, the air drag resistance force F_d, the rolling resistance force F_r, and the gravity force F_g. Furthermore, $m(G)$ is the total inertial mass of the HDV and it is computed as follows:

$$m(G) = m + \frac{J_w + \gamma_G^2 \gamma_f^2 \eta_G \eta_f J_e}{r_w^2}, \tag{6}$$

where G is the active gear, m is the mass of the vehicle, J_w and J_e represent the engine and wheel inertia, respectively, γ_G and γ_f are the gearbox and final-drive ratios, η_G and η_f denote the gearbox and final-drive ratio efficiencies, and r_w is the wheel radius. In order to compute the force F_e and also the fuel consumption, the inverse dynamics of (5) is considered, where the total external forces are combined with the requested acceleration from the vehicle's controller to calculate the required torque from the engine. The fuel consumption rate is determined by interpolation of the torque-RPM-fuel map given the required torque and the engine speed. The total fuel consumption is then calculated by integrating the fuel consumption rate over time. Note that, for any given value of the engine speed, a certain maximum torque can be derived from the map just mentioned. The

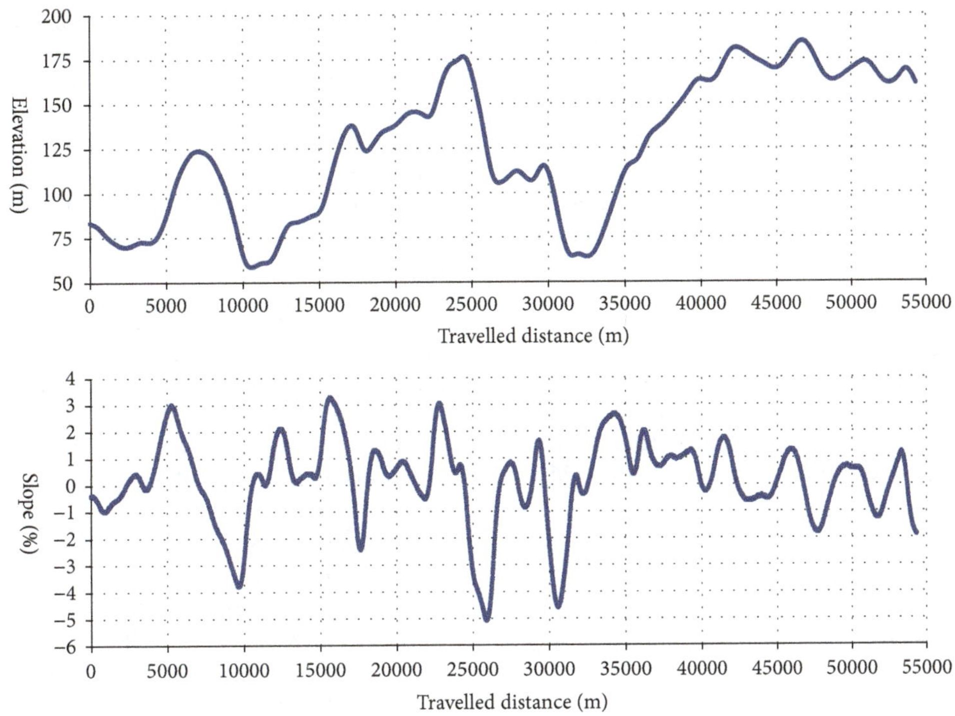

FIGURE 2: Top panel: the road between the two cities of Göteborg and Borås in Sweden. Bottom panel: the slope variation of the same road.

actual torque applied is limited by this value. As for the braking, a simple model has been used [6, 22], in which it is assumed that the truck can deliver any negative acceleration down to a given limit, here conservatively set to -2.5 m/s^2 [22].

The air drag resistance force, F_d, depends on the vehicle's speed. The rolling resistance, F_r, and the gravitational force, F_g are determined by the road slope. Unlike the rolling resistance force, the gravitational force can be either resistive (during uphill driving) or propulsive (during downhill driving). For a detailed description of the model and how these forces are calculated, see [16]. One should note that although the longitudinal model implemented here might seem somewhat simple, the fuel savings results are reproducible in real HDVs, as was shown in our earlier paper [17].

3.2. Air Drag Reduction. The air drag resistance acting on a single vehicle is expressed as

$$F_d = \frac{1}{2} c_D A \rho_a v^2 \tag{7}$$

where c_D is the air drag coefficient, A is the frontal area of the vehicle, ρ_a is the air density, and v is the vehicle's speed. The air drag resistance reduces when the vehicles driving in a platoon, which is described as

$$F_d = \frac{1}{2} \Phi(d) c_D A \rho_a v^2 \tag{8}$$

where $\Phi(d)$ is a coefficient (air drag ratio) that depends on the intervehicle distance. Here, $\Phi(d)$ is modeled as in [10, 20]

$$\Phi(d_i) = \left(1 - \frac{C_{D,1}}{C_{D,2} + d_i}\right) \tag{9}$$

where d_i is the ith vehicle's distance to its preceding vehicle (here, the general form of the equations (applicable to a platoon with any number of vehicles) will be given, even though, in the simulations presented below, the number of vehicles will be equal to 2, as described in Section 3.4).

$C_{D,1}$ and $C_{D,2}$ are constants, obtained by a least-squares fit to the experimental data from [23]. The experimental data and the fitted curve used here are shown in Figure 3 in [10]. As can be seen from that figure, the fit is not very good for small intervehicle distances (less than 10 m) but, nevertheless, it has been used here in order to allow a direct comparison with [10, 20].

In this model, the air drag reduction is considered only for the follower vehicle, since the effect on the preceding vehicle is small in comparison.

3.3. Baseline Case. In order to evaluate the P-SPO platooning approach (described below), a baseline case is included for comparison, namely, the combination of standard cruise control (CC) and standard adaptive cruise control (ACC). In this setup, the lead vehicle aims to maintain a constant speed (using CC) while the follower vehicles attempt to maintain a constant distance to the preceding vehicle (using ACC).

3.3.1. Standard Cruise Control. Here, the CC function is implemented using a simple PID controller with error signal $e(t) = v_s - v(t)$, where v_s is the reference speed (set as a constant for CC) and $v(t)$ is the HDV's instantaneous speed. The control output, $u(t)$, is expressed as follows:

$$u(t) = K_p e(t) + K_i \int_0^t e(\tau)\, d\tau + K_d \dot{e}(t) \tag{10}$$

where K_p, K_i, and K_d are hand-tuned proportional gain, integral gain, and derivative gain. The tuning of the PID controller was carried out in connection with the work for [17]. The gain parameters were set so that the simulated vehicle could follow the speed profile as similarly as possible with respect to the real HDV considered in that work. The requested vehicle acceleration, a_R, is then computed by dividing the control output, $u(t)$, by the total inertial mass, $m(G)$.

3.3.2. Standard Adaptive Cruise Control. For the baseline case, a standard ACC is used for controlling the intervehicle distance between the vehicles within a platoon. ACC allows a vehicle to maintain a desired distance to the preceding vehicle, while following the same speed trajectory. The requested acceleration is then given by the following expression [24]:

$$a_R = k_1 \left(d_{i,i-1} - d_0\right) + k_2 \left(v_{i-1} - v_i\right) \tag{11}$$

where $d_{i,i-1} = x_{i-1} - x_i$ is the intervehicle distance, where $x_i = x_i(t)$ represents the ith vehicle's longitudinal position, $v_i = v_i(t)$ is the ith vehicle's speed, k_1 and k_2 are hand-tuned gains, and d_0 is the desired intervehicle distance.

3.4. Platooning Based on SPO. Consider the problem of moving a platoon of N HDVs from a given starting point to a given end point, in a case where the road profile (here represented by composite Bézier curve) is known in advance. The motion of the vehicles of the platoon, then, can be formulated as desired speed profiles, $v_{d,i}(s)$ defined for the ith vehicle of the platoon, *a priori*, over the complete road profile through which the platoon is driving. The desired speed profiles provide a varying set speed for each vehicle in the platoon, which is used as input to its PID controller that attempts to reach the set speed; see Section 3.3.1. Thus, in this method, henceforth referred to as the *Platooning SPO* (P-SPO) method, the problem of minimizing the fuel consumption of a platoon of HDVs is reduced to finding the optimal speed profiles of each vehicle. One should note that unlike the other approaches mentioned in Section 1, in this approach [16, 17], each vehicle receives and follows its own optimized speed profile *independently* of the other vehicles. In other words, there is no need for using a spacing policy for the following vehicles in a platoon.

In this approach, the safety of the platoon is guaranteed by construction: in the speed profile generation procedure, the safe distances (see Section 3.4.2 below) between vehicles are considered during the evaluation of a speed profile. Therefore, only speed profiles that keep the vehicles within the safe distance are returned by the optimization method. It is common to assume that the vehicles of a platoon are not prevented (by other vehicles) from following their speed profiles, so that the the motion prescribed by P-SPO can be robustly executed, using the PID controller. As they follow their individual speed profiles, the distance between the two HDVs will vary, as shown in the bottom panels of Figures 3 and 4, but with a minimum distance that never drops below the minimum safe distance (see Section 3.4.2 below), However, in the event that one HDV would fail to follow its speed profile, for whatever reason, the ACC function would, in our method, be activated if the distance between the two HDVs drops below the minimum safe distance. Note that the speed profiles are functions of position along the road. Thus, if an HDV is unable to follow its profile for some time, it can later resume following that profile at its current position along the road. The impact of cut-ins by other vehicles (potentially disrupting the platoon's motion) is further discussed in Section 5.

In this work, as in many other works on platooning [9, 10, 25, 26] platoons of two vehicles have been considered, since this is generally sufficient to capture most of the benefits of platooning (except the variability in air drag characteristics between different consecutive vehicles in a platoon). However, the analysis presented here can of course be extended to platoons consisting of more than two vehicles.

3.4.1. Evaluation. Assuming that a set of speed profiles is available, the HDVs of a platoon are simulated following their speed profiles, so that their fuel consumptions can be measured. A dedicated simulation environment was written (in C#.NET), implementing the HDV model and the baseline case for platooning, as described in Sections 3.1 and 3.3, as well as the road and speed profiles described in Section 2 and the P-SPO method just described.

In order to evaluate a set of speed profiles for a platoon over a given road profile, the simulation proceeds as follows: At each time step, the current longitudinal position of the ith vehicle is used for calculating the desired speed from its speed profile. Since the speed profiles are defined in advance, the desired speed of the ith vehicle at any given position can be obtained easily, using the speed profile as a lookup table. The calculated desired speed is then fed to the ith vehicle's PID controller as its reference speed. Note that the reference speed typically varies over a profile. Thus, due to the controller's (nonzero) response time, the vehicle will not actually track the reference speed exactly, and neither is it required to do so in the method presented here. The procedure is repeated (for all the vehicles of the platoon) until the last vehicle passes the end point of the road profile.

3.4.2. Optimization Method. In this work, the P-SPO method is carried out in simulation using a fairly standard genetic algorithm (GA) [19, 27]. GAs, which are search and optimization techniques guided by the principles of Darwinian evolution, have been successfully applied to optimization problems in various fields, but have only recently been used in the case of vehicle platooning [16, 17]. The objective function, to be minimized, was taken simply as the total fuel consumption of all considered vehicles. The GA was implemented as an integral part of the simulation environment described above. SPO was tested for the single vehicle fuel minimization problem in [16, 17]. For an HDV *platoon*, P-SPO is used to generate a speed profile for each HDV separately. The speed profiles that are *generated together* must fulfill the following constraints: (i) the average speed of the entire platoon should not be below a certain threshold $\overline{v}_{\min}$; (ii) the instantaneous speed of all vehicles, $v_{i,\max}$ must be

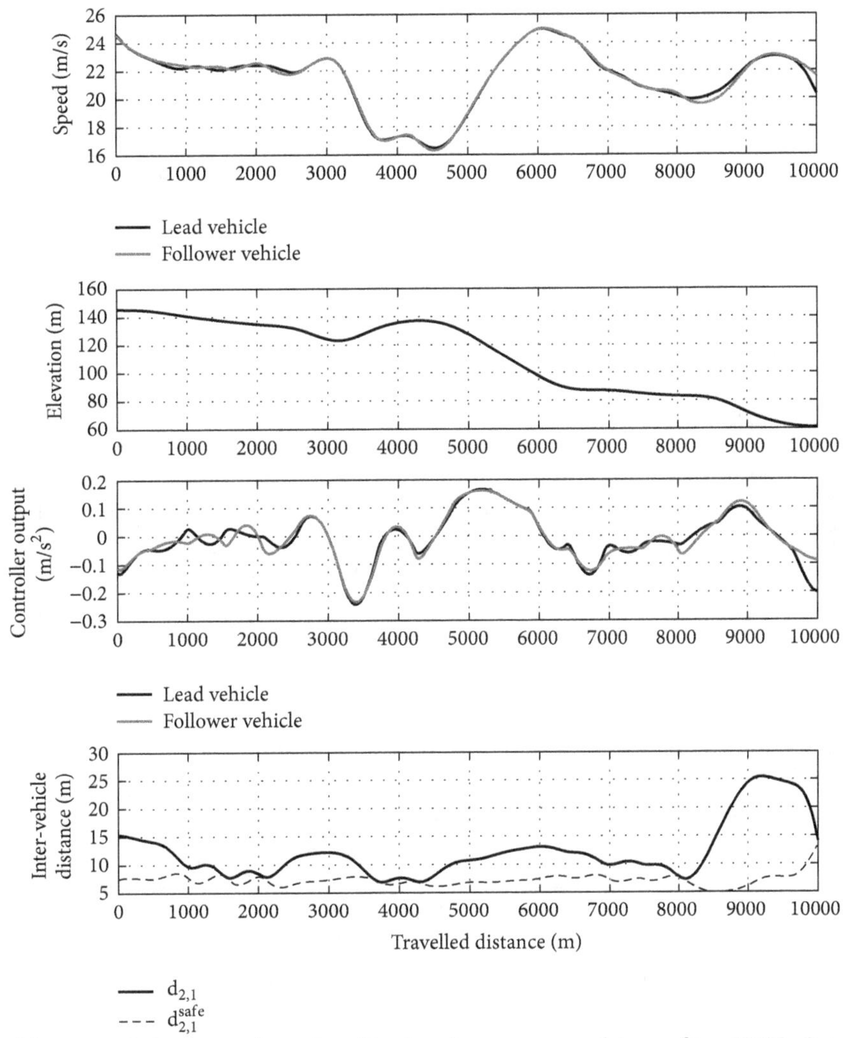

FIGURE 3: Top panel: one of the ten optimized sets of speed profiles for a homogeneous platoon of two HDVs. Second panel: the road profile. Third panel: the HDV's controller output u along the road, normalized by the total inertial mass $m(G)$. Bottom panel: the intervehicle distance $d_{i,i-1}$, (solid line) and the corresponding safe distance, $d_{i,i-1}^{\text{safe}}$, and (dashed line) computed using (12).

below an upper limit (the speed limit); (iii) the instantaneous minimum speed of all vehicles, $v_{i,\min}$, should be above a user-defined threshold. Moreover, in order to make sure that the platoon remains coherent, the intervehicle distance between consecutive vehicles must be below a certain threshold (40 m) at all times.

In order to ensure the safety of the platoon, the intervehicle distances must always remain above the safe distance that, at each time step, is calculated as follows [13]:

$$d_{i,i-1}^{\text{safe}}(t) = d_0 + h(t)\, v_i(t) \tag{12}$$

where $d_{i,i-1}^{\text{safe}}(t)$ is the safe distance between the ith and the $(i-1)^{\text{th}}$ vehicle in the platoon, d_0 is the minimum safe distance (5 m here), $v_i(t)$ is the ith vehicle's speed, and $h(t)$ is the variable time headway, expressed in the following form:

$$h(t) = h_0 - c_h v_r(t) \tag{13}$$

where $h_0 > 0$ (constant) is the minimum time headway, $c_h > 0$ is a constant, and $v_r(t) = v_{i-1}(t) - v_i(t)$ is the relative speed. For safety reasons, the variable time headway $h(t)$ is not allowed to become negative, while very large headways are undesirable as they increase the intervehicle distances. Thus, the variable time headway value is limited to the interval [0, 1] (s). For this work, the value of h_0 and c_h are set to 0.1 and 0.2, respectively, as was proposed in [13]. During the evaluation of a set of speed profiles, if any instantaneous intervehicle distance (between consecutive vehicles) goes below the safe distance defined in (12), the speed profile set is discarded immediately.

The optimization algorithm keeps a population of M individuals (here usually 50-100), in this case each defining N speed profiles (one for each vehicle in the platoon). The individuals encode the optimizable parameters (i.e., the decision variables) in chromosomes consisting of floating-point numbers such that each number corresponds to the second (v) component (here denoted as $P_{i,j}$, $j = 0,\ldots,3$) of the jth two-dimensional control point of the ith spline in

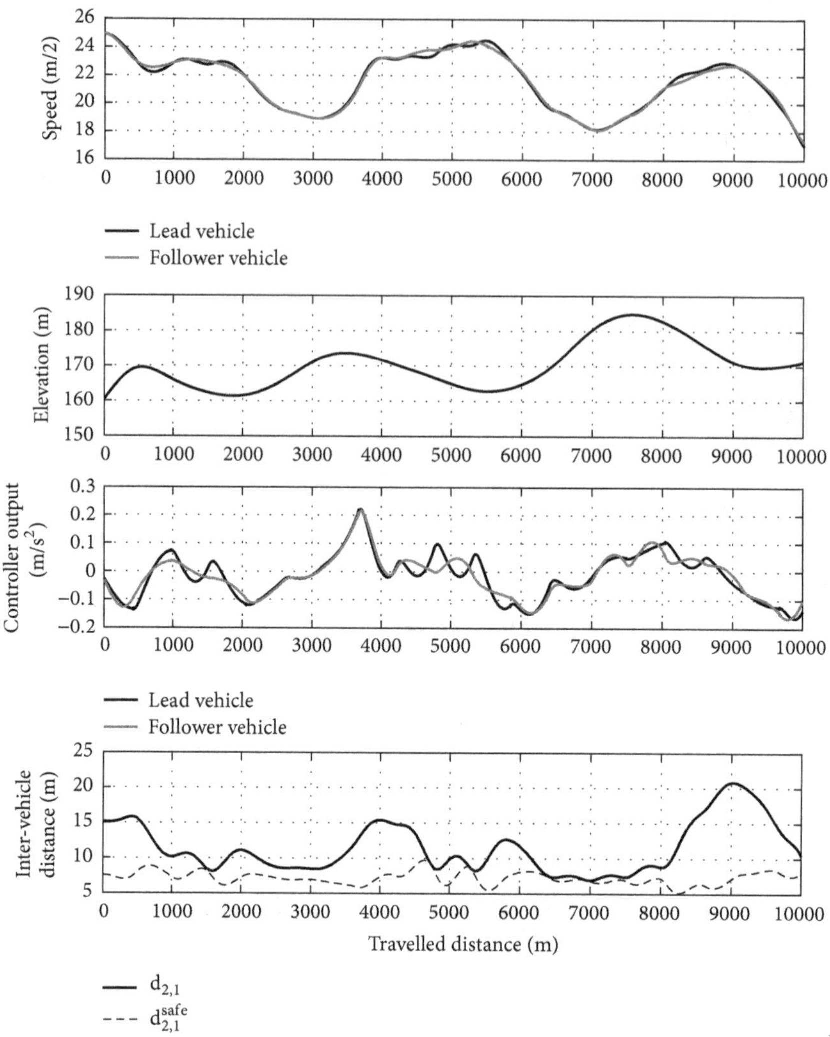

FIGURE 4: Top panel: one of the ten optimized sets of speed profiles for a heterogeneous platoon of two HDVs. Second panel: the road profile. Third panel: the HDV's controller output u along the road, normalized by the total inertial mass $m(G)$. Bottom panel: the intervehicle distance $d_{i,i-1}$, (solid line) and the corresponding safe distance, $d^{\text{safe}}_{i,i-1}$, and (dashed line) computed using (12).

a set of speed profiles (see Section 2.2). There are, however, two additional requirements that should be considered when encoding and decoding a set of speed profiles. Since a speed profile is a composite Bézier curve consisting of several splines, positional (C0) and derivative (C1) continuity of the speed profile must be taken into account during encoding and decoding. C0 and C1 continuity ensure that, after each iteration of the optimization, the decoded chromosome forms a set of speed profiles in which *each* speed profile is continuous and smooth over the entire road profile. In order to make sure that these constraints are met, the following conditions must be hold:

$$P_{i,3} = P_{i+1,0} \quad \text{for } i = 0, \ldots, n-1 \tag{14}$$

and

$$P_{i,3} - P_{i,2} = P_{i+1,1} - P_{i+1,0} \quad \text{for } i = 0, \ldots, n-1 \tag{15}$$

Considering these requirements, the number of parameters (i.e. the length of the chromosome) will be $L = N(4 + 2(n - 1)) = N(2n + 2)$ where n is the number of splines. Note that the speed profile optimization method only tweaks (during mutation) the second component of the control points, thus leaving s unchanged.

For a given individual, the formed set of speed profiles from the decoding procedure is then evaluated as described in Section 3.4.1. The evaluation results in a single number (fitness value), namely, the inverse of the cumulative fuel consumption of the entire platoon at the point where the last vehicle of the platoon reaches the end of the road profile. However, if the intervehicle distance falls below the safe distance at any point, the fitness value is set to 0. Next, if any of the constraints described above are violated, the fitness value is multiplied by a penalty term smaller than 1. For instance, the penalty term for a case in which the instantaneous intervehicle distance exceeds its maximum allowed value is calculated as follows:

$$p_{d_{i,i-1}}(d) = e^{-c_d(d_{i,i-1}/d_{\max}-1)^2}, \tag{16}$$

where $d_{i,i-1}$ is the maximum intervehicle distance between two consecutive vehicles of the platoon during the evaluation,

c_d is the penalty constant, and $d_{\max}$ is the maximum allowed intervehicle distance.

Standard tournament selection is used for selection, during the optimization, with a given tournament size S_t (typically 2–5) and a tournament selection probability of p_t (around 0.7 to 0.8). Single-point crossover is applied to each pair of selected individuals with a probability of p_c (around 0.7 to 0.9). Once two new individuals have been generated, mutation is applied with probability of p_m (with relative probability of $1/L$ where L is the chromosome length), either full-range (with probability of 0.5) or real-number creep mutations (also with probability of 0.5). Finally, elitism is used to preserve the current best individual and pass it to the next generation.

4. Simulation Results

The method described in Section 3.4 was applied, in simulation, to each of the 10 km long road profiles described in Section 2 both with a homogeneous and a heterogeneous platoon. For the homogeneous platoon, all the vehicles had the same mass, set to 30 tonnes here. For the heterogeneous platoon, two general mass configurations were considered, namely, (i) lighter HDV in front and heavier HDV in the back and (ii) heavier HDV in front and lighter HDV in the back. First, however, the platoons' fuel consumption (both homogeneous and heterogeneous) were measured when driving with the baseline case, i.e. the combination of CC+ACC, over all 10 road profiles. For the baseline case, the standard CC speed reference was set to 80 km/h and the desired distance for ACC was set to the minimum allowed value computed using (12) at each time step. A more difficult case (for comparison) was also considered, which was proposed and tested in our earlier work [16], where the lead vehicle of the platoon drives according to its optimized speed profile while the follower vehicle uses ACC in an attempt to follow the same speed trajectory (referred to as SPO+ACC). In this case, as in [16], the desired distance was set to the minimum allowed safe distance; see (12) and (13).

4.1. Optimization Setup. Both in homogeneous and heterogeneous platoons, at the early stages of the optimization, the P-SPO method is used to optimize a single speed profile for the entire platoon. This procedure is used to speed up the optimization progress, by avoiding the possibility of having to discard profiles that result in collisions between HDVs. Next (after a certain number of generations, typically a few hundred), the P-SPO method is used to optimize each vehicle's speed profile, rather than optimizing a single profile for the entire platoon, discarding any set of profiles that result in collisions. Therefore, when the optimization is finished, each vehicle will have its own speed profile to follow that (slightly) differs from the other vehicles' speed profiles. With this approach, the fuel saving potential of driving at close intervehicle distances is fully exploited without compromising the platoon's safety.

During the optimization, the constraints on the average speed, the minimum instantaneous speed, and the maximum allowed speed were set to 80 km/h, 60 km/h, and 90 km/h, respectively. Moreover, the minimum allowed intervehicle distance, at each time step, was calculated based on (12) and the maximum allowed intervehicle distance was set to 40 meters. The initial spacing of the vehicles as they entered a road profile was 15 meters.

TABLE 1: The normalized fuel savings obtained in the simulations for a homogeneous platoon of two 30 tonne HDVs over all the road profiles. The middle column shows the fuel savings obtained with SPO+ACC, whereas the right-most column shows the fuel savings obtained with P-SPO. The normalization was carried out with respect to the baseline case, i.e., CC+ACC where the lead vehicle drives at the constant speed of 80 km/h and the intervehicle distance is set to the allowed minimum.

	Saving (%)	
Road profile	SPO + ACC	P-SPO
Profile 1	11.93	12.41
Profile 2	12.38	12.26
Profile 3	25.70	25.83
Profile 4	8.78	8.81
Profile 5	18.21	18.29
Profile 6	11.76	12.18
Profile 7	16.57	16.91
Profile 8	8.04	8.03
Profile 9	25.05	25.69
Profile 10	17.52	17.54
Average saving (%):	15.59	15.79

4.2. Homogeneous Platoon. The results obtained from the simulations for a homogeneous platoon of 30 tonne HDVs are given in Table 1. In this table, the middle column shows the fuel savings, relative to CC+ ACC, obtained when the lead vehicle follows an optimized speed profile generated using the SPO method, while the follower vehicle tracks the same speed profile using ACC with the desired intervehicle distance set to the minimum allowed safe distance (at each time step). The right-most column of the table shows the fuel savings obtained when each vehicle follows its own optimized speed profile. As can be seen from the table, using SPO leads to large fuel savings in comparison to the case of CC+ACC. However, for a homogeneous platoon, the additional savings obtained when the speed profiles are allowed to differ between vehicles are rather minor. In both cases, the fuel consumption of the entire platoon was reduced by around 16% on average, compared to CC+ACC. An example of the optimized set of speed profiles optimized for road profile 9 when using the P-SPO method is shown in Figure 3, along with the consecutive vehicles intervehicle distances. As can be seen from the figure and the obtained fuel savings from Table 1, by using the P-SPO method that does not force the intervehicle distance to remain constant, the fuel saving can be further improved. In other words, individual speed profile optimization can improve the fuel consumption even though in most cases the effect is very small. The reason is that, by individual optimization, the vehicles are not forced to accelerate or brake, during uphill or downhill driving, to maintain their distance to the preceding vehicle.

However, one should note that, for the SPO+ACC case, the intervehicle distance (set to its minimum allowed value, as mentioned above) was much smaller than in the case of P-SPO. Driving all the time at the small intervehicle distances required (in this comparison) by SPO+ACC is undesirable for the drivers, though, since errors in reading distance and relative speed may lead to dangerous situations; see also the second paragraph in Section 5.

4.3. Heterogeneous Platoons. Similar to the case of homogeneous platoon, for a heterogeneous platoon, the optimization starts by assigning identical speed profiles to all vehicles for a number of generations (in this case, 1000), as mentioned in Section 4.1. After that, the optimizer is allowed to assign different speed profiles for each vehicle, which are optimized together. This procedure will generally result in different speed profiles for each vehicle of a platoon once the optimization has been completed. During the optimization, the constraints on the average speed, the minimum instantaneous speed, the maximum instantaneous speed, and the maximum allowed intervehicle distance were set to the same value as in the homogeneous platoon described above. Moreover, the safe distance was calculated using (12).

In this paper, different heterogeneous platoons configurations of two general forms were considered. In the first configuration, the lead vehicle is lighter (referred to as Configuration I), while in the second configuration (Configuration II), the heavier vehicle is in the lead position. The masses range from 30 to 40 tonnes, and the following configurations have been taken into account: (i) 30 tonnes (lead vehicle) 35 tonnes (follower vehicle), (Configuration I-a); 30 tonnes 40 tonnes (Configuration I-b); 35 tonnes 30 tonnes (Configuration II-a); and 40 tonnes 30 tonnes (Configuration II-b).

The results obtained from the simulations for different mass configurations defined above are given in Table 2. In this table, the middle column shows the average fuel savings obtained (over all road profiles) when the lead vehicle follows an optimized speed profile and the follower vehicle uses ACC function to track the preceding vehicle's speed profile while maintaining a safe distance, calculated using (12). The right-most column of the table shows the average fuel savings obtained (over all the road profiles) when each vehicle follows its own optimized speed profile. From Table 2, it is evident that the SPO, both in SPO+ACC and P-SPO, leads to large fuel savings, ranging from 15.22% to 17.43% for different mass configurations within the platoon, relative to CC+ACC. Furthermore, the fuel savings reported in Table 2 show that the P-SPO method for platooning can further improve, by up to 1.8 percentage points (see the third row in Table 2) the obtained savings compared to SPO+ACC by allowing each vehicle's speed profile to be optimized separately. An example of an optimized set of speed profiles for a heterogeneous platoon is shown in Figure 4. As can be seen from the figure, with the P-SPO approach, the intervehicle distance can vary based on the road topography and the mass of the HDVs, thus, allowing the platoon to drive through a road with varying topography without requiring the vehicles to follow an specific spacing policy and, therefore, achieving higher fuel savings for a heterogeneous platoon relative to SPO+ACC.

Table 2: The average fuel savings obtained in the simulations for a heterogeneous platoon of two HDV, with different mass configurations, over all the road profiles. The middle column shows the fuel savings obtained when the lead vehicle follows the optimized speed profile and the follower vehicles follow the same profile while maintaining the desired intervehicle distance (using ACC). The desired intervehicle distance is set to the minimum allowed safe distance; see (12). The right-most column shows the fuel savings obtained when each vehicle follows its own optimized speed profile. All the fuel savings are normalized with respect to the baseline case, i.e., CC+ACC where the lead vehicle drives at the constant speed of 80 km/h and the inter-vehicle distance is set to the allowed minimum.

	Average saving (%)	
Platoon configuration	SPO + ACC	P-SPO
Configuration I-a	15.26	16.76
Configuration I-b	15.75	16.94
Configuration II-a	15.22	17.05
Configuration II-b	17.23	17.43

5. Discussion

The results presented in Section 4 show that the proposed P-SPO method for platooning does work well in both heterogeneous and homogeneous platoons. In simulations, the P-SPO method reduced the fuel consumption of a homogeneous platoon of two HDVs by around 16% on average for the entire platoon relative to CC+ACC; see Table 1. For heterogeneous platoons, with different mass configurations, the P-SPO method reduced the fuel consumption of the entire platoon by around 17% (on average); see Table 2.

For homogeneous platoons, there is only a small improvement in fuel savings when P-SPO is used compared to the SPO+ACC method used in our paper [16], something that was expected given that the HDVs are identical in this case, resulting in similar optimized speed profiles for all vehicles even when they are optimized individually, see Figure 3. However, if the desired intervehicle distance is set to a fixed value of 15 meters, which is more realistic and safe (in case of communication failure, for example) compared to the minimum allowed intervehicle distance (see Section 4), the fuel savings obtained by SPO+ACC decrease by 1 percentage point. Moreover, one should note that the P-SPO method is more suitable for driving through steep downhill or uphill segments of a road since, in this method, there is no requirement on following a specific spacing policy. The P-SPO method allows the vehicles to exploit a variable intervehicle distance more efficiently than in ACC.

Turning now to heterogeneous platoons, the positive effect of not controlling the intervehicle distance directly becomes more evident. From Table 2, one can see that with the P-SPO method, the fuel savings increased by up to 1.8 percentage points compared to SPO+ACC. Furthermore, unlike other approaches for platooning, the P-SPO method does not require communication between the vehicles within a platoon during driving since, in this method, there is no need for online (i.e., during driving) iterative optimization of the speed profiles. Evidently, the optimized speed profiles must, however, be uploaded to the HDVs when the platoon is formed (before driving). If no profile is available for a given stretch of road, the optimization procedure must thus be completed before the platoon is launched. However, one can of course generate an offline database of speed profiles for commonly used roads and common vehicle masses, so that the profile for the lead vehicle can be made available directly. While this approach may not be feasible for a general situation, there are many cases where such a database could be built, for instance, in the case of a hauling company using a set of similar trucks that frequently carry similar loads, over a fixed set of highway routes.

One can then either use SPO+ACC, which still provides good fuel savings, or run P-SPO (before driving) starting from the available speed profile, in order to generate a set of individual speed profiles for all the vehicles of the platoon, thus achieving a few extra percentage points of fuel savings.

It is interesting to note that the results obtained in this work, with the P-SPO and SPO+ACC methods, compare favorably to the existing methods based on the MPC framework. For example, Murgovski et al. [20] achieved average fuel savings of 9.2%, which is still below our 16%, on driving cycles of 20 km stretches of highway for a homogeneous platoon of four HDVs where each vehicle's mass was 44 tonnes. In their work, a two-layer approach was used, where the top layer is responsible for generating fuel-optimal speed trajectories whereas the bottom layer is used to track the generated trajectory. Since both the simulation setup and the road sections were different compared to ours the results in [20] cannot be compared in detail. For instance, the average set speed used in [20] was 90 km/h with the allowed speed range of ±10 km/h which is different from the setup used here. However, one should note that the vehicle's masses are higher in [20] which, consequently, increases the fuel saving potential of the platoon in general. However, the altitude difference in each road section considered in [20] was at least 45 m and at most around 100 m which is comparable the relative altitude range (23.7 m to 114.7 m) for the road sections used here.

In a more direct comparison between the proposed P-SPO method for platooning and MPC-based approaches, the P-SPO method was applied on a 45 km stretch of highway road between the two Swedish cities of Mariefred and Eskilstuna, for which the fuel savings results obtained with an MPC-based controller was available from an earlier work by Turri et al. [10], where the details of the MPC-based controller can be found. Using the same normalization as in [10], the proposed P-SPO method obtained fuel savings of around 16% (on average) for different platoon configurations (both homogeneous and heterogeneous platoons), which exceeds those obtained in [10] by roughly 3 percentage points (on average) for the entire platoon.

As mentioned above, in MPC-based approaches to platooning, the problem of controlling a platoon of HDVs is generally divided into two layers, in order to reduce the computational complexity of the problem so it can be solved in real time: in the first layer a fuel-optimal speed trajectory is generated while the second layer is responsible for tracking this speed trajectory. In the first layer it is typically required that the vehicles should track the generated speed trajectories precisely, thus the need for handling the speed trajectory tracking in the second layer. In the proposed P-SPO method, by contrast, it is *not* required that the vehicles should track the speed profile precisely, as described in Section 3.4.1; therefore, the simple PID controller used here is sufficient for this method. In our earlier paper [17], by thorough testing with a real HDV it was shown that the results obtained from the SPO method, on which the proposed method P-SPO is based, are transferable to real HDVs, despite the simplicity of the HDV model and the controller.

Moreover, MPC-based methods to date have been implemented in an online fashion, in which it is *required* that the dynamic programming problem should return a solution every few seconds, something that limits the length of the horizon that can be used, due to the exponential growth of the number of states when the horizon is extended. By contrast, the P-SPO method is generally run offline and can comfortably handle horizons of 10 km or more (the entire stretch of road for instance). Alternatively, if the P-SPO method were to be used online, the optimized speed profiles for an upcoming road section could easily be generated while traversing the current road section.

The choice of profile length used here (10 km) is of course somewhat arbitrary; other profile lengths could certainly have been considered. However, that would raise the question of algorithmic (run-time) complexity, which, in this context, can be formulated as follows: How would the optimization algorithm perform as a function of the profile length? Here, one should keep in mind that the number of optimizable parameters, defining the splines for the road profiles, is a linear function of the profile length, as is also the available optimization time, given a certain average speed (see below). The issue of run-time complexity was investigated by randomly extracting 5 road profiles (from the 55 km road considered here) for each of four different profile lengths (here denoted λ), namely 5 km, 10 km, 15 km, and 20 km. For each of these $5 \times 4 = 20$ road profiles, a long offline run was carried out, thus generating speed profiles yielding low fuel consumption. Note that, for simplicity, in this analysis a single truck was considered, rather than a platoon of two trucks. Next, for each of the 20 road profiles, a total of 30 runs were carried out, where the running time was set to a value just below the time required to traverse the road profile in question at the speeds considered here. Specifically, for the 5 km road profiles, the running time for optimization was set to 3.5 minutes, whereas for the 10 km road profiles it was set to 7 minutes etc. For the $5 \times 30 = 150$ runs for each road profile length λ, the ratio $r_\lambda = f_{\text{on}}/f_{\text{off}}$ between the fuel consumption in the online and offline runs was measured. Then averages were formed over the 150 runs for each profile length, thus resulting in four values $\overline{r}_5, \overline{r}_{10}, \overline{r}_{15}$, and $\overline{r}_{20}$. These ratios, normalized by the ratio $\overline{r}_{10}$ (corresponding to the profile length used in this work), are shown in Table 3. As

TABLE 3: The average ratio between the fuel consumption obtained in an online run and that obtained in a long offline run, for four different road profile lengths λ (5, 10, 15, and 20 km), normalized by the value obtained for the 10 km case. The running time for the online runs was 3.5 minutes in the case of 5 km road profiles, 7 minutes for the 10 km road profiles, and so on.

Road profile length	5	10	15	20
$\bar{r}_\lambda/\bar{r}_{10}$	0.996	1.000	0.972	0.974

can be seen in the table, the results obtained for longer profile lengths are slightly better, but the difference is quite small, showing that the algorithm's performance is not very sensitive to the road profile length.

As mentioned in Section 3.4 the safety of the platoon is guaranteed by construction. In other words, when generating a set of speed profiles for the entire platoon, only those profiles that do not violate the safety constraints imposed on the platoon will be considered. However, an important assumption made in this work, which is also common in other work, is the exclusion of external traffic during the simulations. External traffic can endanger the safety of the platoon, e.g., when another vehicle cuts in between two HDVs. Of course, such problems are not specific to this particular method but can affect the performance of any platooning method. However, given that HDVs typically are among the slowest vehicles on the road (especially in the case of highways), cut-ins from other vehicles that last long enough to affect the motion of the platoon are, in fact, rather rare, something that was also observed in our onroad experiments with HDVs in [17]. Here, in order to handle cut-in situations, as soon as the distance of an HDV to its preceding vehicle drops below the safe intervehicle distance defined in (12), the vehicle's ACC function will be activated to control the intervehicle distance. Once the disturbance disappears, the HDV will resume following its (already available) speed profile, *still saving a significant amount of fuel, whether or not the platoon remains coherent* [17]. For cases where the disturbance lasts long enough to disrupt the coherence of the platoon, an online decision-making system could be implemented in the HDVs in order to decide whether or not catching up with the platoon, once the disturbance is gone, can improve the expected fuel savings, over and above the savings expected from just following the HDV's own speed profile. This is a topic for future work.

6. Conclusion

The main contributions of this paper are as follows: first, a new framework (P-SPO) for platooning has been proposed where in contrast to the commonly used leader-follower approaches, each vehicle of the platoon receives its own optimized speed profile to follow. Thus, the problem of controlling the intervehicle distance within a platoon is removed. Furthermore, since the safety of the platoon is explicitly considered during the speed profile optimization procedure, in the proposed approach the safety of the platoon is guaranteed by construction as long as the vehicles are able to follow their speed profiles and by the fallback ACC function in cases where the vehicles cannot do so.

It has been shown that the proposed P-SPO method outperforms the baseline case, i.e., the combination of CC+ACC, both in homogeneous and heterogeneous platoons, by reducing their fuel consumption by 15.8% for a homogeneous platoon, and between 16.8% and 17.4% for heterogeneous platoons with different mass configurations, over road profiles of 10 km length with varying topography. Moreover, the proposed P-SPO method, where the speed profiles are optimized *a priori*, outperforms the common MPC-based approaches by around 3 percentage points in similar settings over the same road profile. Additionally, the P-SPO method requires only a simple PID controller, thus removing the need of having a two-layer control architecture.

Conflicts of Interest

The authors declare that there are no conflicts of interest regarding the publication of this paper.

Acknowledgments

The authors would like to thank the project partners in the SERET project. This work was supported by Vinnova/FFI [Grant no. 2013-04724].

References

[1] European Commission, *Eu Transport in Figures - Statistical Pocketbook*, 2016.

[2] H. Fritz, "Longitudinal and lateral control of heavy duty trucks for automated vehicle following in mixed traffic: Experimental results from the CHAUFFEUR project," in *Proceedings of the 1999 IEEE International Conference on Control Applications (CCA) and IEEE International Symposium on Computer Aided Control System Design (CACSD)*, pp. 1348–1352, August 1999.

[3] J. K. Hedrick, D. McMahon, V. Narendran, and D. Swaroop, "Longitudinal vehicle controller design for IVHS system," in *Proceedings of the 1991 American Control Conference*, pp. 3107–3112, June 1991.

[4] K. S. Chang, J. K. Hedrick, W. Zhang, P. Varaiya, M. Tomizuka, and S. E. Shladover, "Automated highway system experiments in the path program," *I V H S Journal*, vol. 1, no. 1, pp. 63–87, 1993.

[5] L. Alvarez and R. Horowitz, "Safe platooning in automated highway systems," *California Partners for Advanced Transit and Highways (PATH)*, 1997.

[6] A. Alam, *Fuel-Efficient Heavy-Duty Vehicle Platooning [Ph.D. thesis]*, KTH Royal Institute of Technology, Stockholm, May 2014.

[7] T. Valerio, *Fuel-Efficient and Safe Heavy-Duty Vehicle Platooning through Look-Ahead Control [Licentiate thesis]*, KTH Royal Institute of Technology, 2015.

[8] E. Hellström, M. Ivarsson, J. Åslund, and L. Nielsen, "Look-ahead control for heavy trucks to minimize trip time and fuel consumption," *Control Engineering Practice*, vol. 17, no. 2, pp. 245–254, 2009.

[8] E. Hellström, M. Ivarsson, J. Åslund, and L. Nielsen, "Look-ahead control for heavy trucks to minimize trip time and fuel consumption," *Control Engineering Practice*, vol. 17, no. 2, pp. 245–254, 2009.

[9] A. Alam, J. Martensson, and K. H. Johansson, "Look-ahead cruise control for heavy duty vehicle platooning," in *Proceedings of the 2013 16th International IEEE Conference on Intelligent Transportation Systems: Intelligent Transportation Systems for All Modes, ITSC 2013*, pp. 928–935, Netherlands, October 2013.

[10] V. Turri, B. Besselink, and K. H. Johansson, "Cooperative look-ahead control for fuel-efficient and safe heavy-duty vehicle platooning," *IEEE Transactions on Control Systems Technology*, pp. 1–17, 2016.

[11] B. Besselink and K. H. Johansson, "Control of platoons of heavy-duty vehicles using a delay-based spacing policy," *IFAC-PapersOnLine*, vol. 28, no. 12, pp. 364–369, 2015, 12th IFAC Workshop on Time Delay Systems (TDS).

[12] D. Yanakiev and L. Kanellakopoulos, "Nonlinear spacing policies for automated heavy-duty vehicles," *IEEE Transactions on Vehicular Technology*, vol. 47, no. 4, pp. 1365–1377, 1998.

[13] D. Yanakiev and I. Kannellakopoulos, "Variable time headway for string stability of automated heavy-duty vehicles," in *Proceedings of the 1995 34th IEEE Conference on Decision and Control. Part 1 (of 4)*, pp. 4077–4081, December 1995.

[14] E. Hellström, A. Fröberg, and L. Nielsen, "A real-time fuel-optimal cruise controller for heavy trucks using road topography information," *SAE Technical Papers*, 2006.

[15] M. Henzler, M. Buchholz, and K. Dietmayer, "Optimal parameter selection of a model predictive control algorithm for energy efficient driving of heavy duty vehicles," in *Proceedings of the IEEE Intelligent Vehicles Symposium, IV 2015*, pp. 742–748, Republic of Korea, July 2015.

[16] L. Caltagirone, S. Torabi, and M. Wahde, "Truck platooning based on lead vehicle speed profile optimization and artificial physics," in *Proceedings of the 18th IEEE International Conference on Intelligent Transportation Systems, ITSC 2015*, pp. 394–399, Spain, September 2015.

[17] S. Torabi and M. Wahde, "Fuel consumption optimization of heavy-duty vehicles using genetic algorithms," in *Proceedings of the 2017 IEEE Congress on Evolutionary Computation, CEC 2017*, pp. 29–36, Spain, June 2017.

[18] S. Torabi, *Fuel-Efficient Truck Platooning Using Speed Profile Optimization. [Licentiate thesis]*, Chalmers University of Technology, 2017.

[19] J. H. Holland, *Adaptation in natural and artificial systems: an introductory analysis with applications to biology, control, and artificial intelligence*, University of Michigan Press, Oxford, UK, 1975.

[20] N. Murgovski, B. Egardt, and M. Nilsson, "Cooperative energy management of automated vehicles," *Control Engineering Practice*, vol. 57, pp. 84–98, 2016.

[21] H. Richard, J. C. Bartels, and A. Brian, *An Introduction to Splines for Use in Computer Graphics & Geometric Modeling*, Morgan Kaufmann Publishers Inc., San Francisco, CA, USA, 1987.

[22] K. Santhanakrishnan and R. Rajamani, "On spacing policies for highway vehicle automation," *IEEE Transactions on Intelligent Transportation Systems*, vol. 4, no. 4, pp. 198–204, 2003.

[23] W.-H. Hucho, *Aerodynamics of Road Vehicles: From Fluid Mechanics to Vehicle Engineering*, Elsevier Science, 2013.

[24] Q. Deng and X. Ma, "A fast algorithm for planning optimal platoon speeds on highway," in *Proceedings of the 19th IFAC World Congress on International Federation of Automatic Control, IFAC 2014*, pp. 8073–8078, South Africa, August 2014.

[25] A. Al Alam, A. Gattami, and K. H. Johansson, "An experimental study on the fuel reduction potential of heavy duty vehicle platooning," in *Proceedings of the International IEEE Conference on Intelligent Transportation Systems*, pp. 306–311, 2010.

[26] C. Bonnet and H. Fritz, "Fuel consumption reduction in a platoon: Experimental results with two electronically coupled trucks at close spacing," *SAE Technical Papers*, 2000.

[27] D. E. Goldberg, "Genetic algorithms in search, optimization, and machine learning," *Choice Reviews Online*, vol. 27, no. 02, pp. 27-0936–27-0936, 1989.

10

A Pattern Analysis of Daily Electric Vehicle Charging Profiles: Operational Efficiency and Environmental Impacts

Ranjit R. Desai, Roger B. Chen, and William Armington

Golisano Institute for Sustainability, Rochester Institute of Technology, One Lomb Memorial Dr., Rochester, NY 14623-5608, USA

Correspondence should be addressed to Roger B. Chen; rbcgis@rit.edu

Academic Editor: Jing Dong

Plug-in Electric Vehicles (PEVs) are considered one solution to reducing GHG emissions from private transport. Additionally, PEV adopters often have free access to public charging facilities. Through a pattern analysis, this study identifies five distinct clusters of daily PEV charging profiles observed at the public charging stations. Empirically observed patterns indicate a significant amount of operational inefficiency, where 54% of the total parking duration PEVs do not consume electricity, preventing other users from charging. This study identifies the opportunity cost in terms of GHG emissions savings if gasoline vehicles are replaced with potential PEV adopters. The time spent in parking without charging by current PEV users can be used by these potential PEV users to charge their PEVs and replace the use of gasoline. The results suggest that reducing inefficient station use leads to significant reductions in emissions. Overall, there is significant variability in outcomes depending on the specific cluster membership.

1. Introduction and Background

In the United States, the newly registered number of Battery Electric Vehicles (BEVs) and Plug-in Electric Vehicles (PEVs) increased by 159,616 in the year 2016 alone [1]. Though this is not a significant market share, the PEV penetration rate, BEVs and PEVs considered together, is considerably high in the private transportation market. This has led policy-makers to reconsider infrastructure planning to accommodate increasing PEVs and assess sustainability implications of infrastructural development. Recently, a report prepared for the Clinton Climate Initiative [2] outlined the incumbent and future issues of increasing penetration of PEVs. At the forefront there are the adequacy and level-of-service (LOS) of charging stations; these issues are further underscored by the increasing need for out-of-home PEV charging opportunities. A critical component for management and decision-making regarding out-of-home charging decisions is the demand for these services, governed by drivers' charging behaviors. Understanding the timing and duration of PEV charging decisions supports making infrastructure and operational policies that meet economic, operational efficiency, and environmental objectives.

The literature on assessing PEV charging has grown significantly in the past decade, generating studies that investigate both quantitative and qualitative issues. The literature can be roughly segmented into three areas: (i) consumer adoption and use; (ii) infrastructure performance and evaluation; and (iii) operational issues of stations.

1.1. Consumer PEV Adoption. The literature on consumer adoption and use has focused on identifying groups of PEV drivers with respect to their technology ownership and interaction with the infrastructure. One study [3, 4] identifies three groups based on adopters and potential PEV owners in China: (a) early adopters, (b) shapeable groups, and (c) late adopters, each consisting of two motivations that have their own set unique behaviors. For example, early adopters have behaviors labeled as trendy greens and running cost sensitive. Another study [5] assesses PEV charging behavior by applying a "user-battery interaction style" metric developed originally for small electronic devices to find out similarities in device use. In another study [6] PEV drivers were interviewed and found to generally manage without public infrastructure, with the battery still containing plenty of range when station recharging was initiated.

Although PEV range is typically adequate for completing most daily home-based tours without intermediate charging [7], continued PEV adoption, as technological forecasts indicate, can benefit from continued installation and planning of out-of-home charging [8], in addition to further considering PEV driver's perceptions and experiences at stations. Although consumer interviews and analogous insights from the literature are helpful for describing charging behaviors, a study defining user groups or "market segments" based on a pattern analysis of empirical time-dependent charging data from out-of-home charging stations is absent in the literature. Such a study could contribute greatly to helping effective policy implementation from both an operational and sustainability impact perspective.

1.2. Infrastructure Performance and Evaluation. PEV adoption will also impact the electric grid and shift environmental impact from point-source vehicle to electricity generator emissions due to increasing electricity demand. The literature has produced work on network optimization models that quantify and assess infrastructure and environmental impacts at the network level. These studies consider developing public charging infrastructure [9, 10] by optimizing social welfare relative to impacts on the electrical grid. Other studies assess the electric grid impacts from widespread charging through conventional travel surveys, where travel logs of PEV drivers [11] from the NHTS are used to predict electricity consumption and load profiles. Other studies further expand on these by applying network charging scheduling [12] and time-of-use rates [13] to optimize and assess electrical loads on the infrastructure. In another study, large-scale vehicle survey data [14] are used to model charging location decisions by maximizing PEV driving miles to jointly maximum environmental benefit. An activity-based modeling approach [15] was used in another study to assess PEV environmental impacts, suggesting that public charging facilities allow charging during the daytime and can potentially reduce emissions. Although observed travel data is used in these studies, the effects of individual charging behaviors at the stations themselves have not been examined and operational inefficiencies in station turnaround are not considered.

1.3. Operational Issues at Charging Stations. Operational issues have also been considered in the literature. Interviews [16] of PEV owners conducted assessed perceptions of charging etiquette at public and workplace charging locations, concluding that no common charging guidelines evolved unless a pathway of communication between users exists, which ultimately leads to inefficiency in infrastructure use. Another study [17] addresses these inefficiencies by considering occupancy rates, infrastructure costs, and parking premiums to assess the economic feasibility of deploying charging stations using an economic model. This study also finds that policies that promote more infrastructure development and increased station usage are beneficial to decrease range-anxiety issues and charging premiums from the user perspective [17]; however, no empirical data was used. It is important to note that the PEV charging behavior at the public charging stations is not studied before, and therefore the uniqueness of this analysis is; this study brings into light the insights to charging behavior characteristics using pattern analysis of daily charging profiles observed at public charging stations and is supported by dataset.

The overarching goal of this study is to characterize the observed PEV daily charging patterns and assess the environmental impacts. One main task is to identify potential market segments based on the observed patterns of actual PEV adopters at existing charging stations. The focus of this study is to analyze the charging behavior of PEV users at the public charging stations. However, the fundamental unit of this analysis is not a PEV user but a charging profile observed in a single day (as shown later in Figure 2). A second task is to calculate the opportunity cost for improving operational efficiency and environmental impacts. The identified segments are used as a basis to accomplish this scenario analysis. The efficient operations scenario calculates the GHG emissions savings from reducing the duration of parking without charging, allowing other potential PEV drivers and adopters to charge. This displaces gasoline miles that potential PEV drivers would otherwise have incurred from inability to use their PEVs. The dataset lacks information about the PEV users who could not get the access to charging stations if they were occupied (denoting latent demand); the smart charging stations do not have a mechanism to keep track of this parameter. Furthermore, determining latent demand for public infrastructure, such as roads, is in its own right a difficult problem [18]. Similarly, the demand for public charging stations is difficult to determine. In case of the City of Rochester, authorities have confirmed in the personal discussion that there have been instances when PEV users could not get access to public charging stations because all the charging stations at the particular location were occupied. Furthermore, the authors have noted that for extended periods the data shows stations occupied continuously, where a PEV occupies the station just as one PEV departs. Further investigations into this queueing phenomena are underway by the authors.

The City of Rochester, for a population of 208,880 [18]—as of July 1, 2016—has 28 level 2 [19] (with J1772 connector) public charging stations installed and operating through ChargePoint network [20, 21]. Each level 2 charging station can accommodate two PEVs at a time. Therefore, if all the charging stations in the City of Rochester are occupied at a time, there should be 56 PEVs plugged-in to charging stations. In the upstate New York region, Buffalo, NY (population 256,902 [18]), has 29 [19] level 2 and Syracuse, NY (population 143,378 [18]), has 27 public charging stations through ChargePoint network. The data used for this analysis is collected from charging event logs collected at these 28 [19] level 2 smart public charging stations installed in the City of Rochester. The charging stations were installed as part of a New York State Energy Research and Development Authority (NYSERDA) grant to the City of Rochester and are managed by ChargePoint Inc. [21]. For this work, these charging stations are heuristically grouped—as explained in the next section—into three main locations (Figure 1).

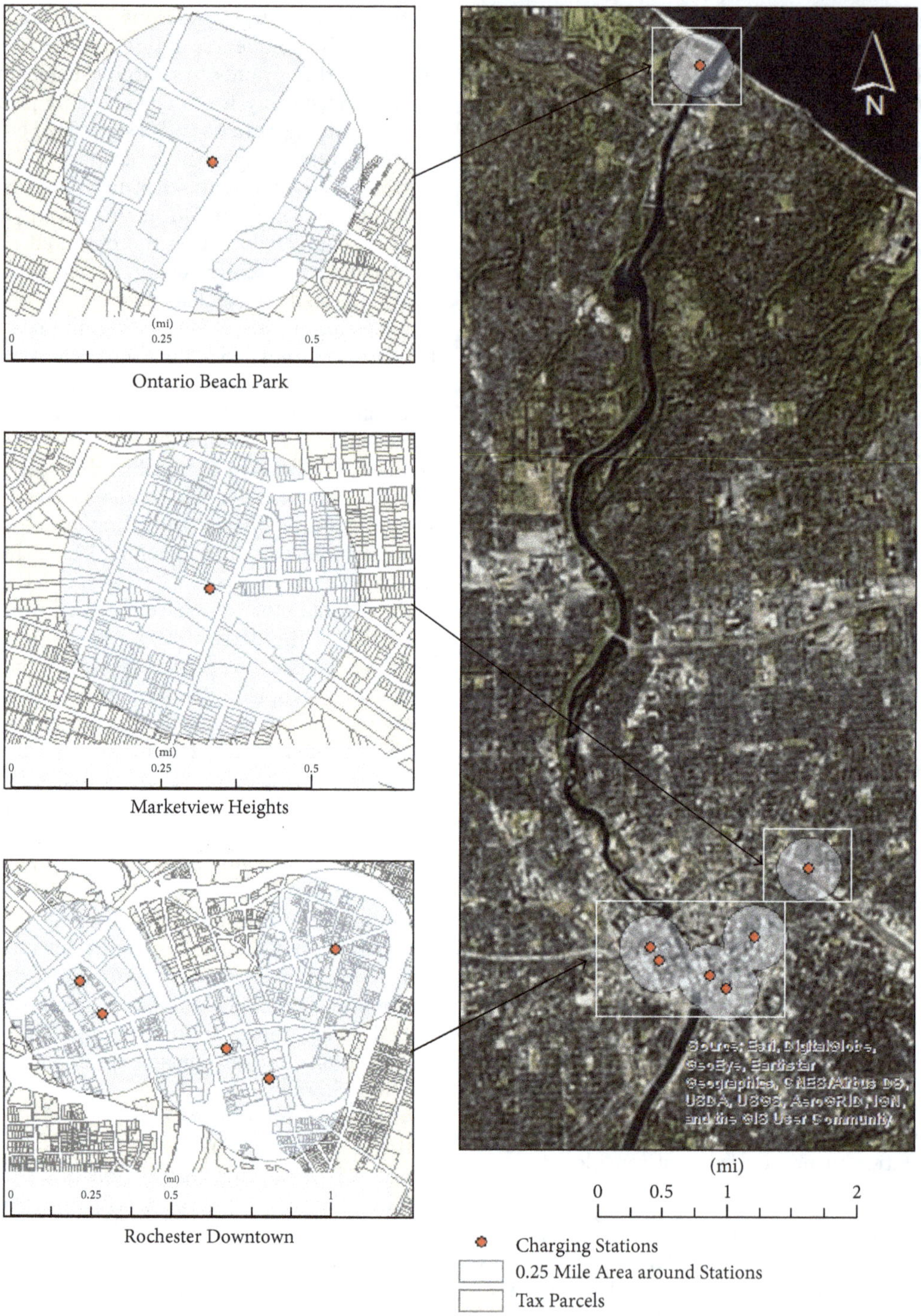

FIGURE 1: Location of charging stations in the City of Rochester (with 1/4 mile radius).

2. Dataset

The City of Rochester currently owns and maintains seven public smart charging stations that log the charging activities of vehicles using the stations. Types of data collected through these stations include (i) timestamps of charging events, such as the time PEVs plug-in at the stations; (ii) computed performance metrics, such as energy consumed (kWh); and (iii) other station information. The data used in the analysis were collected over a period of more than two years between March 2014 and May 2016. Charging station location coordinates were extracted and charging stations were grouped heuristically into three distinct areas: (i) Rochester Downtown (RDT), (ii) Marketview Heights (MVH), and (iii) Ontario Beach Park (OBP). The RDT extent which is defined by the area outlined by the Rochester Downtown Development

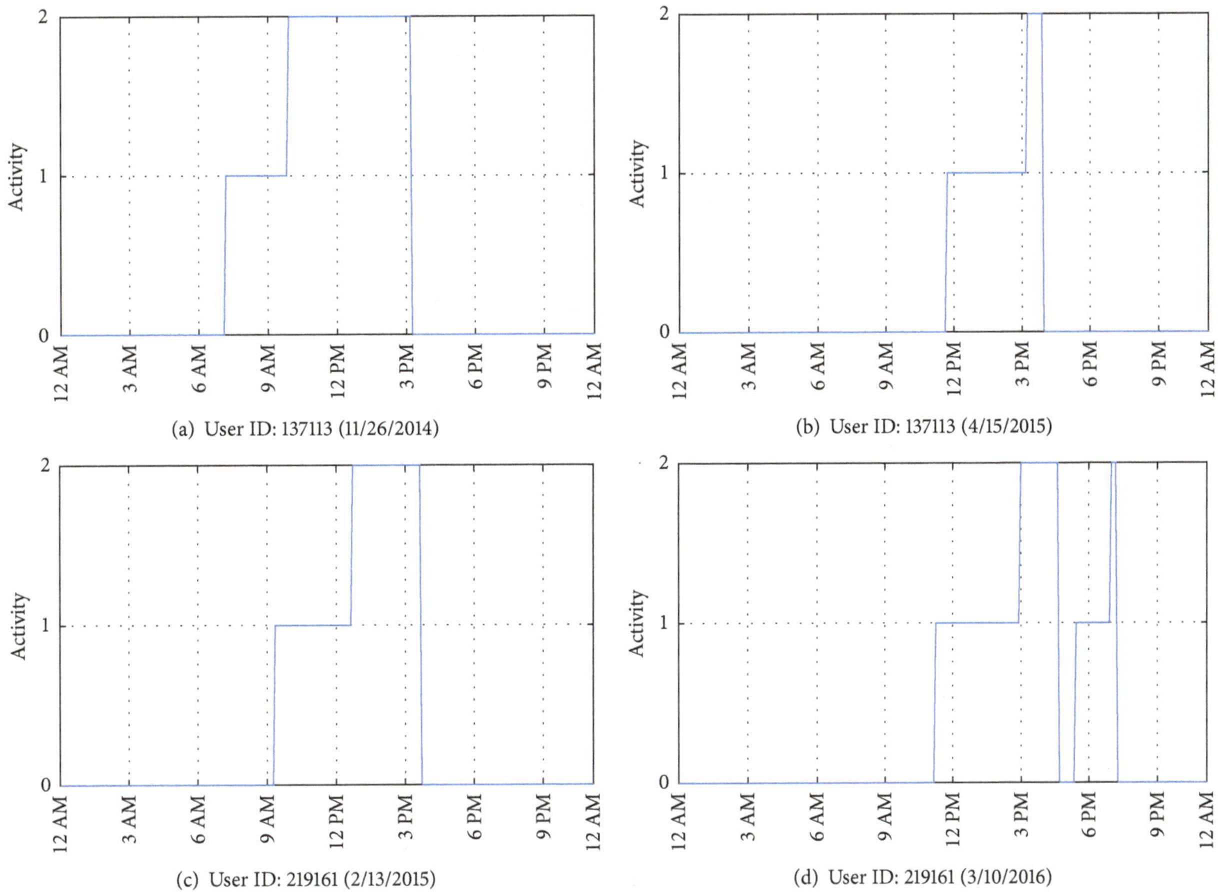

FIGURE 2: Observed charging patterns of PEV drivers in the sample ("0"—no activity, "1"—charging, and "2"—parking without charging).

TABLE 1: Land use around public charging stations [25, 26] (for 1/4 mile radius).

	Rochester Downtown (RDT)	Marketview Heights (MVH)	Ontario Beach Park (OBP)
Population density (per sq. mile)	6200	7700	3500
Unemployment rate	11%	26%	7%
Residential land use	4%	38%	34%
Commercial land use	58%	31%	7%

Corporation [22] and can be considered as a business district with 58% (Table 1) of the land use around the charging stations is occupied for commercial use. The MVH is defined by the area defined in the Rochester Public Market Master Plan Report prepared by Market Ventures Inc. for the City of Rochester in February 2012 [23]. The MVH area has 31% occupancy for commercial use and 38% for residential use. Further, the OBP encompasses the areas as defined by Monroe County Parks Department [24], and 34% of properties around the area are occupied for residential use. Given these guidelines, five stations are located within the RDT area, and one station is located at each of the MVH and OBP extents. Figure 1 shows the locations of charging stations and Table 1 shows the land use of the neighborhood around these public charging stations. The figure marks the locations of charging stations and not actual number of charging stations. For example, a map of parking lots marks locations parking lots and not the capacity of total number of vehicles the lots can accommodate. Similarly, Figure 1 depicts the seven locations of charging stations. Therefore, each location may contain more than one charging station and can accommodate at least two vehicles at a time. Further, Table 2 summarizes data from the charging stations at these locations.

2.1. Data Processing. The unit for this pattern analysis was *one daily PEV charging pattern for a single vehicle observed at public charging stations*. Consequently, drivers who used the stations on multiple days had multiple daily patterns in the sample considered for analysis. PEVs reside in one of three

TABLE 2: Dataset characteristics based on observed plug-in events.

	Sample size: 8929 plug-in events (different from daily patterns)											
	Total parking duration (hours)				Parking with charging (hours)				Parking without charging (hours)			
	Mean	Std. dev.	Min	Max	Mean	Std. dev.	Min	Max	Mean	Std. dev.	Min	Max
Whole network	3.9	3.1	0.0	64.1	1.8	1.1	0.0	12.1	2.1	2.8	0.0	61.7
RDT ($n = 7051$)	4.5	3.2	0.0	64.1	1.9	1.2	0.0	12.1	2.6	2.9	0.0	61.7
OBP ($n = 1316$)	2.3	1.8	0.0	17.4	1.5	0.9	0.0	4.7	0.7	1.4	0.0	14.4
MVH ($n = 562$)	1.5	1.3	0.0	9.1	1.3	1.1	0.0	5.4	0.2	0.6	0.0	7.2
Seasons												
Winter (Dec., Jan., Feb.) ($n = 2250$)	4.3	3.1	0.0	26.8	2.0	1.2	0.0	10.6	2.3	2.7	0.0	17.8
Spring (Mar., Apr., May.) ($n = 2410$)	3.9	2.9	0.0	23.3	1.8	1.1	0.0	10.5	2.1	2.6	0.0	22.2
Summer (Jun., Jul., Aug.) ($n = 2086$)	3.8	3.1	0.0	33.5	1.6	1.0	0.0	12.1	2.2	2.9	0.0	33.0
Autumn (Sep., Oct., Nov.) ($n = 2183$)	3.7	3.2	0.0	64.1	1.7	1.1	0.0	11.4	2.0	2.9	0.0	61.7

states: (i) not parked (no activity); (ii) parked with charging; and (iii) parked without charging, each assigned a numerical value of 0, 1, or 2, respectively. Since only a nominal scale for activities is required for pattern analysis and classification, the actual values of each state are inconsequential. Examples of observed charging patterns are shown in Figure 2. The resulting daily pattern was further discretized into 256 data points taken at 5.625-minute intervals. The final resolution (number of points) is bounded by the requirements of the Walsh-Hadamard Transform [27] used in the pattern analysis, which requires the number of data points (n) to be of the order of 2^n.

Errors in the logged data led to incomplete observations which were omitted from analysis. Charging events that contained total parking durations of two minutes or less were also omitted as these events are from the user (a) opening a charging session without inserting the plug into their car and then the system automatically closes the session to allow other users to use the station, or (b) the original user opens a session and decides to switch charging to an alternate port at the station (these causes were confirmed by the ChargePoint help desk staff). Events with missing or zero station user IDs were also omitted. The original dataset contained a total of 9,680 unique charging events between March 2014 and May 2016. After omitting these events based on previously stated criteria, 8,929 unique charging events were used for subsequent analysis. Using these 8,929 unique charging events, 7,554 unique daily charging patterns were generated by "stitching them together" for a particular vehicle and day for the pattern analysis.

3. Methodological Approach

In order to characterize the daily PEV charging patterns of drivers, a pattern analysis is applied to identify homogenous segments or clusters of patterns. Charging patterns are quite complex, varying over time and geography. Example patterns observed in the sample are given in Figure 2. Each of these patterns represents a daily profile that starts at 12:00 AM and ends at 11:59 PM (24-hour period) for a particular PEV driver. For example, the first pattern (a) (User ID: 137113) shows that the driver started his/her activity around 7:30 AM and used the facility for charging until 10 AM and continued to park without charging until 3:30 PM. This PEV driver, thus, used the facility for 8 hours but charged the PEV for only 2.5 hours. Daily patterns can also consist of multiple charging events. Pattern (d) shows a PEV driver using the station twice (two plug-in events) in a single day. In the case of multiple plug-in events within a single day (24 hours) for the same PEV driver, events were stitched together into a single daily charging pattern. In pattern (d), the driver used the station for charging for a total of 5.5 hours out of 7.25 hours of parking.

Given the complexity and time-varying nature of PEV charging patterns, a pattern analysis helps identify clusters of similar patterns to facilitate discussion and subsequent analysis. Apart from uncovering similarities across patterns, this approach can also simplify and summarize a large sample of pattern data. Past studies use pattern analysis to classify travel patterns [28] and characterize vehicle travel patterns and to examine the feasibility of PEV operation [29]. This study adopts a similar approach towards classifying daily charging patterns of PEV drivers into homogenous segments with respect to their daily patterns.

Pattern analysis has been applied to a wide range of fields from voice recognition to image analysis, all of which aim to classify patterns into sensible segments. The clusters do not group the PEV users but the daily charging profiles of the PEV users observed at the public charging stations. The same PEV user can have multiple patterns of usage and can be observed to use the charging stations at different locations. Importantly, pattern analysis can provide different and a few in number clusters of homogeneous daily charging profiles, and the identified clusters can be used for policy making where a particular charging behavior can be targeted. Furthermore, the observed operational inefficiency depends on the patterns of charging behavior, and along with the total parking without charging time, it is important to understand how this inefficiency is distributed. The pattern analysis provides the specific number of clusters of daily profiles which can further understand the causes of the operational inefficiency.

The overall methodological framework is shown in Figure 3. Conventionally, a sequential three-stage process of pattern analysis consists of (i) pattern specification, (ii) feature extraction, and (iii) clustering. First, plug-in event data is taken from the sample and stitched together to form daily patterns. The process for this was described previously in Section 2. Given the resulting sample of daily charging patterns, features are extracted from these daily patterns for the subsequent clustering stage. The output of clustering and consequently the entire pattern analysis process is a set of homogenous segments of patterns, with patterns within the same segment being similar to each other, while patterns in different segments are dissimilar to patterns in other segments. These identified segments serve as the basis for subsequent scenario analysis to investigate the potential of policy interventions and technology options. The next section describes the pattern analysis process and discusses the clusters that are an outcome of this process.

3.1. Pattern Analysis Process

3.1. Pattern Analysis Process. Conventionally, pattern analysis consists of a sequential three-stage process: (i) pattern specification, (ii) feature extraction, and (iii) clustering. These stages and the overall methodological framework are described further in the next section.

3.1.1. Pattern Specification. First, plug-in event data are taken from the sample and stitched together to form daily patterns. In this stage, all the plug-in events are converted into a "signal" consisting of three possible states that refer to three activities, "0" for no activity, "1" for charging activity, and "2" for parking without charging activity, as explained in the previous section. Daily time-varying EV charging patterns are generated or "stitched" together in chronological order based on the timestamp of charging events logged by the

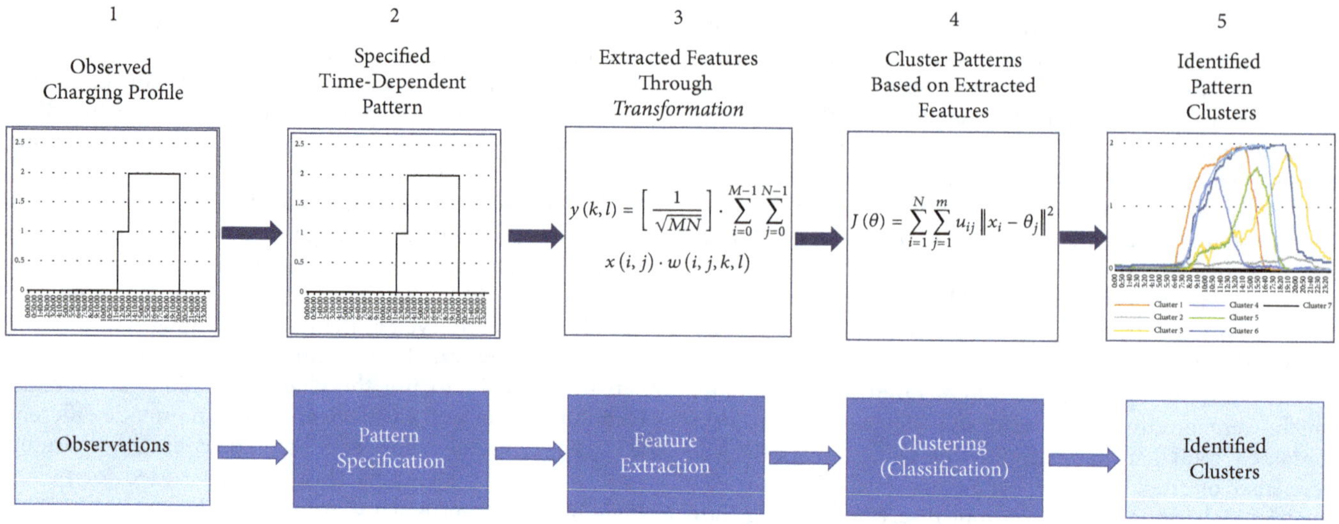

FIGURE 3: Methodological framework.

smart charging stations. Examples of patterns specified and outputted in this stage were shown in Figure 2.

3.1.2. Feature Extraction. In this stage, features are extracted from the specified patterns for subsequent clustering. Observed events such as charging (state = 1) and parking without charging (state = 2) are time-dependent. The main goal of feature extraction is to extract statistically independent "features" from the observed patterns that are best suited for subsequent cluster analysis in the classification stage. These features serve to summarize the patterns and capture the most important pattern characteristics. The feature extractor used is the Walsh-Hadamard Transformation (WHT), though others such as the Karhunen-Loeve or Haar could also have been used [27]. The WHT extracts features referred to as Walsh coefficients [29] that are used in the cluster analysis. These extracted features can conceptually be considered as "building blocks" of the observed patterns. An analogous process would be a principle components analysis on a digital image. In pattern analysis, a feature extraction allows identifying separate orthogonal features from the image, or in this case time-dependent pattern. These features, in this case Walsh coefficients, are used in subsequent clustering of patterns as "attributes" of this observed pattern.

3.1.3. Pattern Classification. The extracted features from the previous stage are used in a clustering algorithm to identify homogenous segments of patterns. The output of clustering and consequently the entire pattern analysis process is a set of pattern segments, with patterns within the same segment being similar or homogenous to each other, while patterns in different clusters are dissimilar to each other. Conceptually, each cluster contains patterns (which are time-dependent) that have similar "features." In the literature, several clustering algorithms exist; this study uses the k-means clustering algorithm.

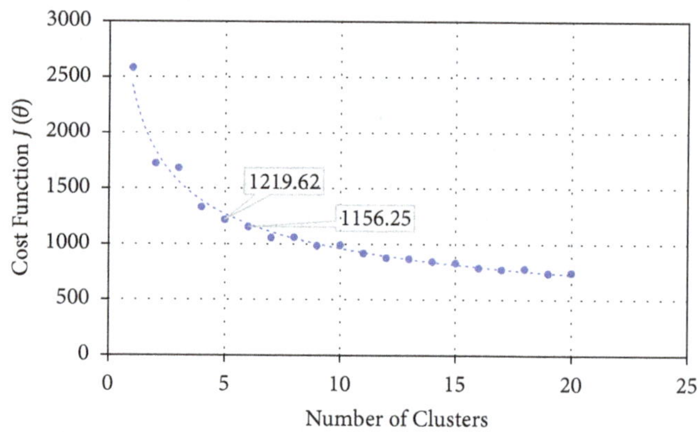

FIGURE 4: Cost function value versus number of clusters assumed a priori.

The k-means algorithm is a cost-based algorithm [30], in which the cost function $J(\theta)$ is given by

$$J(\theta) = \sum_{i=1}^{N} \sum_{j=1}^{k} u_{ij} \left\| x_i - \theta_j \right\|^2, \quad (1)$$

where θ_j is the centroid of cluster j, u_{ij} is a binary indicator that equals 1 if point x_i is nearest to θ_j, N is the total number of points, and k is the number of clusters. For the k-means algorithm, the number of clusters is specified a priori. As the number of clusters increases the marginal reduction in the cost function decreases until a negligible value, indicating the correct number of clusters, is specified. A plot of the cost function ($J(\theta)$) as a function of number of clusters (k) is given in Figure 4. It shows the variation of cost function with respect to increasing number of clusters. The final number of clusters for a particular sample of patterns is determined based on identifying the point of marginal return on the cost function.

4. Identified Clusters of Daily Charging Profiles

The pattern analysis approach described in the previous section identified 5 distinct clusters of daily charging profiles from the sample. This section examines the clusters and their members with respect to (i) charging, (ii) parking without charging, and (iii) parking durations. Additionally, inverse Walsh-Hadamard Transform on the feature centroids of each cluster is used to reconstruct a composite image for each cluster and to facilitate interpretation. These images are shown in Figure 5.

4.1. Characterization of Identified Clusters. Each identified cluster is characterized with respect to charging and parking activities. Figure 5 shows the reconstructed profiles from the inverse Walsh-Hadamard for each cluster. Note that the profile represents a composite of all activities at a specific time. For example, only if *all events* at a specific time were charging (state = 1) would the profile distinctly indicates "charging." For this reason, the distribution of patterns within each cluster residing in the three different states considered is also plotted over the one-day period.

4.1.1. Cluster 1 (CL-1, 772 Patterns; 10% of the Samples of Patterns). The reconstructed composite profile of CL-1 indicates that most charging activities start at 6:30 AM and peak (60% of patterns) at 8:45 AM. The daily profiles in this cluster are observed to have a peak (90% patterns) of parking without charging activity at about 1:45 PM. The patterns from CL-1 are mostly (85%) observed at RDT on weekdays (95%).

4.1.2. Cluster 2 (CL-2, 3,438 Patterns; 46% of the Samples of Patterns). The composite profile of CL-2 has a distribution characterized by a low peak (20% of events) at 2:15 PM. Note again that the profile represents a composite of all activities at a specific time, making it difficult to tease out specific activities. The peak of parking without charging activity is observed at 3:30 PM with only 12% of patterns. Unlike CL-1, the patterns from CL-2 are observed at all the locations and have the highest (30%) contribution of patterns observed on weekends. The patterns observed at MVH have the highest contribution of 14% in CL-2 compared to the rest of the clusters. As CL-2 is largely comprised of shorter plug-in events, and, by MVH being the location near a public market, it can be inferred that the PEV drivers who access the public charging stations at the MVH and have charging profiles belonging to the CL-2 tend to use the charging stations for shorter durations.

4.1.3. Cluster 3 (CL-3, 1,355 Daily Patterns; 18% of the Samples of Patterns). These patterns show a peak (68%) of charging activity at 8:45 AM and a very high level of parking without charging activity with a peak (98.45% of patterns) at 4:15 PM. The patterns in this cluster indicate more parking without charging than parking with charging. CL-3 is constituted of 98% patterns observed at RDT and 97% patterns observed on weekdays. Similar to CL-1, the charging activity of such patterns coincides with a typical working day schedule around the business district of Rochester.

4.1.4. Cluster 4 (CL-4, 1,500 Patterns; 20% of the Samples of Patterns). The parking with charging activity for these patterns peaks (70% of patterns) at 9:00 AM. The parking without charging activities peaks (60% of patterns) at 11:30 AM. CL-4 does not have higher peak of parking without charging activity than parking with charging activity. Further, the patterns observed at RDT are 84% and 92% are observed on weekdays.

4.1.5. Cluster 5 (CL-5, 489 Patterns; 6% of the Samples of Patterns). The parking with charging of these patterns peaks (55% of patterns) at 9:00 AM. The parking without charging peaks (90% of patterns) at 5:30 PM. The patterns observed at OBP have the highest contribution of 27% in CL-5 compared to the rest of the clusters. As OBP can be considered as a leisure place, a typical PEV driver may stay at OBP for the entire day, which can be inferred with the charging activity ranging from 7:30 AM to 6:20 PM.

Importantly, the charging profiles observed in CL-3, CL-1, and CL-5 show charging behavior characteristics which are likely to be observed on a working day at a public charging station located in business district, in this case RDT.

4.2. Comparison of Identified Clusters. Figure 6 shows distribution of three dimensions, across patterns within each identified cluster: (a) parking duration (PD), (b) parking duration with charging (PDWC), (c) parking duration without charging (PDWOC), and (d) total sum of parking without charging durations of all patterns across different clusters. For PD, CL-5 while being the smallest cluster with only 489 daily patterns, had the highest median PD (more than 10 hours). However, CL-2, the largest cluster with 3,438 daily patterns, has the lowest median PD (2 hours). Irrespective of the cluster sizes, the median for PD varies from 2 hours to 10 hours. In contrast, the PDWC does not vary significantly across clusters. The median for PDWC varies from 1.5 hours to 2.5 hours across clusters. Figure 6(c) also shows the distribution of PDWOC for all clusters. CL-5 has the highest median around 7.5 hours and CL-2 has the lowest (almost negligible) median. This characteristic of a charging behavior indicates efficient (or inefficient) operation of charging stations. Furthermore, Figure 6(d) shows that CL-3 has the highest total sum duration of hours spent parking without charging, followed by CL-1 and CL-5. These total hours spent parking without charging activity indicate the inefficiency of user turnaround at stations, likely due to the currently cost-free situation. Figure 6 also shows that, except for CL-2 and CL-4, the remaining clusters exhibit a higher median for PDWOC, relative to PDWC. However, due to their membership sizes, CL-2 and CL-4 still show a significant *total sum duration* of parking without charging.

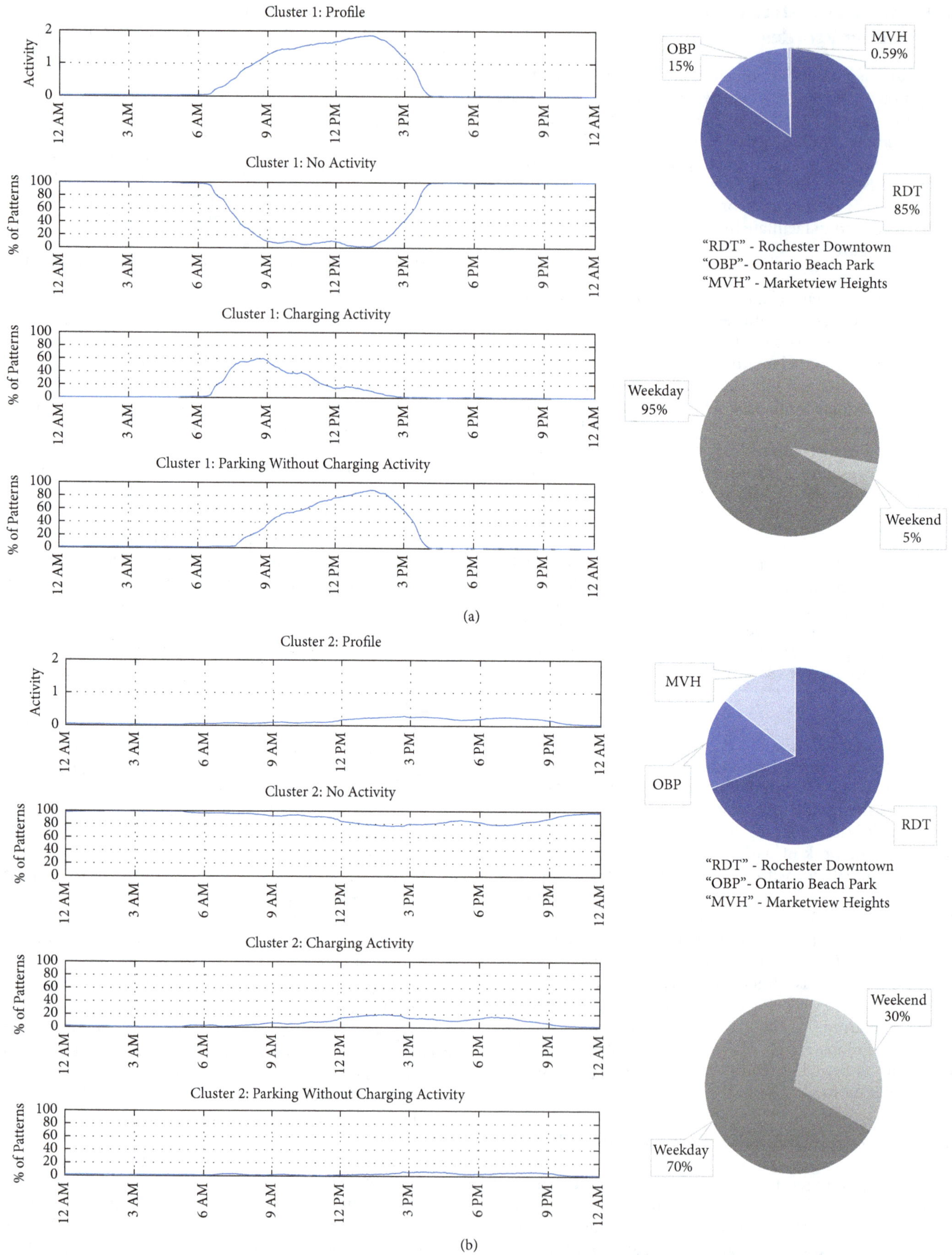

Figure 5: Continued.

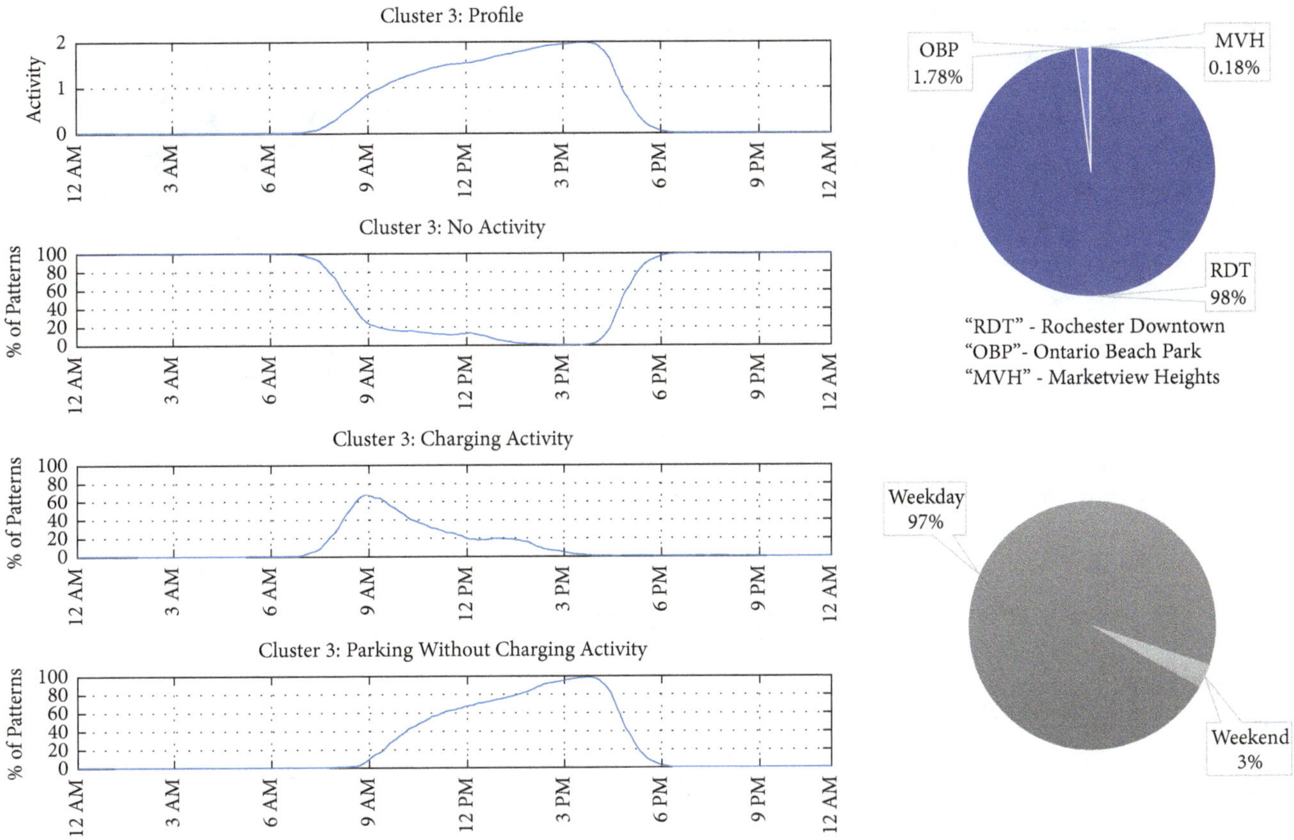

(c)

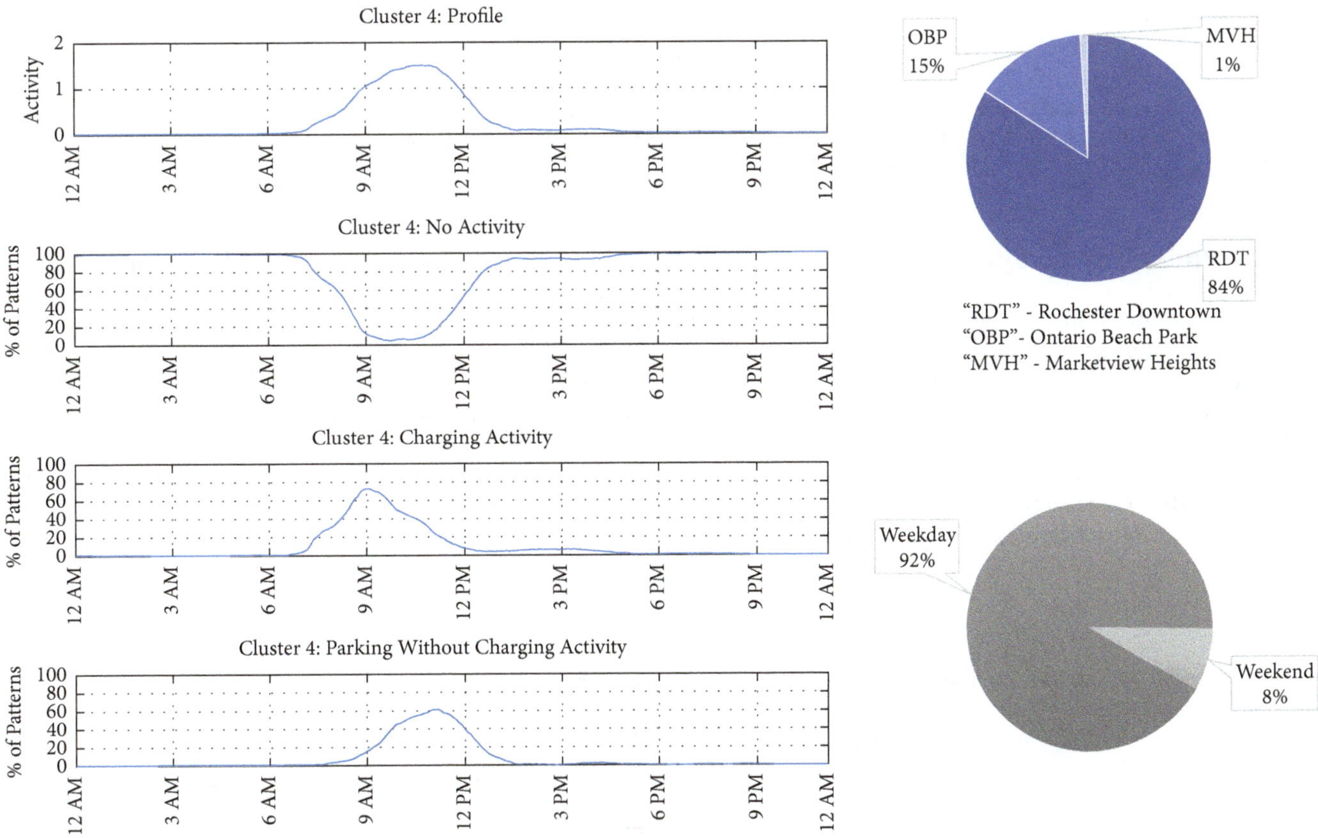

(d)

FIGURE 5: Continued.

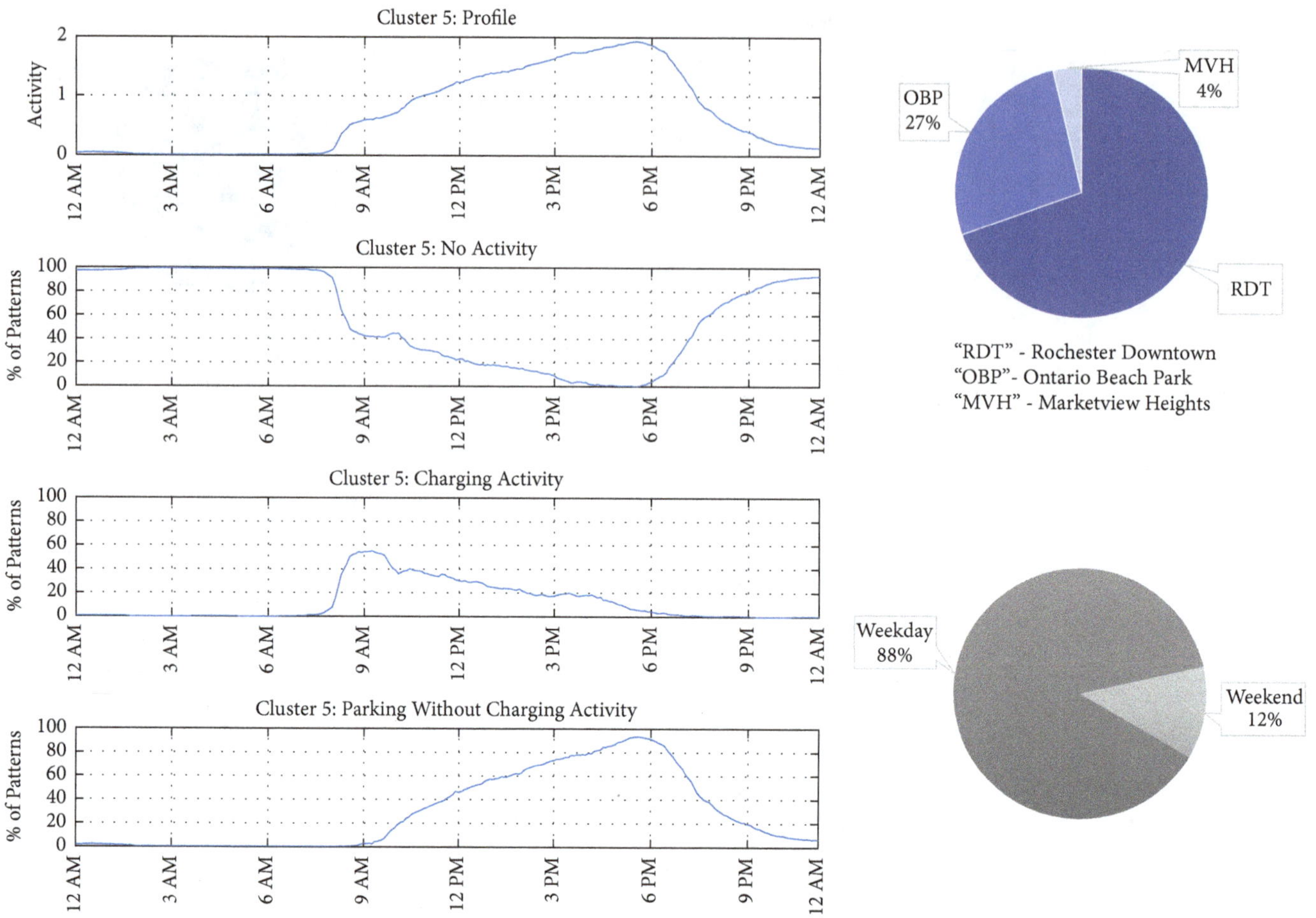

(e)

FIGURE 5: Cluster profiles and distribution of patterns in activities "0"—no activity, "1"—parking with charging, and "2"—parking without charging along with distribution of patterns with respect to location and type of day the pattern was observed (e.g., weekday of weekend) (a) Cluster 1, (b) Cluster 2, (c) Cluster 3, (d) Cluster 4, and (e) Cluster 5.

5. GHG Emissions Analysis

The pattern analysis in the previous section uncovered five distinct clusters of daily charging profiles, each with a large frequency and total sum duration of parking without charging across daily profiles of PEV drivers. User operational inefficiency at charging stations is an issue that needs to be addressed to efficiently allocate institutional resources. Additionally, this inefficiency may demotivate potential PEV adopters who see out-of-home charging locations constantly occupied.

This scenario examines the impact of reducing inefficient station use through reducing the parking without charging duration. By allowing more efficient use and turnaround of charging stations, more PEV drivers can use the stations, thus potentially displacing gasoline miles incurred otherwise with electricity. These drivers could be potential PEV adopters or current PEV drivers who see the occupied station and turn away discouraged. The main assumption is that 100% time spent in parking without charging can be replaced with PEV drivers who need to charge. The GHG emissions savings from this displacement of inefficient station sue are determined as follows:

$$\begin{aligned} \text{GHG Emissions Savings} &= \text{Emissions displaced by replacing gasoline} \\ &= \left\{ \frac{\left\{ \text{Avg.}_{\text{PWOC}} \times P_{\text{avg.}} \times \eta_e \right\}}{\eta_{\text{gas}}} \times \begin{pmatrix} \text{Emissions per gallon of gasoline} \\ (\text{MT CO}_2 \text{ eq./gallon}) \end{pmatrix} \right\} \\ &\quad - \left\{ E \times \begin{pmatrix} \text{Emissions per 1 kWh of electricity generation in NYUP} \\ (\text{MT CO}_2 \text{ eq./kWh}) \end{pmatrix} \right\}, \end{aligned} \tag{2}$$

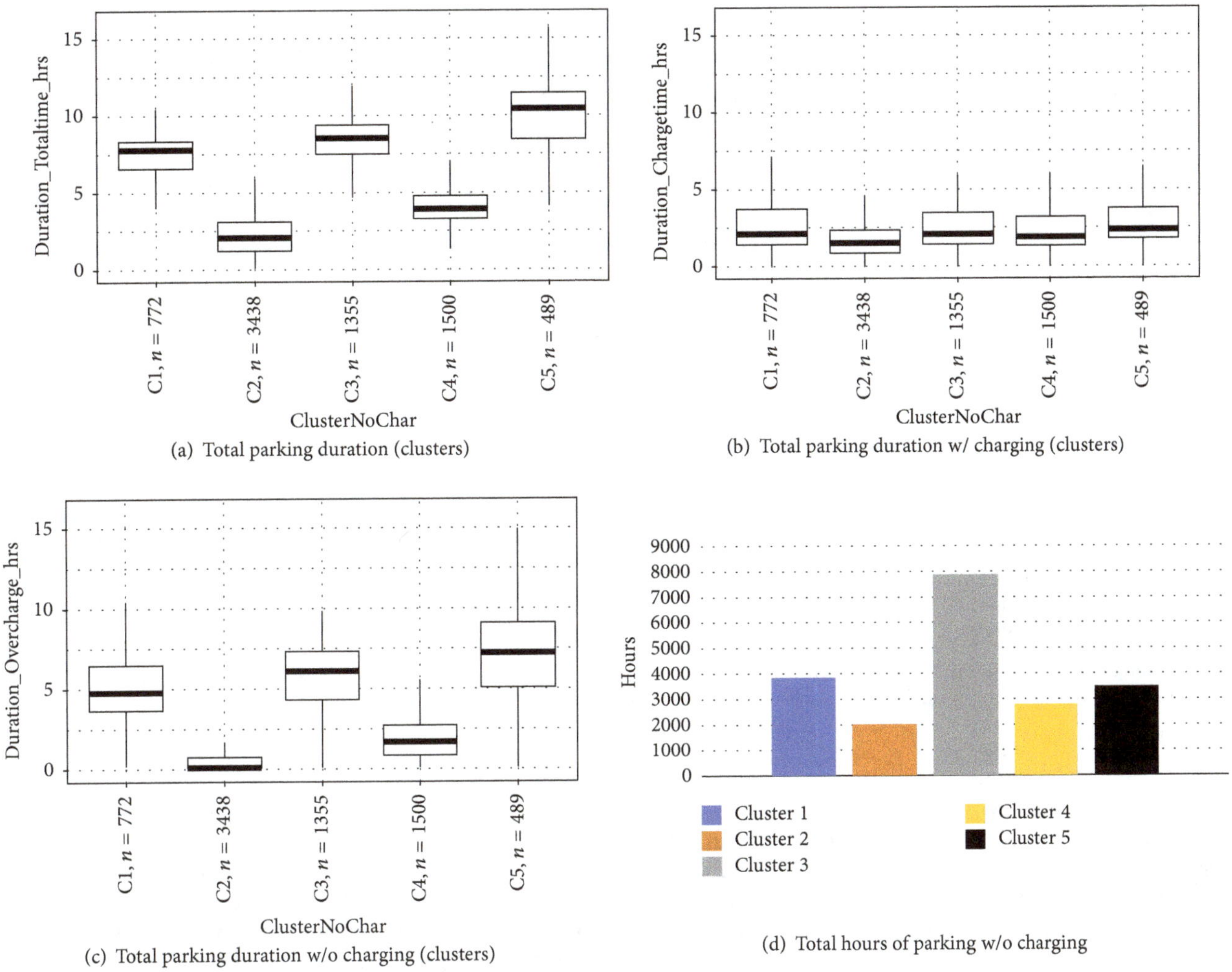

(a) Total parking duration (clusters)

(b) Total parking duration w/ charging (clusters)

(c) Total parking duration w/o charging (clusters)

(d) Total hours of parking w/o charging

FIGURE 6: Variation of activities across clusters: (a) parking duration (hours), (b) parking duration with charging (hours), (c) parking duration without charging (hours), and (d) total (i.e., sum of all the daily patterns in the respective cluster). Hours of parking without charging across clusters.

where $Avg._{PWOC}$ is the average time spent in parking without charging in hours per daily profile in respective cluster, $P_{avg.}$ is the assumed average charging rate of 3 kW, which is averaged rate across all the individual plug-in events in the observed sample in (kW), and η_e is the electrical efficiency of PEVs in (miles/kWh) and is assumed based on the literature: 3 miles/kWh [31]—for increase and decrease in electrical efficiency, the GHG emissions savings vary in proportion and E is electricity consumed per average profile of respective cluster in (kWh). Given that consumption of 1 gallon of gasoline emits 8.887×10^{-3} metric tons (MT) of CO_2 eq. [32]—for gasoline, CO_2 and CO_2 equivalent being the same because gasoline combustion results in only CO_2 and H_2O—the total GHG emissions displaced by allowing more station charging can be calculated. The analysis was carried out for each individual cluster identified.

To determine the GHG emissions from the electricity consumed, hourly electricity consumption is assumed to correspond to the number of PEV drivers using the facility for charging at that particular hour. Additionally, for upstate New York, the electricity production mix changes at each hour depending upon the demand for electricity [33], resulting in varying environmental impact each hour per kW-hr of charge. Figure 7 depicts the global warming potential (GWP) in terms of kg of CO_2 equivalent for one unit (1 kWh) of electricity production for a varying grid mix of upstate New York [32] for a typical day.

Figure 7 shows that every 1 kWh of electricity produced from 12:00 AM until 8:00 AM is cleaner compared to the rest of the day and the GWP stays below 0.5 kg of CO_2 equivalent. However, after 8:00 AM, the GWP of electricity production starts increasing and around 6:00 PM it has the highest GWP of 0.56 kg of CO_2 eq. because at this hour the electricity grid mix has most of the supply from fossil fuel power plants to meet the additional peak demand. Therefore, every 1 kWh of electricity consumed by the PEV drivers during a day has GWP directly proportional to the GWP shown in Figure 7 at that hour. Figure 7 further shows the GHG emissions for

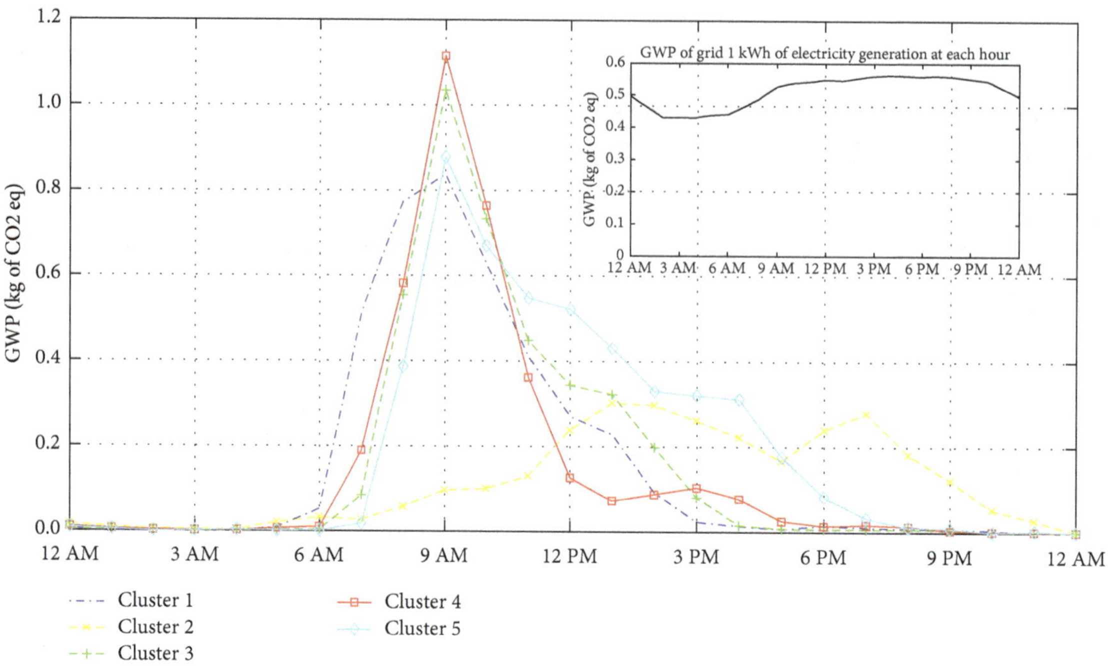

FIGURE 7: Global warming potential (GWP) in terms of kg of CO_2 eq. for electricity consumed by an average daily profile of respective cluster at each hour and emissions from grid for generation of 1 kWh at each hour.

an average profile belonging to each cluster. CL-4 at 9:00 AM shows the highest GHG emissions, followed by CL-3 and CL-5, respectively. Note this peak occurs at a specific time.

Looking at average daily profiles for each cluster and their GHG emissions savings indicate opportunities for significant emissions savings. Overall, if efficiency in PEV user operations is improved, that is, the PDWOC is reduced, significant GHG emissions savings can be realized. For an average daily profile, CL-5 has the maximum total GHG emissions savings potential across the day, largely because profiles in this cluster have the highest average parking without charging time over the day. CL-3 and CL-1 have the second and the third largest GHG emissions savings potential for similar reasons based on the entire daily GHG savings. Finally, CL-2 has the least GHG emission savings, due to low PDWOC, over the day, relative to the other clusters (Figure 6(c)). Overall, the results suggest that displacing inefficient parking at charging stations with PEV charging from additional PEVs can lead to significant GHG emission savings, even in the case of CL-2 which has a lower median of PDWOC relative to the other clusters.

As stated earlier, in Section 4.2, the CL-3 has the highest PDWOC (Figure 6(d)), with respect to the sum of all the daily charging profile in the cluster; proportionately this cluster has 7880 hours of total parking without charging duration, which is the highest among all five clusters. However, a daily charging profile belonging to CL-5 exhibits the longest PWOC activity (starting at 9:00 AM until 12:00 AM, shown in Figure 5(e)). This can also be referred to in Figure 5(c); CL-5 has the highest median of PDWOC per daily profile, followed by CL-1 and CL-3, respectively. If all the clusters were of the same membership size, the CL-5 would have had the largest total parking duration without charging and hence the maximum GHG emissions savings potential. If the hour-specific electricity consumption is considered as a criterion, a CL-4 profile has the highest GHG emissions at 9:00 AM (Figure 7), followed by CL-3 and CL-5, respectively, which corresponds to the charging activity of an average profile belonging to a particular cluster at 9:00 AM.

As this dataset sample may not represent the actual population of charging events, we cannot draw inferences at the population level. However, based on this sample, for realizing GHG emissions savings, we can target the daily profile represented in CL-5, as they exhibit the highest PDWOC on a typical day, or we can target the daily profile represented in CL-3, as the CL-3 has the highest total parking without charging duration (in total hours). If the calculations of GHG savings are performed considering efficiency of gasoline vehicles, with respect to the projections of expected improvements in miles per gallon (mpg) as per CAFE standards [34], the amount of GHG emissions displaced varies in proportion to the gasoline vehicle efficiencies. For higher gasoline engine efficiency, the emission savings are less and vice versa.

6. Conclusion

Our results indicate that the daily charging profiles of PEV drivers observed at the public charging stations can be clustered into five different groups based on their charging behavior. The five identified patterns display distinct PEV charging behaviors with respect to parking with charging activity and parking without charging activity. These five clusters indicate that though the PEV drivers have variability in their charging profiles, the median for charging durations are similar (1.5 hours to 2.5 hours). However, they have significant durations of parking without charging, which can be conceptualized as "inefficient operation of charging stations." This inefficiency is about 54% of the total usage time of the charging stations. The GHG emissions analysis suggests this inefficiency, if reduced, can have a positive environmental impact (in terms of GHG emissions savings) resulting from an assumed displacement of gasoline vehicles.

Several extensions to this work are envisioned. First, the analysis precluded consideration of behavioral changes. To address this, a queueing simulation model of station operations could be implemented based on this data, providing insights into shifts in station turnaround under different policies. The queueing simulation can further be used to analyze a policy of maximum allowed time to use public charging stations to reduce this inefficiency. Second, empirical data on charging prior to reaching the stations could be collected providing a clearer picture of charging needs at the station and subsequently could be used to reduce the duration of parking without charging.

Conflicts of Interest

The authors declare that they have no conflicts of interest.

References

[1] International Energy Agency, "Global EV outlook 2017 two million and counting," 2017, https://www.iea.org/publications/freepublications/publication/GlobalEVOutlook2017.pdf .

[2] R. Philip and A. Wiederer, *Policy Options for Electric Vehicle Charging Infrastructure in C40 Cities*, 2010.

[3] J. Woetzel, S. Sha, and H. Zhang, "Electric Vehicles in Megacities—Shanghai Charges up," 2010.

[4] International Energy Agency, *E-Mobility Technology Roadmap*, Singapore, 2014.

[5] T. Franke and J. F. Krems, "Understanding charging behaviour of electric vehicle users," *Transportation Research Part F: Traffic Psychology and Behaviour*, vol. 21, pp. 75–89, 2013.

[6] L. Bunce, M. Harris, and M. Burgess, "Charge up then charge out? drivers' perceptions and experiences of electric vehicles in the UK," *Transportation Research Part A: Policy and Practice*, vol. 59, pp. 278–287, 2014.

[7] N. S. Pearre, W. Kempton, R. L. Guensler, and V. V. Elango, "Electric vehicles: how much range is required for a day's driving?" *Transportation Research Part C: Emerging Technologies*, vol. 19, no. 6, pp. 1171–1184, 2011.

[8] J. Bailey, A. Miele, and J. Axsen, "Is awareness of public charging associated with consumer interest in plug-in electric vehicles?" *Transportation Research Part D: Transport and Environment*, vol. 36, pp. 1–9, 2015.

[9] F. He, D. Wu, Y. Yin, and Y. Guan, "Optimal deployment of public charging stations for plug-in hybrid electric vehicles," *Transportation Research Part B: Methodological*, vol. 47, no. 2013, pp. 87–101, 2013.

[10] F. He, Y. Yin, and J. Zhou, "Deploying public charging stations for electric vehicles on urban road networks," *Transportation Research Part C: Emerging Technologies*, vol. 60, pp. 227–240, 2015.

[11] J. C. Kelly, J. S. MacDonald, and G. A. Keoleian, "Time-dependent plug-in hybrid electric vehicle charging based on national driving patterns and demographics," *Applied Energy*, vol. 94, pp. 395–405, 2012.

[12] L. Gan and S. Low, "Optimal decentralized protocol for electric vehicle charging," *Power*, vol. 6, no. 1, pp. 1–10, 2007.

[13] B. M. Davis and T. H. Bradley, "The efficacy of electric vehicle time-of-use rates in guiding plug-in hybrid electric vehicle charging behavior," *IEEE Transactions on Smart Grid*, vol. 3, no. 4, pp. 1679–1686, 2012.

[14] N. Shahraki, H. Cai, M. Turkay, and M. Xu, "Optimal locations of electric public charging stations using real world vehicle travel patterns," *Transportation Research Part D: Transport and Environment*, vol. 41, pp. 165–176, 2015.

[15] J. E. Kang and W. W. Recker, "An activity-based assessment of the potential impacts of plug-in hybrid electric vehicles on energy and emissions using 1-day travel data," *Transportation Research Part D: Transport and Environment*, vol. 14, no. 8, pp. 541–556, 2009.

[16] N. Caperello, K. S. Kurani, and J. TyreeHageman, "Do you mind if i plug-in my car? how etiquette shapes PEV drivers' vehicle charging behavior," *Transportation Research Part A: Policy and Practice*, vol. 54, pp. 155–163, 2013.

[17] M. V. Faria, P. C. Baptista, and T. L. Farias, "Electric vehicle parking in European and American context: economic, energy and environmental analysis," *Transportation Research Part A: Policy and Practice*, vol. 64, pp. 110–121, 2014.

[18] T. Litman, "Generated traffic and induced travel: implications for transport planning," *ITE Journal*, vol. 71, no. 4, pp. 38–47, 2001.

[19] United Stated Census Bureau, "U.S. department of commerce," 2017, https://www.census.gov/.

[20] ChargePoint Inc, "ChargePoint Inc," 2017, https://na.chargepoint.com/charge_point.

[21] City of Rochester, "City of Rochester," City of Rochester, 2016, http://www.cityofrochester.gov/evcharging/.

[22] Rochester Downtown Development Corporation, "Rochester Down Town," City of Rochester, 2016, http://www.rochesterdowntown.com/.

[23] Market Ventures Inc., "Rochester public market renovation," 2012, http://www.marketventuresinc.com/index.php/business-planning.

[24] Monroe County New York, "Monroe County-Ontario Beach Park," Monroe County, 2015, http://www2.monroecounty.gov/parks-ontariobeach.php.

[25] GIS Data, "Monroe County," 2017, https://www2.monroecounty.gov/gis-Data.php.

[26] American FactFinder, "United States Census Bureau," U.S. Department of Commerce, 2017, https://factfinder.census.gov/.

[27] S. Theodoridis and K. Koutroumbas, *Pattern Recognition*, Academic Press, 4th edition, 2008.

[28] W. W. Recker, M. G. McNally, and G. S. Root, "Travel/activity analysis: pattern recognition, classification and interpretation," *Transportation Research Part A: General*, vol. 19, no. 4, pp. 279–296, 1985.

[29] J. L. Walsh, "A closed set of normal orthogonal functions," *American Journal of Mathematics Published by: The Johns Hopkins University Press*, vol. 45, no. 1 (Jan., 1923), pp. 5–24, 2016.

[30] J. B. MacQueen, "*K*means Some methods for classification and analysis of multivariate observations," *Berkeley Symposium on Mathematical Statistics and Probability*, vol. 1, no. 333, pp. 281–297, 1967.

[31] Idaho National Laboratory, "Advanced vehicle testing activity INL/MIS-11-22490," 2010.

[32] US-EPA, "GHG equivalencies calculator—calculations and references," 2016, https://www.epa.gov/energy/ghg-equivalencies-calculator-calculations-and-references.

[33] http://www.nyiso.com/public/index.jsp, NYISO (Markets & Operations - Market Data - Graphs), 2016, http://www.nyiso.com/public/markets_operations/market_data/graphs/index.jsp.

[34] U.S.-EPA, "EPA and NHTSA propose to extend the national program to reduce greenhouse gases and improve fuel economy for cars and trucks," 2011.

11

Innovative Bike-Sharing in China: Solving Faulty Bike-Sharing Recycling Problem

Shan Chang, Rui Song, Shiwei He, and Guo Qiu

MOE Key Laboratory for Urban Transportation Complex Systems Theory and Technology, Beijing Jiaotong University, Beijing, China

Correspondence should be addressed to Rui Song; rsong@bjtu.edu.cn

Academic Editor: Samiul Hasan

In China, based on the mobile Internet technology and global positioning system (GPS), innovative bike-sharing is different from traditional bike-sharing system with docking station, for its flexibility and convenience. However, innovative bike-sharing system faces operational challenges, especially in faulty bike-sharing recycling (FBSR) problem. In this paper, a framework is designed based on the optimization method to solve the FBSR problem so that it can minimize the total recycling costs by taking the route optimization and loading capacity ratio as constraints. The FBSR method combines the K-means method for clustering faulty bike-sharing with planning recycling route for operational decisions. Moreover, CPLEX solver is used to obtain the desired result of the FBSR model. Finally, a case study based on a certain area in Beijing, China, is used to verify the validity and applicability of the model. The results show that the value of loading capacity ratio and the number of clustering points greatly affect the results of FBSR problem. Four vehicles are designated to execute FBSR tasks required by different clustering points. This study is of considerable significance for the bike-sharing promotion in the last-mile situation to the real problems arising in the initial period.

1. Introduction

Bike-sharing systems are becoming increasingly popular in cities around the world because they are cheap, efficient, healthy, and green. In recent years, with the development of mobile Internet and global positioning system (GPS) becoming increasingly affordable [1], an innovative bike-sharing system without docking station emerged in 2015 in China. Bike-sharing system can effectively solve the last-mile problem in multimodal transportation; in addition, many Chinese cities show more cycling than the other international cities [2], which leads to wide spread of bike-sharing systems across China now. Spatial distribution of main dockless bike-sharing systems in China is shown in Figure 1.

Compared with traditional bike-sharing [3, 4], dockless bike-sharing is more flexible and more convenient for the users. Bicycles of bike-sharing systems are widely distributed, thus reducing travel distance from traveler to bicycle docking station. It is convenient for users to use bike-sharing; when users use bike-sharing for the first time, first, users should register directly in the app and scan their identification and give a deposit. Then, the smart bike-sharing app will show a map of all the bicycles around users and locate the closest bike. After that, users scan the Quick Response code with the app and within 10 seconds they can hear the click and hit on the road. However, with all the benefits of this innovative bike-sharing system, here come operational challenges.

It is worth noting that there are a large number of people cycling every day. It is unavoidable that bike-sharing may malfunction in its routine use, and accidents might take place due to its faults. According to some statistics, there are more than 10 million bicycles in bike-sharing systems, and faulty bicycles rate slightly less than 1%. If we calculate with rate of 1% faulty bicycles, there will be 100,000 faulty bicycles in China. Meanwhile, lots of bicycles will be scrapped usually in three years in China. These factors could cause the following: (a) the presence of the faulty bike-sharing severely threatens users' safety; (b) the service quality of bike-sharing system will affect the reputation of the companies; (c) faulty bicycles in the city have a very

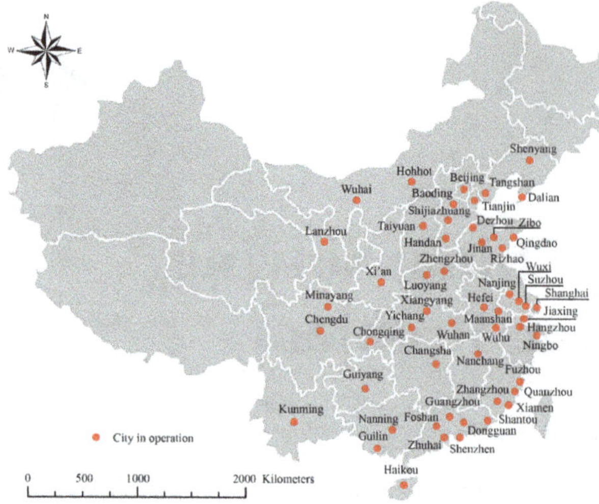

FIGURE 1: Spatial distribution of main dockless bike-sharing systems in China.

bad effect on the city appearance. Therefore, faulty bike-sharing recycling (FBSR) is a very significant issue that should be solved. Moreover, the distribution of bicycles of innovative bike-sharing system is rather dispersed since bike-sharing can park without docking station, whereas for traditional bike-sharing system, faulty bicycles are readily discovered and easily processed at bicycle docking station. Comparatively speaking, it is more difficult for the innovative bike-sharing system to recycle faulty bicycle than traditional bike-sharing. Thus, solving the FBSR problem is the key to reduce recycling costs for the authorities and operators with the means.

Therefore, this paper presents a framework design and optimization method to solve FBSR problem. Firstly, the K-means clustering method is used to divide the faulty bike-sharing into different service points. Then, a FBSR model is established to minimize the total recycling costs with loading capacity ratio as a constraint. At last, this method is verified based on a case study in Beijing, China.

The remaining part of this paper is organized as follows. The relevant literature is reviewed in Section 2. The framework design of faulty bike-sharing recycling is presented in Section 3. Section 4 describes the modeling methodologies and model specifications. In Section 5, the empirical results of the models and the effects of the explanatory variables are presented. The final section provides a summary of the research findings with possible extensions of the research.

2. Literature Review

Since a bike-sharing system was first introduced in the 1960s [5], lots of bike-sharing systems have been implemented throughout the world [6, 7], and now the bike-sharing system has become a necessary for city transportation. According to research by different scholars, the bike-sharing system has potentials due to the following reasons: (a) reducing traffic congestion and fuel use [8]; (b) behavioral shifts towards increased bicycle use for daily mobility [9]; (c) a growing perception of the bicycle as a convenient transportation mode [1].

The existing literature on bike-sharing systems is mainly focused on the traditional bike-sharing system with docking stations including development policies and safety issues [10–12], operation scales and the operator[8, 13], and benefit on the environment [14–18]. In order to improve the user experience for bicycle use and reduce the operating costs, many researchers study rebalancing operations. Chemla et al. [19] were the first to introduce rebalancing operations for bike-sharing system and proposed tabu search algorithms for solving it. Schuijbroek et al. [20] presented unified dual-bounded service level constraints that add inventory flexibility and vehicle routing for static rebalancing in bike-sharing systems. Ghosh et al. [21] developed an optimization formulation to reposition bikes using vehicles while considering the routes for vehicles and future expected demand to improve bike-sharing availability and reduce the usage of private vehicles. Pfrommer et al. [22] proposed a heuristic algorithm for solving the dynamic rebalancing problem with multiple vehicles and presented a dynamic pricing strategy that encouraged users to return bikes to empty stations. However, the bike-sharing system in a city does not have the restriction of fixed docking stations. So the existing studies cannot solve the FBSR in China.

Recently, the new bike-sharing system without docking station has attracted attention from a few scholars. Reiss and Bogenberger [23] created a demand model to obtain an optimal distribution of bikes within the operating area, based on a detailed GPS-data analysis for the bike-sharing system. Caggiani et al. [24] suggested a method to use the revenue collected by a congestion price policy to implement a bike-sharing system. Pal and Zhang [25] proposed a hybrid nested large neighborhood search with variable neighborhood descent algorithm, which can solve static rebalancing problems for bike-sharing system. Caggiani et al. [26] proposed an operator-based bike redistribution methodology that started from the prediction of the number and position of bikes over operating area and ended with a decision support system for the relocation process. Cheng and Gao [27] studied the revenue changes of the platform after the bike-sharing platform adopted the monthly strategy and obtained the best monthly subscription pricing to maximize the platform revenue. However, compared with bike-sharing scale in China, these studies were based on case studies that are of much smaller scale. Therefore, FBSR problem deserves more attention.

In addition, Cervero et al. [28] found that the single strongest predictor for bicycle use is the availability of a bike. The FBSR is one of the important means to ensure the availability of bike-sharing. However, up to now, there is no research on faulty bike-sharing recycling in the literature. Thus, this study bridges this gap by reporting a solution algorithm that addresses this issue and helps bike-sharing companies to reduce operating costs particularly in Chinese cities.

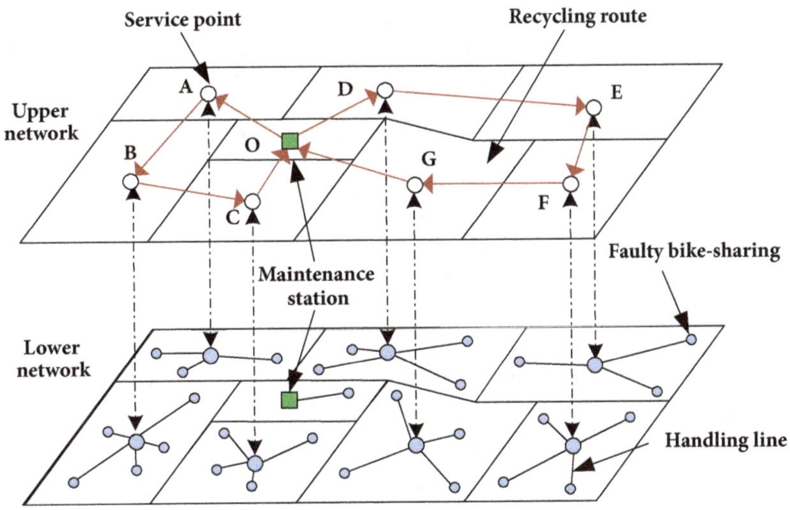

FIGURE 2: Multilevel faulty bike-sharing recycling network.

3. Framework Design of FBSR

3.1. Faulty Bike-Sharing Definition and Classification. In this paper, three types of bicycle as faulty bicycles are defined as follows.

(i) Fault of cycling: it happens when the users unlock the bicycles and find that the bicycle cannot be used; then the users will send the information through the mobile phone app. When the bike-sharing system server receives the information of the faulty bicycle status and location, the bike-sharing company will have it repaired.

(ii) Fault of communication: it happens when bicycles are identified as faulty bike-sharing owing to GPS equipment damage, and the system server will prohibit users from using this bicycle (ignoring GPS equipment lost contact while cycling).

(iii) End of the service life: if a bicycle has reached the state's mandatory write-off standard (generally three years in China), the bike-sharing system server will identify the bicycle as faulty bike-sharing. Then this bicycle will be forced to scrap, prohibiting users from cycling.

3.2. Recycling Rules. Multilevel faulty bike-sharing recycling network is shown in Figure 2, which is divided into upper network and lower network. The lower network shows that faulty bike-sharing in multiple locations within a certain area will be concentrated at clustering service point in the area, where blue dots indicate faulty bike-sharing in some locations and the lines among some dots stand for handing line. The upper network indicates that the clustering service points of each area are obtained, where dots indicate service point, red line shows the line direction, and the green one represents maintenance station. The FBSR model can be applied to the best routes so that total recycling costs are minimized.

For the description, ground rules are established as follows.

(1) A number of maintenance stations are established in a city. Each maintenance station is in a fixed location. The maintenance station provides faulty bicycles maintenance service in a certain area, and faulty bicycles are recycled once a day. It should be noted that the faulty bicycles will not be classified. The faulty bike-sharing information is collected at a fixed time every day; if there are any updates about faulty bike-sharing, they will be dealt with the next day.

(2) Within each local area, faulty bicycles are clustered for different service points according to their locations, which are made in lower network such as point A in Figure 2. To perform a task, a service vehicle will be parked in the clustering service point where the faulty bicycles are carried back by hands. Then service vehicle will move on to the next service point such as point B in Figure 2 after the task is completed at point A. Finally, the faulty bike-sharings of each clustering service point are sent to maintenance point along the planned best line.

(3) There are a number of recycling vehicles in maintenance station. These vehicles traverse all service points from maintenance station in order to recycle all faulty bicycles. It is worth noting that the service point is not fixed, and the service point should have enough space for recycling vehicle to park.

Therefore, there are two key issues to deal with to complete one FBSR in an area:

(i) How to determine the number and the location of bike-sharing service points that are available for recycling vehicles.

(ii) How to determine routes for recycling vehicles to traverse all service points from maintenance station.

3.3. Recycling Framework. Based on the analysis of faulty bike-sharing definition, classification, and recycling rules, the flowchart of faulty bike-sharing recycling is shown in Figure 3. There are four steps in the framework of FBSR, which are the data collection and processing, the faulty bike-sharing clustering, the planning of recycling route, and the performing of recycling task. The details of recycling framework can be described as follows.

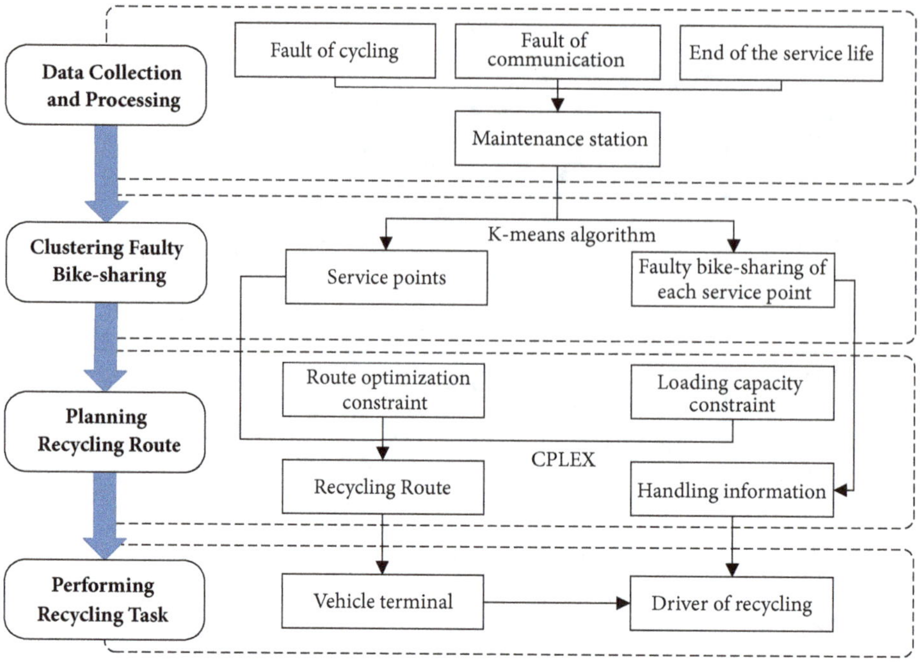

FIGURE 3: Flowchart of faulty bike-sharing recycling.

Step 1 (data collection and processing). There are three types of faulty bike-sharing in the section above. For type 1, the system server will directly set it as faulty bike-sharing to be processed. For type 2, when the location information of bike-sharing is lost, the system needs to record the last received location information. Type 1 and type 2 bicycles will be stored in the faulty bike-sharing database. For type 3, when a bike-sharing company launches bicycles in the city every time, the system server will set the time of the bicycles' service life, after which the time of service life of these bicycles will be stored in the faulty bike-sharing database by system server. By confirmation and processing of three types above, the bike-sharing company will have these faulty bikes fixed.

Step 2 (faulty bike-sharing clustering). Faulty bike-sharing clustering is represented by lower network as shown in Figure 2. The system will use the K-means algorithm to cluster the faulty bicycles to different service points in each area according to the result of data collection and processing. Based on the different service point, the faulty bicycles are carried to this location by manual handling and wait for company trucks to carry them in each clustering service point along the planned route driving to maintenance point.

Step 3 (planning recycling route). Planning recycling route is represented by the upper network as shown in Figure 2. Every recycling route is planned based on the result of faulty bike-sharing clustering. Meanwhile, recycling route optimization tries to achieve minimum total recycling costs with route optimization and vehicle capacity as constraints. The recycling vehicles start from the maintenance station, collect all the faulty bicycles, and finally return to the station.

Step 4 (performing recycling task). According to the result of Step 2 and Step 3, the recycling route and the faulty bike-sharing location information are sent to the vehicle terminal so that every driver of recycling vehicle can perform the task of recycling faulty bicycles.

It is worth noting that Step 2 and Step 3 are the key scientific problems. The FBSR aims at recycling all faulty bicycles and traversing all service points while optimizing the routes for recycling vehicles in order to reduce the costs of recycling.

4. Model Formulation

Based on framework design of faulty bike-sharing recycling, the FBSR model in this paper mainly consists of two parts: faulty bike-sharing clustering and recycling route modeling.

4.1. Faulty Bike-Sharing Clustering. K-means algorithm [27] is a method commonly used to automatically partition a dataset into k groups. We use it to cluster the faulty bicycles to different service points in an area. Suppose that a data set $\{x_1, ..., x_N\}$ consists of N observations of a random 2-dimensional Euclidean variable x. The objective is to partition the dataset into some number K of clusters, given the value of K. First, introduce a set of 2-dimensional vectors c_k, where $C = \{c_k, k = 1, ..., K\}$; c_k represents the center of the kth cluster. Second, find an assignment of data points to clusters, so that the sum of the squares of the distances of each data that points to its closest vector c_k is a minimum.

$$\min \quad SSE = \sum_{n=1}^{N}\sum_{k=1}^{K} r_{nk} \left\| x_n - c_k \right\|^2 \tag{1}$$

$$r_{nk} = \begin{cases} 1, & x_n \in c_k \\ 0, & x_n \notin c_k \end{cases} \quad (2)$$

where SSE (sum of squares for error) is the sum of the distances from the data points to the clustering points, binary indicator variables $r_{nk} \in \{0, 1\}$, and k describes which cluster the data point x_n is assigned to, so that if x_n is assigned to cluster k, then $r_{nk} = 1$, and $r_{nj} = 0$ for $j \neq k$.

4.2. Recycling Route Modeling. We develop a recycling route model, where the decision-maker (the transit authority or operator) wishes to determine recycling route while minimizing total recycling costs. It is assumed that the result of faulty bike-sharing clustering remains unchanged throughout that period. Consider a network rooted at depot (the maintenance station) named {0} and let the depot be denoted as node 0; S_0 is the set of nodes including the maintenance station and service points; that is, $S = S_0 \backslash \{0\}$. We suppose that travel time of recycling vehicles retains their characteristics throughout that period. Consider a maintenance station that consists of H vehicles, $h = 1, \ldots, H$. Identical recycling vehicles with capacity cap_h serve the route for all service points, $\forall i \in S_0, \forall j \in S_0, i \neq j$, starting from depot ($i = 0$) and ending at depot ($j = 0$). The travel time (ignoring dwell time) of each vehicle between successive points i and j is denoted by t_{ij}. In order to model FBSR problem, we make assumptions as follows:

(i) All route cycles are repeated with identical characteristics.

(ii) The round-trip distance between the two service points is the same.

(iii) The location of the faulty bicycle without positioning is accurate.

Thus, the optimization problem is

$$x_{ijh} \begin{cases} 1 & \text{if vehicle } h \text{ traverses arc}\,(i, j) \\ 0 & \text{otherwise} \end{cases}$$
$$y_{ih} \begin{cases} 1 & \text{if vehicle } h \text{ serves node } i \\ 0 & \text{otherwise} \end{cases} \quad (3)$$

$$\text{Minimize} \quad C_m \sum_{i\in S_0} \sum_{j\in S_0} \sum_{k\in H} t_{ij} x_{ijh} \quad (4)$$

$$\text{Subject to} \quad \sum_{i\in S_0} x_{ijh} = y_{hj} \quad \forall j \in S_0,\ h \in H \quad (5)$$

$$\sum_{j\in S_0} x_{ijh} = y_{hi} \quad \forall i \in S_0,\ h \in H \quad (6)$$

$$\sum_{k\in H} y_{hi} = 1 \quad \forall i \in S \quad (7)$$

$$\sum_i y_{hi} d_i \leq cap_h \quad h \in H \quad (8)$$

$$y_{hi} \in \{0, 1\} \quad \forall i \in S_0,\ h \in H \quad (9)$$

$$x_{ijh} \in \{0, 1\} \quad \forall i, j \in S_0,\ h \in H \quad (10)$$

The objective function minimizes the total recycling costs. It is multiplied by C_m, which indicates recycling vehicles operating costs per unit time. d_i indicates the number of bicycles handled by hands at node i.Constraint (6) states that if a vehicle h decides to serve node j which is serviced by only one vehicle, then it should travel along the arc from that node i, that is, arc(i, j). Constraint (7) states that a vehicle h has traversed arc(i, j) from node i which is serviced by only one vehicle; then it should travel along the arc leading to node j. Constraint (6) and constraint (7) ensure that the path of each vehicle is successive. Constraint (8) ensures that every node is serviced by only one vehicle, whereas constraint (9) ensures that the vehicle capacity is not exceeded. In order to model the entire process of FBSR, the formulation can easily accommodate fixed costs by modification of the objective function as follows:

$$C_m \sum_{i\in S_0} \sum_{j\in S_0} \sum_{h\in H} t_{ij} x_{ijh} + C_n \sum_{i\in S_0} \bar{t} n_i \quad (11)$$

The objective includes two components. The first refers to the objective function (5). The second refers to the expectation of the sum of manual handling costs, where C_n indicates faulty bicycles manual handling costs per unit time, $\bar{t}$ is the average time of handling each faulty bicycle, and, based on formula (1), n_i is the number of faulty bicycles in the ith service points.

Let us define

$$\bar{t} = \frac{2 \times SSE}{\bar{v} \sum_{i\in S} n_i} \quad (12)$$

where $\bar{v}$ is the average velocity of handling each faulty bicycle. It is multiplied by 2 because each vehicle must return to the clustering points after service. Objective function (11) is now

$$C_m \sum_{i\in S_0} \sum_{j\in S_0} \sum_{h\in H} t_{ij} x_{ijh} + C_n \frac{2 \times SSE}{\bar{v}} \quad (13)$$

In order to improve the utilization rate of recycling vehicles, we propose a constraint to restrict load capacity as follows:

$$\sum_{i\in S} y_{hi} n_i \geq \lambda cap_h \quad \forall h \in H \quad (14)$$

where λ is the coefficient of load capacity ratio.

Besides, in order to avoid the loop in the route, it is sufficient that problem includes the following bonding constraint:

$$U_{ih} - U_{jh} + |S| \cdot x_{ijh} \leq |S| - 1 \quad \forall i, j \in S_0,\ h \in H \quad (15)$$

where U_{ih} is the auxiliary variable to eliminate the constraint of loop for hth vehicle.

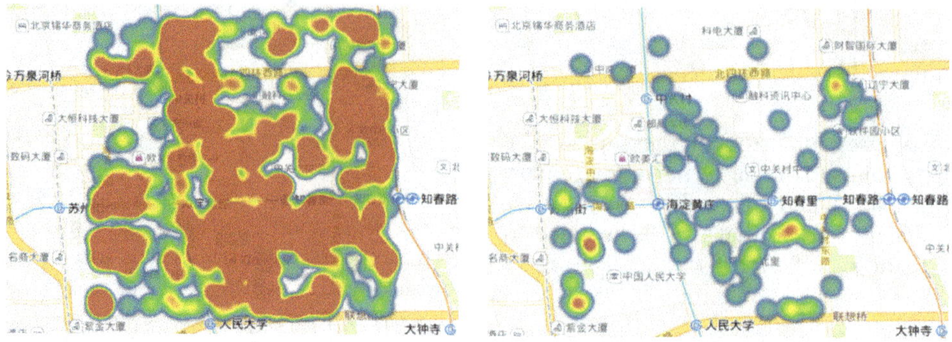

(a) (b)

FIGURE 4: Heat map of bike-sharing.

In the process of problem solving, the recycling route model belongs to 0-1 integer programming model. Considering the scale of the problem, it can be solved by CPLEX solver with branch and bound approach for its advantages of the direct and concise input and strong computation ability.

5. Case Study

In this section, a real-world bike-sharing from an area of Beijing, China, was selected as a case study. The test area is a square area that has a side length of 2.6 km and Haidian Huang Zhuang subway station is selected as areal maintenance station. We set the number of recycling vehicles in a maintenance station as 4, and the capacity of each vehicle is 30 bicycles. There are 1900 bicycles (Figure 4(a)) in this area, and we randomly select 95 bicycles (5% of the total) as faulty bike-sharing, which is shown in Figure 4(b). Red indicates that faulty bicycles are dense, followed by orange, and blue means fewer bicycles in heat map.

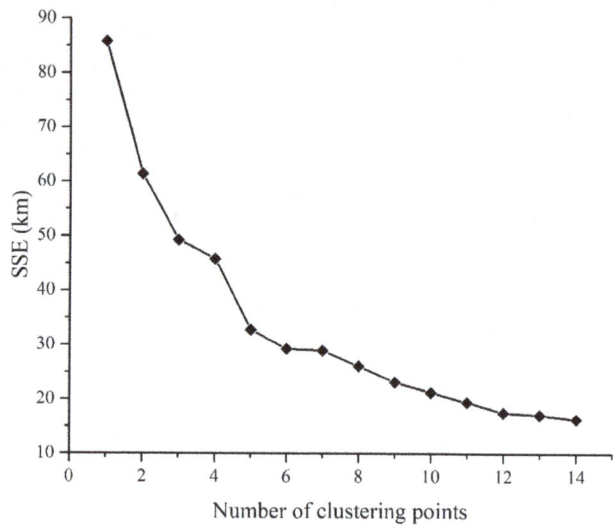

FIGURE 5: The SSE graph of K-means.

5.1. Result of Faulty Bike-Sharing Clustering. We set up 1 to 14 groups clustering experiment. The times of iterations of each clustering algorithm are not more than 20, which proves that the computation is highly efficient. The *SSE* graph of K-means is shown in Figure 5. We show different clustering points in the x-axis and *SSE* of different clustering points in the y-axis. From the chart, it can be seen that the *SSE* of the clustering is stable when we select 12 clustering points, which is 17.531 km. We first select 12 clustering points for experimental analysis.

Latitude and longitude coordinates of the clustering centers are obtained by the K-means algorithm and their locations on a map are shown in Figure 6. From the diagram, it can be seen that the clustering points distribute relatively equally in this area, due to the fact that faulty bikes are distributed relatively evenly.

5.2. Recycling Route Analysis. Considering the result of faulty bike-sharing clustering by K-means algorithm of 12 service points, $|S| = 12$. Travel time t_{ij} between every clustering point is obtained by Google Maps (see Tables 4, 5, and 6), which is calculated based on the shortest time of driving. The coefficient of load capacity ratio λ is 0.7. The recycling vehicle operating costs C_m are 5.5 Renminbi (RMB) per minute, and the faulty bicycles manual handling costs C_n are 0.5 RMB per minute. The average velocity of handling each faulty bicycle $\overline{v}$ is 60 m/min. The optimal results of the FBSR model are solved with the CPLEX solver, as shown in Table 1.

In this example, we can see that the total costs are 677 RMB from the FBSR model. In Table 1, travel time is derived from the recycling route model; carrying time is derived from the clustering results for each service point. It can be seen clearly that each route of total time is between 112 min and 206 min, which demonstrates that the results of faulty bike-sharing clustering and recycling route are reasonable. In addition, the number of bicycles loaded is 24, 23, 25, and 23 for each vehicle; capacity ratio of each vehicle is more than 76.7%. According to the FBSR model, each vehicle achieves an optimized recycling route of FBSR with the vehicle load capacity constraint.

The comparison experiment for the coefficient of load capacity ratio was conducted. Figure 7 shows the number of bicycles loaded for the different capacity ratio, and the higher the coefficient of load capacity ratio is, the more average

TABLE 1: The optimal result of FBSP model.

Vehicle	Capacity ratio	Route	Travel time/min	Carrying time/min	Total time/min	Total costs/RMB
1	80%	$0 \rightarrow 1 \rightarrow 4 \rightarrow 2 \rightarrow 10 \rightarrow 0$	17	95	112	141
2	76.7%	$0 \rightarrow 5 \rightarrow 9 \rightarrow 7 \rightarrow 0$	17	176	193	181.5
3	83	$0 \rightarrow 11 \rightarrow 3 \rightarrow 12 \rightarrow 0$	24	182	206	223
4	76.7%	$0 \rightarrow 6 \rightarrow 8 \rightarrow 0$	12	131	143	131.5
	Total		70	584	654	677

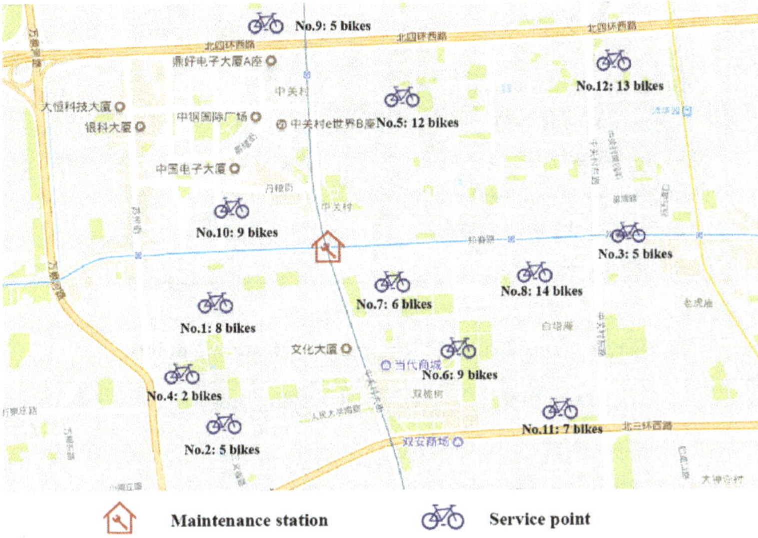

FIGURE 6: The graph of clustering centers by K-means.

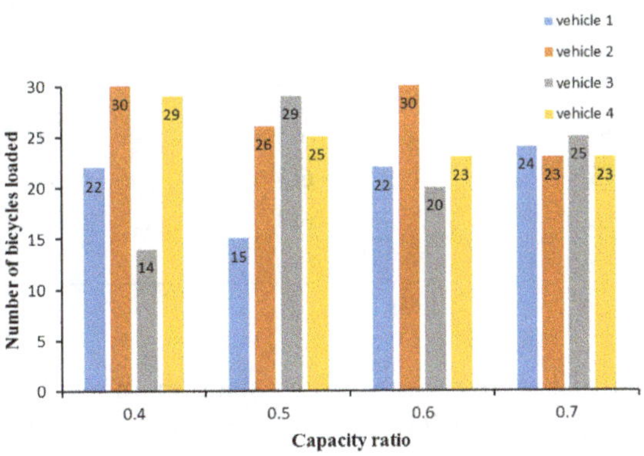

FIGURE 7: The number of bicycles loaded for the coefficient of load capacity ratio.

number of bicycles is loaded. It means that the load capacity constraint has an impact on the result of the FBSR model. In Table 2, we may observe travel time, carrying time, and total cost in several coefficients of load capacity ratios, which shows the optimal result of FBSP model for the different capacity ratio. We can find that travel time and capacity ratio have a positive correlation; that is, the lower the value of the capacity ratio is, the lower the travel time is and the lower the costs are. The lowest value of the capacity ratio is 0.4, which responds to the lowest travel time, 67, and the lowest costs, 660.5. And the highest value of the capacity ratio is 0.7, which responds to the highest travel time, 70, and the highest costs, 677. However, the value of capacity ratio will affect the fairness of the recycling task and the utilization rate of the recycling vehicles; as a result, decision-makers can consider adding the maximum coefficient as objective to FBSR model according to the actual situation of recycling.

Table 3 shows that different clustering points are selected for contrast experiments. According to the experimental results, we can see that the total costs of 10-node clustering points are always the highest with different load capacity rates. Then the results show that the total costs of 12 nodes are more than the total costs of 14 nodes when the coefficient of load capacity ratio is no more than 0.6, and the total costs of 14 nodes are more than those of 12 nodes when the coefficient of load capacity ratio is equal to 0.7. When the number of clustering points is less, the increase in manual operation time will lead to high cost. In addition, the increase in service points will result in an increase of the recycling distance and the higher costs of recycling vehicles operating. Hence, the appropriate recycling scheme should be selected according to the costs changes of C_m and C_n.

TABLE 2: The optimal result of FBSP model for the different capacity ratio.

λ	Vehicle	route	Travel time/min	Carrying time/min	Total costs/RMB
0.4	1	$0 \to 11 \to 2 \to 4 \to 1 \to 0$	67	584	660.5
	2	$0 \to 5 \to 12 \to 3 \to 0$			
	3	$0 \to 10 \to 9 \to 0$			
	4	$0 \to 8 \to 6 \to 7 \to 0$			
0.5	1	$0 \to 1 \to 4 \to 2 \to 0$	68	584	666
	2	$0 \to 5 \to 9 \to 10 \to 0$			
	3	$0 \to 8 \to 6 \to 7 \to 0$			
	4	$0 \to 12 \to 3 \to 11 \to 0$			
0.6	1	$0 \to 1 \to 4 \to 2 \to 11 \to 0$	69	584	671.5
	2	$0 \to 5 \to 12 \to 3 \to 0$			
	3	$0 \to 10 \to 7 \to 9 \to 0$			
	4	$0 \to 8 \to 6 \to 0$			
0.7	1	$0 \to 1 \to 4 \to 2 \to 10 \to 0$	70	584	677
	2	$0 \to 5 \to 9 \to 7 \to 0$			
	3	$0 \to 11 \to 3 \to 12 \to 0$			
	4	$0 \to 8 \to 6 \to 0$			

TABLE 3: The effects of the number of clustering points.

Network	λ	Travel time /min	Carrying time /min	Total time /min	Total costs /RMB
10-node	0.4	65	710	775	712.5
	0.5	65	710	775	712.5
	0.6	65	710	775	712.5
	0.7	65	710	775	712.5
12-node	0.4	67	584	651	660.5
	0.5	68	584	652	666
	0.6	69	584	653	671.5
	0.7	70	584	654	677
14-node	0.4	69	546	615	652.5
	0.5	71	546	617	663.5
	0.6	72	546	618	669
	0.7	74	546	620	680

6. Conclusions

This paper introduces the framework design and optimization method to solve FBSR problem. In China, the number of bike-sharing systems grows rapidly due to its flexibility and convenience for the users. However, they face operational challenges such as FBSR problem. For this reason, we presented a framework design of FBSR to solve the problem of recycling faulty bike-sharing. Then, we propose FBSR optimization model that is able to minimize the total recycling costs through the K-means clustering method that is used to divide the faulty bike-sharing into different service points. It can provide bike-sharing company's managers with good insight into the design of a FBSR problem. The comparison among different service points can help decision-makers to choose the best solution. In addition, the vehicle capacity restriction should be included in a FBSR optimization model to improve the number of bicycles loaded for each vehicle.

A typical example solved by CPLEX solver was created to illustrate the proposed model. The results of a case study show that the model works well. It is indicated that the best route recycles the faulty bicycles according to the clustering points. Compared with the values of capacity ratio models in the FBRS model, we can find that travel time and capacity ratio have positive correlation. And the comparison experiment has shown that the number of clustering points has a significant impact on total recycling costs, which needs to be considered carefully in the FBSR model.

TABLE 4: 10-node travel time matrix for each point (unit: min).

t_{ij}	0	1	2	3	4	5	6	7	8	9	10
0	0	2	6	9	12	4	7	4	7	3	4
1	2	0	4	14	2	6	6	5	12	6	7
2	6	4	0	17	1	12	8	9	16	13	8
3	9	14	17	0	13	7	8	10	7	4	12
4	12	2	1	13	0	12	9	11	13	11	7
5	4	6	12	7	12	0	8	6	8	4	7
6	7	6	8	8	9	8	0	5	4	7	6
7	4	5	9	10	11	6	5	0	7	6	5
8	7	12	16	7	13	8	4	7	0	9	11
9	3	6	13	4	11	4	7	6	9	0	6
10	4	7	8	12	7	7	6	5	11	6	0

TABLE 5: 12-node travel time matrix for each point (unit: min).

t_{ij}	0	1	2	3	4	5	6	7	8	9	10	11	12
0	0	2	6	7	12	4	4	3	5	5	4	5	11
1	2	0	4	13	2	6	5	4	6	12	7	6	16
2	6	4	0	15	1	12	6	13	13	14	8	6	19
3	7	13	15	0	17	11	5	9	4	7	14	3	5
4	12	2	1	17	0	12	12	11	12	12	7	10	16
5	4	6	12	11	12	0	6	6	5	4	7	9	5
6	4	5	6	5	12	6	0	3	3	9	8	6	8
7	3	4	13	9	11	6	3	0	5	6	5	6	12
8	5	6	13	4	12	5	3	5	0	8	7	7	9
9	5	12	14	7	12	4	9	6	8	0	7	8	12
10	4	7	8	14	7	7	8	5	7	7	0	7	11
11	5	6	6	3	10	9	6	6	7	8	7	0	10
12	11	16	19	5	16	5	8	12	9	12	11	10	0

TABLE 6: 14-node travel time matrix for each point (unit: min).

t_{ij}	0	1	2	3	4	5	6	7	8	9	10	11	12	13	14
0	0	2	6	8	12	4	4	2	4	5	4	6	6	8	7
1	2	0	4	13	2	6	11	10	12	12	7	11	15	17	16
2	6	4	0	14	1	12	8	8	13	12	8	8	16	18	17
3	8	13	14	0	17	11	6	9	6	7	14	5	9	5	4
4	12	2	1	17	0	12	9	8	13	11	5	9	15	17	16
5	4	6	12	11	12	0	5	6	5	4	7	8	4	5	4
6	4	11	8	6	9	5	0	3	3	8	8	5	7	10	7
7	2	10	8	9	8	6	3	0	4	8	4	5	8	10	9
8	4	12	13	6	13	5	3	4	0	8	7	7	8	8	7
9	5	12	12	7	11	4	8	8	8	0	6	8	3	6	6
10	4	7	8	14	5	7	8	4	7	6	0	6	9	10	10
11	6	11	8	5	9	8	5	5	7	8	6	0	10	9	8
12	6	15	16	9	15	4	7	8	8	3	9	10	0	3	2
13	8	17	18	5	17	5	10	10	8	6	10	9	3	0	2
14	7	16	17	4	16	4	7	9	7	6	10	8	2	2	0

In the future, with bike-sharing systems growing in China, we will present a solution model for dealing with dynamic faulty bike-sharing recycling. In addition, random searching for faulty bicycle without GPS should also be considered by the researchers.

Conflicts of Interest

The authors declare that they have no conflicts of interest.

Acknowledgments

This research is supported by the National Natural Science Foundation of China (no. U1434207 and no. U1734204).

References

[1] E. Fishman, "Bikeshare: a Review of Recent Literature," *Transport Reviews*, vol. 36, no. 1, pp. 92–113, 2016.

[2] Y. Tang, H. Pan, and Q. Shen, "Bike-sharing systems in Beijing, Shanghai and Hangzhou and their impact on travel behaviour," in *Proceedings of the Transportation Research Board 90th Annual Meeting*, 2010.

[3] S. A. Shaheen, H. Zhang, E. Martin, and S. Guzman, "China's Hangzhou Public Bicycle: Understanding early adoption and behavioral response to bikesharing," *Transportation Research Record*, no. 2247, pp. 33–41, 2011.

[4] Q. Chen and T. Sun, "A model for the layout of bike stations in public bike-sharing systems," *Journal of Advanced Transportation*, vol. 49, no. 8, pp. 884–900, 2015.

[5] P. Demaio, "Bike-sharing: history, impacts, models of provision, and future," *Journal of Public Transportation*, vol. 12, no. 4, pp. 41–56, 2009.

[6] P. J. Demaio, "Smart bikes: Public transportation for the 21st century," *Transportation Quarterly*, vol. 57, no. 1, pp. 9–11, 2003.

[7] S. A. Shaheen, S. H. Guzaman, and H. Zhang, "Bikesharing in Europe, the Americas, and Asia past, present, and future," *Transportation Research Record*, vol. 2143, Article ID 1316350, pp. 159–167, 2010.

[8] A. Audikana, E. Ravalet, V. Baranger, and V. Kaufmann, "Implementing bikesharing systems in small cities: Evidence from the Swiss experience," *Transport Policy*, vol. 55, pp. 18–28, 2017.

[9] Y. T. Hsu, L. Kang, and Y. H. Wu, "user behavior of bikesharing systems under demand–supply imbalance," *Transportation Research Record*, vol. 2587, pp. 117–124, 2016.

[10] L. Aultman-Hall and M. G. Kaltenecker, "Toronto bicycle commuter safety rates," *Accident Analysis & Prevention*, vol. 31, no. 6, pp. 675–686, 1999.

[11] E. Fishman, S. Washington, and N. Haworth, "Bike Share: a synthesis of the literature," *Transport Reviews*, vol. 33, no. 2, pp. 148–165, 2013.

[12] K. Martens, "Promoting bike-and-ride: The Dutch experience," *Transportation Research Part A: Policy and Practice*, vol. 41, no. 4, pp. 326–338, 2007.

[13] H. Nakamura and N. Abe, "Evaluation of the hybrid model of public bicycle-sharing operation and private bicycle parking management," *Transport Policy*, vol. 35, pp. 31–41, 2014.

[14] K. Martens, "The bicycle as a feedering mode: experiences from three European countries," *Transportation Research Part D: Transport and Environment*, vol. 9, no. 4, pp. 281–294, 2004.

[15] A. Kaltenbrunner, R. Meza, J. Grivolla, J. Codina, and R. Banchs, "Urban cycles and mobility patterns: Exploring and predicting trends in a bicycle-based public transport system," *Pervasive and Mobile Computing*, vol. 6, no. 4, pp. 455–466, 2010.

[16] J. Pucher, R. Buehler, and M. Seinen, "Bicycling renaissance in North America? An update and re-appraisal of cycling trends and policies," *Transportation Research Part A: Policy and Practice*, vol. 45, no. 6, pp. 451–475, 2011.

[17] S. Kaplan, F. Manca, T. A. S. Nielsen, and C. G. Prato, "Intentions to use bike-sharing for holiday cycling: an application of the theory of planned behavior," *Tourism Management*, vol. 47, pp. 34–46, 2015.

[18] S.-Y. Chen, "Green helpfulness or fun? Influences of green perceived value on the green loyalty of users and non-users of public bikes," *Transport Policy*, vol. 47, pp. 149–159, 2016.

[19] D. Chemla, F. Meunier, and R. W. Calvo, "Bike sharing systems: solving the static rebalancing problem," *Discrete Optimization*, vol. 10, no. 2, pp. 120–146, 2013.

[20] J. Schuijbroek, R. C. Hampshire, and W.-J. van Hoeve, "Inventory rebalancing and vehicle routing in bike sharing systems," *European Journal of Operational Research*, vol. 257, no. 3, pp. 992–1004, 2017.

[21] S. Ghosh, P. Varakantham, Y. Adulyasak, and P. Jaillet, "Dynamic repositioning to reduce lost demand in bike sharing systems," *Journal of Artificial Intelligence Research*, vol. 58, pp. 387–430, 2017.

[22] J. Pfrommer, J. Warrington, G. Schildbach, and M. Morari, "Dynamic vehicle redistribution and online price incentives in shared mobility systems," *IEEE Transactions on Intelligent Transportation Systems*, vol. 15, no. 4, pp. 1567–1578, 2014.

[23] S. Reiss and K. Bogenberger, "GPS-Data Analysis of Munich's Free-Floating Bike Sharing System and Application of an Operator-based Relocation Strategy," in *Proceedings of the 18th IEEE International Conference on Intelligent Transportation Systems*, pp. 584–589, September 2015.

[24] L. Caggiani, R. Camporeale, and M. Ottomanelli, "Planning and design of equitable free-floating bike-sharing systems implementing a road pricing strategy," *Journal of Advanced Transportation*, vol. 2017, Article ID 3182387, 18 pages, 2017.

[25] A. Pal and Y. Zhang, "Free-floating bike sharing: Solving real-life large-scale static rebalancing problems," *Transportation Research Part C: Emerging Technologies*, vol. 80, pp. 92–116, 2017.

[26] L. Caggiani, R. Camporeale, M. Ottomanelli, and W. Y. Szeto, "A modeling framework for the dynamic management of free-floating bike-sharing systems," *Transportation Research Part C: Emerging Technologies*, vol. 87, pp. 159–182, 2018.

[27] X. Cheng and Y. Gao, "The optimal monthly strategy pricing of free-floating bike sharing platform," *Modern Economy*, vol. 9, no. 2, pp. 318–338, 2018.

[28] R. Cervero, O. Lsarmiento, E. Jacoby, L. F. Gomez, A. Nei-Man, and G. Xue, "Influences of built environments on walking and cycling: lessons from Bogotá," *Urban Transport of China*, vol. 3, no. 4, pp. 203–226, 2016.

12

Uncertainty Analysis for Natural Gas Transport Pipeline Network Layout: A New Methodology based on Monte Carlo Method

Jun Zeng,[1,2] Chaoxu Sun,[3] Zhenjun Zhu,[1,4] Jiangling Wu,[5] and Hongsheng Chen[6]

[1] *School of Transportation, Southeast University, Nanjing 211189, China*
[2] *Department of Civil, Architectural and Environmental Engineering, The University of Texas at Austin, Austin, TX 78712, USA*
[3] *Zhejiang Provincial Natural Gas Development Co. Ltd., Hangzhou 310052, China*
[4] *Department of City and Regional Planning, University of California, Berkeley, Berkeley, CA 94720, USA*
[5] *School of Civil Engineering and Architecture, Henan University, Kaifeng 475004, China*
[6] *School of Architecture, Southeast University, Nanjing 210096, China*

Correspondence should be addressed to Zhenjun Zhu; shangjingqu@163.com

Academic Editor: Zhi-Chun Li

Natural gas plays an increasing important role in the China's energy revolution. The rapid market development and refined government regulation demand improvements in the natural gas transport pipeline network. Therefore, it is of great theoretical and practical significance to conduct a study regarding the layout of pipeline networks. To reflect the comprehensive benefits of pipeline projects and obtain global optimal solution, this study introduces the dominance degree model (DDM). Aiming at optimizing the layout of natural gas transport pipeline networks, this paper studies the uncertainty of the DDM and the corresponding method for network layout. This study proposes an uncertainty analysis based on the Monte Carlo method to quantify the uncertainty of the DDM and its influential factors. Finally, the methodology is applied to the real case of a natural gas transport pipeline project in Zhejiang Province, China. The calculation results suggest that the methodology appropriately addresses the problem of pipeline network layout for natural gas transport. This has important implications for other potential pipeline networks not only in the Zhejiang Province but also throughout China and beyond.

1. Introduction

Natural gas is an efficient and clean energy source that can be utilised in the production of low-carbon energy consumption [1]. According to China's energy development strategic plan and natural gas transport pipeline network plan, the share of natural gas in the primary energy consumption will continue to increase up to 10% by 2020, while the total length of natural gas pipelines is planned to reach 104,000 km. Driven by the continued growth of consumption and infrastructure strengthening, the trunk pipeline coverage will be further expanded and the regional gas transport pipeline network will be improved.

The natural gas transport pipeline network is tasked with gas distribution, which plays a significant role in improving the overall regional socioeconomic conditions. Although pipeline projects require a large investment and have a long payback period, the gas market is exhibiting a dynamic development trend [2]. Therefore, when planning the pipeline network, the benefits from investments on pipeline projects, construction costs, other influential factors, and the impact of uncertainty should be considered to determine the optimal layout scheme.

Previous studies on the layout of natural gas transport pipeline networks have proposed several methods based on pipeline network topology, such as graph theory, dynamic programming, neural networks, genetic algorithms, and complex methods [3, 4]. These methods essentially solve the network structure to satisfy a given criterion and focus on mathematical optimization [5, 6]. The minimum spanning

TABLE 1: Advantages and limitations of existing methods.

Methods	Advantages	Limitations
Graph Theory	Easy access to computer program processing; higher computing efficiency; effectively solving the shortest tree problem within a large scale network	Only solves the network layout among known fixed points; project investment costs are not considered; solutions can only be used as the initial network layout
Dynamic Programming	Solving optimization problems with multiple decision-making variables	Not suitable for dealing with large-scale network systems, dimension obstacles exist in solution process
Neural Network Method	Solving optimization problems with multiple decision-making variables	Only obtaining the local optimal solution
Genetic Algorithm	Higher adaptability that can overcome the difficulties of solving nonlinear optimization	Lower computing efficiency; no effective quantitative analysis concerning algorithm precision, feasibility and computational complexity
Complex Method	The algorithm is simple and suitable for dealing with constrained optimization problems	Unable to deal with multi-variable, multi-constraint optimization problems

tree method (MSTM) and dynamic programming (DP) are the most commonly used solutions. MSTM abstracts the pipeline network into an undirected network, including the classic solutions of the Dijkstra, Kruskal, and Steiner algorithms [7, 8]. Compared to traditional graph theory solutions, these algorithms are implemented by computer programs for processing and have a relatively higher operational efficiency. The Steiner algorithms are effective in solving the shortest path problem of a large-scale network [9–11]. However, the three algorithms mentioned above do not consider the investment costs of pipeline projects, and their results can only be regarded as the initial pipeline network layout. DP can deal with the optimization problem of multiple decision-making variables. However, dimension obstacles exist during the solution process. Specifically, the computation will increase exponentially as the number of variables grows. When the dimension of this problem increases to a certain extent, the problem cannot be solved [12]. Thus, the current commonly used methods have certain advantages and limitations, as presented in Table 1. In summary, the optimization of the natural gas transport pipeline network is a multiobjective nonlinear programming problem, which should consider the uncertainty caused by the gas market and costs. However, this problem cannot be solved easily and effectively by using the abovementioned methods.

With regard to uncertainty and network layout, previous studies have mainly focused on transportation and logistics [13–15]. For example, regarding uncertainty and traffic network layout, Partriksson used a stochastic bilevel programming model to solve the optimal transportation network layout scheme based on the uncertainty of demand [16]. Yin et al. [17, 18] studied the urban road network layout methods under the impact of demand uncertainty and proposed sensitivity-based, scenario-based, and min-max optimization models. Zhang et al. [19] investigated the joint optimization problem of the green vehicle scheduling and routing problem in time-varying traffic networks and developed a corresponding joint optimization model. However, extensive studies on uncertainty associated with and the layout of the natural gas transport pipeline network have not been reported.

Therefore, this study uses the dominance degree model (DDM) of pipeline projects and the corresponding layout method, which considers socioeconomic benefits and construction costs. The layout method based on the DDM is simple: it provides a global optimal solution to obtain the comprehensive benefits of the pipeline network. Therefore, by analyzing the uncertain influential factors of the DDM, this study proposes an innovative uncertainty analysis of the natural gas transport pipeline network layout based on the Monte Carlo method. This proposed method uses the Monte Carlo method and sensitivity analysis to determine the impact of uncertainty factors on the model results. This can quantify the uncertainty and its influence and thus strengthen the practicability of the DDM and function as a future reference for the optimal layout of the pipeline network. Finally, to verify the validity of the methodology, natural gas transport pipeline projects in Zhejiang Province, China, are taken as a case study.

2. Dominance Degree Model and Its Layout Method

The dominance degree model (DDM) of pipeline projects is a new method that was developed to optimize the pipeline network layout [20]. This method used the dominance degree to reflect the comprehensive benefits of transport pipeline projects by combining the potential model (PM) and economic potential theory (EPT) to build the dominance degree model (DDM) for pipeline projects.

The DDM of pipeline projects embodies the comprehensive socioeconomic benefits of the projects. By comparing the dominance degree of pipeline projects when applying the DDM to the layout of natural gas transport pipeline network, the optimal layout scheme and construction sequences are determined, which will leverage the advantages of pipeline projects to obtain the maximum socioeconomic benefits of the pipeline network.

TABLE 2: Calibration for the value of K_i.

P (10,000 People)	K_i
<20	1
20~50	1.25
50~100	1.5
100~300	2.0
300~500	2.5
500~1000	3.0

2.1. Dominance Degree Model. This study combined the potential model (PM) and economic potential theory (EPT) to establish the dominance degree model (DDM) of the pipeline projects. The following assumptions and methods were used when developing the model:

(1) Pipeline projects will generate economic potential, which can be regarded as an index reflecting the city scale.

(2) Construction costs of pipeline projects will directly affect their economic benefits; therefore, construction costs were taken as a parameter reflecting the impedance in the PM.

(3) The influence of pipeline projects on economic potential of the city is closely related to its development level. Pipeline projects will directly affect the transformation of economic potential.

Therefore, some specific indicators can be selected to reflect these influential factors. Above all, the DDM of gas transport pipeline projects is established as follows:

$$A_{ij} = K_i \frac{\mu_i I_i}{VG\beta a_{ij}^2 + a_{ij} \sum_{k=1}^{4} m_k w_k}, \quad (1)$$

where A_{ij} is the dominance degree of pipeline projects between city i and gas transport station j. K_i is the dielectric constant, which is closely related to the urban functional orientation of cities and towns. For the DDM of natural gas transport pipeline projects, the city scale P is selected to reflect this parameter. Considering the adjustment in 2014 of the classification criteria for city scales, K_i can be calibrated as shown in Table 2.

μ_i is related to the richness of urban resources, market development level, living standards of residents, and existing state of the industry. In this study, the macroeconomic indicator of GDP per capita is used to represent this parameter.

I_i is the gas demand of city i, which is related to the local energy consumption structure and policies. To facilitate the analysis of this question, the energy consumption per unit GDP is selected to reflect the status of energy consumption and saving. To some extent, it also reflects the current gas demand and development potential of natural gas.

a_{ij} is the distance between city i and gas transport station j, which can be expressed by a straight-line distance.

V is the pipeline price per unit weight. G is the pipeline weight per unit length. The values of V and G can be obtained from pipelines of the same diameter, material, and thickness, as a standard. β is the terrain correction coefficient, which is used to modify the pipeline route length. Different topography conditions have different values, which can be determined by referring to engineering data within the study area. m_k is the number of crossings of the kth type, and w_k is the corresponding construction cost. $k = 1, 2, 3, 4$ represent a river, grade highway, substandard highway, and railway, respectively.

2.2. Layout Methods Based on DDM. By analyzing and comparing the dominance degree of pipeline projects within the natural gas transport pipeline network, the layout methods based on DDM can not only determine the optimal layout scheme, but also obtain construction sequences of pipeline projects. Thus, the socioeconomic benefits of the pipeline network will be maximized. The basic application process is shown in Figure 1.

According to the layout process, the layout method for a natural gas transport network based on DDM consists of four steps, as follows.

Step 1 (abstract pipeline network in the regional area). First, abstract gas transport stations and valve chests into spatial nodes and record a set of nodes $j = \{1, 2, \ldots, m\}$. Determine their locations on the map by using a geographic information system (GIS). Then, connect these nodes based on pipeline alignments and obtain a spatial structure diagram of the regional pipeline network.

Step 2 (calculate the dominance degree of pipeline projects). Similarly, abstract the cities and towns uncovered by the gas transport pipeline network into a set of nodes $i = \{1, 2, \ldots, n\}$. Calibrate model parameters and use the DDM to calculate the dominance degree between node i and node j sequentially. Then, sort the calculation results in descending order.

Step 3 (update the pipeline network in the regional area). According to the calculation results, take the pipeline project of maximum dominance degree and connect its nodes at both ends to form a new pipeline route. Then, the corresponding nodes i are incorporated into a set of node j: that is, $j = \{1, 2, \ldots, m + 1\}$. For example, assume that $i = \{1, 2, 3, 4\}$ and $j = \{1, 2\}$, the calculated maximum dominance degree A_{21}, which means $i = 2$ is removed from a set of node i, subsumed into a set of node j and denoted as $j = 3$; then, $i = \{1, 3, 4\}$ and $j = \{1, 2, 3\}$. Finally, the updated spatial structure diagram of the regional pipeline network is obtained.

Step 4 (calculate cyclically and output the results). Repeat Steps 2 and 3 until the set of nodes $i = \{1, 2, \ldots, n\}$ is an empty set, while the set of nodes $j = \{1, 2, \ldots, m + n\}$. At this time, all cities and towns have been covered by the pipeline network and the layout scheme of pipeline network is obtained.

3. Uncertainty of Dominance Degree Model

3.1. Source of Model Uncertainty. The uncertainty of the DDM arises from a series of factors, including wrong settings of the model, imperfect input information, and the inherent randomness of events and behaviors [21]. The PM and EPT are combined to establish the DDM, which fully embodies the comprehensive socioeconomic benefits of pipeline projects.

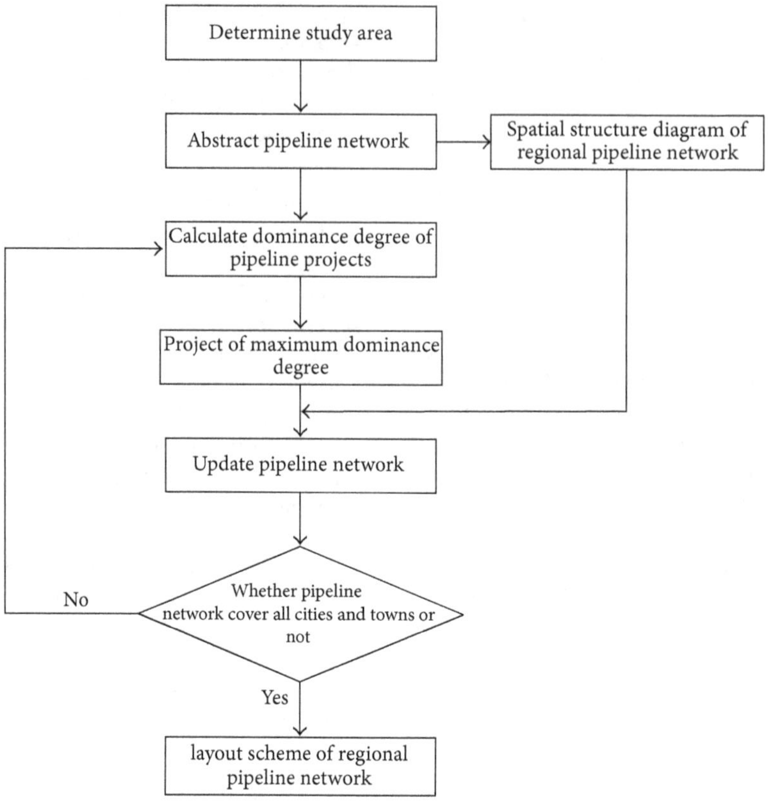

FIGURE 1: Natural gas transport pipeline network layout process.

However, the natural gas market is subject to economic development, gas pricing, policy adjustments, and other factors, whose development trend shows a certain irregularity. In addition, the construction costs of pipeline projects vary when the market prices and construction costs of steel pipelines change. When applying the DDM to obtain the layout scheme of a pipeline network, the inputs that reflect the market scale and construction costs should be determined based on statistical data or similar empirical engineering data. Therefore, there is a certain degree of uncertainty in the DDM, which is mainly derived from the inputs.

3.2. Uncertainty of Inputs. Model inputs mainly refer to data that describes the basic situation and related events and factors that affect the model performance [22]. The inputs of the DDM are obtained through related statistical survey data and by reference to similar empirical engineering data, which have a certain uncertainty. The survey data are the basic data, which are mainly used to characterize the urban socioeconomic conditions, energy consumption, and related policies along the pipeline project. Generally, there are errors in the surveys, although such data errors are somewhat limited to a certain extent and do not expand in the layout loop computation thanks to modern statistical methods. Therefore, the impact of the survey data error on the model uncertainty is relatively small. However, the inputs determined by similar empirical engineering data, such as the terrain correction coefficient and construction costs, will have a significant impact on model uncertainty. If, owing to conditional constraints, the survey of geographical environment conditions along the pipeline project is not detailed, the data will be more subjective. There are inevitable differences in geographical and environmental conditions along different natural gas transport pipeline projects, and thus, analogous data from similar projects may differ from the actual project. Therefore, the uncertainty of the model is likely to be dominated by the uncertainty of these inputs.

4. Uncertainty Analysis of Model

4.1. Quantitative Analysis of Uncertainty. The method of moments and Monte Carlo method are two main methods used to evaluate the model outputs [23]. The method of moments requires that the outputs are specified as an explicit, single function of the inputs, where higher order derivatives are used to ensure computational accuracy [24]. From the perspective of model setting, owing to the computational complexity inherent to the DDM and its layout method, it is difficult to calculate the outputs by the method of moments. Therefore, this study uses the Monte Carlo method, since it can obtain more accurate results to quantify the uncertainty of the DDM. According to the general process of Monte Carlo method [25], based on the characteristics of the DDM, the quantitative analysis of uncertainty includes the following three aspects.

4.1.1. Determine the Probability Distribution of Inputs. The appropriate probability distribution is selected based on the characteristics of inputs of the DDM. In the DDM, many variables vary within the nonnegative range. To avoid the occurrence of negative numbers in the process of generating random numbers, the logarithmic normal distribution (LND) is used to represent the probability distribution of the DDM. The probability density function of the LND can be calculated as follows:

$$f(t) = \frac{1}{\sqrt{2\pi}\sigma_{\text{Int}}} \exp\left[-\frac{1}{2}\left(\frac{\text{Int} - \mu_{\text{Int}}}{\sigma_{\text{Int}}}\right)^2\right], \quad (2)$$

where μ_{Int} is the location parameter, which represents the logarithmic mean of the probability distribution. σ_{Int} is the shape parameter, which represents the logarithmic standard deviation.

The mean and variance of random variable t can be represented as follows:

$$E(t) = \exp\left(\mu_{\text{Int}} + \frac{1}{2}\sigma_{\text{Int}}^2\right) \quad (3)$$

$$D(t) = \left(e^{\sigma_{\text{Int}}^2} - 1\right)\exp\left(2\mu_{\text{Int}} + \sigma_{\text{Int}}^2\right). \quad (4)$$

From (3) and (4), the coefficient of variation (CoV) can be obtained as follows:

$$\text{CoV} = \frac{\sqrt{D(t)}}{E(t)} = \frac{\sqrt{\left(e^{\sigma_{\text{Int}}^2} - 1\right)\exp\left(2\mu_{\text{Int}} + \sigma_{\text{Int}}^2\right)}}{\exp\left(\mu_{\text{Int}} + (1/2)\sigma_{\text{Int}}^2\right)} = \sqrt{e^{\sigma_{\text{Int}}^2} - 1}. \quad (5)$$

4.1.2. Determine the CoV of Inputs. The CoV can be used to compare the degree of variation among two or more units [26]. Therefore, the CoV can be regarded as the expression of parameters of uncertainty. Considering that the CoV of some inputs is difficult to be determined directly, the CoV of inputs can be assumed based on the relevant research results to analyze its impact on the outputs of the DDM [27]. Moreover, the standard deviation can be calculated by multiplying the CoV and mean value of the inputs.

4.1.3. Simulation Analysis of Dominance Degree Model. The MATLAB software was used to program and solve the Monte Carlo simulation analysis of the DDM. According to the probability distribution of inputs, random samples are generated. Then, the results of outputs of DDM and their distribution can be calculated.

4.2. The Analysis Method of Uncertainty. This study uses a multivariate sensitivity analysis method to quantify the uncertainty of the DDM. Based on the linear regression of inputs and outputs, the impact of inputs on output uncertainty is analyzed. This method provides assistance in determining this impact and uses the normalized regression coefficient to represent the influence degree [28, 29].

4.2.1. Multiple Linear Regression Analysis. The multiple linear regression analysis is a statistical method to investigate the relationship between a dependent variable and multiple independent variables [30]. Assume that there is a number of independent variables $x_1, x_2, \ldots, x_p$ and a dependent variable Y. The model between them can be expressed as follows:

$$Y = \beta_0 + \beta_1 x_1 + \beta_2 x_2 + \cdots + \beta_p x_p + \varepsilon, \quad (6)$$

where $\beta_0, \beta_1, \ldots, \beta_p$ represents the regression coefficient. ε represents the error term that obeys normal distribution $N(0, \sigma^2)$.

By observing the variables $x_1, x_2, \ldots, x_p, Y$ for n times, the observations of $(x_{i1}, x_{i2}, \ldots, x_{ip}, y_i)$ of n groups are obtained. The least squares method is used to calculate $\beta_0, \beta_1, \ldots, \beta_p$, and then, the estimated value $\widehat{\beta}_0, \widehat{\beta}_1, \ldots, \widehat{\beta}_p$ can be obtained.

When carrying out the model uncertainty analysis, the outputs are set as dependent variables and the inputs are set as independent variables. Combined with the Monte Carlo simulation results, the regression coefficients were obtained by regression analysis.

4.2.2. Significance Tests of Regression Coefficients. To test if a regression coefficient β_j $(1 \le j \le p)$ is zero, it is equivalent to test whether its corresponding value x_j has an impact on Y. Based on the linear regression model, we assume that H_0 : $\beta_j = 0,\ 1 \le j \le p$, and use $\widehat{\beta}_j$ as a test to obtain the statistics T of Student's distribution.

$$T = \frac{\widehat{\beta}_j}{\sqrt{c_{jj}}\sqrt{Q/(n-p-1)}}$$
$$Q = \sum_{i=1}^{n}(Y_i - \widehat{y}_i)^2, \quad (7)$$

where c_{jj} represents the jth element on the main diagonal of matrix $C = A^{-1}$ and matrix A is expressed as follows:

$$A = \begin{bmatrix} n & \sum_i x_{i1} & \sum_i x_{i2} & \cdots & \sum_i x_{ip} \\ \sum_i x_{i1} & \sum_i x_{i1}^2 & \sum_i x_{i1}x_{i2} & \cdots & \sum_i x_{i1}x_{ip} \\ \vdots & \vdots & \vdots & & \vdots \\ \sum_i x_{ip} & \sum_i x_{ip}x_{i1} & \sum_i x_{ip}x_{i2} & \cdots & \sum_i x_{ip}^2 \end{bmatrix}. \quad (8)$$

Given a significant level α, if $|T| \ge t_{\alpha/2}(n-p-1)$, H_0 is rejected, and β_j is significantly not equal to zero. If $|T| < t_{\alpha/2}(n-p-1)$, H_0 is accepted, and β_j is significantly equal to zero. When making a significant test of regression coefficients, the value of T and p of β_j can be obtained. If $p < \alpha$, it can also be determined that it is not equal to zero; namely, x_j has a significant impact on model outputs. If $p > \alpha$, x_j has no impact on model outputs, and thus, the variable can be excluded.

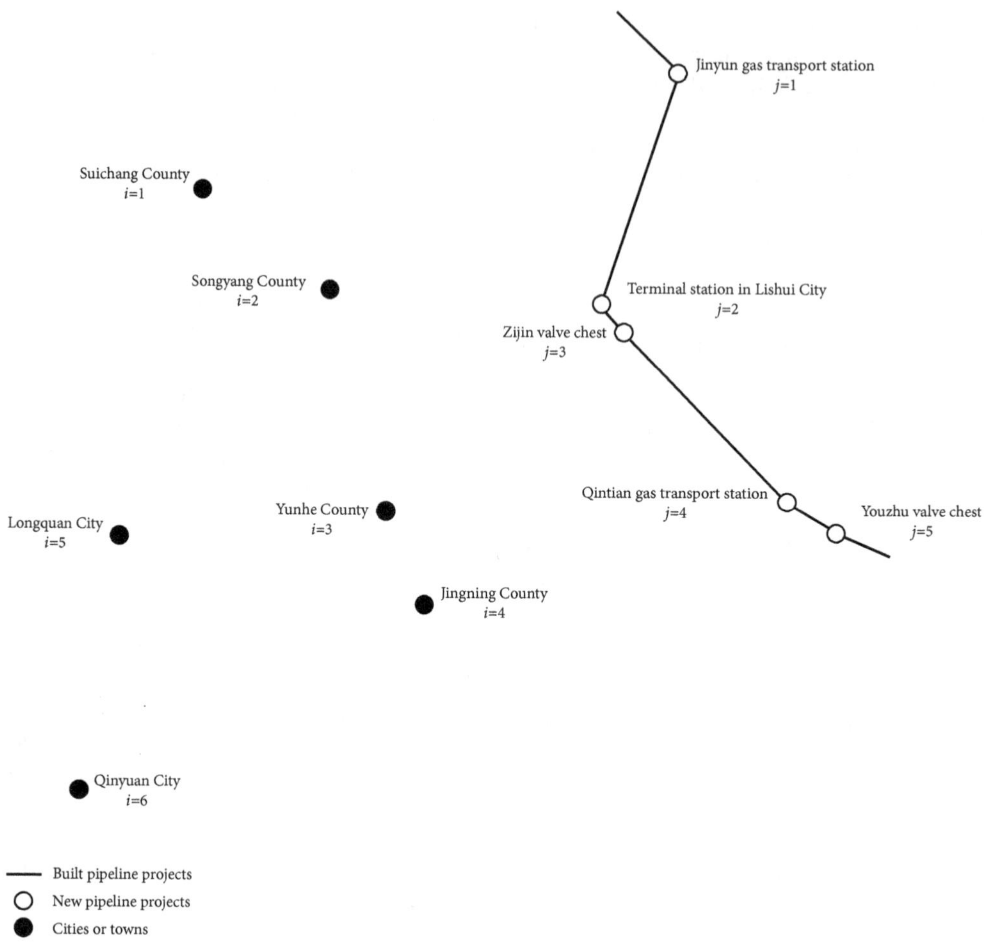

FIGURE 2: Natural gas transport pipeline network.

4.2.3. Sensitivity Analysis. A nonparametric statistics method is proposed to examine sensitivity [31]. After the linear regression analysis and significance test of model outputs and inputs, the standardized regression coefficient SRC_i, which can be used to determine the impact of inputs on the uncertainty of outputs, can be calculated as follows:

$$\mathrm{SRC}_i = \frac{\beta_i \times \sigma_i}{\sigma_y}, \quad (9)$$

where β_i represents the regression coefficient. σ_i represents the standard deviation of independent variables. σ_y represents the standard deviation of dependent variables.

5. Case Study

The Lishui region along the Jinhua-Lishui-Wenzhou gas transport trunk pipeline engineering in Zhejiang Province, China, is taken as a case study. DDM and uncertainty analysis are used to determine the optimal layout scheme.

5.1. Pipeline Network and Model Inputs. GIS is used to determine the locations of the three natural gas transport stations and two valve chests. Then, the pipeline network in the regional area is schematized as shown in Figure 2.

According to the Statistical Yearbook of Lishui City of 2015, and energy consumption statistics of each county of Lishui City (Mátyás 1998), input parameters are as listed in Table 3.

By reference to related data of the Lishui section of the Jinhua-Lishui-Wenzhou pipeline engineering project, the L450M steel is used for standard pipelines. The value of β is 1.032. V is 0.77 ten thousand yuan/tons. G is 222.3 tons/km. w_1, w_2, w_3, w_4 are, respectively, 250, 15, 4, and 25 thousand yuan/crossing.

5.2. Model Calculation Results. By assuming that the model inputs obey the LND, the CoV is assigned a value of 0.3 based on related studies [32, 33]. Using the proposed methodology, the two layout schemes and project construction sequence of the gas pipeline network in the Lishui area are obtained, as shown in Figures 3 and 4.

The results present in Table 4 are obtained by setting the significant level $\alpha = 0.05$ and using the multivariate sensitivity analysis method to quantify uncertainty of the DDM.

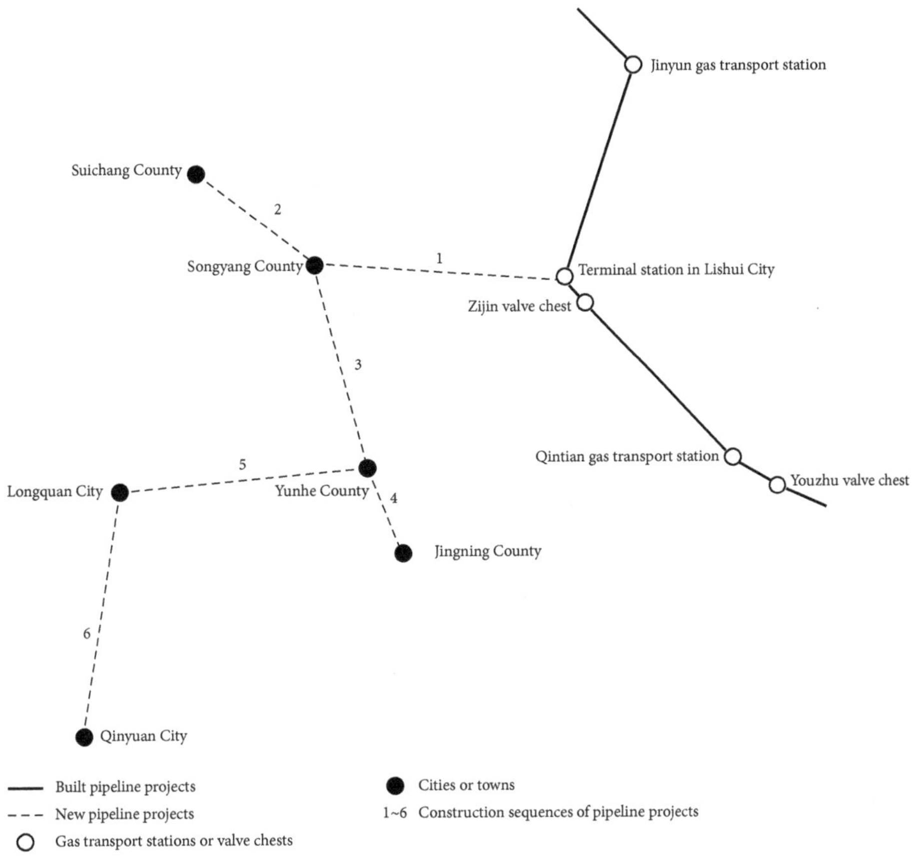

FIGURE 3: Layout scheme of natural gas transport pipeline network (Scheme 1).

TABLE 3: Input parameters of cities and towns.

Name	K	M (10,000 yuan/person)	I (Tons of standard coal/10,000 yuan)
Suiyang County	1.25	46.596	0.470
Songyang County	1.25	44.037	0.456
Yunhe County	1	45.627	0.532
Jingning County	1	38.541	0.458
Longquan City	1.25	39.546	0.491
Qinyuan County	1.25	43.943	0.490

TABLE 4: Results of sensitivity analysis.

Inputs	A	
	SRC	P
μ	3.050	0.000
K	1.739	0.001
I	0.379	0.030
a	−3.326	0.000
w_1	−1.594	0.000
w_4	−0.750	0.001
w_2	−0.279	0.001
β	−0.221	0.025
V	−0.167	0.046

The results reveal that the uncertainty of model inputs directly affects the pipeline network layout schemes and optimal construction sequences. The inputs having a significant impact on the uncertainty of the DDM are the GDP per capita μ, the dielectric constant K, the distance between city and gas transport station a, and the corresponding construction costs of crossing a river and railway w_1, w_4.

Natural gas consumption and city scale are closely related to economic development level, which determine the market development potential and growth rate and become the main influential factors that affect the natural gas market. Thus, the uncertainty of μ and K has a significant impact on model results. The pipeline project construction cost is related to the length of pipelines a and the corresponding construction costs of crossings, which have a direct impact on the project

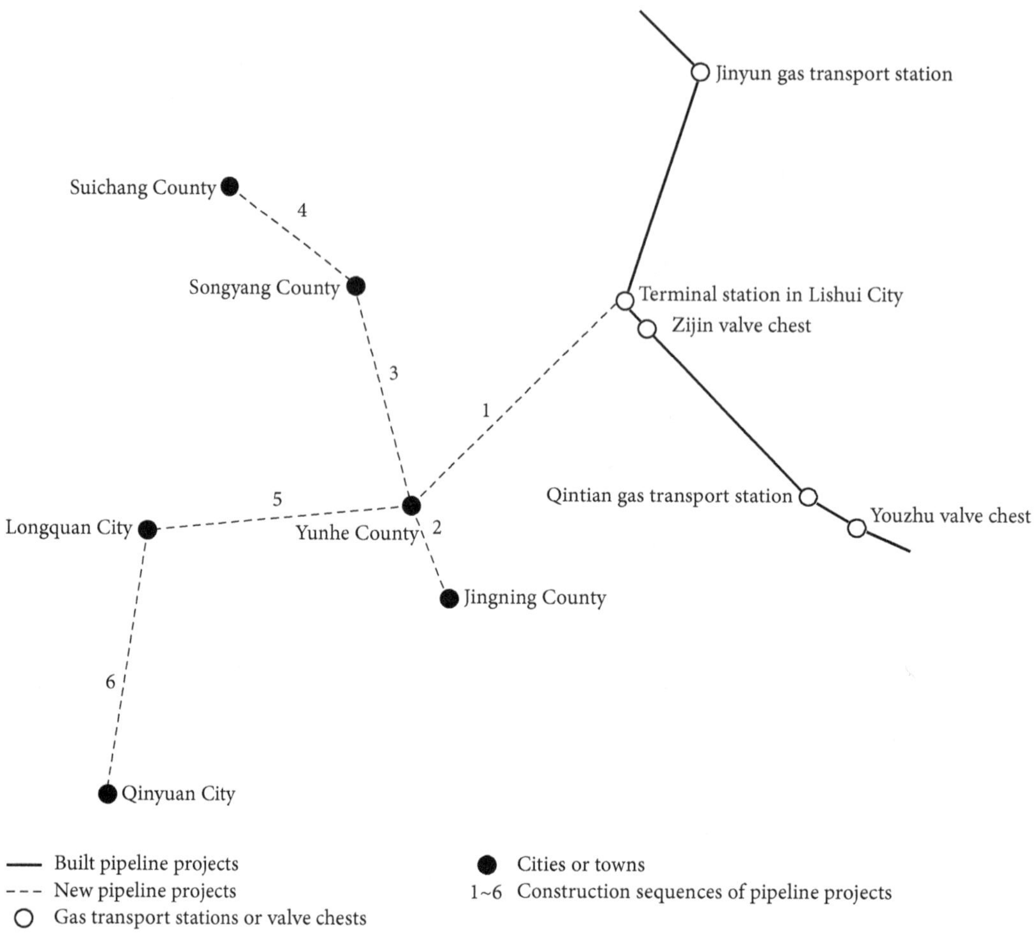

FIGURE 4: Layout scheme of natural gas transport pipeline network (Scheme 2).

investment and income. The higher costs of w_1 and w_4 directly affect construction costs. Thus, the uncertainty of a, w_1, and w_4 has a significant impact on the results of the model.

In summary, when the DDM is applied to determine the layout of a natural gas transport pipeline network, we should consider the impact of input uncertainty, so as to try and reduce this impact and improve the accuracy of the model results. Accordingly, this will support the best decision regarding the layout of the natural gas transport pipeline network.

6. Conclusions and Future Studies

The optimization of the layout of a natural gas transport pipeline network is a scientific and forward-looking topic with socioeconomic benefits. It is also a complex problem, which is affected by multiple factors including uncertainty. This study investigated the DDM and its layout method and analyzes the source and impact of model uncertainty and then proposed a new methodology for uncertainty analysis based on the Monte Carlo method. The Jinhua-Lishui-Wenzhou gas transport trunk pipeline engineering project at the Lishui region in Zhejiang Province, China, was taken as a case study, and the corresponding pipeline network layout scheme and the uncertainty analysis results were obtained.

Uncertainty is one type of limitation that exists almost in every model. The uncertainty analysis method that was developed for the DDM is a method that can quantify the uncertainty of the model and examine the influence of different uncertainty factors on the layout of a regional pipeline network. It was observed from the case study results that the uncertainty of the model directly affected the layout optimization for the gas transport pipeline network and provided us with different results. Therefore, in practical applications of the DDM for regional network layouts, it is necessary to conduct an uncertainty analysis and quantify uncertainty effects of the model. Through the uncertainty analysis, we were able to determine the factors that significantly affect the model uncertainty and take measures to improve the accuracy of these factors by reducing the influence of uncertainty and obtaining the optimal layout of gas transport pipeline network, which could be used as a reference and provide technical support to decision-making with regard to an optimal pipeline network layout.

However, there are several limitations in the present study. First, this study set the CoV to reflect changes in model inputs, while various differences existed between the set based on the empirical data and the actual situation. Thus, the method to determine the value of the CoV should be further studied. Second, a multivariate sensitivity analysis was used to analyze the impact of specific inputs on model uncertainty, while in reality, the model uncertainty may be caused by the combined impact of multiple inputs. Therefore, the impact of uncertainty derived from combinations of multiple inputs on the model results should be further studied.

Conflicts of Interest

The authors declare that there are no conflicts of interest regarding the publication of this paper.

References

[1] M. F. Chávez-Rodríguez, L. Dias, S. Simoes et al., "Modelling the natural gas dynamics in the Southern Cone of Latin America," *Applied Energy*, vol. 201, pp. 219–239, 2017.

[2] X. Ma and Z. Liu, "Application of a novel time-delayed polynomial grey model to predict the natural gas consumption in China," *Journal of Computational and Applied Mathematics*, vol. 324, pp. 17–24, 2017.

[3] S. Sanaye and J. Mahmoudimehr, "Optimal design of a natural gas transmission network layout," *Chemical Engineering Research and Design*, vol. 91, no. 12, pp. 2465–2476, 2013.

[4] H. Üster and Ş. Dilaveroğlu, "Optimization for design and operation of natural gas transmission networks," *Applied Energy*, vol. 133, pp. 56–69, 2014.

[5] R. R. Singh and P. K. S. Nain, "Optimization of natural pipeline design and its total cost using GA," *International Journal of Science and Research Publications*, vol. 2, no. 8, pp. 1-10, 2012.

[6] F. D. S. Alves, J. N. M. D. Souza, and A. L. H. Costa, "Multi-objective design optimization of natural gas transmission networks," *Computers & Chemical Engineering*, vol. 93, pp. 212–220, 2016.

[7] J. An and S. Peng, "Layout optimization of natural gas network planning: Synchronizing minimum risk loss with total cost," *Journal of Natural Gas Science and Engineering*, vol. 33, pp. 255–263, 2016.

[8] Z. Shariat, A. Movaghar, and M. Hoseinzadeh, "A learning automata and clustering-based routing protocol for named data networking," *Telecommunication Systems*, vol. 65, no. 1, pp. 9–29, 2017.

[9] Y. Cuan and X. Chen, "Application of improved Dijkstra algorithm in selection of gas source node in gas network," in *Proceedings of the 2012 International Conference on Industrial Control and Electronics Engineering, ICICEE 2012*, pp. 1558–1560, chn, August 2012.

[10] D.-M. Han and J.-H. Lim, "Design and implementation of smart home energy management systems based on ZigBee," *IEEE Transactions on Consumer Electronics*, vol. 56, no. 3, pp. 1417–1425, 2010.

[11] H. J. Prömel and A. Steger, *The Steiner Tree Problem: A Tour through Graphs, Algorithms, And Complexity*, Advanced Lectures in Mathematics, Friedr. Vieweg & Sohn, Braunschweig, 2002.

[12] R. E. Bellman, S. E. Dreyfus, R. Bellman et al., *Applied Dynamic Programming*, Princeton University Press, Princeton, NJ, USA, 2015.

[13] T. M. Masaud, R. D. Mistry, and P. K. Sen, "Placement of large-scale utility-owned wind distributed generation based on probabilistic forecasting of line congestion," *IET Renewable Power Generation*, vol. 11, no. 7, pp. 979–986, 2017.

[14] M. M. J. Knoope, A. Ramírez, and A. P. C. Faaij, "The influence of uncertainty in the development of a CO_2 infrastructure network," *Applied Energy*, vol. 158, pp. 332–347, 2015.

[15] X.-F. Xu, J. Hao, Y.-R. Deng, and Y. Wang, "Design optimization of resource combination for collaborative logistics network under uncertainty," *Applied Soft Computing*, vol. 56, pp. 684–691, 2017.

[16] M. Patriksson, "Robust bi-level optimization models in transportation science," *Philosophical Transactions of the Royal Society A: Mathematical, Physical and Engineering Sciences*, vol. 366, no. 1872, pp. 1989–2004, 2008.

[17] Y. Yin, S. Lawphongpanich, and Y. Lou, "Estimating investment requirement for maintaining and improving highway systems," *Transportation Research Part C: Emerging Technologies*, vol. 16, no. 2, pp. 199–211, 2008.

[18] Y. Yin, S. M. Madanat, and X.-Y. Lu, "Robust improvement schemes for road networks under demand uncertainty," *European Journal of Operational Research*, vol. 198, no. 2, pp. 470–479, 2009.

[19] D. Zhang, X. Wang, S. Li, N. Ni, Z. Zhang, and X. Ma, "Joint optimization of green vehicle scheduling and routing problem with time-varying speeds," *PLoS ONE*, vol. 13, no. 2, article e0192000, 2018.

[20] Z. Zhu, C. Sun, J. Zeng, and G. Chen, "Optimization of natural gas transport pipeline network layout: a new methodology based on dominance degree model," *Transport*, vol. 33, no. 1, pp. 143–150, 2018.

[21] K. Hirano and J. H. Wright, "Forecasting with model uncertainty: representations and risk reduction," *Econometrica*, vol. 85, no. 2, pp. 617–643, 2017.

[22] W. Xie, B. L. Nelson, and R. R. Barton, "Multivariate input uncertainty in output analysis for stochastic simulation," *ACM Transactions on Modeling and Computer Simulation (TOMACS)*, vol. 27, no. 1, Art. 5, 22 pages, 2016.

[23] S. Mohajeryami and M. Doostan, "Investigating the lightning effect on compact transmission lines by employing Monte Carlo method," in *Proceedings of the 2016 IEEE/PES Transmission and Distribution Conference and Exposition*, May 2016.

[24] S. Lefrancois, C. Husko, A. Blanco-Redondo, and B. J. Eggleton, "Nonlinear silicon photonics analyzed with the moment method," *Journal of the Optical Society of America B: Optical Physics*, vol. 32, no. 2, pp. 218–226, 2015.

[25] R. Y. Rubinstein and D. P. Kroese, *Simulation and the Monte Carlo Method*, John Wiley & Sons, Inc., 2008.

[26] W. C. Yeong, M. B. Khoo, S. L. Lim, and M. H. Lee, "A direct procedure for monitoring the coefficient of variation using a variable sample size scheme," *Communications in Statistics—Simulation and Computation*, vol. 46, no. 6, pp. 4210–4225, 2017.

[27] S.-P. Wan, "Multi-attribute decision making method based on possibility variance coefficient of triangular intuitionistic fuzzy numbers," *International Journal of Uncertainty, Fuzziness and Knowledge-Based Systems*, vol. 21, no. 2, pp. 223–243, 2013.

[28] E. Borgonovo and E. Plischke, "Sensitivity analysis: a review of recent advances," *European Journal of Operational Research*, vol. 248, no. 3, pp. 869–887, 2016.

[29] M. Lamboni, H. Monod, and D. Makowski, "Multivariate sensitivity analysis to measure global contribution of input factors in dynamic models," *Reliability Engineering & System Safety*, vol. 96, no. 4, pp. 450–459, 2011.

[30] G. Grégoire, "Multiple linear regression," *EAS Publications Series*, vol. 66, Article ID 1466005, pp. 45–72, 2015.

[31] A. Saltelli and J. Marivoet, "Non-parametric statistics in sensitivity analysis for model output: a comparison of selected techniques," *Reliability Engineering & System Safety*, vol. 28, no. 2, pp. 229–253, 1990.

[32] W. L. Teoh, M. B. Khoo, P. Castagliola, W. C. Yeong, and S. Y. Teh, "Run-sum control charts for monitoring the coefficient of variation," *European Journal of Operational Research*, vol. 257, no. 1, pp. 144–158, 2017.

[33] M. E. Calzada and S. M. Scariano, "A synthetic control chart for the coefficient of variation," *Journal of Statistical Computation and Simulation*, vol. 83, no. 5, pp. 851–865, 2013.

13

Multiple Equilibrium Behaviors considering Human Exposure to Vehicular Emissions

Zhanbo Sun, [1,2] **Yu Tan**, [1,2] **Rui Ma**, [3] **Xia Yang**, [4] **and Jie Zhang** [5]

[1] *National Engineering Laboratory of Integrated Transportation Big Data Application Technology, Chengdu, China*
[2] *School of Transportation and Logistics, Southwest Jiaotong University, Chengdu, China*
[3] *Department of Civil and Environmental Engineering, University of California, Davis, CA, USA*
[4] *Department of Civil Engineering, SUNY Polytechnic Institute, Utica, NY, USA*
[5] *School of Information Science and Technology, Southwest Jiaotong University, Chengdu, China*

Correspondence should be addressed to Zhanbo Sun; zhanbo.sun@home.swjtu.edu.cn

Academic Editor: Zhi-Chun Li

Emissions produced by urban transportation activities are harmful to people's health and they also affect people's trip-making decisions. In this paper, we explore the multiple equilibrium behaviors considering human exposure to vehicular emissions. We assume that a portion of transportation users are environmental advocates and their route decisions are based on some composite cost functions comprise of a travel time component and an emission exposure component. We then study the multiple equilibrium behaviors with multiple types of users on a traffic network. The multiple equilibrium problems are further converted into variational inequality (VI) problems and they are solved using a method of successive average- (MSA-) based diagonalization method. Per the specific network setting, we find that as travelers become more concerned about their exposure to vehicular emissions, the system emission exposure, travel time, and the total cost get reduced; i.e., Pareto improving solutions are achieved. By analyzing the multiple equilibrium behaviors, we find that the system gets better if more users become environmental advocates. And the change of a small percentage of users should already lead to a good system improvement.

1. Introduction

Transportation activities account for about 27% carbon dioxide, 59% nitrogen oxide, and 54% carbon monoxide of the atmospheric emissions [1]. Recently, transportation emissions and their negative impacts have been increasingly noticed. Long time exposure to air pollution can cause short-term and long-term health problems [2]. According to the US Environmental Protection Agency [3], living close to congested urban location has lifetime cancer risks greater than 100 in million, which is about 10 times higher than the national average. This evidence motivated researchers to develop mechanisms to quantify and decrease traffic emissions [4–7]. In the network modeling field, efforts are spent to model and control the network emissions, usually in an aggregated manner, without paying much attention to the impacts of vehicular emissions on people. Travelers are directly exposed to vehicular emissions. And vehicular emissions, in turn, affect traveler's trip-making decisions. Such interrelationship between trip-making decisions and human exposure to vehicular emissions has not been fully studied.

Modeling travelers' trip making behavior is challenging. Taking route choice behavior as an example. The most well-known theories are the Wardropian principles, including user equilibrium (UE) and system optimum (SO). UE assumes that users always choose the shortest path to minimize their individual cost so that no one can unilaterally change routes to reduce his/her own cost. SO assumes that all users are altruistic to minimize the total system cost. Researchers have recognized there are other users which are neither fully competitive (UE) nor fully cooperative (SO), e.g., the groups of users that follow the Cournot-Nash (CN) behaviors [8–12]. CN users are like drivers in a same transport company, who try to minimize their group total costs. And different CN users are fully competitive. The problems that UE, CN, and sometimes SO players coexist in the network are referred to as the mixed equilibrium or multiple equilibrium problems [12]. The work on this topic mostly considered single model objective such as travel time cost [8, 9, 11]. Zhang et al. [10] established a unified framework that simultaneously considered travel time cost and users' time-irrelevant costs. He et al. [13] formulated tradable credit schemes based on the mixed equilibrium network with consideration of users' transaction cost. Yang et al. [12] investigated the mixed equilibrium behaviors with general common constraints and took traffic emissions as a general common constraint. Little work has been done in the network-modeling field to explicitly consider human's emission exposure in the modeling framework.

To capture human exposure to vehicular emissions, it is necessary to describe emission modeling and exposure in an integrated framework (see [14]). Such a framework should encapsulate three components, including (i) a traffic flow model; (ii) an emission model that interprets the source of emission as a function of traffic flow; and (iii) a model that describes how the produced tailpipe emissions spread away from the source and affect the neighborhood. With respect to (i) and (ii), microscopic tools such as Comprehensive Modal Emissions Model (CMEM; [15, 16]), VT-Micro [17, 18], and Motor Vehicle Emission Simulator (MOVES; [19]) can estimate emissions with fine resolution by breaking down the physical behavior of vehicles into different operating modes (e.g., acceleration, deceleration, idle, and cruising). These models, however, require extensive input data such as vehicle type and trajectory information (see [6]) from component (i) and usually do not have analytical forms. These models may not be suitable for real time network-scale applications (route choice modeling). Alternatively, mesoscopic and macroscopic approaches [4, 20–24] can estimate the source of emissions as a function of a set of traffic variables, which are more appropriate for network traffic analysis. In the field of atmospheric environment, component (iii) is known as emission dispersion modeling. Emission dispersion models include Gaussian Plume model [25], Box model [26], Lagrangian model [27], and others [28], among which Gaussian dispersion model is most widely used due to its simplistic nature. Gaussian model-based software packages such as HIWAY2 and California Line Source Dispersion Model version 4 (CALINE4) have been widely adopted to estimate emission concentration and exposure near freeways and major arterials [29–33].

In this study, an integrated modeling framework is applied to analytically estimate vehicular emissions and concentrations within the traffic network. The proposed method adopts the BPR function and a macroscopic emission model to describe the source of emissions. Gaussian dispersion model is then used to capture the dispersion of emissions from the source. The framework can be used to calculate travelers' exposure to emissions, which is considered as a part of the generalized travel cost. We further assume that a portion of transportation users are environmental advocates and their route choice decisions are based on some composite cost functions comprise of a travel time component and an emission exposure component. We then study the multiple equilibrium behaviors with multiple types of users on a traffic network. The multiple equilibrium problems are further converted into variational inequality (VI) problems and they are solved using a method of successive average- (MSA-) based diagonalization method. The model framework proposed in the paper forms the basis for analyzing the more practically meaningful multiple equilibrium behaviors with considerations of people's perception and reaction to emission exposures.

2. Methodology

It assumes that multiple types of players may exist on a multiple-origin and multiple-destination network. The first type of users follows Wardropian user equilibrium, and it is referred to as a common UE player who seeks to minimize the individual travel time cost. The second type of users also follows the UE principle, but it considers both travel time and the emission exposure cost, and this type of users is referred to as an environmental advocacy UE player. The third type of users is assumed to be controlled by CN player(s) whose objective is to minimize the group cost. The equilibrium behaviors under such user composition are referred to as multiple equilibrium hereafter in this paper. Similar definitions can be found in Harker [8], Yang et al. [9], Yang et al. [34], Zhang et al. [10], and Yang et al. [12]. The multiple equilibrium behaviors are studied in the paper to analyze the effects of some environmental-cost–based routing models (e.g., smartphone apps) that provide routing services considering the emission exposure of the transportation users. The notations used in the paper are similar to the ones used in Yang et al. [12].

Assume that all users have been transformed into equivalent passenger car units. We use Ω^U to represent the feasible set for the common UE player, Ω^E to represent the feasible set of the environmental advocacy UE player, and Ω^k, $\forall k \in K$ to represent the feasible set of each environmental advocacy CN player.

For the common UE player

$\Omega^U = \{v^U \mid v^U$ statisfies the following constraints (1)–(3)}:

$$\sum_{p \in P_w} f_p^U = D_w^U \quad \forall w \in W^U \tag{1}$$

$$v_a^U = \sum_{p \in P^U} \delta_{ap} f_p^U \quad \forall a \in A \tag{2}$$

$$f_p^U \geq 0 \quad \forall p \in P^U \tag{3}$$

For the environmental advocacy UE player

$\Omega^E = \{v^E \mid v^E$ statisfies the following constraints (4)–(6)}:

$$\sum_{p \in P_w} f_p^E = D_w^E \quad \forall w \in W^E \tag{4}$$

$$v_a^E = \sum_{p \in P^E} \delta_{ap} f_p^E \quad \forall a \in A \tag{5}$$

$$f_p^E \geq 0 \quad \forall p \in P^E \tag{6}$$

For the environmental advocacy CN players

$\Omega^k = \{v^k \mid v^k$ statisfies the following constraints (7)–(9)}, $\forall k \in K$:

$$\sum_{p \in P_w} f_p^k = D_w^k \quad \forall w \in W^k \tag{7}$$

$$v_a^k = \sum_{p \in P^k} \delta_{ap} f_p^k \quad \forall a \in A \tag{8}$$

$$f_p^k \geq 0 \quad \forall p \in P^k \tag{9}$$

2.1. Emission Exposure Model. A macroscopic model is applied to estimate CO emission as a function of travel time, as in (10), where $e_a(v_a)$ is the CO emission in grams per vehicle per hour, l_a is the length of link a in kilometers, and $t_a(v_a)$ is the link travel time in minutes, which takes the form of BPR function (Bureau of Public Roads 1964), as in (11). Similar emission models have been discussed in Wallace et al. [20] and Yin & Lawphongpanich [4].

$$e_a(v_a) = 0.2038 t_a(v_a) e^{0.7962 l_a / t_a(v_a)} \tag{10}$$

$$t_a(v_a) = t_a^0 \left(1 + \alpha \left(\frac{v_a}{c_a}\right)^\beta\right) \tag{11}$$

For the ease of emission dispersion modeling, we further convert the CO emission from grams per vehicle per hour to milligram (mg) per meter per second; see (12). Therefore, link a can be treated as a homogeneous line with a linear source strength of q_a, in mg/s.

$$q_a(v_a) = 0.2038 \left(\frac{v_a}{3600 l_a}\right) t_a(v_a) e^{0.7962 l_a / t_a(v_a)} \tag{12}$$

Turner [35] proposed the Gaussian equation for a continuous point source, as (13), which is applied to compute the emission concentration resulted from a point source.

$$c(x, y, z) = \frac{q}{2\pi \sigma_y \sigma_z u} \exp\left(-\frac{y^2}{2\sigma_y^2}\right) \cdot \left[\exp\left(-\frac{(z - H_p)^2}{2\sigma_z^2}\right) + \exp\left(-\frac{(z + H_p)^2}{2\sigma_z^2}\right)\right] \tag{13}$$

Here $c(x, y, z)$ is the mean concentration at point (x, y, z), in mg/m^3; x is an along-wind coordinate measured in wind direction from the source; y is a cross-wind coordinate measured perpendicular to the wind direction; z is an elevation coordinate measured from the ground; u is the mean wind velocity, in m/s; q is the strength of the point source, in mg/s; H_p is the elevation of plume where emissions are released; σ_y and σ_z are the standard deviations of the horizontal and vertical dispersion and are dependent upon the atmospheric stability, among other factors. The parameterization of the Gaussian dispersion parameters is out of the scope of the study. Details can be found in Liu et al. [36]. In this paper, we assume that CO disperses in the near ground, i.e., $z = 0, H_p = 0$. Therefore, the previous model can be simplified as equation (14a) for any downwind location. And $c(x, y, z = 0) = 0$ for any upwind as in (14b).

$$c(x, y, z = 0) = \frac{q}{\pi \sigma_y \sigma_z u} \exp\left(-\frac{y^2}{2\sigma_y^2}\right), \quad x \geq 0 \tag{14a}$$

$$c(x, y, z = 0) = 0, \quad x < 0 \tag{14b}$$

Based on the point emission concentration, the next step is to extend the point source to a line source. Bensen [29] presented a method to decompose each line source (e.g., link a) into a number of short elements and every element is an equivalent finite line source (FLS) whose center is the element midpoint. Since the points along the equivalent FLS have the same x-coordinate, the emission concentration $C_e(x, y)$ caused by short element (or equivalently, FLS) e can be computed by integrating the continuous point source along the length of the FLS, as equation (15). Note that y_e^1 and y_e^2 represent the y-coordinate distances between the two endpoints of equivalent FLS e and the receptor, which can be easily determined given the length of the short element (l_e) and the angle (φ_e) between the element and the wind direction, i.e., $y_e^1 = y - l_e \sin \varphi_e / 2$ and $y_e^2 = y + l_e \sin \varphi_e / 2$, $0° \leq \varphi_e \leq 90°$. The Gaussian dispersion from a line source or its equivalent finite line source (FLS) is illustrated in Figure 1. Further details can be found in Turner [35].

$$C_e(x, y) = \int_{y_e^1}^{y_e^2} c(x, y, z = 0)\, dy = \frac{q_e}{\pi \sigma_y \sigma_z u \sin \varphi_e} \int_{y_e^1}^{y_e^2} \exp\left(-\frac{y^2}{2\sigma_y^2}\right) dy \tag{15}$$

The emission concentration at any location s with coordinate (x_s, y_s) caused by a FLS $e \in F$ can be described using (16), where F is the set that contains all the finite line sources of the network; q_e is the emission concentration of FLS e; and obviously $q_e = q_a$ if $e \in F_a$, where F_a is the set that contains all FLS of link a. Supposing that $p = y / \sigma_y$, (16) can be derived, where $\Phi(\bullet)$ is the cumulative probability density function of a standard normal distribution.

$$C_e(x_s, y_s) = \frac{q_e}{\pi \sigma_z u \sin \varphi_e} \int_{y_e^1/\sigma_{y_s}}^{y_e^2/\sigma_{y_s}} \exp\left(-\frac{p^2}{2}\right) dp = \frac{\sqrt{2} q_e}{\sqrt{\pi} \sigma_z u \sin \varphi_e} \left[\Phi\left(\frac{y_e^2}{\sigma_{y_s}}\right) - \Phi\left(\frac{y_e^1}{\sigma_{y_s}}\right)\right] \tag{16}$$

In this paper, we are particularly interested in modeling the emission exposure of transportation users. The emission exposure of a transportation user on link a can be expressed as (17), which can be considered as the emission concentration multiplies the duration of exposure, where t_s denotes

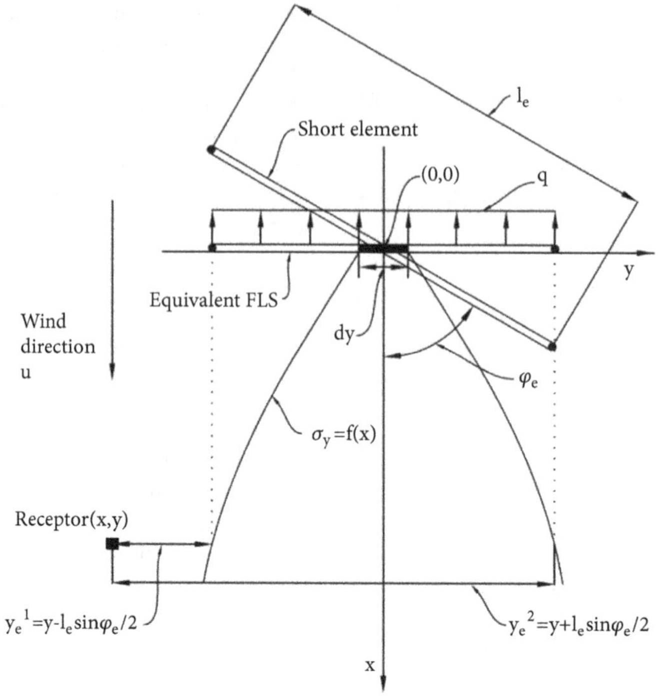

FIGURE 1: Dispersion from a short element/equivalent FLS.

the segment travel time of a finite line source $s \in F_a$, and $\sum_{s\in F_a} t_s = t_a$.

$$\tau_a = \sum_{s\in F_a}\sum_{e\in F} C_e(x_s, y_s)\, t_s \tag{17}$$

Assuming link a is decomposed evenly and the link travel time is uniformly distributed along the link, then $t_s = t_a/|F_a|$, where $|F_a|$ represents the cardinality of set F_a. To make the cost function separable, we further assume that the emission concentration on link a is only dependent upon the flow on this link. Therefore the emission exposure cost can be simplified as (18).

$$\tilde{\tau}_a = \tilde{\tau}_a(v_a) \equiv \frac{t_a}{|F_a|}\sum_{s\in F_a} C_s(x_s, y_s) \tag{18}$$

2.2. Multiple Equilibrium considering Human Exposure. In this section, we consider a generalized cost function for the transportation users, as in (19). The composite function comprises a travel time component t_a and an emission exposure component $\tilde{\tau}_a$; coefficient λ is used to leverage the tradeoff between travel time and emission exposure.

$$g_a = t_a + \lambda\tilde{\tau}_a \tag{19}$$

Therefore, the model for exposure of different players on the traffic emission can be formulated as below.

For the common UE player

$$\min_{v^U\in\Omega^U}\sum_{a\in A}\int_0^{v_a^U} t_a\left(v_a^E + v_a^K + x\right)dx \tag{20}$$

The function $t_a(v_a)$ is increasing, or $\partial t_a(v_a)/\partial v_a > 0$, $\forall a \in A$ and the function is convex $\partial^2 t_a(v_a)/\partial t_a^2 \geq 0$, $\forall a \in A$, so the minimization problem has a unique solution.

For the environmental advocacy UE player

$$\min_{v^E\in\Omega^E}\sum_{a\in A}\int_0^{v_a^E}\left[t_a\left(v_a^U + v_a^K + x\right) + \lambda\tilde{\tau}_a\left(v_a^U + v_a^K + x\right)\right]dx \tag{21}$$

Note that the emission exposure cost $\tilde{\tau}_a(v_a)$ is conditionally increasing and convex. In fact, it is easy to prove that the minimization problem has a unique solution under the condition of travel speed which is less than 150 km/h, which is true in most urban areas. The proof is provided in the appendix.

For the environmental advocacy CN players, $\forall k \in K$

$$\begin{aligned}\min_{v^k\in\Omega^k}\sum_{a\in A}&\left[t_a\left(v_a^U + v_a^E + v_a^k + v_a^{k-}\right)\right.\\&\left.+\lambda\tilde{\tau}_a\left(v_a^U + v_a^E + v_a^k + v_a^{k-}\right)\right]v_a^k\end{aligned} \tag{22}$$

Where the first derivative $t_a(v_a) + \lambda\tilde{\tau}_a(v_a) + v_a^k(t_a'(v_a) + \lambda\tilde{\tau}_a'(v_a)) > 0$ and the second derivative $2(t_a'(v_a) + \lambda\tilde{\tau}_a'(v_a)) + v_a^k(t_a''(v_a) + \lambda\tilde{\tau}_a''(v_a)) > 0$, $\forall a \in A$, so the minimization problem also has a unique solution.

Therefore, the minimization problems in (20), (21), and (22) can be characterized by a set of variational inequalities (VI):

$$\begin{aligned}&\sum_{a\in A} f_a^{\varphi}\left(v_a^{\varphi^*}, v_a^{-\varphi}\right)\left(v_a^{\varphi} - v_a^{\varphi^*}\right) \geq 0,\\&\qquad\qquad\forall v^{\varphi} \in \Omega^{\varphi},\ \forall\varphi \in \Phi\end{aligned} \tag{23}$$

TABLE 1: Network specifications.

Link	(1, 3)	(2, 4)	(1, 5)	(5, 6)	(2, 5)	(6, 3)	(6, 4)
t_a^0(min)	3	3	1.5	1.5	1.5	1.5	1.5
c_a(veh/hr)	4000	4000	4000	8000	4000	4000	4000
l_a(km)	4	4	2	2	2	2	2

Here $\Phi \equiv U \cup E \cup K$, $v_a^{-\varphi}$ is the total flow on link $a \in A$ assigned by all players in Φ except φ, $\varphi \in \Phi$. Note that $f_a^{\varphi}(v_a^{\varphi}, v_a^{-\varphi})$ is the perceived link cost of player $\varphi \in \Phi$, which is given as below:

$$f_a^{\varphi}\left(v_a^{\varphi}, v_a^{-\varphi}\right) = \begin{cases} t_a\left(v_a^{-\varphi} + v_a^{\varphi}\right), & \text{if } \varphi \in U \\ g_a\left(v_a^{-\varphi} + v_a^{\varphi}\right), & \text{if } \varphi \in E \\ g_a\left(v_a^{-\varphi} + v_a^{\varphi}\right) + v_a^{\varphi} g_a'\left(v_a^{-\varphi} + v_a^{\varphi}\right) & \text{if } \varphi = k, \forall k \in K \end{cases} \quad (24)$$

Since the objective functions (20)-(22) are increasing and convex in most urban traffic settings, the VI problem (23) has a unique solution, and similar proof can be found in Yang et al. [9].

2.3. Solution Algorithm. In this paper, we applied a diagonalization method with nested method of successive averages (MSA) to solve the aforementioned VI problems. The solution approach is provided below.

Step 0. For iteration number $n = 0$; use t_a^0 as the initial perceived link cost for all players and perform all-or-nothing assignment to get the initial total link flow $v_a^{(0)}$;

Step 1. Calculate the perceived cost of the common UE player, $f_a^{(n+1,1)} = f_a(v_a^{(n)})$. Perform all-or-nothing assignment for the common UE player based on the perceived cost, and obtain a set of updated link flow $y_a^{(n+1,1)}$. Use step size (γ); i.e., $\gamma = 1/n$ to update the traffic flow on the network, $v_a^{(n+1,1)} = v_a^{(n)} + \gamma(y_a^{(n+1,1)} - v_a^{(n)})$;

Step 2. Calculate the perceived link cost of the environmental advocacy UE player, $f_a^{(n+1,2)} = f_a(v_a^{(n+1,1)})$. Perform all-or-nothing assignment for the environmental UE player based on the perceived cost, and obtain a set of updated link flow $y_a^{(n+1,2)}$. Update the traffic flow on the network, $v_a^{(n+1,2)} = v_a^{(n+1,1)} + \gamma(y_a^{(n+1,2)} - v_a^{(n+1,1)})$;

Step 3. Calculate the perceived link cost for each environmental advocacy CN player, $f_a^{(n+1,k+2)} = f_a(v_a^{(n+1,k+1)})$, $\forall k = 1, 2, 3 \cdots \|K\|$. Perform all-or-nothing assignment for each environmental advocacy CN player, and obtain a set of updated link flow $y_a^{(n+1,k+2)}$. Update the total traffic flow on the network, $v_a^{(n+1,k+2)} = v_a^{(n+1,k+1)} + \gamma(y_a^{(n+1,k+2)} - v_a^{(n+1,k+1)})$;

Step 4 (convergence criteria). The convergence index is $\varepsilon_{\varphi}^{n} = \sqrt{\sum_{a\in A}(v_a^{\varphi(n+1)} - v_a^{\varphi(n)})^2} / \sum_{a\in A} v_a^{\varphi(n)}$, $\varphi \in \Phi$. If $\varepsilon_{\varphi}^{n}$ satisfies the convergence condition, stop; otherwise, make $v_a^{(n+1)} = v_a^{(n+1,\|K\|+2)}$, $n = n+1$, and proceed to step 1.

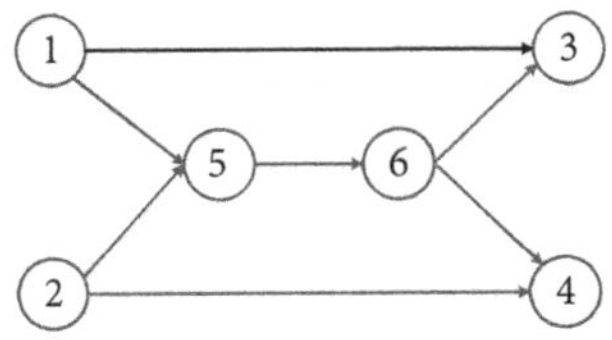

FIGURE 2: Study network.

3. Numerical Experiment

An idealistic six-node, seven-link transportation network, as shown in Figure 2, has been used to test the proposed modeling framework. There are two Origin-Destination (OD) pairs (1, 3) and (2, 4), both with the demand of 6000 vehicles per hour. The free speed of traffic flow is 80 km/h for all links on the idealistic traffic network. The mean wind velocity is 0.5 m/s, and we use $\alpha = 0.15, \beta = 4$ to parameterize the BPR function. The network specifications and some model parameters are summarized in Table 1.

3.1. Environmental Advocacy UE (Case 1). In the first case, we assume that all transportation users follow the first Wardropian principle and all the users take the same composite cost function. There are two paths (e.g., path 1-3 and path 1-5-6-3) for OD pair (1, 3).

Figure 3 depicts the change of traffic flow, travel time, and emission exposure on the two paths with respect to the change of parameter λ. It is found that when $\lambda = 0$, the travel time costs of the two paths are identical, which satisfy the UE principle. It is found that when λ increases, people become more concerned about their emission exposure, as a result, some flow has switched from path 1-3 to path 1-5-6-3 and the emission exposure cost as well as the travel time on path 1-3 decreased. Noteworthy the travel time on path 1-5-6-3 only slightly increases as the capacity on this path is much larger than the flow, therefore the marginal effects of flow is quite small. The results for path 2-4 and 2-5-6-4 are not shown here since the network and OD volumes are symmetric. The results show that the route choice patterns have been changed by considering the emission exposure cost in the cost function.

Figure 4 shows the system total cost (represented by $\sum_a [t_a(v_a) + \widetilde{\tau}_a(v_a)]$), travel time, and emission exposure with respect to the change of λ. Per the specific network setting, we find the total cost decreased as λ increasing. This suggests that as travelers become more concerned about their exposure to vehicular emissions, the system emission exposure, travel

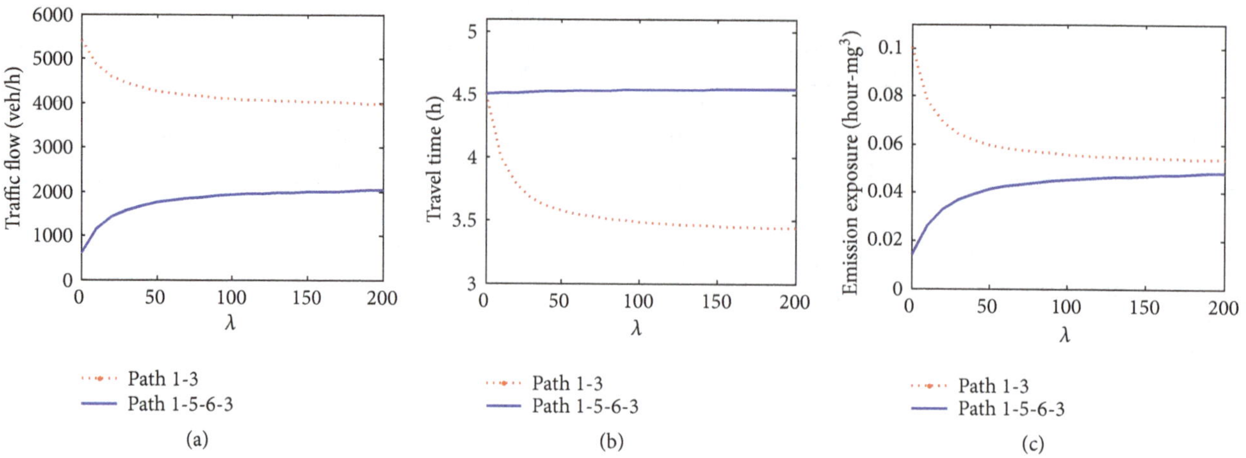

FIGURE 3: Path-based results of the environmental advocacy UE: (a) traffic flow; (b) travel time; (c) emission exposure.

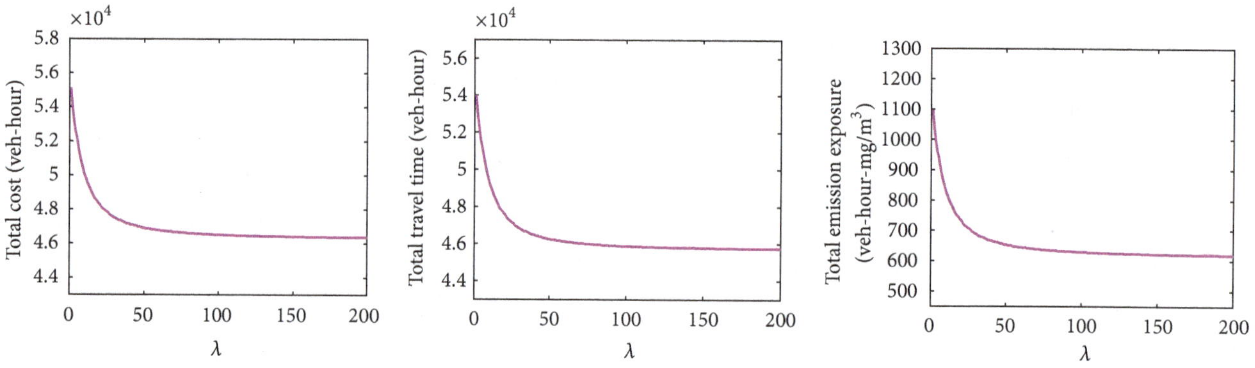

(a) $\sum_a [t_a(v_a) + \tilde{\tau}_a(v_a)]$ (b) $\sum_a t_a(v_a)$ (c) $\sum_a \tilde{\tau}_a(v_a)$

FIGURE 4: System costs of case 1: (a) total cost; (b) travel time; (c) emission exposure.

time, and the total cost all get reduced, i.e., Pareto improving solutions are achieved.

3.2. Multiple Equilibrium with a Common UE and an Environmental Advocacy UE (Case 2). In the second case, we assume there are a common UE player and an environmental advocacy UE player. We have only two types of players in the case; therefore the proposed approach simplified can be to tackle this case. By setting different user proportions, we observe the change of system total cost, travel time, and emission exposure with respect to λ; the results are provided in Figure 5. Here θ represents the percentage of the environmental advocacy UE users, and $(1 - \theta)$ represents the percentage of the common UE users.

When $\theta = 0$, since only the common UE users exist, the system total cost, travel time, and emission exposure remain unchanged with the change of λ. In Figure 5(a), the total cost first reduces when a small percentage of users become environmental advocacy UE users. The cost remains stable when the percentage of environmental advocacy UE users becomes larger than 20%. Similar trends are discovered for the system travel time and emission exposure. This suggests that the system gets better if more users become environmental advocators. In an operation perspective, the change of a small percentage of users should already lead to good system improvement.

3.3. Multiple Equilibrium with a Common UE and an Environmental Advocacy CN (Case 3). We further look at the scenario with a common UE player and an environmental advocacy CN player on the network. This scenario is practically meaningful since there may be some environmental-cost based routing models (e.g., smartphone apps) that provide routing services to minimize individual/group-wide emission exposure.

The results of this case are in general similar to case 2. As shown in Figure 6(a), as θ increasing, the total cost first decreases and becomes stable when θ becomes larger than 60%. Figures 6(b) and 6(c) reveal similar trends for the travel time and emission exposure cost.

The results from case 3 are compared to the ones in case 2, as indicated in Figure 7. Figure 7(a) shows that when 50% users are environmental advocacy users, the total cost of case 2 is larger than case 3; when $\lambda < 50$, the result is consistent with common sense that cooperative behavior leads to reduced system cost. But as the users seriously concern their

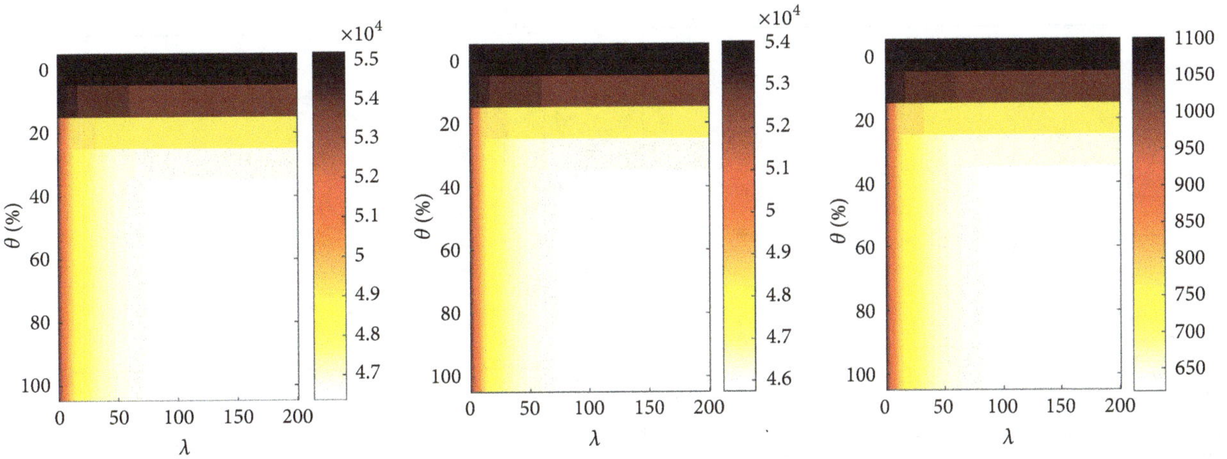

(a) $\sum_a [t_a(v_a) + \tilde{\tau}_a(v_a)]$ (b) $\sum_a t_a(v_a)$ (c) $\sum_a \tilde{\tau}_a(v_a)$

FIGURE 5: System costs with a common UE and an environmental advocacy UE: (a) total cost; (b) travel time; (c) emission exposure.

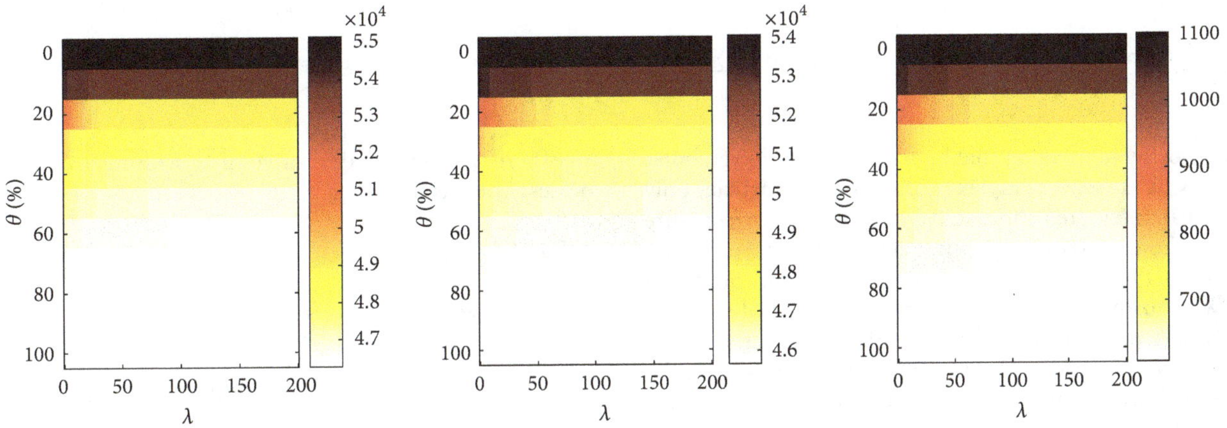

(a) $\sum_a [t_a(v_a) + \tilde{\tau}_a(v_a)]$ (b) $\sum_a t_a(v_a)$ (c) $\sum_a \tilde{\tau}_a(v_a)$

FIGURE 6: System costs with a common UE and an environmental advocacy CN; (a) total cost; (b) travel time; (c) emission exposure.

emission exposure (e.g., $\lambda \geq 50$), the cooperative behavior (CN) results in worse system performance. Figure 7(b) shows that when 70% or more users are environmental advocacy users, the cooperative behavior (CN) is always better than the competitive behavior, and such effect is more significant at lower λ value.

4. Concluding Remarks

In this study, an integrated modeling framework was applied to analytically estimate vehicular emission and concentration within the traffic network. The framework was used to calculate travelers' exposure to emissions, which was considered as a part of the generalized travel cost. The experimental results showed that the route choice behaviors have been changed by considering the emission exposure cost in the cost function. Per the specific network setting, we found that as travelers become more concerned about their exposure to vehicular emissions, the system emission exposure, travel time, and the total cost all get reduced; i.e., Pareto improving solutions are achieved. By analyzing the multiple equilibrium behaviors, we found the system gets better if more users become environmental advocates. For example, in the case that all users follow common UE principle, when 20% of users become environmental advocates, the total system cost reduces about 11%. And the change of a small percentage of users should already lead to good system improvement. By comparing the case with a common UE and an environmental advocacy UE to the case with a common UE and an environmental advocacy CN, we found that cooperative behaviors (CN) lead to better system performance when people's concerns of emission exposure remain at a rational level. The findings of the paper clearly show the necessity and importance of providing exposure-based routing applications to reduce human exposures and to improve the overall performance of transportation networks. As travelers are becoming more concerned of their emission exposure, such eco-routing models are more behavioral sound compared to the conventional emission-based routing models.

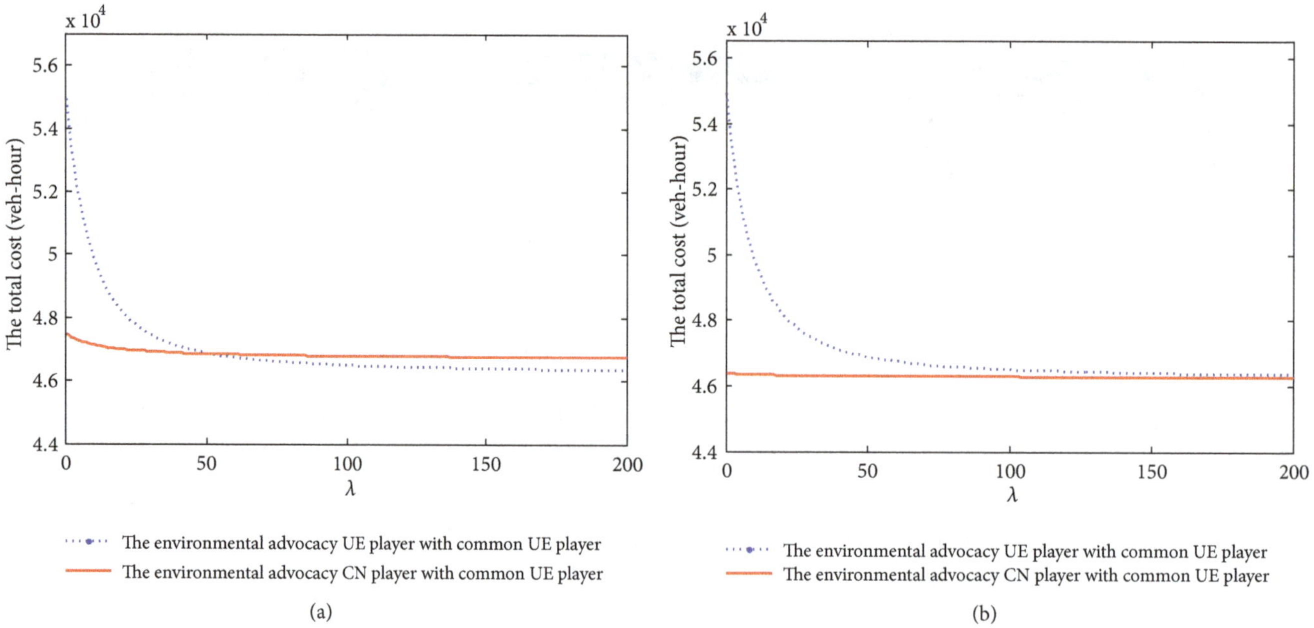

FIGURE 7: Comparison of case 2 and case 3 in total cost $((\sum_a [t_a(v_a) + \tilde{\tau}_a(v_a)]))$: (a) total cost when $\theta = 0.5$ and (b) total cost when $\theta = 0.7$.

It is worthy to mention that human exposure to vehicular emissions is a very complex problem. This article only provided an integrated approach to capture human exposure in route choice. Many modeling details need to be further studied. In the future, the authors will test the proposed framework using more realistic traffic flow and dispersion models and using real traffic networks. Another possible extension is to consider dynamic and elastic demand in the proposed multiple equilibrium. More results and findings will be provided in our subsequent work.

Appendix

$$\begin{aligned}\tilde{\tau}_a(v_a) &= \frac{t_a}{|F_a|}\sum_{s\in F_a} C_s(x_s, y_s) = \frac{t_a}{|F_a|} \\ &\cdot \sum_{s\in F_a}\frac{\sqrt{2}}{\sqrt{\pi}\sigma_z u \sin\varphi_s}\left(\Phi\left(\frac{y_s^2}{\sigma_y}\right) - \Phi\left(\frac{y_s^1}{\sigma_y}\right)\right) q_s \\ &= 0.2038\frac{t_a}{|F_a|} \\ &\cdot \sum_{s\in F_a}\frac{\sqrt{2}}{\sqrt{\pi}\sigma_z u \sin\varphi_s}\left(\Phi\left(\frac{y_s^2}{\sigma_y}\right) - \Phi\left(\frac{y_s^1}{\sigma_y}\right)\right) \\ &\cdot \frac{v_a t_a}{3600 l_a} e^{0.7962 l_a/t_a} = \frac{0.2038}{3600 l_a |F_a|} \\ &\cdot \sum_{s\in F_a}\frac{\sqrt{2}}{\sqrt{\pi}\sigma_z u \sin\varphi_s}\left(\Phi\left(\frac{y_s^2}{\sigma_y}\right) - \Phi\left(\frac{y_s^1}{\sigma_y}\right)\right) \\ &\cdot v_a t_a^2 e^{0.7962 l_a/t_a} = const * v_a t_a^2 e^{0.7962 l_a/t_a}\end{aligned} \tag{A.1}$$

Therefore the first derivative of $\tilde{\tau}_a$ with respect to v_a can be described as in

$$\begin{aligned}\tilde{\tau}'_a(v_a) &= \frac{d\tilde{\tau}_a}{dv_a} \\ &= const \\ &* e^{0.7962 l_a/t_a}\left(t_a^2 + \frac{dt_a}{dv_a}(2v_a t_a - 0.7962 l_a v_a)\right)\end{aligned} \tag{A.2}$$

Since $t_a^2 \geq 0$ and $dt_a/dv_a \geq 0$, we have $\tilde{\tau}'_a(v_a) \geq 0$ if $l_a/t_a \leq (2 * 60)/0.7962 = 150.7$km/h. That is, $\tilde{\tau}_a(v_a)$ is an increasing function if link speed is less than 150.7km/h.

The second derivative of $\tilde{\tau}_a$ with respect to v_a can be calculated as in

$$\begin{aligned}\tilde{\tau}''_a(v_a) &= \frac{d\tilde{\tau}^2_a}{dv_a^2} = const * e^{0.7962 l_a/t_a}\left(t_a^2\right. \\ &\left.+ \frac{dt_a}{dv_a}(2v_a t_a - 0.7962 l_a v_a)\right) = const \\ &* e^{0.7962 l_a/t_a}\left(2t_a\frac{dt_a}{dv_a} + \frac{dt_a}{dv_a}(2t_a - 0.7962 l_a)\right. \\ &\left.+ 2v_a\left(\frac{dt_a}{dv_a}\right)^2 + \frac{dt^2_a}{dv_a^2}(2v_a t_a - 0.7962 l_a v_a)\right) \\ &- const * e^{0.7962 l_a/t_a}\frac{0.7962 l_a}{t_a^2}\frac{dt_a}{dv_a}\left(t_a^2\right. \\ &\left.+ \frac{dt_a}{dv_a}(2v_a t_a - 0.7962 l_a v_a)\right) = const \\ &* 2e^{0.7962 l_a/t_a}\frac{dt_a}{dv_a}(2t_a - 0.7962 l_a) + const\end{aligned}$$

$$* e^{0.7962 l_a/t_a} \frac{dt^2_a}{dv_a^2} (2v_a t_a - 0.7962 l_a v_a) + const$$

$$* e^{0.7962 l_a/t_a} v_a \left(\frac{dt_a}{dv_a} \right)^2 \left(1 + \left(1 - \frac{0.7962 l_a}{t_a} \right)^2 \right) \quad (A.3)$$

Similarly, we have $\tilde{\tau}''_a(v_a) > 0$ if $l_a/t_a \leq (2*60)/0.7962 = 150.7$km/h. Therefore $\tilde{\tau}_a(v_a)$ is strictly convex if link speed is less than 150.7km/h,. This completes the proof.

Notations

N: The set of nodes of the network
A: The set of links of the network
U: The common UE player on the network
E: The environmental advocacy UE player on the network
K: The set of environmental advocacy CN players on the network, $k \in K$
W^U: The set of OD pairs of pertaining to the common UE player
W^E: The set of OD pairs pertaining to the environmental advocacy UE player
W^k: The set of OD pairs pertaining to the environmental advocacy CN player(s)
W: $W \equiv W^U \cup W^E \cup_{k\in K} W^k$
D^U_w: The demand between OD pair $w \in W^U$
D^E_w: The demand between OD pair $w \in W^E$
D^k_w: The demand between OD pair $w \in W^k,\ k \in K$
P_w: The set of paths between OD pair $w \in W$
P^U: $P^U \equiv \cup_{w\in W^U} P_w$
P^E: $P^E \equiv \cup_{w\in W^E} P_w$
P^K: $P^K \equiv \cup_{w\in W^k, k\in K} P_w$
P: $P \equiv P^U \cup P^E \cup P^K$
f^U_p: The flow on path p pertaining to the common UE player
f^E_p: The flow on path p pertaining to the environmental advocacy UE player
f^k_p: The flow on path p pertaining to the environmental advocacy CN player k
δ_{ap}: 1 if link $a \in A$ is on route $p \in P$ and 0 otherwise
v^U_a: The flow of link a that follows the common UE principle
v^E_a: The flow of link a controlled by the environmental advocacy UE player
v^k_a: The flow of link a controlled by environmental advocacy CN player k
v^K_a: $v^K_a \equiv \sum_{k\in K} v^k_a$
v_a: Total flow on link a, $v_a \equiv v^U_a + v^E_a + v^K_a$
v^U: $v^U \equiv (\cdots, v^U_{a-1}, v^U_a, v^U_{a+1}, \cdots)$
v^E: $v^E \equiv (\cdots, v^E_{a-1}, v^E_a, v^E_{a+1}, \cdots)$
v^k: $v^k \equiv (\cdots, v^k_{a-1}, v^k_a, v^k_{a+1}, \cdots)$
v^K: $v^K \equiv (\cdots, v^{k-1}, v^k, v^{k+1}, \cdots)$
τ_a: Emission exposure on link a
$\tilde{\tau}_a$: Simplified emission exposure on link a
$t_a(v_a)$: Travel time on link a
$e_a(v_a)$: CO emission rate on link a
$q_a(v_a)$: CO emission source strength on link a
F: The set that contains all finite line sources of the network
F_a: The set that contains all finite line sources of link a
g_a: A generalized cost function of link a.

Conflicts of Interest

The authors declare that they have no conflicts of interest.

Acknowledgments

The paper was improved and extended based on a conference version presented at CICTP 2018. Information of the conference paper can be found at http://www.cota-home.org/Docs/Conference/CICTP2018/TechSession.pdf. The authors are supported by the National Natural Science Foundation of China (NSFC) via grant 71701173. Any opinions, findings, and conclusions or recommendations expressed in this paper are those of the authors and do not necessarily reflect the views of the NSFC.

References

[1] US Environmental Protection Agency (EPA). (2017). Air Pollutant Emissions Trends Data, Retrieved (May 23, 2018) from https://www.epa.gov/transportation-air-pollution-and-climate-change/accomplishments-and-success-air-pollution-transportation.

[2] X. Tian, H. Dai, Y. Geng et al., "Economic impacts from PM2.5 pollution-related health effects in China's road transport sector: A provincial-level analysis," *Environment International*, vol. 115, pp. 220–229, 2018.

[3] US Environmental Protection Agency (EPA). (2011). National-scale air toxics assessment. Retrieved (July 23, 2016) from https://www.epa.gov/national-air-toxics-assessment/2011-national-air-toxics-assessment.

[4] Y. Yin and S. Lawphongpanich, "Internalizing emission externality on road networks," *Transportation Research Part D: Transport and Environment*, vol. 11, no. 4, pp. 292–301, 2006.

[5] A. Skabardonis, N. Geroliminis, and E. Christofa, "Vehicle emissions estimation under oversaturated conditions along signalized arterials," in *Proceedings of the 91th TRB Annual Meeting (CD)*, Washington, DC, USA, 2012.

[6] Z. Sun, P. Hao, X. Ban, and D. Yang, "Trajectory-based vehicle energy/emissions estimation for signalized arterials using mobile sensing data," *Transportation Research Part D: Transport and Environment*, vol. 34, pp. 27–40, 2015.

[7] W. Zhu, S. Ma, and J. Tian, "Optimizing congestion and emissions via tradable credit charge and reward scheme without initial credit allocations," *Physica A: Statistical Mechanics and its Applications*, vol. 465, pp. 438–448, 2017.

[8] P. T. Harker, "Multiple equilibrium behaviors on networks," *Operations Research Society of America. Transportation Science Section. Transportation Science*, vol. 22, no. 1, pp. 39–46, 1988.

[9] H. Yang, X. Zhang, and Q. Meng, "Stackelberg games and multiple equilibrium behaviors on networks," *Transportation Research Part B: Methodological*, vol. 41, no. 8, pp. 841–861, 2007.

[10] X. Zhang, H. Yang, and H. Huang, "Multiclass multicriteria mixed equilibrium on networks and uniform link tolls for system optimum," *European Journal of Operational Research*, vol. 189, no. 1, pp. 146–158, 2008.

[11] M. Meng, C. F. Shao, Y. D. Wong, and J. Zhang, "Multimodal Network Equilibrium with Stochastic Travel Times," *Mathematical Problems in Engineering*, vol. 2014, Article ID 136872, 8 pages, 2014.

[12] X. Yang, X. J. Ban, and R. Ma, "Mixed equilibria with common constraints on transportation networks," *Networks and Spatial Economics*, vol. 17, no. 2, pp. 547–579, 2017.

[13] F. He, Y. Yin, N. Shirmohammadi, and Y. M. Nie, "Tradable credit schemes on networks with mixed equilibrium behaviors," *Transportation Research Part B: Methodological*, vol. 57, pp. 47–65, 2013.

[14] A. Alzuhairi, M. Aldhaheri, Z.-B. Sun, J.-S. Oh, and V. Kwigizile, "Vehicular emissions and concentrations in school zones: A case study," *Journal of Central South University*, vol. 23, no. 7, pp. 1778–1785, 2016.

[15] M. Barth, G. Scora, and T. Younglove, "Estimating emissions and fuel consumption for different levels of freeway congestion," *Transportation Research Record*, no. 1664, pp. 47–57, 1999.

[16] G. Scora and M. Barth, *Comprehensive Modal Emissions Model (CMEM), version 3.1, User's Guide. 2006. LIST OF TABLES TABLE, 1*, 2014.

[17] H. Rakha, K. Ahn, and A. Trani, "Development of VT-Micro model for estimating hot stabilized light duty vehicle and truck emissions," *Transportation Research Part D: Transport and Environment*, vol. 9, no. 1, pp. 49–74, 2004.

[18] D. Llopis-Castelló, A. M. Pérez-Zuriaga, F. J. Camacho-Torregrosa, and A. García, "Impact of horizontal geometric design of two-lane rural roads on vehicle co2 emissions," *Transportation Research Part D: Transport and Environment*, vol. 59, pp. 46–57, 2018.

[19] US Environmental Protection Agency (EPA), (2014). MOVES2014 a user guide, Retrieved (July 23, 2016) from https://www3.epa.gov/otaq/models/moves/documents/420b15095.pdf.

[20] C. Wallace, K. Courage, M. Hadi, and A. Gan, *TRANSYT-7F User's Guide*, University of Florida, Gainesville, FL, USA, 1998.

[21] K. Boriboonsomsin, M. J. Barth, W. Zhu, and A. Vu, "Eco-routing navigation system based on multisource historical and real-time traffic information," *IEEE Transactions on Intelligent Transportation Systems*, vol. 13, no. 4, pp. 1694–1704, 2012.

[22] A. Skabardonis, N. Geroliminis, and E. Christofa, "Prediction of vehicle activity for emissions estimation under oversaturated conditions along signalized arterials," *Journal of Intelligent Transportation Systems: Technology, Planning, and Operations*, vol. 17, no. 3, pp. 191–199, 2013.

[23] R. Ma, X. Ban, and W. Y. Szeto, "Emission modeling and pricing in dynamic traffic networks," *Transportation Research Procedia*, vol. 9, pp. 106–129, 2015.

[24] R. Saedi, R. Verma, A. Zockaie, M. Ghamami, and T. Gates, "A Framework for Incorporating the Network-Wide Fundamental Diagram into Large-Scale Emission Estimation," *Journal of Transport Health*, vol. 9, pp. S54–S55, 2018.

[25] M. R. Beychok, *Fundamentals of Stack Gas Dispersion*, Newport Beach, CA, USA, 1994.

[26] S. P. Arya, *Air Pollution Meteorology and Dispersion*, Oxford University Press, New York, NY, USA, 1999.

[27] T. Chai, R. Draxler, and A. Stein, "Source term estimation using air concentration measurements and a Lagrangian dispersion model - Experiments with pseudo and real cesium-137 observations from the Fukushima nuclear accident," *Atmospheric Environment*, vol. 106, pp. 241–251, 2015.

[28] N. Sharma, K. K. Chaudhry, and C. V. Chalapati Rao, "Vehicular pollution prediction modelling: A review of highway dispersion models," *Transport Reviews*, vol. 24, no. 4, pp. 409–435, 2004.

[29] P. E. Benson, "CALINE4 – A dispersion model for predicting air pollutant concentrations near roadways," Report No. FHWA/CA/TL-84-15, California Department of Transportation, Sacramento, CA, USA, 1984.

[30] P. Benson, *CALINE3 line source dispersion model, FHWA-CA-TL-84-15*, California Department of Transportation, Sacramento, CA, USA, 1984.

[31] J. Levitin, J. Härkönen, J. Kukkonen, and J. Nikmo, "Evaluation of the CALINE4 and CAR-FMI models against measurements near a major road," *Atmospheric Environment*, vol. 39, no. 25, pp. 4439–4452, 2005.

[32] S. Samaranayake, S. Glaser, D. Holstius et al., "Real-Time Estimation of Pollution Emissions and Dispersion from Highway Traffic," *Computer-Aided Civil and Infrastructure Engineering*, vol. 29, no. 7, pp. 546–558, 2014.

[33] T. L. Alderete, R. B. Jones, Z. Chen et al., "Exposure to traffic-related air pollution and the composition of the gut microbiota in overweight and obese adolescents," *Environmental research*, vol. 161, pp. 472–478, 2018.

[34] H. Yang and X. Zhang, "Existence of anonymous link tolls for system optimum on networks with mixed equilibrium behaviors," *Transportation Research Part B: Methodological*, vol. 42, no. 2, pp. 99–112, 2008.

[35] D. B. Turner, *Workbook of atmospheric dispersion estimates: an introduction to dispersion modeling*, CRC Press, London, UK, 2nd edition, 1994.

[36] X. Liu, A. Godbole, C. Lu, G. Michal, and P. Venton, "Optimisation of dispersion parameters of Gaussian plume model for CO_2 dispersion," *Environmental Science and Pollution Research*, vol. 22, no. 22, pp. 18288–18299, 2015.

14

A Sparse Optimization Approach for Energy-Efficient Timetabling in Metro Railway Systems

Ziyan Luo,[1] Xiaoyu Li,[2] and Naihua Xiu[2]

[1] *State Key Laboratory of Rail Traffic Control and Safety, Beijing Jiaotong University, Beijing 100044, China*
[2] *Department of Mathematics, Beijing Jiaotong University, Beijing 100044, China*

Correspondence should be addressed to Xiaoyu Li; 16121607@bjtu.edu.cn

Academic Editor: Luca Pugi

In this paper, we propose a sparse optimization approach to maximize the utilization of regenerative energy produced by braking trains for energy-efficient timetabling in metro railway systems. By introducing the cardinality function and the square of the Euclidean norm function as the objective function, the resulting sparse optimization model can characterize the utilization of the regenerative energy appropriately. A two-stage alternating direction method of multipliers is designed to efficiently solve the convex relaxation counterpart of the original NP-hard problem and then to produce an energy-efficient timetable of trains. The resulting approach is applied to Beijing Metro Yizhuang Line with different instances of service for case study. Comparison with the existing two-step linear program approach is also conducted which illustrates the effectiveness of our proposed sparse optimization model in terms of the energy saving rate and the efficiency of our numerical optimization algorithm in terms of computational time.

1. Introduction

With the rapid increase in industry and population in cities, the metro railway system has been playing a more and more important role in urban economic and social development for metropolis due to its high loading capacity and low pollution. Despite the inherent efficiency, the energy consumption of metro railway systems is of a huge magnitude and the efficient energy management for such systems is then of global importance. It is known that the energy consumption of trains is largely influenced by the driving strategy and the arrivals and departures in the timetable, which inspires the extensive study on energy-efficient train control and energy-efficient train timetabling. A careful treatment on such topics can be found in the nice survey paper by Scheepmaker et al. in [1], with focuses on reviewing the mathematical problem formulations and solution approaches in the literature. Related work on energy optimization for rail traffic systems such as tramway systems has also been well studied; see, e.g., [2, 3].

Among all the involved energy saving strategies in metro railway systems, an energy-efficient timetable with regenerative braking which schedules the arrival time and the departure time of each train to and from the platforms it visits for the minimization of energy consumption and the maximization of produced regenerative energy utilization, has gained considerable academical attention in recent years. In 2012, Peña-Alcaraz et al. [4] proposed a large-scale mixed integer programming model to maximize the utilization of regenerative braking energy by means of maximizing the total duration of all possible synchronization processes of acceleration and regenerative braking between all possible train pairs. The simulation model is applied to line three of the Madrid metro system and the energy savings of the optimized timetable are reported to be about 7% on average. By merely choosing those train pairs suitable for regenerating energy transfer other than all possible train pairs in [4], Das Gupta et al. developed a more tractable mixed integer programming model in [5]. In 2013, Yang et al. [6] presented a cooperative integer programming model to maximize the time overlaps of nearby accelerating and braking trains and applied a genetic algorithm to approximately solve the resulting optimization model. For the case study on Beijing Metro Yizhuang Line, they reported 22.1% and 15.2% increase of effective time overlaps during peak hours and off-peak hours, respectively. A two-objective timetable optimization model for maximizing the regenerative braking energy with consideration of the passenger waiting time was developed by Yang et al. in [7] where the genetic algorithm was again adopted to solve the optimization model. An integrated energy-efficient operation model was proposed by Li & Lo in [8] with the speed profile part to minimize the net energy consumption and the timetabling part with headway constraints to synchronize the accelerating and regenerative braking for the reuse of regenerative energy. About 25% energy savings were reported for Beijing Metro Yizhuang Line with the headway between trains being 90 seconds. Li & Lo further developed an integrated energy-efficient timetable and speed profile optimization model to determine the cycle time, the headway time, and the speed profiles for a metro line to minimize the energy consumption [9]. The resulting mathematical model is a strictly convex quadratic programming problem with simple bound constraints and an analytical approach based on the first-order optimality condition (i.e., the KKT condition). An integrated timetable and speed profile optimization model with Taylor approximation were also considered by Yang et al. in [10] where the active set method was employed to handle the resulting strictly convex quadratic programming with linear constraints. Under the assumption that trains are operating in the optimal speed profiles, Yang et al. [11] proposed a mixed integer programming model aiming at synchronizing the arrivals and departures of the trains such that regenerative braking energy can be used as much as possible during the whole day service. A two-step linear programming model with the first step to minimize the total energy consumed by all trains and the second step to maximize the utilization of regenerative energy produced by braking trains for a whole day service in metro railway networks was proposed by Das Gupta et al. in [12]. To ensure the transfer of maximum possible regenerative energy from braking trains to accelerating trains, a time overlap vector was introduced which requires to be as close as possible to the zero vector. To effectively quantify such a requirement, the ℓ_0-norm and the ℓ_1-norm were employed in [12]. By adopting the ℓ_1-norm heuristic for the original ℓ_0-norm (i.e., the number of nonzero components in a vector), together with epigraph approach that equivalently transforms the convex ℓ_1-norm minimization to a linear programming, the resulting relaxation problem in the second step was efficiently solved by Gurobi Optimizer. Numerical experiments on different instances of service PES2-SFM2 of line 8 of Shanghai Metro network spanning a full service period of one day were conducted and a significant reduction in effective energy consumption with the worst case being 19.27% was reported by applying the proposed approach.

The original ℓ_0-norm minimization problem proposed in [12] is well-known to be NP-hard due to the discontinuity and nonconvexity of the ℓ_0-norm. Such a type of optimization problems is coined by sparse optimization in the optimization community and has a wide range of applications in compressed sensing, signal processing, statistical regression, financial portfolio, and so on. Various tractable relaxation strategies are proposed to replace the ℓ_0-norm, such as the popular convex surrogate ℓ_1-norm [13], and some weighted ℓ_1-norm variants [14]. Numerical algorithms are also developed for the resulting relaxation models, such as the greedy-type methods [15], the first-order optimization methods [16], and second-order methods for some special structured models [17].

Inspired by the development of sparse optimization methods, in this paper, a sparse optimization model with the ℓ_0-norm and the squared Euclidean norm as the objective function is built to maximize the utilization of the regenerative braking energy. Motivated by the adaptive LASSO model in the high-dimensional statistical regression [14], we propose a weighted ℓ_1-norm relaxation scheme. A two-stage alternating direction method of multipliers (TSADMM) is then proposed with the first stage for calculating an appropriate weight vector and an initial point for the iterative algorithm in the second stage. To illustrate the effectiveness of our new mathematical model, the linear programming model in [12], a quadratic model merely considering the squared Euclidean norm, and our weighted-ℓ_1 companied with the squared Euclidean norm are applied to Beijing Metro Yizhuang Line for comparison. The numerical results on energy savings show that our model is the best among them. To demonstrate the efficiency of our proposed TSADMM, the academical version of CVX (http://cvxr.com/cvx/) is called to solve our model as well. The numerical results in computation time show that our algorithm outperforms that used in CVX.

The rest of this paper is organized as follows. In Section 2, the sparse optimization model for energy-efficient timetabling in metro railway systems is built. In Section 3, the weighted ℓ_1 norm relaxation model is proposed and a two-stage alternating direction method of multipliers (TSADMM) is designed to get an approximate optimal solution. Theoretical global convergence is established as well. By applying our proposed approach to Beijing Metro Yizhuang Line, we demonstrate the effectiveness of our proposed model and the efficiency of our proposed TSADMM by the numerical results in Section 4. Conclusions and future research are given in Section 5. For reference, the appendix provides the proof of the global convergence theorem for our algorithm.

2. Model Formulation

2.1. Notation. The notation that will be used throughout the paper is listed in Table 1 for the convenience of the subsequent model formulation.

2.2. Constraints. The set of constraints in the metro train networks including constraints of the trip time, the dwell time, the headway time, the total travel time, and the domain of event times will be described in this subsection, following from [12].

(i) Trip time constraint: it includes the trip time constraints with tracks and with crossing-overs. For the former one, when a train $t \in \mathcal{T}$ has a trip from platform $i \in \mathcal{N}^t$ to platform $j \in \mathcal{N}^t$ along the track $(i, j) \in \mathcal{K}^t$, the departure time d_i^t and the arrival

TABLE 1: A list of notations.

a_i^t	The arrival time of train t at platform i
d_i^t	The departure time of train t from platform i
$\mathcal{T}$	The set of all trains
$\lvert\cdot\rvert$	The cardinality of a set, i.e., the number of elements in a set
$[\underline{tr}_{ij}^t, \overline{tr}_{ij}^t]$	The trip time window for train t from platform i to j
ϕ	The set of all the crossing-overs
$\mathcal{O}_{ij}$	The set of all train pairs associated with the crossing-over (i, j)
$[\underline{\kappa}_{ij}^{tt'}, \overline{\kappa}_{ij}^{tt'}]$	The trip window for train t on the crossing-over (i, j)
$[\underline{dw}_i^t, \overline{dw}_i^t]$	The dwell time window for train t at platform i
$\mathcal{S}$	The set of all platform pairs at the same interchange station
$\mathcal{C}_{ij}$	The set of all connection train pairs for a platform pair (i, j)
$[\underline{c}_{ij}^{tt'}, \overline{c}_{ij}^{tt'}]$	The connection window between the train t at the platform i and the train t' at platform j
$\mathcal{H}_{i,j}$	The set of train-pairs that move along the track (i, j)
$[\underline{h}_i^{tt'}, \overline{h}_i^{tt'}]$	The headway window between train t and t' arrival at or departure from platform i
$[\underline{tt}^t, \overline{tt}^t]$	The window of total travel time constraint for train t
f_{ij}	Energy consumption associated with the trip $(i, j) \in \mathcal{A}_{tr}$
$\mathcal{P}$	The set of all platform pairs powered by a same electrical substation
$\overrightarrow{t}$	Temporally closest train to the right of train t
$\overleftarrow{t}$	Temporally closest train to the left of train t
∇_i^t	The relative distance from a_i^t to regenerative alignment point
Δ_i^t	The relative distance from d_i^t to the consumption alignment point
$\overrightarrow{\varepsilon}$	A set of ε containing elements of the form $(i, j, t, \overrightarrow{t})$
$\overleftarrow{\varepsilon}$	A set of ε containing elements of the form $(i, j, t, \overleftarrow{t})$
$\mathcal{N}$	The set of all platforms
$\mathcal{K}$	The set of all tracks
$\mathcal{K}^t$	The set of tracks visited by a train t
$\mathcal{N}^t$	The set of platforms visited by a train t
$\mathcal{K}_{tr}$	The set of all arcs associated with trip time constraints
$\delta_S(\cdot)$	The indicator function with respect to the set S
$\Pi_S(\cdot)$	The projection operator onto the set S
$\text{sign}(\cdot)$	The sign function
$\circ$	The componentwise product, also known as the Hadamard product
A^T	The transpose of the matrix A
$vec(X)$	The column vector generated by stacking all columns of the matrix X

time a_j^t are bounded by $\underline{tr}_{ij}^t$ and $\overline{tr}_{ij}^t$, respectively. The corresponding constraints are

$$\underline{tr}_{ij}^t \leq a_j^t - d_i^t \leq \overline{tr}_{ij}^t, \quad \forall t \in \mathcal{T},\ \forall (i, j) \in \mathcal{K}^t. \tag{1}$$

For the latter one, one knows that a crossing-over connecting two train-lines is a special track, where a train-line is a path with some nonopposite platforms and tracks. The crossing-over is to connect the terminal platform of a train-line with the first platform of another. A train t turns around by a crossing-over and departs from the first platform to travel through the train-line after it arrives at the final platform at another train-line, and then the same train is labelled as t' in order to treat it as two different trains [20]. Let ϕ represent the set of all crossing-overs. Consider a crossing-over $(i, j) \in \phi$, we denote the set of all train pairs associated with a same physical train through the crossing-over (i, j) by $\mathcal{O}_{ij}$. The trip time to travelling through the crossing-over $(i, j) \in \phi$ is the time from the departure from the platform i (train is labelled as t) to the arrival at the platform j (train is labelled as t'), which has to stay within a window $[\underline{\kappa}_{ij}^{tt'}, \overline{\kappa}_{ij}^{tt'}]$. And the corresponding constraints can be described as below:

$$\underline{\kappa}_{ij}^{tt'} \leq a_j^{t'} - d_i^t \leq \overline{\kappa}_{ij}^{tt'}, \quad \forall (i, j) \in \phi,\ \forall (t, t') \in \mathcal{O}_{i,j}. \tag{2}$$

(ii) Dwell time constraint: when a train $t \in \mathcal{T}$ arrives at a platform $i \in \mathcal{N}^t$, it should dwell there in a time interval denoted by $[\underline{dw}_i^t, \overline{dw}_i^t]$ to make sure that passengers can get off and get on the train before it departures from the platform i. The dwell time constraint can be written as follows:

$$\underline{dw}_i^t \leq d_i^t - a_i^t \leq \overline{dw}_i^t, \quad \forall t \in \mathcal{T}, \ \forall i \in \mathcal{N}^t. \quad (3)$$

(iii) Connection constraint: an interchange station used by connecting trains is to make passengers transfer between trains on different lines. Any platform pair $(i, j) \in \mathcal{S}$ consists of two platforms at a same interchange station. For a connection train pair $(t, t') \in \mathcal{C}_{ij}$, we assume that the train t arrives at platform i and another train t' departs from platform j. The difference of both times can be in a window $[\underline{c}_{ij}^{tt'}, \overline{c}_{ij}^{tt'}]$ to maintain that passengers can get off from the former train and get on the latter. The constraint of connection can be described as follows:

$$\underline{c}_{ij}^{tt'} \leq d_j^{t'} - a_i^t \leq \overline{c}_{ij}^{tt'}, \quad \forall (i, j) \in \mathcal{S}, \ \forall (t, t') \in \mathcal{C}_{ij}. \quad (4)$$

(iv) Headway time constraint: similar to the trip time constraints, the headway time constraints includes those with tracks and with crossing-overs. In any subway system, a minimum interval between the departures and arrivals of sequenced trains on the same track is maintained, called headway time. To meet the needs of the passengers, railway networks in many cities designate an upper bound between the departures and arrivals of consecutive trains, in order to ensure passengers do not wait for a long time before the arrival of next train. Define $\mathcal{H}_{ij}$ as the set of pairs of successive trains that pass by track $(i, j) \in \mathcal{K}$. Consider $(t, t') \in \mathcal{H}_{ij}$, and let $[\underline{h}_i^{tt'}, \overline{h}_i^{tt'}]$ and $[\underline{h}_j^{tt'}, \overline{h}_j^{tt'}]$ be the time intervals that maintained the departures and arrivals of consecutive trains t and t' from and to platform i and platform j, respectively. For any $(i, j) \in \mathcal{K}$ between platform i and platform j, headway time constraint with tracks can be described as follows:

$$\begin{aligned} &\underline{h}_i^{tt'} \leq d_i^t - d_i^{t'} \leq \overline{h}_i^{tt'}, \\ &\underline{h}_j^{tt'} \leq a_j^t - a_j^{t'} \leq \overline{h}_j^{tt'}, \\ &\forall (i, j) \in \mathcal{K}, \ (t, t') \in \mathcal{H}_{ij}. \end{aligned} \quad (5)$$

In the other case, consider two consecutive trains t_1 and t_2 that go through a crossing-over $(i, j) \in \phi$ after departing from the final platform i of a train-line. They will be labelled as t_1' and t_2' when they arrive at the first platform j of some other train-line. Let $\widetilde{\mathcal{O}}_{ij}$ be the set of all such two-train pairs $((t_1, t_1'), (t_2, t_2'))$. There is an interval time window $[\underline{h}_i^{t_1t_2}, \overline{h}_i^{t_1t_2}]$ between the departures of trains t_1 and t_2 from platform i and a window $[\underline{h}_j^{t_1't_2'}, \overline{h}_j^{t_1't_2'}]$ between the arrivals of trains t_1' and t_2' at platform j. We can then write the headway constraints with crossing-overs as follows:

$$\begin{aligned} &\underline{h}_i^{t_1t_2} \leq d_i^{t_1} - d_i^{t_2} \leq \overline{h}_i^{t_1t_2}, \\ &\underline{h}_j^{t_1't_2'} \leq a_j^{t_1'} - a_j^{t_2'} \leq \overline{h}_j^{t_1't_2'}, \\ &\forall (i, j) \in \phi, \ \forall ((t_1, t_1'), (t_2, t_2')) \in \widetilde{\mathcal{O}}_{ij}. \end{aligned} \quad (6)$$

Besides, the headway time is related to the passenger demand and the number of trains. Let $\overline{m}'$ be the number of trains that are in service per hour, c_t be the train capacity, u be the train utility rate, and D be the passenger demand. Then, $D = c_t \times u \times \overline{m}'$ [8]. Let h be the headway time. Since $h = 3600/\overline{m}'$, we have

$$h = \frac{3600 \times c_t \times u}{D}. \quad (7)$$

Here we assume that u and c are constant while the passenger demand varies. In other words, the headway time of trains changes at different periods in a day.

(v) Total travel time constraint: when a train $t \in \mathcal{T}$ has a trip, traversing all platforms and tracks in chronological order, it can have the total travel time between $\underline{tt}^t$ and $\overline{tt}^t$, to meet the needs of passengers in subway systems. Total travel time constraints can be written as follows:

$$\underline{tt}^t \leq d_{\mathcal{N}^t(|\mathcal{N}^t|)}^t - a_{\mathcal{N}^t(1)}^t \leq \overline{tt}^t, \quad \forall t \in \mathcal{T}, \quad (8)$$

(vi) Domain of event times: we set zero second as the time of the departure of the first train from a platform in a day when the railway is in service. By adding the maximum possible values of all trip times and dwell times, we can set an upper bound for arrival of the last train at the final platform. And we denote the upper bound by $m_T \in Z_{++}$. So, we can write the domain of the event time constraints as follows:

$$\begin{aligned} &0 \leq a_i^t \leq m_T, \\ &0 \leq d_i^t \leq m_T, \\ &\forall t \in \mathcal{T}, \ \forall i \in \mathcal{N}^t. \end{aligned} \quad (9)$$

2.3. Minimizing the Energy Consumption. The major energy consumption of a train is from the acceleration phase. To efficiently minimize the total energy consumption of trains in a metro railway network, Das Gupta [12] proposed an optimization model with linear inequality constraints to get a feasible timetable. The involved optimization model takes the form of

$$\begin{aligned} \min \quad & \sum_{(i,j) \in \mathcal{K}_{tr}, t \in \mathcal{T}} f_{ij}(a_j^t - d_i^t) \\ \text{s.t.} \quad & (1)\text{–}(9) \end{aligned} \quad (10)$$

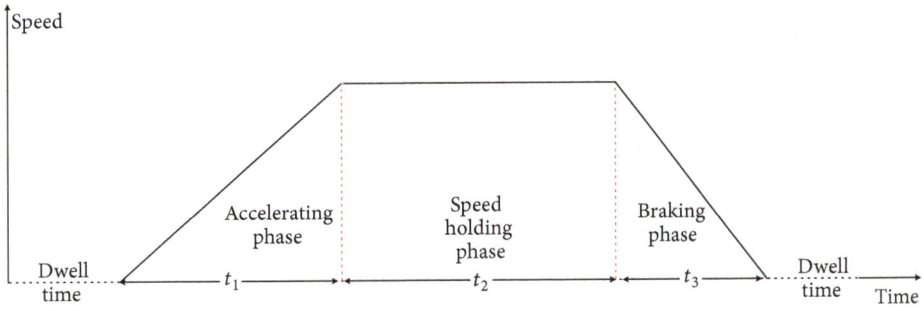

FIGURE 1: Three working phases (accelerating-cruising-braking) of speed profile.

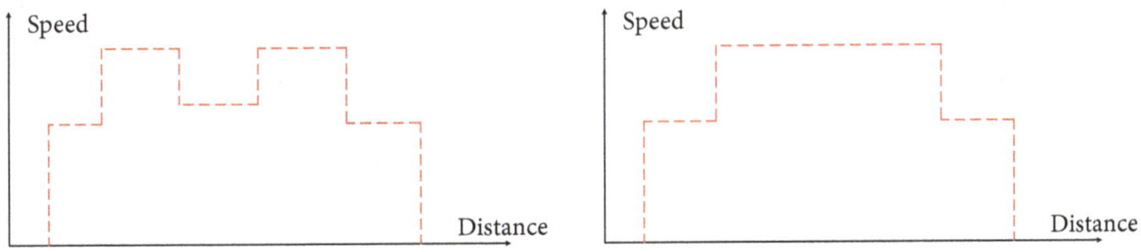

FIGURE 2: The speed limit of a train with a multiphase section.

where $f_{ij} : \mathbf{R}_{++} \longrightarrow \mathbf{R}_{++}$ with argument $(a_j^t - d_i^t)$ being the energy consumption function that is unknown in advance and some best possible affine approximate function in the sense of least-squares by using the measured energy in practice is then utilized. By denoting $x := \text{vec}(X^T)$ with

$$X = \begin{pmatrix} a_1^1 & a_1^2 & \cdots & a_1^{|\mathcal{T}|-1} & a_1^{|\mathcal{T}|} \\ d_1^1 & d_1^2 & \cdots & d_1^{|\mathcal{T}|-1} & d_1^{|\mathcal{T}|} \\ a_2^1 & a_2^2 & \cdots & a_2^{|\mathcal{T}|-1} & a_2^{|\mathcal{T}|} \\ d_2^1 & d_2^2 & \cdots & d_2^{|\mathcal{T}|-1} & d_2^{|\mathcal{T}|} \\ \vdots & \vdots & \ddots & \vdots & \vdots \\ a_{|\mathcal{N}|}^1 & a_{|\mathcal{N}|}^2 & \cdots & a_{|\mathcal{N}|}^{|\mathcal{T}|-1} & a_{|\mathcal{N}|}^{|\mathcal{T}|} \\ d_{|\mathcal{N}|}^1 & d_{|\mathcal{N}|}^2 & \cdots & d_{|\mathcal{N}|}^{|\mathcal{T}|-1} & d_{|\mathcal{N}|}^{|\mathcal{T}|} \end{pmatrix} \quad (11)$$

the approximate counterpart of (10) can be formulated by the following linear programming:

$$\begin{aligned} \min \quad & \sum_{(i,j)\in\overline{\mathcal{K}}_{tr}} c_{ij}\left(x_j - x_i\right) \\ \text{s.t.} \quad & l_{ij} \le x_j - x_i \le u_{ij}, \quad (i,j) \in \overline{\mathcal{K}} \\ & 0 \le x_i \le m_T, \quad i = 1, 2, \ldots, n, \end{aligned} \quad (12)$$

where $n = 2|\mathcal{T}||\mathcal{N}|$, $\overline{\mathcal{K}}$ is a subset of $\{(i,j) \mid i, j = 1, 2, \ldots, n\}$ that collects all those indices according to constraints (1)-(8), and $\overline{\mathcal{K}}_{tr}$ is a subset of $\overline{\mathcal{K}}$ that collects all those indices according to constraints (1)-(2). By utilizing its optimal solution $\overline{x}$, we can get $\overline{a}_j^t$'s and $\overline{d}_j^t$'s such that

$$a_j^t - d_i^t = \overline{a}_j^t - \overline{d}_i^t, \quad \forall t \in \mathcal{T}, \ \forall (i,j) \in \mathcal{K}^t. \quad (13)$$

$$a_j^{t'} - d_i^t = \overline{a}_j^{t'} - \overline{d}_i^t, \quad \forall (i,j) \in \phi, \ \forall \left(t, t'\right) \in \mathcal{O}_{i,j}. \quad (14)$$

For all other constraints, lower and upper bounds are allowed to vary as described in (3)-(9).

2.4. Maximizing the Energy Regeneration: A Sparse Optimization Model. We adopt the speed profile by maximum accelerating, speed holding, maximum braking strategy from [18] with a multi-phase-speed-limit section, as described in Figures 1 and 2.

Let t_1, t_2, and t_3 be the time of maximum accelerating, speed hold, and maximum braking phases, and a_{acc} and a_{bra} are the accelerating rate of the accelerating and braking phase. It is known that, with regenerative braking, kinetic energy can be converted into electricity which can be fed back to the power supply system to be used by other nearby trains in the same electric substation. Denote the conversion factor from electricity to kinetic energy as θ_{acc} and the conversion factor from kinetic energy to regenerative electricity as θ_{bra} and the resistance rate at the speed holding phase as γ. For any two adjacent stations, we adopt the formula from [18] to calculate the energy consumption for accelerating and the regenerative

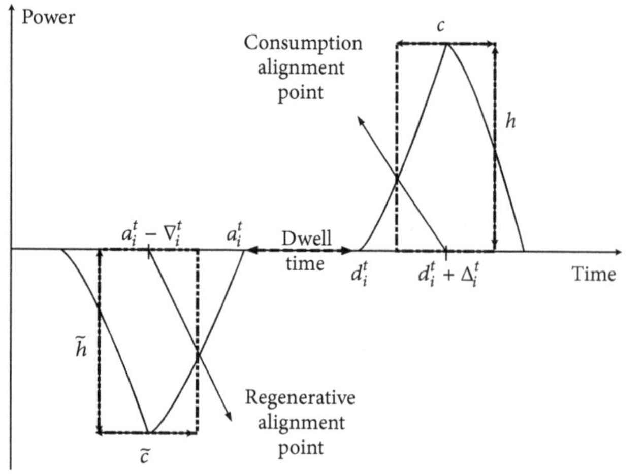

FIGURE 3: Use rectangles instead of energy consumption and regenerative energy.

energy produced during braking, denoted by E^{con} and E^{reg}, respectively, in the way as below:

$$\begin{aligned} E^{con} &= \frac{a_{acc}^2 t_1^2}{2\theta_{acc}} + \int_{t_1}^{t_1+t_2} \frac{\gamma^2 (t-t_1)}{\theta_{acc}} dt \\ &= \frac{a_{acc}^2 t_1^2}{2\theta_{acc}} + \frac{\gamma^2 t_2^2}{2\theta_{acc}} \end{aligned} \quad (15)$$

$$\text{and } E^{reg} = \frac{\theta_{bra} a_{bra}^2 t_3^2}{2}.$$

Here E^{con} and E^{reg} are both for the unit mass. A simple heuristic method is applied to approximate the power graph as shown in Figure 3, where c and $\tilde{c}$ are the width of the rectangles for accelerating and braking and h and $\tilde{h}$ are heights of the rectangles. Similar to those in [12], the midpoint of the width in the rectangle is called the regenerative or consumptive alignment point, where Δ_i^t represents the distance between d_i^t and the regenerative alignment point and ∇_i^t represents the distance between a_i^t and the consumption alignment point.

Here, we align the distance between regenerative and consumption alignment points to maximum the utilization. Before introducing such a distance, we recall the suitable train pairs as introduced in [12]. Assume that platform pairs are opposite to each other at the same electrical substation. Let $\mathcal{P}$ be the set of all platform pairs. For any pair $(i, j) \in \mathcal{P}$ and the set $\mathcal{T}_i \subseteq \mathcal{T}$ contains all trains that arrives at, dwells, departs from platform i. When the train $t \in \mathcal{T}_i$ departs from the platform i, it is a best choice to find a train $\overrightarrow{t}$ at the platform j which are producing the regenerative braking energy using the following definition.

Definition 1 (see [12]). Consider any $(i, j) \in \mathcal{P}$. For every train $t \in \mathcal{T}_i$, the train $\overrightarrow{t} \in \mathcal{T}_j$ is called the temporally closest train to the right of t if

$$\overrightarrow{t} = \mathop{\arg\min}_{t' \in \{x \in \mathcal{T}_j : 0 \le (\overline{a}_j^x + \overline{d}_j^x)/2 - (\overline{a}_i^t + \overline{d}_i^t)/2 \le \beta\}} \left\{ \left| \frac{\overline{a}_i^t + \overline{d}_i^t}{2} - \frac{\overline{a}_j^{t'} + \overline{d}_j^{t'}}{2} \right| \right\} \quad (16)$$

where β is an empirical parameter determined by the timetable designer and is much smaller than the time horizon of the entire timetable [12]. Similarly, when a train $t \in \mathcal{T}_i$ arrives at platform i, it produces the regenerative energy. To find a suitable train $\overleftarrow{t}$ departing from platform j for transferring the regenerative braking energy, we recall the following definition from [12].

Definition 2 (see [12]). Consider any $(i, j) \in \mathcal{P}$. For every train $t \in \mathcal{T}_i$, the train $\overleftarrow{t} \in \mathcal{T}_j$ is called the temporally closest train to the left of t if

$$\overleftarrow{t} = \mathop{\arg\min}_{t' \in \{x \in \mathcal{T}_j : 0 \le (\overline{a}_i^t + \overline{d}_i^t)/2 - (\overline{a}_j^x + \overline{d}_j^x)/2 \le \beta\}} \left\{ \left| \frac{\overline{a}_i^t + \overline{d}_i^t}{2} - \frac{\overline{a}_j^{t'} + \overline{d}_j^{t'}}{2} \right| \right\}. \quad (17)$$

Denote the sets $\overleftarrow{\varepsilon}$ and $\overrightarrow{\varepsilon}$ whose components are $(i, j, t, \overleftarrow{t})$ and $(i, j, t, \overrightarrow{t})$, respectively. Consider any $(i, j, t, \overleftarrow{t}) \in \overleftarrow{\varepsilon}$. To maximize the transfer of the regenerative energy from the braking train $\overleftarrow{t}$ by the accelerating train t, we attempt to make the term $(d_i^t + \Delta_i^t - a_j^{\overrightarrow{t}} + \nabla_j^{\overrightarrow{t}})$ to be zero or as close to zero as possible. Similarly, for $(i, j, t, \overrightarrow{t}) \in \overrightarrow{\varepsilon}$, our goal is to make the term $(d_j^{\overleftarrow{t}} + \Delta_j^{\overleftarrow{t}} - a_i^t + \nabla_i^t)$ to be zero or as close to zero as possible. Following from [12], we define an auxiliary variable $y \in \mathbf{R}^n$ with $n = 2\overline{n}\,\overline{m}$, $\overline{n} = |\mathcal{N}|$ and $\overline{m} = |\mathcal{T}|$ by

$$y = \left(\left(d_i^t + \Delta_i^t - a_j^{\overrightarrow{t}} + \nabla_j^{\overrightarrow{t}} \right)_{(i,j,t,\overrightarrow{t}) \in \overrightarrow{\varepsilon}}, \left(d_j^{\overleftarrow{t}} + \Delta_j^{\overleftarrow{t}} - a_i^t + \nabla_i^t \right)_{(i,j,t,\overleftarrow{t}) \in \overleftarrow{\varepsilon}} \right). \quad (18)$$

The components of y represent difference of the time between the alignment points of accelerating and braking. Once the running time is fixed, Δ_i^t and ∇_i^t are two constants.

The ultimate energy consumption can be approximately calculated as follows (see Figure 4):

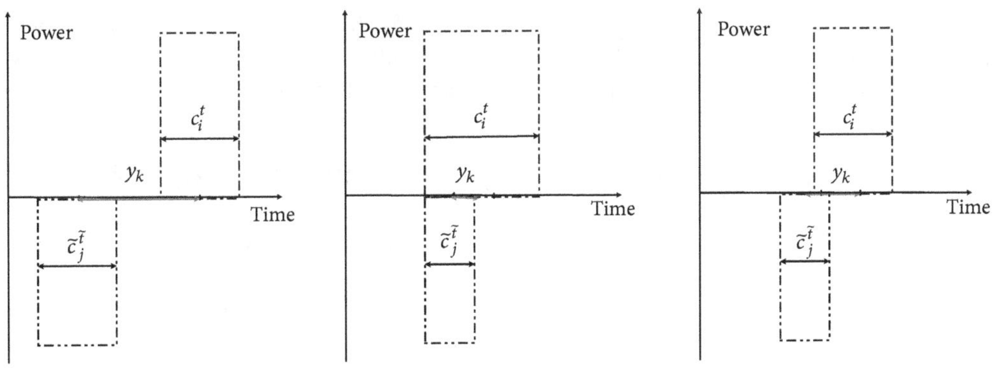

FIGURE 4: Three cases for net energy consumption function.

$$E_{(i,j,t,\tilde{t})} = \begin{cases} E_k^{t,con}, & \text{if } |y_k| \geq \dfrac{\left(\tilde{c}_j^{\tilde{t}} + c_i^t\right)}{2}, \\ E_k^{t,con} - \min\left\{\tilde{c}_j^{\tilde{t}}, c_i^t\right\} * \min\left\{h_i^t, \tilde{h}_j^{\tilde{t}}\right\}, & \text{if } |y_k| \leq \left|\dfrac{\left(c_i^t - \tilde{c}_j^{\tilde{t}}\right)}{2}\right|, \\ E_k^{t,con} - \left(\dfrac{\left(\tilde{c}_j^{\tilde{t}} + c_i^t\right)}{2} - y_k\right) * \min\left\{h_i^t, h_j^{\tilde{t}}\right\}, & \text{otherwise.} \end{cases} \tag{19}$$

where train t is accelerating and train $\tilde{t}$ is braking.

By sorting all possible consecutive platforms for braking energy absorbing, a 0-1 vector $\alpha \in \mathbf{R}^n$ is introduced to indicate all those locations in y with regeneration energy usage when the components of α take the value 1. In order to make the utilization of the regenerative energy as much as possible, we then turn to minimizing the number of nonzero elements of $\alpha \circ y$ which corresponds to minimizing $\|\alpha \circ y\|_0$ and making nonzero elements in $\alpha \circ y$ as close to zero as possible which corresponds to minimizing $\|\alpha \circ y\|_2^2$. The resulting sparse optimization model turns out to be

$$\begin{aligned} \min_{x,y} \quad & \lambda_1 \|\alpha \circ y\|_0 + \frac{\lambda_2}{2} \|\alpha \circ y\|_2^2 \\ s.t. \quad & (3)\text{–}(9),\ (13),\ (14),\ (18). \end{aligned} \tag{20}$$

where $x, y \in \mathbf{R}^n$ are decision variables. Similar to (12), (20) can be rewritten as

$$\min_{x,y} \left\{ \lambda_1 \|\alpha \circ y\|_0 + \frac{\lambda_2}{2} \|\alpha \circ y\|_2^2 : y - Ax - b = 0,\ Fx - g = 0,\ f - Ex \geq 0. \right\} \tag{21}$$

where $A \in \mathbf{R}^{n\times n}$, $E \in \mathbf{R}^{m\times n}$, $F \in \mathbf{R}^{l\times n}$ are

$$E = \begin{pmatrix} E_1 \\ -E_1 \\ E_2 \\ -E_2 \\ E_3 \\ -E_3 \\ E_4 \\ -E_4 \end{pmatrix},$$

$$F = \begin{pmatrix} O_{\bar m\times\bar m} & \widehat{E}_1 & O_{\bar m\times 2\bar m} & \cdots & O_{\bar m\times 2\bar m} & O_{\bar m\times\bar m} \\ O_{\bar m\times\bar m} & O_{\bar m\times 2\bar m} & \widehat{E}_2 & \cdots & O_{\bar m\times 2\bar m} & O_{\bar m\times\bar m} \\ \vdots & \vdots & \vdots & \ddots & \vdots & \vdots \\ O_{\bar m\times\bar m} & O_{\bar m\times 2\bar m} & \cdots & O_{\bar m\times 2\bar m} & \widehat{E}_{\bar n-1} & O_{\bar m\times\bar m} \end{pmatrix} \tag{22}$$

with

$$E_1 = \begin{pmatrix} \widehat{E}_1 & O_{\bar m\times 2\bar m} & \cdots & O_{\bar m\times 2\bar m} \\ O_{\bar m\times 2\bar m} & \widehat{E}_2 & \cdots & O_{\bar m\times 2\bar m} \\ \vdots & \vdots & \ddots & \vdots \\ O_{\bar m\times 2\bar m} & O_{\bar m\times 2\bar m} & \cdots & \widehat{E}_{\bar n} \end{pmatrix},$$

$$\widehat{E}_i = \left(-I_{\bar m\times\bar m} \quad I_{\bar m\times\bar m}\right),$$

$$E_2 = \left(-I_{\bar m\times\bar m} \quad O_{\bar m\times(2\bar n\bar m - 2\bar m)} \quad I_{\bar m\times\bar m}\right),$$

$$E_3 = I_{2\bar n\bar m\times 2\bar n\bar m},$$

$$E_4 = \begin{pmatrix} \widetilde{E}_1 & O_{(\overline{m}-1)\times\overline{m}} & \cdots & O_{(\overline{m}-1)\times\overline{m}} \\ O_{(\overline{m}-1)\times\overline{m}} & \widetilde{E}_2 & \cdots & O_{(\overline{m}-1)\times\overline{m}} \\ \vdots & \vdots & \ddots & \vdots \\ O_{(\overline{m}-1)\times\overline{m}} & O_{(\overline{m}-1)\times\overline{m}} & \cdots & \widetilde{E}_{2\overline{n}} \end{pmatrix},$$

$$\widetilde{E}_i = \begin{pmatrix} -1 & 1 & 0 & \cdots & 0 \\ 0 & -1 & 1 & \cdots & 0 \\ \vdots & \vdots & \ddots & \ddots & \vdots \\ 0 & 0 & \cdots & -1 & 1 \end{pmatrix}, \tag{23}$$

$m = 2(2\overline{n}(\overline{m}-1) + \overline{n}\overline{m} + \overline{m} + n)$ and $l = (\overline{n}-1)\overline{m}$, b, g, and f are the corresponding column vectors from constraints (18), (13), (14), and (3)-(9). Apparently, $y - Ax - b = 0$ is the reformulation of the constraint (18), $Fx - g = 0$ is the condensed form of constraints (13) and (14), and $f - Ex \geq 0$ is the representation of all the inequality constraints distributed in (3)-(9).

3. Model Relaxation and Solution Algorithm

3.1. Relaxation Model. Problem (21) is generally NP-hard due to the combinatorial property of the involved ℓ_0-norm [21]. By adopting the well-known convex surrogate ℓ_1-norm initiated in compressed sensing [13, 22], we get the following $\ell_1 + \ell_2^2$ relaxation model:

$$\min_{x,y} \left\{ \lambda_1 \|\alpha \circ y\|_1 + \frac{\lambda_2}{2} \|\alpha \circ y\|_2^2 : y - Ax - b = 0,\ Fx - g = 0,\ f - Ex \geq 0. \right\} \tag{24}$$

Inspired by Oracle property beneficial from the adaptive lasso strategy proposed by Zou et al. [14], we will further utilize a reweighted ℓ_1-norm term instead of the ℓ_1-norm in (24) and get the following relaxation counterpart:

$$\min_{x,y} \left\{ \lambda_1 \|\alpha \circ w \circ y\|_1 + \frac{\lambda_2}{2} \|\alpha \circ y\|_2^2 : y - Ax - b = 0,\ Fx - g = 0,\ f - Ex \geq 0. \right\} \tag{25}$$

where w is some appropriate positive column vector to encourage a low sparsity of y.

3.2. A Two-Stage Alternating Direction Method of Multipliers. By introducing an auxiliary vector $z \in \mathbf{R}^m$, (25) can be rewritten as the following equality constrained optimization problem:

$$\min_{x,y,z} \left\{ \lambda_1 \|\alpha \circ w \circ y\|_1 + \frac{\lambda_2}{2} \|\alpha \circ y\|_2^2 + \delta_{\mathbf{R}_+^m}(z) : y - Ax - b = 0,\ Fx - g = 0,\ z - f + Ex = 0, \right\} \tag{26}$$

where $\delta_{\mathbf{R}_+^m}(h)$ is the indicator function with respect to $\mathbf{R}_+^m$ which takes the value 0 if $h \in \mathbf{R}_+^m$, and $+\infty$ otherwise. The augmented Lagrangian function of (26) takes the form of

$$\begin{aligned} L_\sigma^A(x, y, z; u, v, s) &= \lambda_1 \sum_{i=1}^{n} w_i |\alpha_i y_i| + \frac{\lambda_2}{2} \|\alpha \circ y\|_2^2 \\ &+ \delta_{\mathbf{R}_+^m}(z) - \langle u, y - Ax - b \rangle - \langle v, Fx - g \rangle - \langle s, z - f + Ex \rangle + \frac{\sigma}{2} \left(\|y - Ax - b\|_2^2 + \|Fx - g\|_2^2 + \|z - f + Ex\|_2^2 \right), \end{aligned} \tag{27}$$

where $\sigma > 0$ is the penalty parameter, and $u \in \mathbf{R}^n$, $v \in \mathbf{R}^l$, $s \in \mathbf{R}^m$ are Lagrangian multipliers corresponding to those three equality constraint systems in (26). The well-known alternating direction method of multipliers (ADMM) proposed by Glowinski and Marroco [23] and Gabay and Mercier [24] is then adopted for handling (26) with the iterative scheme for $k = 0, 1, \ldots,$

$$\begin{aligned} (y^{k+1}, z^{k+1}) &= \arg\min \left\{ L_\sigma^A(x^k, y, z; u^k, v^k, s^k) \mid y \in \mathbf{R}^n,\ z \in \mathbf{R}^m \right\}; \\ x^{k+1} &= \arg\min \left\{ L_\sigma^A(x, y^{k+1}, z^{k+1}; u^k, v^k, s^k) \mid x \in \mathbf{R}^n \right\}; \\ (u^{k+1}, v^{k+1}, s^{k+1}) &= (u^k, v^k, s^k) - \tau\sigma \left(y^{k+1} - Ax^{k+1} - b,\ Fx^{k+1} - g,\ z^{k+1} - f + Ex^{k+1} \right); \end{aligned} \tag{28}$$

where $\tau \in (0, (1+\sqrt{5})/2)$ is the dual stepsize. Note that $L_\sigma^A(x, y, z; u, v, s)$ is separable of y and z. Thus the first subproblem in (28) can be easily solved by the corresponding first-order optimality condition which leads to the following closed-form solution:

$$y_i^{k+1} = \begin{cases} \left(Ax^k + b + \dfrac{u^k}{\sigma} \right)_i & \text{if } \alpha_i = 0, \\ \operatorname{sign}(q_i) \max \left\{ |q_i| - \dfrac{\lambda_1 w_i}{\lambda_2 + \sigma}, 0 \right\} & \text{otherwise,} \end{cases} \tag{29}$$

with $q_i = (1/(\lambda_2 + \sigma))(u^k + \sigma Ax^k + \sigma b)_i$ for all $i = 1, \ldots, n$, and

$$z^{k+1} = \Pi_{\mathbf{R}_+^m} \left(f - Ex^k + \frac{1}{\sigma} s^k \right). \tag{30}$$

The update for x^{k+1} from the second subproblem can be easily obtained by solving the following linear system:

$$Hx = rhs, \tag{31}$$

where $rhs = A^T(y^{k+1} - b - (1/\sigma)u^k) + F^T(g + (1/\sigma)v^k) + E^T(f - z^{k+1} + (1/\sigma)s^k)$ and $H = A^TA + F^TF + E^TE$. Note that E is full

Require: Choose an initial point $(x_0^0, y_0^0, z_0^0, u_0^0, v_0^0, s_0^0)$, accuracy parameters $\varepsilon_1, \varepsilon_2 > 0$, the parameters $\lambda_1^0, \lambda_2^0, \sigma, r, \epsilon$.
Ensure: (x^*, y^*, z^*);
Stage I Compute $(\hat{x}, \hat{y}, \hat{z}, \hat{u}, \hat{v}, \hat{s})$.
Step 1.1 Set $\lambda_1 = \lambda_1^0, \lambda_2 = 0, (x^0, y^0, z^0, u^0, v^0, s^0) = (x_0^0, y_0^0, z_0^0, u_0^0, v_0^0, s_0^0)$, $w = e$ and $k = 0$;
Step 1.2 Compute $(x^{k+1}, y^{k+1}, z^{k+1}, u^{k+1}, v^{k+1}, s^{k+1})$ by (31), (29), (30) and (28);
Step 1.3 Set $k = k + 1$. If some criterion is satisfied to the accuracy ε_1, then stop. Otherwise, go to Step 1.2.
Set $(\hat{x}, \hat{y}, \hat{z}, \hat{u}, \hat{v}, \hat{s}) = (x^k, y^k, z^k, u^k, v^k, s^k)$ and update $w = (1/(|\hat{y}|^r + \epsilon))$;
Stage II Solve Problem (26) and output (x^*, y^*, z^*).
Step 2.1 Set $\lambda_1 = \lambda_1^0, \lambda_2 = \lambda_2^0, (x^0, y^0, z^0) = (\hat{x}, \hat{y}, \hat{z}), (u^0, v^0, s^0) = 0$ and $k = 0$;
Step 2.2 Compute $(x^{k+1}, y^{k+1}, z^{k+1}, u^{k+1}, v^{k+1}, s^{k+1})$ by (31), (29), (30) and (28);
Step 2.3 Set $k = k + 1$. If some criterion is satisfied to the accuracy ε_2, then set $(x^*, y^*, z^*) = (x^k, y^k, z^k)$ and stop. Otherwise, go to Step 2.2.

ALGORITHM 1: A two-stage alternating direction methods of multipliers (TSADMM).

column rank. Thus, $E^T E$ is positive definition and also is H. This indicates that (31) admits a unique solution:

$$x^{k+1} = H^{-1} rhs. \tag{32}$$

To effectively reduce the computation cost resulting from the large size of the involved matrices in H, the preconditioned conjugate gradient (PCG) method is employed to numerically solve the linear system (31) instead of the direct way as in (32).

It is worth mentioning that we can utilize the above ADMM method to solve an ℓ_1 problem which is exactly (26) by taking $w = e$ and $\lambda_2 = 0$ to a relatively low accuracy. The resulting numerical solution $(\hat{x}, \hat{y}, \hat{z})$ can not only be served as the initial point, but also produce the weight vector w in the way that $w = (1/(|\hat{y}_i|^r + \epsilon))$ with $r \in (0, +\infty)$ and $\epsilon > 0$ sufficiently small for a low sparsity promotion when solving (26). The two-stage ADMM algorithmic framework then can be summarized in Algorithm 1.

3.3. Global Convergence. Algorithm 1 admits a global convergence property as stated in the following theorem whose proof will be established in the appendix.

Theorem 3. *Let $\{(x^k, y^k, z^k, w)\}$ be generated from Algorithm 1. If $\tau \in (0, (1 + \sqrt{5})/2)$, then the sequence $\{(x^k, y^k, z^k)\}$ converges to an optimal solution to (26).*

4. Numerical Experiments

Numerical experiments on Beijing Metro Yizhuang Line will be conducted in this section to evaluate our proposed model and to illustrate the performance our designed TSADMM algorithm. The procedure of the numerical study is shown in Figure 6. All the computational results are obtained by running Matlab (version 2016b) on a windows desktop (Intel(R) Core(TM) i7-6700 CPU @ 3.40GHz 3.40GHz RAM 16.0 G).

Beijing Metro Yizhuang Line, opened on December in 2010, is one of the 20 lines in Beijing Metro Networks. It has two train lines with 14 stations as shown in Figure 5. The total length of Yizhuang Line is 23.23 km, with the average distance of two station being 1.8 km, the minimum distance being 1 km, and the maximum being 2.6 km. In the section, we calculate the energy conversion when a train has a trip under unit mass.

4.1. Inputs and Algorithm Initialization. For (26), $\alpha \in \mathbf{R}^n$ in the objective function is a 0-1 vector defined in Section 2.4. And the weight vector is denoted by $w = (1/(|\hat{y}|^r + \epsilon))$, where $\hat{y}$ is an approximate solution generated by Stage I in Algorithm 1. Here we set $r = 0.5$ and $\epsilon = 10^{-20}$. In the equality constraint $y - Ax - b = 0$, $A \in \mathbf{R}^{n\times n}$ with $n = 2\overline{n}\,\overline{m}$, $\overline{n} = |\mathcal{N}| = 28$, $\overline{m}$ the number of trains in one day service, $b \in \mathbf{R}^n$ with components $(\Delta_i^t + \nabla_j^{\overline{t}})_{(i,j,t,\overline{t})}$ sorting in the appropriate way, where train t is accelerating and $\overline{t}$ is braking, and platforms i and j are opposite to each other, satisfying the equation $i + j = 29$. $Fx - g = 0$ in (26) is an equality constraint about trip times. $F \in \mathbf{R}^{l\times m}$ is the coefficient matrix in (26), $l = (\overline{n} - 1)\overline{m}$. The components of $g \in \mathbf{R}^l$ are the trip times. $E \in \mathbf{R}^{m\times n}$ in $z - f + Ex = 0$ is the coefficient matrix of inequality constraints, and $m = 2(2\overline{n}(\overline{m}-1)+\overline{n}\,\overline{m}+\overline{m}+n)$. The elements in the vector $f \in \mathbf{R}^m$ are upper and lower bounds of dwell time constraints, headway time constraints, total travel time constraints, and domain of event times constraints. For testing purpose, elements of f are given in the following way. The bounds of dwell time constraints are given in Table 2. Let D be the passenger demand for one-day service (16 hours) and $u^D \in \mathbf{R}^{16}$ be the distribution of D for each hour which is presented in Table 3. Let c_t be the capacity of a train, which takes the value 1440, and $u \in \mathbf{R}^{16}$ be the train utility rate for different hours as stated in Table 3 as well. By utilizing formula (7), we can get the headway time for each different period of time (per hour). Specifically, denote the headway time vector by $h \in \mathbf{R}^{16}$. Then $h_k = (3600\times1440\times u_k)/(D\times u_k^D)$ for all $k = 1, 2, \ldots, 16$. Set the windows for the headway time constraints to be $h_i \pm 8$ for all possible train pairs in the ith hour. The total travel time constraints are related

TABLE 2: The windows of the dwell time constraints.

Platforms	SJZ1	XC1	XHM1	JG1	YZQ1	YZWHY1	WYJ1	RJDJ1	RCDJ1	TJNL1	JHL1	CQN1	CQ1	YZHCZ1
Lower bounds	25	25	25	25	25	25	25	25	25	25	25	25	25	25
Upper bounds	45	45	45	45	45	45	45	45	45	45	45	45	45	45
Platforms	SJZ2	XC2	XHM2	JG2	YZQ2	YZWHY2	WYJ2	RJDJ2	RCDJ2	TJNL2	JHL2	CQN2	CQ2	YZHCZ2
Lower bounds	25	25	25	25	25	25	25	25	25	25	25	25	25	25
Upper bounds	45	45	45	45	45	45	45	45	45	45	45	45	45	45

TABLE 3: The train utility rate and distribution of D for one-day (16 hours) services.

	Hours															
Periods	1	2	3	4	5	6	7	8	9	10	11	12	13	14	15	16
u	0.3	0.8	0.8	0.6	0.4	0.4	0.5	0.5	0.3	0.6	0.6	0.8	0.8	0.4	0.3	0.3
u^D	0.03	0.11	0.13	0.07	0.05	0.05	0.04	0.05	0.04	0.05	0.07	0.13	0.12	0.03	0.02	0.02

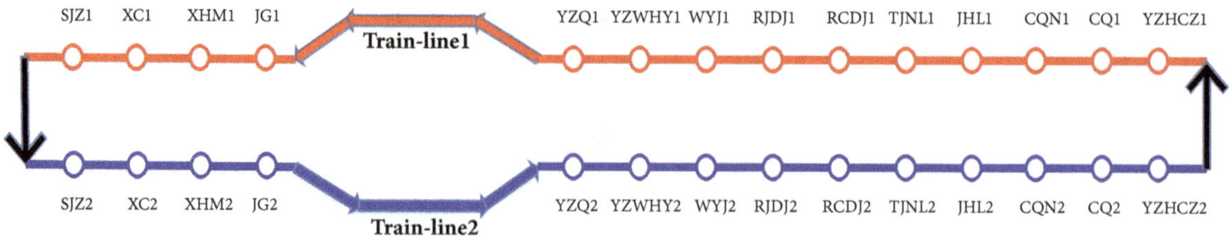

FIGURE 5: Railway network considered for numerical experiment.

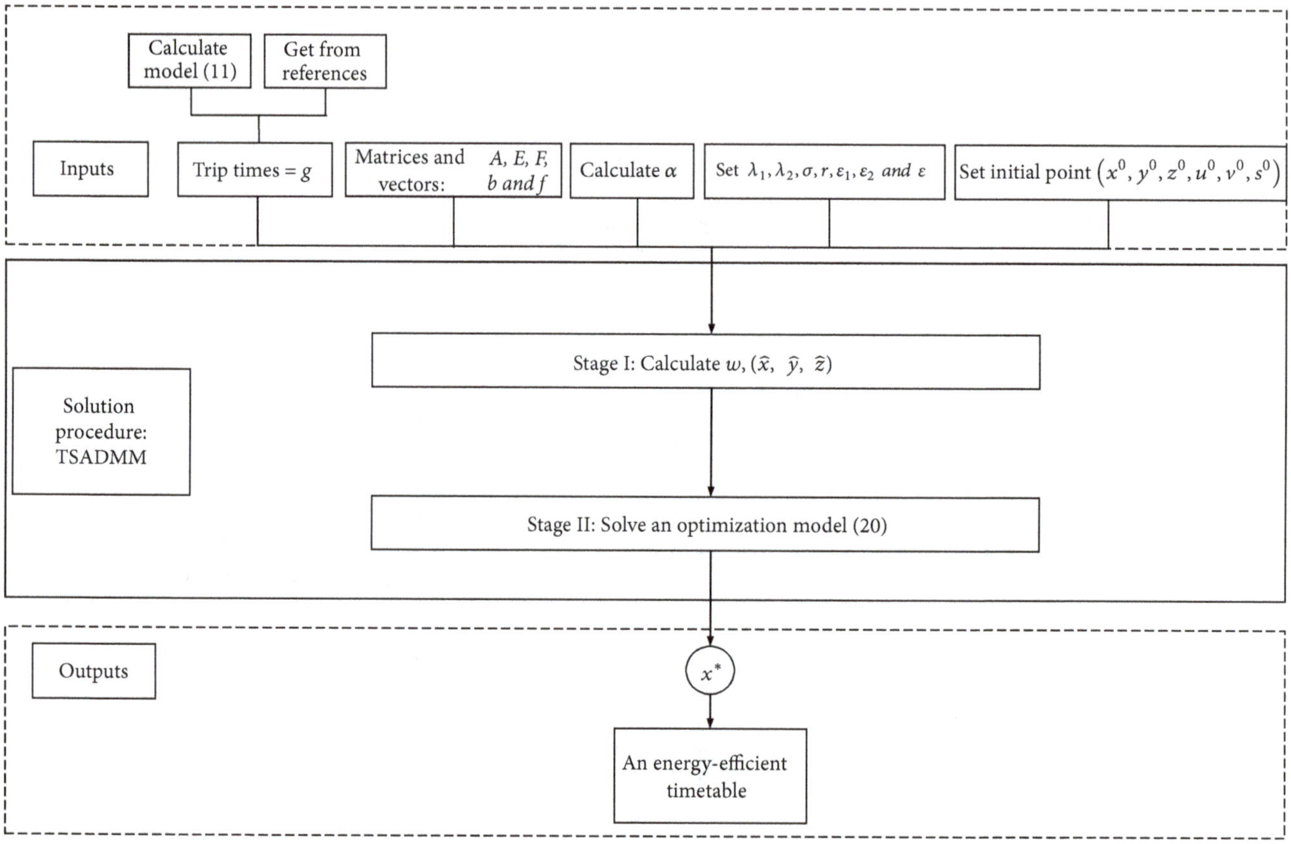

FIGURE 6: The procedure of the numerical experiments.

to the dwell time and trip time. The bounds of trip times are calculated by Table 4 as reference. For each train t, the bounds of trip time are more than $\sum_{q=1}^{5}(s_q/v_q)$ for all sections (ij), $i = 1, 2, \ldots, 27$, $j = i + 1$. The bounds of trip time constraints are presented in Table 5. The following formulae are adopted for calculating the windows of total travel time constraints,

$$\begin{aligned} \underline{tt} &= r_1 \times \left(\sum_{i=1}^{28} \underline{dw}_i + \sum_{\substack{i=1 \\ i \neq 14}}^{27} \underline{tr}_{ij} + \sum_{i=14} \underline{\kappa}_{ij} \right) \\ \overline{tt} &= r_2 \times \left(\sum_{i=1}^{28} \overline{dw}_i + \sum_{\substack{i=1 \\ i \neq 14}}^{27} \overline{tr}_{ij} + \sum_{i=14} \overline{\kappa}_{ij} \right) \end{aligned} \quad (33)$$

where $r_1 = 1.1$ and $r_2 = 0.85$. For the domain of event times constraints, we set $m_T = 16 \times 3600$.

The initial point (u^0, v^0, s^0) in Algorithm 1 is chosen to be 0. To calculate the energy consumption and regenerative energy, we set the maximum acceleration $a_{acc} = 0.5m/s^2$ in accelerating, $a_{bra} = -0.8m/s^2$ for braking, the resistance acceleration $\gamma = -0.05m/s^2$ during the speed holding phase, the conversion factor from electricity to kinetic energy $\theta_{acc} = 0.7$, and the ratio of kinetic energy converting to regenerative energy $\theta_{bra} = 0.5$ as used in [18]. For testing purpose, we approximately calculate the energy consumption for accelerating and the regenerative energy produced during braking in the following way. As presented in Section 2.4, the maximum accelerating, speed holding, maximum braking strategy from [18] is adopted and the parameters t_1, t_2, and t_3 are set in Table 6. The parameters c and $\tilde{c}$ in Figure 3 which represent

Table 4: Speed limits of every section on Beijing Metro Yizhuang Line following from [18].

Sections	Segment		Segment		Segment		Segment		Segment	
	v_1(m/s)	s_1(m)	v_2(m/s)	s_2(m)	v_3(m/s)	s_3(m)	v_4(m/s)	s_4(m)	v_5(m/s)	s_5(m)
SJZ-XC	13.89	0-150	23.61	150-480	18.06	480-1161	23.61	1161-2501	16.67	2501-2631
XC-XHM	16.67	2631-2643	23.61	2643-2797	20.83	2797-3534	23.61	3534-3780	16.67	3780-3905
XHM-JG	16.67	3905-3918	23.61	3918-5808	20.83	5808-6141	23.61	6141-6271	-	-
JG-YZQ	16.67	6271-6281	23.61	6281-8122	16.67	8122-8254	-	-	-	-
YZQ-YZWHY	16.67	8254-8265	23.61	8265-9116	16.67	9116-9246	-	-	-	-
YZWHY-WYJ	16.67	9246-9256	23.61	9259-10655	16.67	10655-10785	-	-	-	-
WYJ-RJDJ	16.67	10785-10797	23.61	10797-11933	16.67	11933-12065	-	-	-	-
RJDJ-RCDJ	16.67	12065-12077	23.61	12077-13289	16.67	13289-13419	-	-	-	-
RCDJ-TJNL	16.67	13419-13431	23.61	13419-14649	20.83	14649-15426	23.61	15426-15624	16.67	15624-15756
TJNL-JHL	16.67	15756-15768	23.61	15768-17891	16.67	17891-18021	-	-	-	-
JHL-CQN	16.67	18021-18033	23.61	18033-19982	16.67	19982-20107	-	-	-	-
CQN-CQ	16.67	20107-20120	23.61	20120-21264	16.67	21264-21394	-	-	-	-
CQ-YZHCZ	16.67	21394-21406	23.61	21406-22569	16.67	22569-22728	-	-	-	-

Table 5: The windows of the trip time constraints.

Sections	1	2	3	4	5	6	7	8	9	10	11	12	13	14
Lower bounds	160	92	150	130	80	100	100	100	140	130	130	100	100	170
Upper bounds	190	110	170	150	110	125	120	120	170	160	145	120	120	200
Sections	15	16	17	18	19	20	21	22	23	24	25	26	27	-
Lower bounds	100	100	130	130	140	100	100	100	80	130	150	92	160	-
Upper bounds	120	120	145	160	170	120	120	125	110	150	170	110	190	-

TABLE 6: Parameters for calculating E^{con} and E^{bra}.

t_1	t_2	t_3	c	$\tilde{c}$
20%	45%	35%	14%	21%

TABLE 7: Results of numerical experiments on Beijing Metro Yizhuang Line: A1 stands for the linear programming model (36) solved by Gurobi; A2 stands for the quadratic programming (37) solved by CVX; A3 stands for our proposed model (26) solved by Algorithm 1.

		$\eta(\%)$	
$\overline{m}$	A1	A2	A3
357	25.54	24.52	**29.92**
368	23.67	23.44	**28.36**
380	25.56	23.01	**28.54**
391	24.50	23.80	**28.49**
403	24.78	24.17	**27.47**
414	27.02	26.57	**30.05**
426	27.73	27.07	**30.52**
437	27.59	26.94	**30.46**
449	28.26	27.54	**30.84**

the width of the rectangles for E^{con} and E^{bra} (in percentage of the trip time) are also given in Table 6 for approximately calculating the energy consumption for accelerating E^{con} and the regenerative energy produced during braking E^{bra}.

4.2. Stopping Criteria. To measure the accuracy of an approximate optimal solution (x, y, z) of Problem (26) from (x, y, z, u, v, s) generated by Stage II in Algorithm 1, the relative primal and dual infeasibilities will be adopted. Specifically, denote

$$\eta_p = \max\left\{\frac{\|y - Ax - b\|}{1 + \|y\|}, \frac{\|Fx - g\|}{1 + \|g\|}, \frac{\|z - f + Ex\|}{1 + \|z\|}\right\}, \quad (34)$$
$$\eta_d = \frac{\|A^T u - F^T w - E^T s\|}{1 + \|A^T u\| + \|F^T w\| + \|E^T s\|}.$$

For a given tolerance $\varepsilon_2 > 0$, we will stop the tested algorithms when both η_p and η_d are less than ε_2. The algorithms will also be terminated when they reach a given maximum number of iterations $MaxIter_2$. Here we set $\varepsilon_2 = 10^{-3}$ and $MaxIter_2 = 10^4$. For Stage I in Algorithm 1, it will be stopped when both η_p and η_d are less than ε_1 or the maximum iteration reaches $MaxIter_1$. Here we set $\varepsilon_1 = 10^{-1}$ and $MaxIter_1 = 10^3$.

4.3. Experiment Results. To evaluate our model with ℓ_0-norm plus ℓ_2^2-norm as the object, the other two possible models are computed for comparison. The first model is the ℓ_0-norm plus ℓ_1-norm proposed by Das Gupta et al. in [12] which takes the form of

$$\min_x \{\|y\|_0 + \lambda \|y\|_1 : y - Ax - b = 0,\ Fx - g = 0,\ Ex \le f\} \quad (35)$$

which is handled by approximately solving its linear programming relaxation problem

$$\min_x \{e^T s : y \le s,\ y \ge -s,\ y - Ax - b = 0,\ Fx - g = 0,\ Ex \le f\}. \quad (36)$$

The second possible model is the squared ℓ_2-norm related quadratic programming,

$$\min_x \left\{\frac{1}{2}\|y\|_2^2 : y - Ax - b = 0,\ Fx - g = 0,\ Ex \le f\right\}, \quad (37)$$

where only the squared Euclidean norm is employed to measure the vanishing of y. Since we focus on the reduction of the effective energy consumption, we define the energy saving rate η as follows:

$$\eta = \frac{\sum_{t=1}^{|\mathcal{T}||\mathcal{N}|} E_{(i,j,t,\tilde{t})}}{\sum_{t=1}^{|\mathcal{T}||\mathcal{N}|} E_t^{con}}. \quad (38)$$

Energy savings by using these three different models are listed in Table 7 with different choices of $\overline{m}$.

As shown in Table 7, the energy savings by using our model are better than those by the other two in all the testing instances with at least 2.58 percentage higher in the energy saving rate η.

Other choices for the initial trip times from [6, 9–11, 19] with $\overline{m} = 368$ are also considered as inputs. The comparison results in terms of the energy saving rate η are stated in Table 8.

TABLE 8: Energy saving rates for $\overline{m} = 368$: A1, A2, and A3 are the same as in Table 7.

		$\eta(\%)$	
Ref.	A1	A2	A3
[19]	19.82	21.34	**24.94**
[9]	18.85	18.83	**23.11**
[10]	18.99	18.88	**23.11**
[11]	19.02	18.85	**23.12**
[6]	18.97	18.78	**23.29**

As one can see in Table 8, the utilization of regenerative energy of our approach is the best among these three models

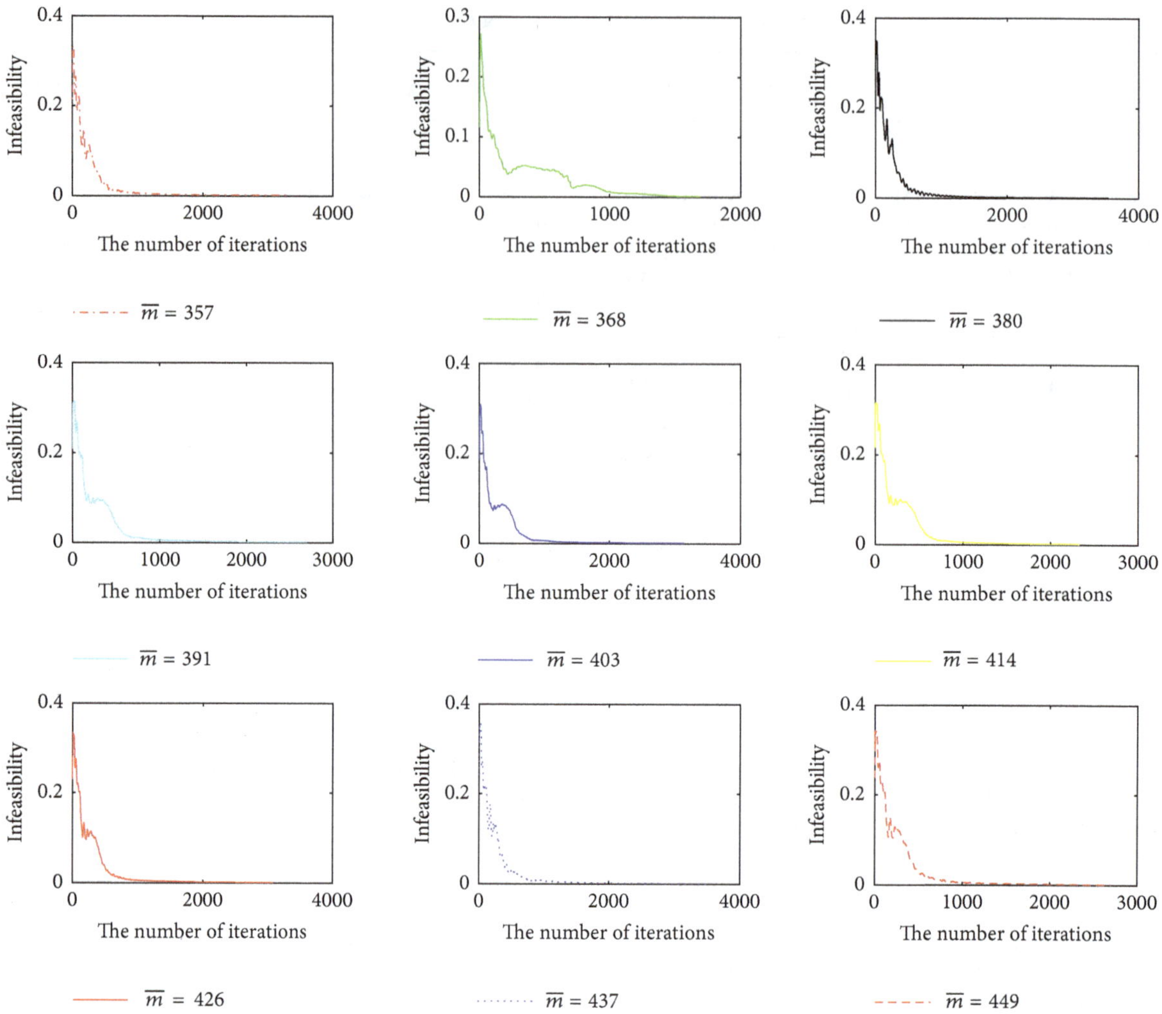

FIGURE 7: The numerical convergence behavior of TSADMM with different choices of $\overline{m}$.

as well with the energy saving rate at least 3.60 percent higher than the other two. These evidently show the effectiveness of our proposed model.

In order to demonstrate the efficiency of our proposed TSADMM algorithm, we call the academical CVX to solve ours model for comparison. The time comparisons for all the aforementioned instances are presented in Tables 9 and 10, respectively.

As can be seen in Tables 9 and 10, our proposed TSADMM outperforms the CVX solver for the model (26) in terms of computation time. Particularly, for the cases of $\overline{m}$ equaling 368 and 449 in Table 9, the computation time for TSADMM is less than one-third of those for CVX. Besides the theoretical global convergence of our TSADMM, the numerical convergence behaviors in terms of the infeasibility measure for all the above 14 instances are presented in Figures 7 and 8. One can see that for each instance, the infeasibility decreases rapidly as the number of iterations increases and it meets the required accuracy 10^{-3} within less than 4000 iterations.

5. Conclusion

In this paper we have proposed a sparse optimization model using ℓ_0-norm and squared ℓ_2-norm as objective function so as to maximize the utilization of regenerative energy in metro railway system. To overcome the NP-hardness resulting from the ℓ_0-norm, the weighted ℓ_1-norm by invoking the adaptive lasso strategy in statistical regression has been introduced to make the relaxation counterpart computationally tractable and numerically effective. A two-stage alternating direction method of multipliers (TSADMM) has been designed to efficiently solve the proposed mathematical model and the global convergence of the algorithm has been established as well. Numerical experiments have been conducted on Beijing Metro Yizhuang Line for case study. The comparison results with the linear programming proposed by Das Gupta et al. and a quadratic programming with least-squares loss function have illustrated the effectiveness of our proposed model in terms of the energy saving rate, and the timing comparison results with the CVX toolbox have demonstrated the efficiency of our proposed TSADMM algorithm.

TABLE 9: Time comparison between CVX and TSADMM with different choices of $\overline{m}$.

Alg.	$\overline{m}$								
	357	368	380	391	403	414	426	437	449
	time(s)								
CVX	47.17	47.06	54.33	56.90	31.09	50.63	56.18	70.72	71.14
TSADMM	24.35	12.80	31.06	21.26	25.39	19.36	27.61	30.80	23.66

TABLE 10: Time comparison between CVX and TSADMM with different trip times from [6, 9–11, 19].

Alg.	Ref.				
	[19]	[9]	[10]	[11]	[6]
			time(s)		
CVX	55.61	57.86	55.45	51.25	47.81
TSADMM	20.36	29.22	28.21	27.64	31.84

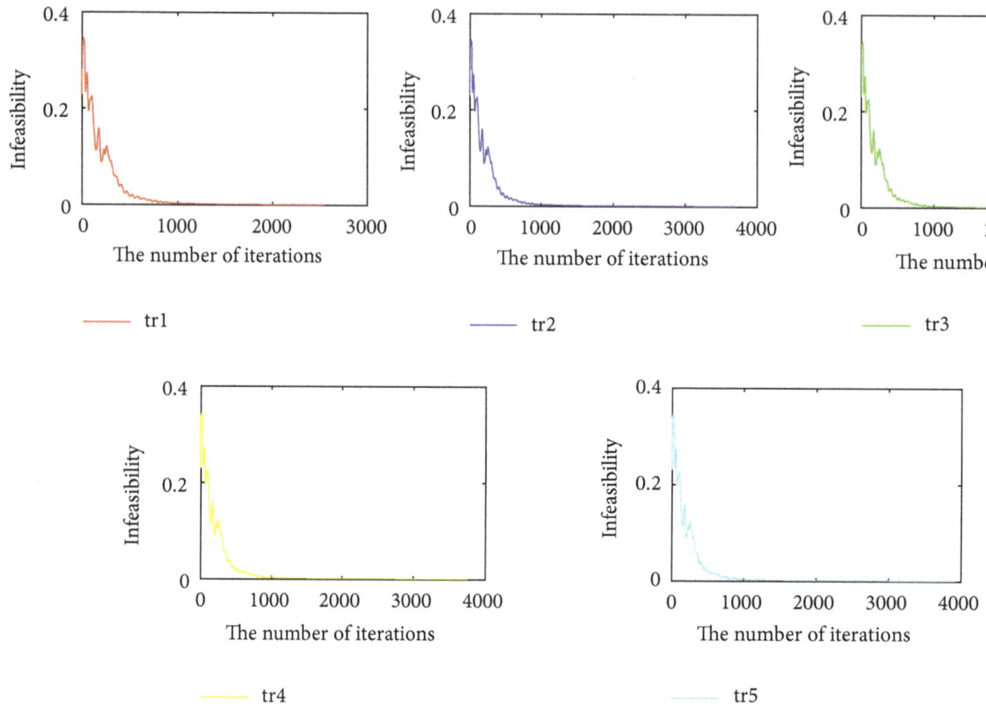

FIGURE 8: The numerical convergence behavior of TSADMM for different trip times: tr1,...,tr5 stand for the cases for different trip times proposed in [6, 9–11, 19].

Since the weighted ℓ_1-norm is a convex approximation of the ℓ_0-norm, the optimal solution to the (26) is a proximal optimal solution to the original sparse optimization model. How to design some more effective nonconvex relaxation scheme to get a better proximal optimal solution so as to get a better energy saving rate will be one of our future work. Meanwhile, the delay perturbations of the metro trains and the randomness of passenger demands have not been considered in our static timetabling. How to use advanced optimization techniques for handling such cases will also be our future research issue.

Appendix

The Proof of Theorem 3

Proof. Let $\overline{x} = ([x^{(1)}]^T, \ldots, [x^{(2\overline{n})}]^T)^T \in \mathbf{R}^n$ be an optimal solution of Problem (12). Then, the optimality and the feasibility of $\overline{x}$ imply that $F\overline{x}-g=0$ and $f-E\overline{x} \geq 0$. By virtue of the expressions of E and F in Section 2.4, together with the time window constraints as stated in f, we can find some $\widetilde{x}$ in a neighborhood $N(\overline{x}, \delta)$ with $\delta < \min_{i: f_i-(E\overline{x})_i>0} f_i-(E\overline{x})$, such that $F\widetilde{x}-g=0$ and $f-E\widetilde{x}>0$. Set $\widetilde{y}=A\widetilde{x}+b$ and $\widetilde{z}=f-E\widetilde{x}$. It is easy to verify that $(\widetilde{x}, \widetilde{y}, \widetilde{z}) \in (\mathbf{R}^n \times \mathbf{R}^n \times \mathbf{R}^m_{++}) \cap P$ with P being the feasible region of Problem (26). By employing [25, Theorem B.1], we can get the desired global convergence result in Theorem 3. □

Conflicts of Interest

The authors declare that they have no conflicts of interest regarding the publication of this paper.

Acknowledgments

This work was partly supported by the State Key Laboratory of Rail Traffic Control and Safety, Beijing Jiaotong University [Grants nos. RCS2017ZJ001 and RCS2016ZZ01], and National Natural Science Foundation of China [Grants nos. 11771038 and 11431002].

References

[1] G. M. Scheepmaker, R. M. Goverde, and L. Kroon, "Review of energy-efficient train control and timetabling," *European Journal of Operational Research*, vol. 257, no. 2, pp. 355–376, 2017.

[2] S. Barsali, P. Bolognesi, M. Ceraolo, M. Funaioli, and G. Lutzemberger, "Cyber-physical modelling of railroad vehicle systems using modelica simulation language," in *Railways*, Corsica, Ajaccio, France, 2014.

[3] A. Capasso, R. Lamedica, A. Ruvio, M. Ceraolo, and G. Lutzemberger, "Modelling and simulation of electric urban transportation systems with energy storage," in *Proceedings of the 16th International Conference on Environment and Electrical Engineering (EEEIC '16)*, Florence, Italy, June 2016.

[4] M. Peña-Alcaraz, A. Fernández, A. P. Cucala, A. Ramos, and R. R. Pecharromán, "Optimal underground timetable design based on power flow for maximizing the use of regenerative-braking energy," *Proceedings of the Institution of Mechanical Engineers, Part F: Journal of Rail and Rapid Transit*, vol. 226, no. 4, pp. 397–408, 2011.

[5] S. D. Gupta, L. Pavel, and J. K. Tobin, "An optimization model to utilize regenerative braking energy in a railway network," in *Proceedings of the American Control Conference (ACC '15)*, pp. 5919–5924, July 2015.

[6] X. Yang, X. Li, Z. Gao, H. Wang, and T. Tang, "A cooperative scheduling model for timetable optimization in subway systems," *IEEE Transactions on Intelligent Transportation Systems*, vol. 14, no. 1, pp. 438–447, 2013.

[7] X. Yang, B. Ning, X. Li, and T. Tang, "A two-objective timetable optimization model in subway systems," *IEEE Transactions on Intelligent Transportation Systems*, vol. 15, no. 5, pp. 1913–1921, 2014.

[8] X. Li and K. Lo Hong, "An energy-efficient scheduling and speed control approach for metro rail operations," *Transportation Research Part B: Methodological*, vol. 64, pp. 73–89, 2014.

[9] X. Li and H. K. Lo, "Energy minimization in dynamic train scheduling and control for metro rail operations," *Transportation Research Part B: Methodological*, vol. 70, pp. 269–284, 2014.

[10] S. Yang, J. Wu, X. Yang, H. Sun, and Z. Gao, "Energy-efficient timetable and speed profile optimization with multi-phase speed limits: theoretical analysis and application," *Applied Mathematical Modelling: Simulation and Computation for Engineering and Environmental Systems*, vol. 56, pp. 32–50, 2018.

[11] X. Yang, A. Chen, X. Li, B. Ning, and T. Tang, "An energy-efficient scheduling approach to improve the utilization of regenerative energy for metro systems," *Transportation Research Part C: Emerging Technologies*, vol. 57, pp. 13–29, 2015.

[12] S. D. Gupta, J. K. Tobin, and L. Pavel, "A two-step linear programming model for energy-efficient timetables in metro railway networks," *Transportation Research Part B: Methodological*, vol. 93, pp. 57–74, 2016.

[13] D. L. Donoho, "Compressed sensing," *IEEE Transactions on Information Theory*, vol. 52, no. 4, pp. 1289–1306, 2006.

[14] H. Zou, "The adaptive lasso and its oracle properties," *Journal of the American Statistical Association*, vol. 101, no. 476, pp. 1418–1429, 2006.

[15] S. S. Chen, D. L. Donoho, and M. A. Saunders, "Atomic decomposition by basis pursuit," *SIAM Review*, vol. 43, no. 1, pp. 129–159, 2001.

[16] A. Beck and M. Teboulle, "A fast iterative shrinkage-thresholding algorithm for linear inverse problems," *SIAM Journal on Imaging Sciences*, vol. 2, no. 1, pp. 183–202, 2009.

[17] X. Li, D. Sun, and K.-C. Toh, "A highly efficient semismooth Newton augmented Lagrangian method for solving lasso problems," *SIAM Journal on Optimization*, vol. 28, no. 1, pp. 433–458, 2018.

[18] S. Yang, J. Wu, H. Sun, X. Yang, Z. Gao, and A. Chen, "Bi-objective nonlinear programming with minimum energy consumption and passenger waiting time for metro systems, based on the real-world smart-card data," *Transportmetrica B*, vol. 4, pp. 1–18, 2017.

[19] Z. Tian, P. Weston, N. Zhao, S. Hillmansen, C. Roberts, and L. Chen, "System energy optimisation strategies for metros with regeneration," *Transportation Research Part C: Emerging Technologies*, vol. 75, pp. 120–135, 2017.

[20] L. W. P. Peeters, *Cyclic Railway Timetable Optimization*, Erasmus University Rotterdam, 2003.

[21] B. K. Natarajan, "Sparse approximate solutions to linear systems," *SIAM Journal on Computing*, vol. 24, no. 2, pp. 227–234, 1995.

[22] E. J. Candes and T. Tao, "Decoding by linear programming," *IEEE Transactions on Information Theory*, vol. 51, no. 12, pp. 4203–4215, 2005.

[23] R. Glowinski and A. Marroco, "Sur lapproximation, par elements finis dordre un, et la resolution, par penalisation-dualite', dune classe de problems de dirichlet non lineares," *Revue Francaise d'Automatique, Informatique, et Recherche Ope'rationelle*, vol. 9, pp. 41–76, 1975.

[24] D. Gabay and B. Mercier, "A dual algorithm for the solution of nonlinear variational problems via finite element approximation," *Computers & Mathematics with Applications*, vol. 2, no. 1, pp. 17–40, 1976.

[25] M. Fazel, T. K. Pong, D. Sun, and P. Tseng, "Hankel matrix rank minimization with applications to system identification and realization," *SIAM Journal on Matrix Analysis and Applications*, vol. 34, no. 3, pp. 946–977, 2013.

15

Modelling Rail-Based Park and Ride with Environmental Constraints in a Multimodal Transport Network

Xinyuan Chen and Inhi Kim

Monash Institute of Transport Studies, Australia

Correspondence should be addressed to Inhi Kim; inhi.kim@monash.edu

Guest Editor: David Z. W. Wang

Environmental sustainability is a significant aspect in the sustainable development of modern urban cities, especially in the road transport system. As traffic demands increase, public transport requires more promotion to accommodate the increasing travel demands while maintaining the environmental quality. Public transport, however, is less attractive in vast suburb areas mainly due to its longer travel distance and waiting time. Therefore, this paper proposes a rail-based Park-and-Ride (RPR) scheme to promote public transport in the multimodal transport network. To remedy the heterogeneous distribution of vehicle pollutants in the network, regulations in environmental sensitive districts are required and studied in this paper. To quantitatively evaluate and analyse this joint RPR and environmental regulation strategy in multimodal transport systems, this paper develops an environmental constrained combined modal split and traffic assignment (EC-CMSTA) model. The proposed formulation adopts the concept of fix-point to reformulate the nonlinear complementarity conditions associated with the combined modal split and user equilibrium conditions, which is subsequently incorporated into a VI formulated nonlinear complementarity conditions associated with environmental constraints. The proposed VI formulation can handle a general constraint structure, which enhances the modelling adaptability and flexibility. The strictly monotone and Lipschitz continuity properties of this model are rigorously proved, giving rise to efficient algorithms for the model. A customized projection based self-adaptive gradient projection (SAGP) algorithm is then developed. Numerical studies demonstrate that the EC-MSTA model could enhance the behavioural modelling of network users' travel decisions and assist in quantitatively evaluating the effectiveness of RPR schemes and environmental regulations.

1. Introduction

Environmental protection is an important component in the sustainable development of modern urban cities and closely related to the transport system. According to Hong Kong Air Pollutant Emission Inventory (2016), the amount of CO generated in the year of 2016 in Hong Kong is 58600 tones and among them, road transport contributes to 31500 tones and shares 54% of the CO emissions. Also, the US's environmental protection agency reported motorized vehicles in the United States contribute to 86% of CO emissions, 62% of NO_x emissions, 12% of PM2.5 emissions, and 7% of SO_2 emissions [1]. Kitthamkesorn et al. [2] pointed out that only about 20% of the world's town residents enjoy good enough air quality as measured by the levels of emissions and over 600 million people in urban areas worldwide were exposed to dangerous levels of traffic-generated emissions from congestion. In addition, these emissions have not only adverse effects on human health but also a source of greenhouse gas that contributes to global warming [3].

With these adverse effects, the promotion of alternative green transport modes is thus pressing for the governments and transportation authorities to accommodate the growing travel demand while maintaining the environmental quality. Public transport aiming to satisfy the travel demand of urban commuters safely, quickly, and affordably is therefore a significant alternative to private transport. However, in vast suburb areas which are far away from an urban centre, public transport becomes less attractive, mainly due to the unreliable time table, tortuous bus itinerary, much lower travel speed, and lack of door to door services [4]. Large-scale deployments of public transport system in the remote areas are hardly seen as an economical available strategy. Park and Ride (P&R), therefore, becomes an important strategy to promote public transport since it enables travellers to drive when the road is not congested and to change a mode to transit accessing to CBD without traffic congestion. Since its first implementation in 1930s, P&R has been widely recognized as a significant traffic management strategy and becomes an indispensable part of majority modern urban cities.

P&R can be categorized into bus-based P&R (BPR) and rail-based P&R (RPR) based on its ride component [5]. In a metropolis with advanced rail system, rail-based P&R is superior to bus-based P&R, due to the faster speed, higher level of safety, comfortability, and its advantages in environmental protection (zero-emission).

Despite the fact that environmental deterioration could somehow be alleviated by promoting public transport through RPR schemes, the environmental requirements are usually not strictly met in certain districts mainly due to the heterogeneous distribution of pollution in the network [3]. To remedy this limitation, specific environmental regulations, therefore, are required to certain district, e.g., the city business district (CBD) with many tall buildings, to protect the environmental quality in this area.

Therefore, we aim to use RPR services to assist mode shift to reduce the overall vehicle travel demands in the road network and then apply area-based regulation scheme to deal with the environmental deterioration caused by heterogeneity of traffic flows. The environmental protection approaches are integrated as a combined strategy in this paper. Previous studies revealed that a systematic study is necessary to properly understand and evaluate the effectiveness of the environmental regulations, as well as the RPR scheme. Thus, this paper focuses on the modelling and solution algorithm development of the proposed environmental protection strategy.

2. Literature Review

Road network equilibrium approaches are commonly preferred for estimating vehicle emissions and noises [3], because the distribution of vehicle pollution heavily relies on the states of traffic flows [6]. Other than the overall pollution level, the spatial distribution level should be controlled since it is heterogeneous.

One line of the research developed models by setting environmental protection as objectives [3], modelling environmental considerations as an objective function. Rilett and Benedek [7] proposed the concept of emission optimum (EO) and emission-oriented user equilibrium (EOUE), in which the former one assumes travellers make route choices with the objective of minimizing the total emissions or noises, and the latter includes the toll charges to represent the impact on the environment. Later, Benedek and Rilett [8] made extension by modelling the assignment of vehicles based on equity which considers the objectives of neighbouring residents, rather than the system operator's or drivers' objectives. Sugawara and Niemeier [9] used the emissions-optimized trip assignment model to estimate the maximum carbon monoxide emission reduction under varying congestion levels.

Although the overall environmental quality could somehow be improved by taking the environmental considerations as the objective, the pollution level is not always strictly met on each link and zone in the road network. Therefore, another line of the research focuses on modelling the emission as environmental constraints. Chen, Zhou and Ryu [10] developed a traffic assignment model by considering various physical and environment restrictions as side constraints, which is formulated as a VI problem and solved by a predictor-corrector (PC) algorithm. Li et al. [11] developed a toll design model by simultaneously considering congestion and environmental externalities. Recently, Xu, Chen, and Cheng [1] reformulated the environmental constrained traffic assignment model through a smooth gap function to enhance the modelling adaptability and flexibility.

As a rational extension of traffic assignment problem, Nagurney, Ramanujam, and Dhanda [12] extended the environmental constrained traffic assignment problem to the case of multimodal traffic network. However, the model presented in this paper simply follow the deterministic user equilibrium (DUE) principle for both modal split and traffic assignment. In order to evaluate the environmental regulation strategy in the multimodal transport network, a better behavioural model is needed. Therefore, the combined modal split and traffic assignment model (CMSTA) is adopted in this paper. In the literature, the CMSTA problems are widely adopted to evaluate the P&R schemes. Fernandez et al. [13] conducted an initial work to model P&R in deterministic case. Later, this work was extended to the case of asymmetric link travel time functions by García and Marín [14] and to the logit-based stochastic model by Li et al. [15] and Lam et al. [16]. Recently, Liu and Meng [5] provided a formulation for the bus-based P&R in a network with the congestion charge. Pineda et al. [17] and Song, He, and Zhang [18] developed an integrated P&R facilities and transit service optimization model. Kitthamkesorn et al. [2] and Kitthamkesorn and Chen [19] considered the influence of "go green" transport modes on environmental protection by taking use of recent advances in logit model. Liu et al. [20] proposed a new concept of P&R, remote park and ride, and developed an exact solution method for the optimal design of parking facilities, following the nonlinear valid inequality technique [21, 22]. Some other studies focus on the P&R schemes in a linear corridor, e.g., Liu et al. [23], Wang and Du [24], and Du and Wang [25]. However, these studies mainly focus on congestion alleviation. As abovementioned, the environmental requirements in certain areas are not considered. Therefore, a more general modelling framework that incorporates environmental requirements is needed. To this end, this paper aims to build an environmental constrained combined modal split and traffic assignment (EC-CMSTA) model to better evaluate and analyse environmental sustainable transportation development strategies.

This paper distinguishes itself from previous studies in threefold aspects. First, we propose EC-CMSTA, CMSTA model with side constraints, which could handle the general environmental constraint structures, including linear or nonlinear, link based, or area-based constraints. Also, the general link/route cost structure is considered here. Second, by incorporating general side constraints into CMSTA, solving is much more complex compared to the traditional side constrained unimodal traffic equilibrium problems. To explicitly consider the general environmental constraints and make use of existing efficient algorithms, we propose a VI formulation by adapting the concept of fix-point. This VI formulation is strictly monotone and Lipschitz continuous, permitting a number of existing efficient algorithms for its solution [10, 26]. Third, even though the solution methods for the traffic assignment problems with side constraints have been well studied with numerous achievements [1, 10, 12, 27], solutions for the multimodal traffic equilibrium with side constraints are relatively limited. Recently, Meng, Liu and Wang [28] proposed a CA incorporated self-adaptive PC algorithms to solve asymmetric stochastic user equilibrium (SUE) problem with elastic demand and link capacity constraints. Inspired by Meng, Liu, and Wang [28], we extend this approach to multimodal traffic equilibrium problem and develop a customized projection based algorithm with a self-adaptive step size scheme which permits the convergence under a much milder assumption of the underlying mapping.

The remainder of this paper is organized as follows: we present assumptions and an EC-CMSTA model for the environmental regulations in the multimodal transport system with RPR scheme in the next section. A general VI formulation and a projection based solution algorithm are provided in Sections 3 and 4. Section 5 provides two numerical studies to validate the proposed methodology. Finally, Section 6 concludes this study.

3. Problem Description

Considering a specific environmental regulation strategy, to analyse its impact in the multimodal transport system, we build a model for the equilibrium flows following environmental requirements.

Given a multimodal transportation network represented by a strongly connected topology graph $G(N, A)$ where N is the set of nodes and A is the set of links. The origins, destinations, and RPR facilities are scattered in the network and represented by certain nodes, $n \in N$. The set of links A can be further categorized into three subsets: set of autolinks A^a; set of rail links A^r; and set of P&R links A^p; $A = A^a \cup A^r \cup A^p$.

3.1. Modelling Environmental Requirements. As abovementioned, the environmental requirements are not strictly met in the road network. Thus, it is necessary to explicit considering environmental protection thresholds to prevent environmental deterioration in certain zones. In the literature, the environmental constraints can be expressed into a general form:

$$g_e(\mathbf{v}) \le 0, \quad \forall e \in E \tag{1}$$

where e is a component of E and E represents a set of environmental sensitivity elements. In the case of link based environmental protection strategy, e represents a link which is assumed in this paper. $\mathbf{v}$ is the vector of link flows; $g_e(\mathbf{v})$ is a function to measure the amount of vehicle pollution and environmental thresholds. Equations (1) ensure the amount of pollution no larger than predetermined thresholds. Note that (1) are general formulation that could measure different types of vehicle pollutants [1]. Herein, we make assumption as follows.

Assumption 1. The environmental threshold measure function $g_e(\mathbf{v}), \forall e \in E$, is a continuously differentiable and strictly monotone increasing function of the link flows $\mathbf{v}$.

Suppose the total number of environmentally sensitive areas in the network is I. The inbound flows of these areas should be restricted according to environmental protection strategies. All the links in the area $i \in I$ are grouped into set E_i, where $E_i \in A$. Then, the specific link based environmental constraints in these areas could be expressed as

$$g_a(\mathbf{v}) \cdot v_a \le \overline{g}_a, \quad \forall a \in E_i,\ i \in I \tag{2}$$

where $g_a(\mathbf{v})$ is the amount of vehicle pollutants per vehicle on link a and $\overline{g}_a$ is the environmental protection threshold on link a imposed by the operators.

3.2. Network Assessment. Due to the fact that the travel decisions of a network user consist of modal choice and route choice, a combined modal split and traffic assignment model is needed. The travel impedance on an auto or rail link is clearly influenced by its flows. For the RPR links, each of them always connects an autolink and a rail link, so that the RPR flows could transfer. The total travel impedance on a RPR link includes the time spent on parking, waiting/boarding the dedicated bus as well as the bus in-vehicle travel time, which is clearly flow-dependent. We then make the following assumption.

Assumption 2. The link travel time $t_a(\mathbf{v}), a \in A$ is a continuously differentiable, monotone increasing of the link flow $\mathbf{v}$.

Let K_m^{od} denote the set of paths under m; $f_k^{m,od}$ denotes the flow on path $k \in K_m^{od}$ for mode m; q_m^{od} denotes the demand of mode m; and $\overline{q}^{od}$ denotes the total travel demand between OD pair (o, d). We have the following flow conservation conditions:

$$\sum_{m \in M_{od}} q_m^{od} = \overline{q}^{od}, \quad \forall od \in W \tag{3}$$

$$q_m^{od} \ge 0, \quad \forall m \in M_{od},\ od \in W \tag{4}$$

$$\sum_{k \in K_m^{od}} f_k^{m,od} = q_m^{od}, \quad \forall m \in M_{od},\ od \in W \tag{5}$$

$$f_k^{m,od} \ge 0, \quad \forall k \in K_m^{od},\ m \in M_{od},\ od \in W \tag{6}$$

$$v_a = \sum_{od \in W} \sum_{m \in M_{od}} \sum_{k \in K_m^{od}} f_k^{m,od} \delta_{a,k}^{m,od}, \quad \forall a \in A \tag{7}$$

where (3) and (5) define demand/flow conservation conditions. Equations (4) and (6) are the nonnegativities. Equations (7) define the link-path flow relationship. Constraints (1), (3)-(7) define the feasible set of Ω of $(\mathbf{v}, \mathbf{f}, \mathbf{q})$ where $\mathbf{v} := (v_a, a \in A)$ is the vector of link flows; $\mathbf{f} := (f_k^{em,od}, k \in K_m^{od}, m \in M^{od}, od \in W)$ is the vector of path flows; and $\mathbf{q} := (q_m^{od}, m \in M^{od}, od \in W)$ is the vector of modal demands.

Let W denote the set of OD pairs; $\overline{q}^{od}$ denote the total travel demand for $(o,d) \in W$; and M_{od} denote the set of modes between OD pair (o,d). The multinomial logit (MNL) model is adopted to analyse the modal split pattern. The probability of choosing any mode m between an OD pair (o,d) equals

$$p_m^{od} = \frac{\exp\left(-\theta \cdot \tau_m^{od}\right)}{\sum_{i\in M_{od}} \exp\left(-\theta \cdot \tau_i^{od}\right)}, \quad \forall m \in M,\ od \in W \tag{8}$$

where $\theta \in [0,+\infty)$ is the dispersion parameter and τ_m^{od} is the deterministic component of the utility of mode m between OD pair (o,d), which can be taken as the equilibrium path travel time of mode m. The satisfaction of MNL, $S_{od}(\tau_m^{od})$, equals $\tau_m^{od} + (1/\theta)\ln(q_m^{od})$, which is concave function [28].

The user equilibrium path flow solution is then formulated by the following conditions:

$$f_k^{m,od}\left(c_k^{m,od} - \varphi_m^{od}\right) = 0, \quad \forall k \in K_m^{od},\ m \in M_{od},\ od \in W \tag{9}$$

$$f_k^{m,od} \geq 0, \quad \left(c_k^{m,od} - \varphi_m^{od}\right) \geq 0, \quad \forall k \in K_m^{od},\ m \in M_{od},\ od \in W \tag{10}$$

where φ_m^{od} is the Lagrange multiplier associated with the constraints (5), which represents the travel impedance of using mode m. At equilibrium, the travel time $c_k^{m,od}$ of a used path $k \in K_m^{od}$ equals φ_m^{od}. All commuters under mode m have identical path travel time, and no traveller can unilaterally reduce his/her travel time by changing routes.

4. Variational Inequality Formulation

In this section, we propose two equivalent VI formulations for the proposed EC-CMSTA model. In order to strengthen the flexibility and adaptability of the proposed model, we further make the following two assumptions.

Assumption 3. The link travel time function is asymmetric, i.e., t_a is not only affected by its own flow v_a, but also flows on some other links, which is expressed as $t_a = t_a(\mathbf{v})$.

Assumption 4. The path travel time c_k is nonadditive to its links, i.e., $c_k \neq \sum_a t_a \delta_{a,k}$, where $\delta_{a,k} = 1$ if path k uses link $a \in A$ and 0 otherwise.

Then, the VI mode for EC-CMSTA is formulated as follows:

$$\sum_{od\in W}\sum_{m\in M_{od}}\sum_{k\in K_m^{od}} c_k^{m,od}\left(\mathbf{f}^*,\mathbf{q}^*\right)\left(f_k^{m,od} - f_k^{m,od*}\right) + \sum_{od\in W}\sum_{m\in M_{od}} \tilde{c}_m^{od}\left(\mathbf{f}^*,\mathbf{q}^*\right)\left(q_m^{od} - q_m^{od*}\right) \geq 0, \quad \forall\left(\mathbf{f}^*,\mathbf{q}^*\right) \in \Omega \tag{11}$$

where $c_k^{m,od}$ is the travel time on route k, where $k \in K_m^{od}$. $\tilde{c}_m^{od} = (1/\theta)\ln(q_m^{od})$ represents the entropy term in travel cost function of mode m, $m \in M_{od}$; $f_k^{m,od*}$ and q_m^{od*} are the optimal solution of this problem; $\mathbf{f}^*$ and $\mathbf{q}^*$ are the vector of optimal solution.

Let $\mathbf{c}$ be the vector of $c_k^{m,od}$ and $\tilde{\boldsymbol{\tau}}$ be the vector of $\tilde{c}_m^{od}$, i.e., $\mathbf{c} := (c_k^{m,od},\ k \in K_m^{od},\ m \in M_{od},\ od \in W)^T$ and $\tilde{\boldsymbol{\tau}} := (\tilde{c}_m^{od},\ m \in M_{od},\ od \in W)^T$. The VI formulation can be simplified as

$$\mathbf{c}\left(\mathbf{f}^*,\mathbf{q}^*\right)^T\left(\mathbf{f}-\mathbf{f}^*\right) + \tilde{\boldsymbol{\tau}}\left(\mathbf{f}^*,\mathbf{q}^*\right)^T\left(\mathbf{q}-\mathbf{q}^*\right) \geq 0, \quad \forall\left(\mathbf{f}^*,\mathbf{q}^*\right) \in \Omega \tag{12}$$

Note that the inclusion of nonlinear environmental constraints makes the feasible set more complex in general. Existing solution algorithms for CMSTA cannot be directly adopted to solve it. Most of existing studies focus on solving the simplified linear link capacity constrained traffic equilibrium problems by a penalty based algorithm. However, due to the complexity of EC-CMSTA (simultaneous achieving equilibrium for both route flow and modal demand), reformulating it into an integrated formulation or directly solving it by a penalty/gradient based algorithm is less effective. Therefore, this paper develops an alternative approach to reformulate the EC-CMSTA which permits existing solution algorithms for its solution.

4.1. Alternative VI Formulation. Daganzo [29] built a fix-point model in terms of link flows for the asymmetric SUE problem with fixed demand and showed availability of the MSA, which was extended by Cantarella [30] to the case of asymmetric SUE problem with elastic demand. Recently, Meng, Liu and Wang [28] further made extension by incorporating link capacity constraints. This problem was formulated as a VI problem by incorporating the fix-point formulated asymmetric SUE problem with elastic demand. Following the same technique, we propose a fixed point reformulated CMSTA model which is expressed as follows:

$$v_a = \sum_{od\in W} \overline{q}^{od} \sum_{m\in M_{od}} p_m^{od}\left(\mathbf{f}\left(\mathbf{v}\right)\right) \times P_a^{m,od}\left(\mathbf{f}\left(\mathbf{v}\right)\right), \quad \forall a \in A \tag{13}$$

where $\mathbf{v}$ is the equilibrium link flow solution, p_m^{od} is the probability of mode choice defined in (8), and $P_a^{m,od}$ is the link usage probability defined by

$$P_a^{m,od}\left(\mathbf{f}\left(\mathbf{v}\right)\right) = \sum_{k\in K_m^{od}} p_k^{m,od}\left(\mathbf{f}\left(\mathbf{v}\right)\right)\delta_{k,a}^{m,od}, \quad \forall a \in A \tag{14}$$

and $p_k^{m,od}$ is the probability of path k perceived as the shortest one among the path set K_m^{od}. In the deterministic case, $p_k^{m,od}$ equals one if path k is the shortest path among alternatives and zero otherwise.

A set of generalized SUE conditions with elastic demand and capacity constraints was proposed by Meng, Liu, and Wang [28] for the optimal solution of SUE problem with elastic demand, asymmetric link travel time function and link capacity constraints. These conditions can be further extended to the multimodal transport system which is the problem addressed in this paper. Let $\mathbf{v}$ denote the equilibrium link flow pattern of the addressed problem if and only if there exists a vector of Lagrangian multipliers corresponding to environmental constraints, denoted by $\mathbf{u}^* = (u_e^*, e \in E)^T$.

$$v_a(\mathbf{u}) = \sum_{od \in W} \sum_{m \in M_{od}} q_m^{od}(\mathbf{f}(\mathbf{v}), \boldsymbol{\lambda}(\mathbf{u}^*)) \times P_a^{m,od}(\mathbf{f}(\mathbf{v}), \boldsymbol{\lambda}(\mathbf{u}^*)), \quad \forall a \in A \tag{15}$$

$$-g_e(\mathbf{v}) \geq 0, \quad \forall e \in E \tag{16}$$

$$u_e^* \cdot (-g_e(\mathbf{v})) = 0, \quad \forall e \in E \tag{17}$$

$$u_e^* \geq 0, \quad \forall e \in E \tag{18}$$

where $\boldsymbol{\lambda}(\mathbf{u}^*)$ is defined by

$$\boldsymbol{\lambda}(\mathbf{u}^*) = \left(\lambda_k^{m,od} = \sum_{e \in E} u_e^* \zeta_{e,k}^{m,od}, \ \forall k \in K_m^{od}, \ m \in M_{od}, \ od \in W \right)^T \tag{19}$$

and $\zeta_{e,k}^{m,od} = 1$ if path $k \in K_m^{od}$ goes through $e \in E$ and $\zeta_{e,k}^{m,od} = 0$, otherwise. Based on the optimal Lagrangian multiplier $\mathbf{u}^*$, the EC-CMSTA is identical to the CMSTA in terms of generalized link travel time functions.

$$\hat{t}_a(\mathbf{v}, \mathbf{u}^*) = \begin{cases} t_a(\mathbf{v}) + u_a^*, & \forall a \in E \\ t_a(\mathbf{v}), & \forall a \in A/E \end{cases} \tag{20}$$

Equations (16)-(18) imply that if there is no environmental constraint on any link $a \in E$, $u_a^* = 0$.

Note that the complementary slackness conditions (see (16)-(18)) could be regarded as a nonlinear complementary problem (NCP) of the vector of Lagrangian multipliers $\mathbf{u}^*$. Due to the feasible set of $\mathbf{u}$ is defined on the whole nonnegative orthant, this NCP problem can be equivalently formulated as a VI model. Hence, the alternative formulation of VI (11) is proposed as follows:

$$\mathbf{g}(\mathbf{u}^*)^T (\mathbf{u} - \mathbf{u}^*) \geq 0, \quad \forall \mathbf{u} \in \mathfrak{R}_+^{|E|} \tag{21}$$

where $\mathfrak{R}_+^{|E|} = (u_e \geq 0, e \in E)^T$ denotes the feasible set. $\mathbf{g}(\mathbf{u})$ is the $|E|$ dimensional vector functions, which measures vehicle pollution:

$$\mathbf{g}(\mathbf{u}) = (-g_e(\mathbf{v}(\mathbf{u})), \ \forall e \in E)^T : \mathfrak{R}_+^{|E|} \longrightarrow \mathfrak{R}^{|E|} \tag{22}$$

$\mathbf{g}(\mathbf{u}^*)$ is thus the optimal solution of the VI problem (21). Herein, $\mathbf{v}(\mathbf{u})$ is a link flow solution for the CMSTA with the generalized link travel time functions (20) (without environmental constraints). In other words, $\mathbf{v}(\mathbf{u})$ is a solution of the fix-point model with the Lagrangian multiplier $\mathbf{u}$:

$$v_a(\mathbf{u}) = \sum_{od \in W} \sum_{m \in M_{od}} q_m^{od}(\hat{\mathbf{t}}(\mathbf{v}, \mathbf{u})) \times P_a^{m,od}(\hat{\mathbf{t}}(\mathbf{v}, \mathbf{u})), \quad \forall a \in A \tag{23}$$

where $\hat{\mathbf{t}}(\mathbf{v}, \mathbf{u}) = (\hat{t}_a(\mathbf{v}, \mathbf{u}), \forall a \in A)$ denotes the vector of generalized link travel time functions.

Note that the feasible set $\mathfrak{R}_+^{|E|}$ of $VI(\mathbf{g}, \mathfrak{R}_+^{|E|})$ is unbounded, which makes it difficult to shown the existence of a solution to VI model $VI(\mathbf{g}, \mathfrak{R}_+^{|E|})$. Fortunately, according to Meng, Liu and Wang [28], one can always find an upper bound $\overline{M}$ for $u_a, \forall a \in E$. Therefore, we define a subset $\overline{\Omega}$ of $\mathfrak{R}_+^{|E|}$ which is bounded by $\overline{M}$. The restricted VI model is subsequently formulated as

$$\mathbf{g}(\mathbf{u}^*)^T (\mathbf{u} - \mathbf{u}^*) \geq 0, \quad \forall \mathbf{u} \in \overline{\Omega} \tag{24}$$

The feasible set $\overline{\Omega}$ is a nonempty, closed, and convex set. To show the existence of solution to $VI(\mathbf{g}, \overline{\Omega})$ as well as the convergence of its solution algorithm, the following properties of vector function $\mathbf{g}(\mathbf{u})$ are proposed. Rigorous proofs of Propositions 5 and 6 are given in the Appendices A and B, respectively.

Proposition 5. *Vector function* $\mathbf{g}(\mathbf{u})$ *is strictly monotone on* $\overline{\Omega}$; *namely,*

$$\left(\mathbf{g}(\mathbf{u}') - \mathbf{g}(\mathbf{u}'')\right)^T (\mathbf{u}' - \mathbf{u}'') > 0, \quad \forall \mathbf{u}', \mathbf{u}'' \in \overline{\Omega} \tag{25}$$

Proposition 6. *Vector function* $\mathbf{g}(\mathbf{u})$ *is Lipschitz continuous on* $\overline{\Omega}$; *namely,*

$$\left\| \mathbf{g}(\mathbf{u}') - \mathbf{g}(\mathbf{u}'') \right\|_2 \leq L \left\| \mathbf{u}' - \mathbf{u}'' \right\|, \quad \forall \mathbf{u}', \mathbf{u}'' \in \overline{\Omega} \tag{26}$$

Though we have Propositions 5 and 6, the existence of a solution to VI model $VI(\mathbf{g}, \overline{\Omega})$ can be ensured. We proceed to prove the equivalence between the VI model $VI(\mathbf{g}, \overline{\Omega})$ and the generalized CMSTA conditions and summarized in Proposition 7. Rigorous proofs are given in Appendix C.

Proposition 7. $\mathbf{u}^* = (u_a^*, \forall a \in E)$ *is a solution of* $VI(\mathbf{g}, \overline{\Omega})$ *if and only if* $\mathbf{u}^*$ *and* $\mathbf{v}(\mathbf{u}^*)$ *fulfil the generalized CMSTA conditions (15)-(18).*

5. Solution Algorithm

As the vector function $\mathbf{g}(\mathbf{u})$ of the VI model $VI(\mathbf{g}, \overline{\Omega})$ is an implicit function, it is burdensome to directly evaluate its gradient. On the other hand, with any given Lagrangian multiplier vector $\mathbf{u}$, the value of vector function $\mathbf{g}(\mathbf{u})$ can be evaluated by solving the CMSTA with generalized travel time with no constraints, which can be efficiently solved by a number of existing solution algorithms.

Projection type algorithms have been thoroughly studied and proven to be efficient in solving traffic equilibrium problems [31–33]. Among them, the gradient projection (GP) is commonly adopted due to its concise form and good performance, for which a recent survey can be found in Ryu, Chen, and Choi [34]. The self-adaptive gradient projection (SAGP) algorithm was originally proposed by He et al. [35], where the gradient projection operation is adopted to establish the descent direction, coupled with a self-adaptive step size scheme to permit the convergence under a much milder assumption of the underlying mapping, and thus can handle general cost mapping structures and generalized side constraints. Later, Han and Sun [36] adopted this method to solve the asymmetric traffic assignment problems. Chen, Zhou, and Xu [32] further used this method to solve the asymmetric traffic assignment problem with nonadditive route cost functions. Inspired by previous studies, this paper presents the SAGP algorithm specifically designed for solving the CMSTA with generalized travel time functions.

As for the Lagrangian multipliers **u**, the projection operation on Ω of the VI model $VI(\mathbf{g}, \overline{\Omega})$ is straightforward, as well. Propositions 5 and 6 imply the monotony and continuity of the VI model $VI(\mathbf{g}, \overline{\Omega})$. Therefore, the projection based algorithm is also suitable to evaluate the Lagrangian multipliers solutions. To this end, this section proposes a nested SAGP algorithm to solve the EC-CMSTA model which is presented as follows.

Step 0 (initiation). Set parameters of SAGP within the given range: $\delta \in (0,1)$, $\mu \in [0.5, 1]$, $\varepsilon > 0$, $\alpha_{\max} > 0$, $\alpha^{(0)} > 0$, $\gamma^{(0)} = \alpha^{(0)}$. Set the iteration counter: $n = 0$, and the initial solution which should be a feasible solution: $\mathbf{u}^{(0)} = (u_a^{(0)} = 0, a \in E)^T$.

Step 1 (self-adaptive scaling procedure). This step aims to find the largest step size $\alpha^{(n+1)}$ that satisfies the given criteria in (27). To this end, let $\alpha^{(n+1)} = \mu^{l^{(n)}} \gamma^{(n)}$, and set nonnegative integer $l^{(n)} = 0$. We perform the following procedures, if the given criterion is satisfied, terminate and go to Step 2; otherwise, let $l^{(n)} = l^{(n)} + 1$, and perform the following procedures again.

Step 1.1. Given vector $\mathbf{u}^{(n)}$, evaluate the CMSTA with generalized travel time functions (23), using SAGP algorithm, and obtain the equilibrium flows $(\mathbf{v}^{(n)}, \mathbf{q}^{(n)})$. The solutions are subsequently used to evaluate the environmental measurement function $\mathbf{g}(\mathbf{u}^{(n)})$.

Step 1.2. Perform the projection operation for $\mathbf{u}^{(n)}$ to obtain a tentative vector $\widetilde{\mathbf{u}}^{(n+1)}$, i.e., $\widetilde{\mathbf{u}}^{(n+1)} = P_{\Omega}[\mathbf{u}^{(n)} - \alpha^{(n+1)} \cdot \mathbf{g}(\mathbf{u}^{(n)})]$.

Step 1.3. Let $\overline{\mathbf{u}}^{(n+1)} = \max[0, \mathbf{u}^{(n)} - \alpha^{(n)} \mathbf{g}(\mathbf{u}^{(n)})]$. If the following inequality is satisfied

$$\begin{aligned} &(2-\delta)\,\alpha^{(n+1)} \left(\mathbf{u}^{(n)} - \widetilde{\mathbf{u}}^{(n+1)}\right)^T \left(\mathbf{g}\left(\mathbf{u}^{(n)}\right) - \mathbf{g}\left(\widetilde{\mathbf{u}}^{(n+1)}\right)\right) \\ &\quad + \left(\alpha^{(n+1)}\right)^2 \left\|\mathbf{g}\left(\mathbf{u}^{(n)}\right) - \mathbf{g}\left(\widetilde{\mathbf{u}}^{(n+1)}\right)\right\|^2 \\ &\quad \geq \max\left\{0, \frac{\left(\alpha^{(n+1)}\right)^2 - \left(\alpha^{(n)}\right)^2}{\left(\alpha^{(n)}\right)^2} \left\|\mathbf{u}^{(n)} - \overline{\mathbf{u}}^{(n+1)}\right\|^2\right\} \end{aligned} \tag{27}$$

go Step 2. Otherwise, let $l^{(n)} = l^{(n)} + 1$, update step size $\alpha^{(n+1)}$, and go back to Step 1.1.

Step 2 (selection of $\gamma^{(n+1)}$). If the following condition is satisfied, let $\gamma^{(n+1)} = \max\{\alpha^{(n+1)}/\mu, \alpha_{\max}\}$.

$$\begin{aligned} &0.5\alpha^{(n+1)} \left(\mathbf{u}^{(n)} - \widetilde{\mathbf{u}}^{(n+1)}\right)^T \left(\mathbf{g}\left(\mathbf{u}^{(n)}\right) - \mathbf{g}\left(\widetilde{\mathbf{u}}^{(n+1)}\right)\right) \\ &\quad + \left(\alpha^{(n+1)}\right)^2 \left\|\mathbf{g}\left(\mathbf{u}^{(n)}\right) - \mathbf{g}\left(\widetilde{\mathbf{u}}^{(n+1)}\right)\right\|^2 \\ &\quad \geq \max\left\{0, \frac{\left(\alpha^{(n+1)}\right)^2 - \left(\alpha^{(n)}\right)^2}{\left(\alpha^{(n)}\right)^2} \left\|\mathbf{u}^{(n)} - \overline{\mathbf{u}}^{(n+1)}\right\|^2\right\} \end{aligned} \tag{28}$$

Step 3 (convergence test). If the convergence criterion is satisfied, terminate; otherwise, let iteration counter: $n = n+1$, go back to Step 1.

Note that the CMSTA with generalized travel time functions in Step 1.1 is also solved by the SAGP algorithm, which is not repeated here for conciseness.

6. Numerical Examples

In this section, two experiments based on a linear network and the Sioux-Falls network are provided to demonstrate the features of the proposed EC-CMSTA and validity of the proposed solution algorithm.

It is well known that carbon monoxide (CO), carbon dioxide (CO_2), sulphur dioxide (SO_2), nitrogen oxides (NO_x), respirable suspended particulates (RSP or PM_{10}), fine suspended particulates (FSP or $PM_{2.5}$), and volatile organic compounds (VOC) are the main types of vehicular emission pollutants that are critical for air quality protection. Among others, CO is taken as an important indicator of the level of vehicle atmosphere pollution, mainly because (1) vehicles contribute to the majority of CO emissions in the atmosphere; (2) CO is one of the most critical pollutants among the various types of vehicular emissions; (3) the emission functions of other pollutants are similar to that of CO [11]. In addition, this consideration has also been used in a number of recent studies, e.g., Yin and Lawphongpanich [37]; Nagurney, Qiang and Nagurney [38]; Chen, Zhou and Ryu [10]; Li et al. [11]; Xu, Chen and Cheng [1]; Yang, Ban and Ma [39]; Kitthamkesorn et al. [2]. Therefore, we use CO as an illustration to model vehicle emission. The nonlinear macroscopic formula of CO emission rate proposed by Wallace et al. [6] is used for analysis as shown below:

$$g_a(t_a) = 0.2038 \cdot t_a \cdot \exp\left(\frac{0.7962 \cdot L_a}{t_a}\right) \tag{29}$$

where $g_a(t_a)$ is the CO emission rate on link a measured in grams per hour. L_a is the length of link a measured in kilometers, and t_a is the link travel time measured in minutes. This nonlinear function can be shown to be convex and continuously differentiable. The Bureau of Public Road (BPR) type function is adopted for the link travel time, i.e., $t_a = t_a^0(1+0.15(v_a/C_a)^4)$ for road links, and $t_a = t_a^0(1+0.1(v_a/C_a)^4)$ for rail and RPR links. The dispersion parameter of MNL mode is set as $\theta = 1.0$.

6.1. A Linear Corridor. As shown in Figure 1, consider a linear corridor $G(N, A)$, with one rail line (dotted line) from suburb o to the urban city area d and another highway (solid line)

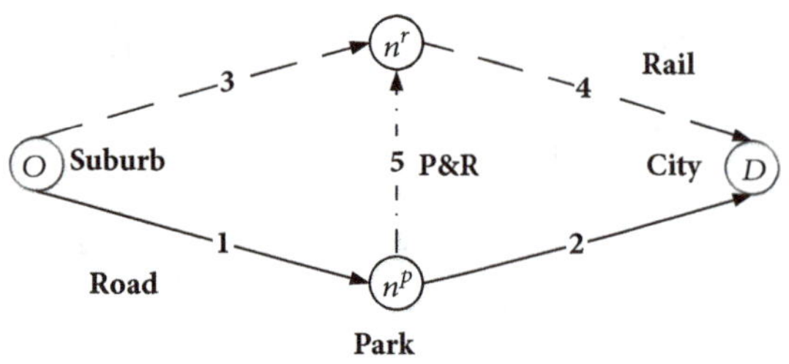

FIGURE 1: A linear corridor with RPR facility.

TABLE 1: Attributes of the linear corridor.

Link Type	Auto		Rail		RPR
Link No.	1	2	3	4	5
Free Flow Travel Time (t_a^0)	6	4	12	5	3
Capacity (C_a)	250	200	350	350	350

TABLE 2: Modal share and equilibrium travel time.

Modes	With RPR scheme	
	Demand	Travel time
Auto	250.45	16.66
Train	150.15	17.16
RPR	99.38	17.58

connecting (o, d). The OD demand is assumed to be 500 per hour. One RPR site is constructed at n^p, adjacent to the train station n^r. Travellers transfer from the RPR site n^p to the train station n^r through the RPR link (link 5). Link 2 passes into an environmentally sensitive urban area, which exerts restriction on the total CO emissions on link 2. Using the CO emission rate function defined by (29), the nonlinear environmental constraint on link a can be expressed as

$$0.2038 \cdot t_a \cdot \exp\left(\frac{0.7962 L_a}{t_a}\right) \cdot v_a - \overline{g}_a \leq 0, \quad \forall a \in A \quad (30)$$

where $\overline{g}_a$ is the CO emission threshold on link a. Attributes of the network are given in Table 1.

Setting the environmental constraints at 500, we first examine the correctness of the results. As expected, the results in Table 2 satisfy the modal demand conversation condition and the MNL based modal choice condition. The Lagrangian multiplier on link 2 equals 1.73, permitting the emission on link 2 subject to the requirements.

We proceed to explore the influence of the environmental constraint on the flow shift of modal demands, link flows, and total system travel time by varying the emission constraint $\overline{g}_2$ between 0 and 800. The corresponding modal demand shift pattern under environmental constraint is given in Figure 2. We can see that as $\overline{g}_2$ increases (relax the CO emission constraints), the auto demand first increases and then becomes stable; meanwhile, the demands of rail and RPR show the crosscurrent. It makes sense because relaxing $\overline{g}_2$ is similar to increase the road capacity. When the emission level is beyond 631.47, constraint (30) is not blinding anymore which means the PRP facilities are able to achieve the environmental

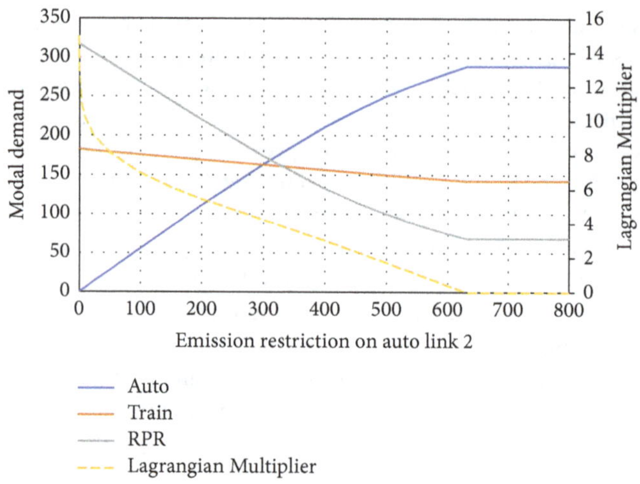

FIGURE 2: Modal split pattern with various environmental constraint level.

protection requirements without additional regulations or restrictions. The dotted line indicates the evolution of Lagrangian multiplier on link 2, u_2^*. As the environmental regulation can be strengthened, u_2^* keeps increasing and trends to be infinity, when $\overline{g}_2 = 0$.

Detailed results of link flows, link travel time, and Lagrangian multiplier are shown in Table 3. As marked by bold and italic text in Table 3 the minimal system travel cost is achieved when $\overline{g}_2$ is increased to around 631.47. In other words, setting $\overline{g}_2 \geq 631.47$ is equivalent to have no CO emission constraint on link 2, which reduces the problem into CMSTA. Meanwhile, the Lagrangian multiplier u_2^* reduces to zero as well indicating that there is no restrictions for vehicles which come into this area.

6.2. Sioux-Falls Network. Another numerical example built from the Sioux-Falls network (Figure 3) is used to further access the performance of the proposed model and solution algorithm for solving the environmental constrained CMSTA problem. This study network consists of 30 nodes and 76 links. The area highlighted by the red ellipse is the CBD district. It has four rail lines between the suburb areas and the city loop. Detail attributes of this network are tabulated in Appendix D. We focus on the travel demands in commuting hours, where the majority of travel demands origin from the suburbs and go into business districts. There are 24 OD pairs and the OD demand data are given in Appendix E. Four sites (nodes 3, 6, 19, and 14) which locate close to rail lines are selected as RPR car parks where RPR users park their car and transfer to rail mode.

The column generation scheme is adopted in this case and incorporated into our proposed solution algorithm to generate paths in evaluating the equilibrium flows. In the multimodal transport network, we first examine the effect of the proposed RPR scheme with the CMSTA model and access the emissions in the road subnetwork. 5 links in the CBD district are subsequently identified as the critical links, and be further grouped into the link set $\overline{A} = \{14, 22, 29, 46, 48\}$. As shown in Figure 4, the provision of RPR facilities significantly

TABLE 3: Link flows and link cost with varied restricted CO emission level.

$\overline{g}_2$	Link 1		Link 2			Link 3		Link 4		Link 5	
	Time	Flow	Time	Flow	u^*	Time	Flow	Time	Flow	Time	Flow
0	8.33	317.29	4.00	0.00	$+\infty$	12.09	182.71	7.08	499.94	3.20	317.23
100	8.55	324.24	4.00	55.32	6.99	12.08	175.76	6.30	444.68	3.10	268.92
200	8.76	330.69	4.06	110.34	5.45	12.07	169.31	5.77	389.66	3.05	220.35
300	8.97	337.09	4.27	163.54	4.24	12.06	162.91	5.43	336.46	3.02	173.55
400	9.21	343.56	4.75	211.36	3.02	12.05	156.44	5.23	288.64	3.01	132.20
500	9.45	349.85	5.48	250.47	1.73	12.04	150.15	5.13	249.53	3.00	99.38
600	9.69	355.77	6.34	281.02	0.41	12.03	144.23	5.08	218.98	3.00	74.75
631.47	***9.77***	***357.57***	***6.62***	***289.22***	***0.00***	***12.03***	***142.43***	***5.07***	***210.78***	***3.00***	***68.35***
700	9.77	357.57	6.62	289.22	0.00	12.03	142.43	5.07	210.78	3.00	68.35
800	9.77	357.57	6.62	289.22	0.00	12.03	142.43	5.07	210.78	3.00	68.35
900	9.77	357.57	6.62	289.22	0.00	12.03	142.43	5.07	210.78	3.00	68.35

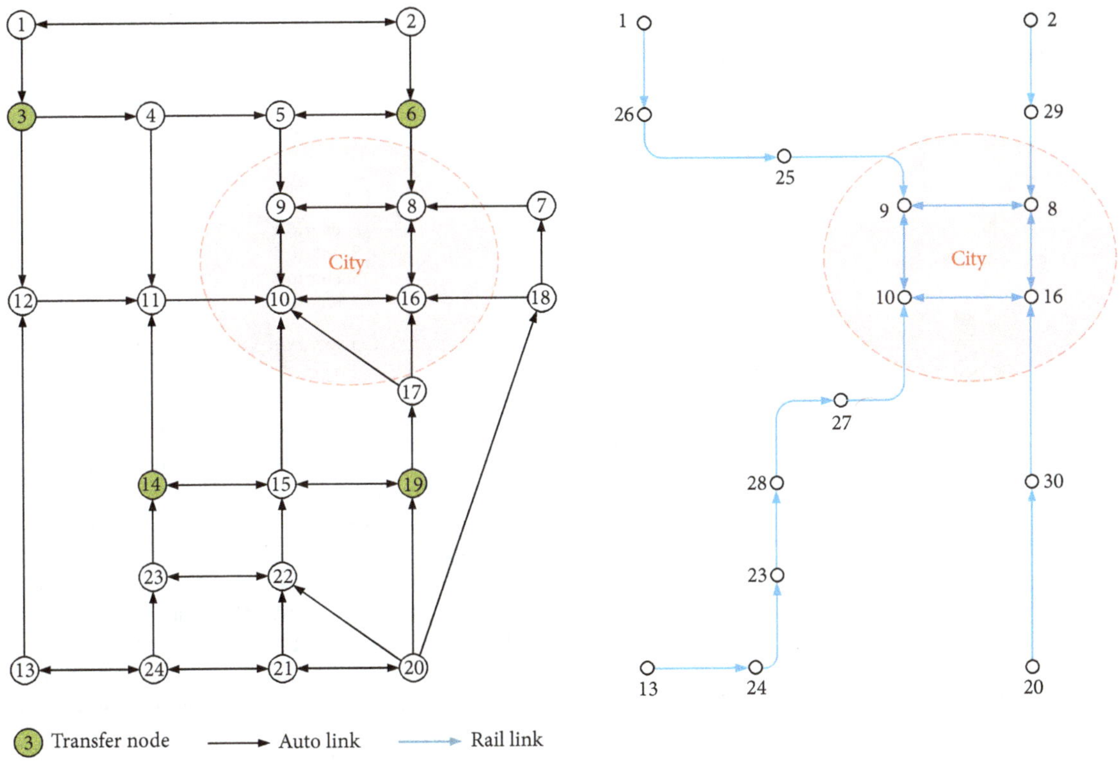

FIGURE 3: Modified Sioux-Falls network with RPR scheme.

influence travellers' mode choice as well as route choice behaviour. 14.03% of previous autodrivers and 5.51% of previous train riders shift to RPR mode, due to the utility provided by RPR services. The amount of emissions on critical links also significantly reduced with the provision of RPR schemes. This phenomenon indicates RPR schemes could alleviate the environmental deterioration by redistributing the flow and modal demand in the network and reflects the rationality of modelling users modal choice and route choice decisions in a combined model, i.e., CMSTA.

We proceed to exert environmental constraints on these critical links. Here, we consider five scenarios of the thresholds and summarize in Table 4. Parameters in the SAGP algorithm are set as follows: $\mu = 0.85$, $\delta = 0.8$, and $\alpha^{(0)} = 1.0$. In order to reflect the convergence trend of the SAGP algorithm apparently, we take the following logarithmic value of the maximum absolute error as a performance index, namely,

$$\varepsilon = \log_{10}\left(\max\left(\left\|T_a - g_a v_a\right\|_2, \forall a \in \overline{A}\right)\right) \tag{31}$$

The SAGP algorithm is then used to solve the environmental constrained CMSTA in terms of the five scenarios provided in Table 4. Figure 5 depicts the logarithmic value of the maximum absolute error versus the number of iterations used for solving scenarios 1, 2, 3, and 4. It clearly shows that the

Table 4: Five scenarios of environmental constraints.

Link ID	Scenario 1 Threshold T_a^1	Scenario 2 Threshold T_a^2	Scenario 3 Threshold T_a^3	Scenario 4 Threshold T_a^4	Scenario 5 Threshold T_a^5	Emission	Link flow
14	500	600	700	800	900	724.16	320.28
22	500	600	700	800	900	1007.05	370.93
29	500	600	700	800	900	796.13	319.51
46	500	600	700	800	900	866.88	351.1
48	500	600	700	800	900	735.61	531.39

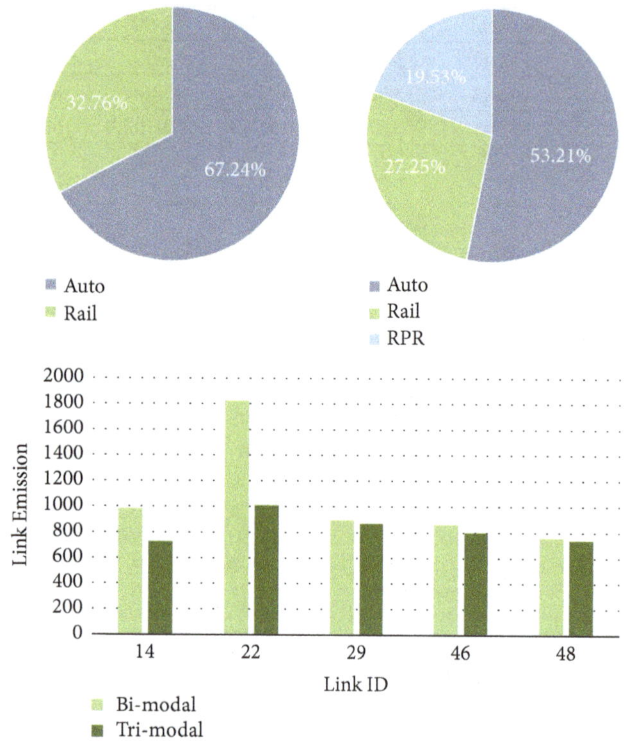

Figure 4: Mode share and emissions with/without RPR scheme.

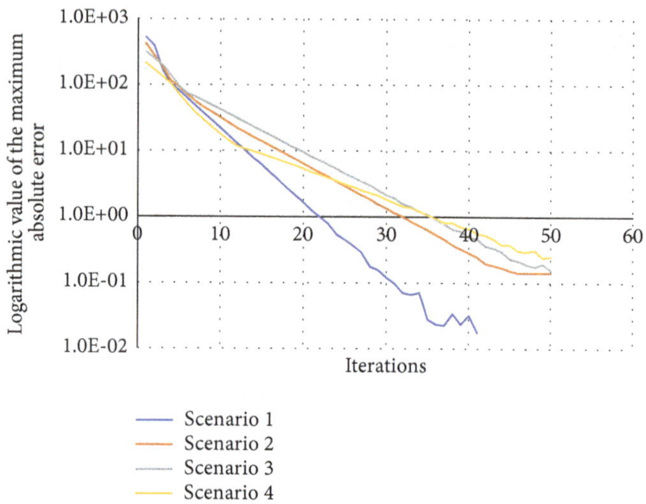

Figure 5: Convergence trend of the SAGP algorithm.

algorithm can converge to a solution of VI model $VI(\mathbf{g}, \overline{\Omega})$. In scenario 5, the solution converges much faster than other scenarios because only link 22 is constrained.

Table 5 gives the resultant link emissions and corresponding Lagrangian multiplier for each scenario. We can observe that the emissions on each link are bounded by the required thresholds (some link emissions slightly excess the requirements caused by the computation error). The corresponding multiplier takes positive value only on the links with emission bounded by the threshold. These two phenomena tally with (16)-(18), which numerically verify the effectiveness of the VI model $VI(\mathbf{g}, \overline{\Omega})$ and the SAGP algorithm. In addition, from scenario 1 to scenario 5, it can be seen that, by tightening the thresholds, the corresponding Lagrangian multipliers become larger, which indicates that it would be more costly to restrict the emission to a much lower level.

Table 6 further investigates the modal shift trend from scenario 1 to scenario 5. As the values of threshold getting smaller, travellers who use automode become less. The environmental regulations in road network could significantly promote the modal shift from auto to "green" travel modes (i.e., Train and RPR). Taking scenario 1, for example, the automode share is depressed about 11.11%, from 53.21% to 42.10%, where 5.38% of travel demand shifts to rail mode and 5.72% rest shifts to RPR mode. This phenomenon indicates that environmental regulations significantly influence not only travellers' route choice behaviour but also mode choice behaviour.

To further see the impacts of demand level on the Lagrangian multipliers, a sensitivity test is carried out on the dispersion parameter on the modal choice function (8). Without loss of generality, the first scenario of threshold setting is taken for this test, and then the value of θ is changed from 0.5 to 0.9 with an interval of 0.1. The Lagrangian multipliers obtained by the SAGP algorithm in these five cases are presented in Table 7. We noted that a larger value of θ results in a larger level of autodemand level and the road network congestion level is higher accordingly. Table 7 indicates the generalized CMSTA conditions can still be fulfilled. As the demand level ascends, the corresponding Lagrangian multipliers also increases, meaning that it is more difficult to maintain the threshold constraints at a higher demand level.

We further compare the network-level performance before and after implementing the set of environmental

TABLE 5: Link emission and corresponding Lagrangian multiplier.

Link ID	Scenario 1		Scenario 2		Scenario 3		Scenario 4		Scenario 5		Without
	Emission	LM	Emission	LM	Emission	LM	Emission	LM	Emission	LM	
14	499.99	1.64	600.06	0.93	700.07	0.22	728.78	0.00	726.51	0.00	724.16
22	500.00	2.29	600.03	1.74	700.10	1.20	800.17	0.64	900.00	0.38	1007.05
29	500.00	0.79	600.00	0.23	700.03	0.10	800.06	0.03	867.92	0.00	866.88
46	500.02	2.32	600.14	1.5	700.09	0.77	800.21	0.15	803.76	0.00	796.13
48	500.03	0.86	600.02	0.71	699.99	0.32	743.59	0.00	736.09	0.00	735.61

∗LM denotes Lagrangian multiplier.

TABLE 6: Mode share with different threshold levels.

Mode	With Threshold					Without Threshold
	Scenario 1	Scenario 2	Scenario 3	Scenario 4	Scenario 5	
Auto	42.10%	45.68%	48.96%	51.53%	52.35%	53.21%
Train	32.64%	30.64%	28.89%	27.61%	27.39%	27.25%
RPR	25.26%	23.68%	22.15%	20.86%	20.27%	19.53%

TABLE 7: Sensitivity test on the dispersion parameter.

link ID	$\theta = 0.5$		$\theta = 0.6$		$\theta = 0.7$		$\theta = 0.8$		$\theta = 0.9$	
	Emission	LM	Emission	LM	Emission	LM	Emission	LM	Emission	LM
14	499.99	1.64	500.00	1.70	499.98	1.73	500.01	1.74	500.02	1.75
22	500.00	2.29	500.00	2.02	500.00	1.82	500.02	1.66	500.01	1.54
29	500.00	0.79	500.01	0.68	500.01	0.59	500.00	0.52	500.01	0.48
46	500.02	2.32	500.03	2.38	500.03	2.41	500.00	2.42	500.06	2.43
48	500.03	0.86	500.01	0.85	499.94	0.84	500.02	0.84	499.96	0.83

∗LM denotes Lagrangian multiplier.

restraints. The first scenario of threshold setting is still taken for this test. Without loss of generality, the total travel time (TTT) and total vehicle distances (TVD) travelled are used in this study to measure the network-wide performance. From Table 8 we can observe that implementation of environmental constraints decreases both TTT and TVD (-13.89% decrease of TTT and -13.19% decrease of TVM), which implies that some travellers must be shifting their modes in order to benefit the air quality of the CBD area.

7. Conclusions

Environmental protection is an important consideration in planning a transportation system. Environmental constraint is a useful means to explicitly reflect environmental protection requirements. In this paper, we studied a combined environmental protection strategy in the multimodal transport network: (1) a rail-based park-and-ride scheme to assist mode shift to reduce the overall vehicle travel demands in the road network and (2) an area-based regulation scheme to deal with the environmental deterioration caused by heterogeneity of traffic flows. To quantitatively evaluate the effectiveness of the proposed scheme, we built an environmental constrained combined modal split and traffic assignment (EC-CMSTA) model. Specifically, we developed an equivalent VI reformulation for the EC-CMSTA. Different from the explicit formulation as a VI or NCP, the proposed reformulation adapts the concept of fix-point to reformulate the CMSTA conditions, which is, later, incorporated into the VI formulated nonlinear complementarity conditions associated with the nonlinear environmental constraints. The proposed VI reformulation can handle a general nonlinear environmental constraint structure and a general link and route cost structure, enhancing the modelling adaptability and flexibility, and in addition permits a number of efficient algorithms for its solution. A customized projection algorithm with self-adaptive step size scheme is adopted in this paper. Numerical examples demonstrated the EC-CMSTA model has the potential to enhance the behavioural modelling of network users' travel decisions and to assist in quantitatively evaluating the effectiveness of RPR schemes and environmental regulations.

Several directions are worthy of further investigation. First, in this paper, we use CO as an indicator to model vehicle emission. Further study will take other types of emission into consideration, e.g., the CO_2. Second, a number of efficient solution algorithms proposed by recent studies, e.g., Ryu, Chen, and Choi [34]; Ryu, Chen, and Choi [40], should be explored to further improve the solution efficiency. Third, we plan to test our model on larger networks with more real world applications. For very large-scale problems, existing algorithms may be not efficient and specific algorithms should be developed.

TABLE 8: Network performance with and without environmental regulations.

Network performance	Without environmental Regulation (a)	With environmental regulation (b)	Relative difference [(b-a)/a]
Total travel time	41442.50	35685.36	-13.89%
Total vehicle distance	38995.10	33851.32	-13.19%

Appendix

A. Proof of Proposition 5

Vector function $\mathbf{g}(\mathbf{u})$ is strictly monotone on $\overline{\Omega}$; namely,

$$\left(\mathbf{g}\left(\mathbf{u}'\right)-\mathbf{g}\left(\mathbf{u}''\right)\right)^T\left(\mathbf{u}'-\mathbf{u}''\right)>0, \quad \forall \mathbf{u}',\mathbf{u}''\in\overline{\Omega} \tag{A.1}$$

Proof. For any two distinct nonnegative Lagrangian multiplier vectors $\mathbf{u}'$ and $\mathbf{u}''$, let $\mathbf{v}(\mathbf{u}')$ and $\mathbf{v}(\mathbf{u}'')$ denote the corresponding CMSTA link flow solutions, and similarly, let $\mathbf{f}(\mathbf{u}')$, $\mathbf{f}(\mathbf{u}'')$, $\mathbf{q}(\mathbf{u}')$, and $\mathbf{q}(\mathbf{u}'')$ denote the corresponding CMSTA path flow and modal demand solutions. We have

$$\mathbf{v}\left(\mathbf{u}'\right)=\Delta\mathbf{f}\left(\mathbf{u}'\right) \tag{A.2}$$

$$\mathbf{v}\left(\mathbf{u}''\right)=\Delta\mathbf{f}\left(\mathbf{u}''\right) \tag{A.3}$$

$$\mathbf{f}\left(\mathbf{u}'\right)=\mathbf{q}\left(\boldsymbol{\tau}'\right)^T\cdot\mathbf{p}\left(\widehat{\mathbf{c}}'\right) \tag{A.4}$$

$$\mathbf{f}\left(\mathbf{u}''\right)=\mathbf{q}\left(\boldsymbol{\tau}''\right)^T\cdot\mathbf{p}\left(\widehat{\mathbf{c}}''\right) \tag{A.5}$$

$$\mathbf{q}\left(\widehat{\boldsymbol{\tau}}'\right)=\overline{\mathbf{q}}^T\cdot\overline{\mathbf{p}}\left(\widehat{\boldsymbol{\tau}}'\right) \tag{A.6}$$

$$\mathbf{q}\left(\widehat{\boldsymbol{\tau}}''\right)=\overline{\mathbf{q}}^T\cdot\overline{\mathbf{p}}\left(\widehat{\boldsymbol{\tau}}''\right) \tag{A.7}$$

where Δ defines the link-path incidence, $\widehat{\mathbf{c}}' = \Delta^T\widehat{\mathbf{t}}(\mathbf{v}(\mathbf{u}'),\mathbf{u}')$, $\widehat{\mathbf{c}}'' = \Delta^T\widehat{\mathbf{t}}(\mathbf{v}(\mathbf{u}''),\mathbf{u}'')$, $\mathbf{p}(\widehat{\mathbf{c}}') = (p_k^{m,od}(\widehat{\mathbf{c}}'),\ \forall k\in K_m^{od},\ m\in M_{od},\ od\in W)^T$, $\mathbf{p}(\widehat{\mathbf{c}}'') = (p_k^{m,od}(\widehat{\mathbf{c}}''),\ \forall k\in K_m^{od},\ m\in M_{od},\ od\in W)^T$, $\widehat{\boldsymbol{\tau}}' = (\tau_m^{\prime od}(\widehat{\mathbf{c}}'),\ \forall m\in M_{od},\ od\in W)^T$, $\widehat{\boldsymbol{\tau}}'' = (\widehat{\tau}_m^{\prime\prime od}(\widehat{\mathbf{c}}''),\ \forall m\in M_{od},\ od\in W)^T$, $\overline{\mathbf{p}}(\widehat{\boldsymbol{\tau}}') = (p_m^{od}(\widehat{\boldsymbol{\tau}}'),\ \forall m\in M_{od},\ od\in W)^T$, $\overline{\mathbf{p}}(\widehat{\boldsymbol{\tau}}'') = (p_m^{od}(\widehat{\boldsymbol{\tau}}''),\ \forall m\in M_{od},\ od\in W)^T$, and $\overline{\mathbf{q}} = (\overline{q}^{od},\ \forall od\in W)^T$.

According to the definition of CMSTA conditions, the modal travel time τ_m^{od} is taken as the equilibrium path travel time of mode m, it follows that

$$\tau_m^{od}\le c_k^{m,od},\quad \forall k\in K_m^{od}; \tag{A.8}$$

together with the flow conservation condition (5), we have

$$\mathbf{q}\left(\boldsymbol{\tau}'\right)^T\boldsymbol{\tau}'\le\mathbf{f}\left(\mathbf{u}'\right)^T\mathbf{c}' \tag{A.9}$$

$$\mathbf{q}\left(\boldsymbol{\tau}''\right)^T\boldsymbol{\tau}''\le\mathbf{f}\left(\mathbf{u}''\right)^T\mathbf{c}'' \tag{A.10}$$

Subtracting (A.9) and (A.10) yields that

$$\mathbf{q}\left(\boldsymbol{\tau}'\right)^T\left(\boldsymbol{\tau}'-\boldsymbol{\tau}''\right)\le\mathbf{f}\left(\mathbf{u}'\right)^T\left(\mathbf{c}'-\mathbf{c}''\right) \tag{A.11}$$

$$\mathbf{q}\left(\boldsymbol{\tau}''\right)^T\left(\boldsymbol{\tau}'-\boldsymbol{\tau}''\right)\le\mathbf{f}\left(\mathbf{u}''\right)^T\left(\mathbf{c}'-\mathbf{c}''\right); \tag{A.12}$$

namely,

$$\begin{aligned}&\left(\mathbf{q}\left(\boldsymbol{\tau}'\right)-\mathbf{q}\left(\boldsymbol{\tau}''\right)\right)^T\left(\boldsymbol{\tau}'-\boldsymbol{\tau}''\right)\\&\quad\le\left(\mathbf{f}\left(\mathbf{u}'\right)-\mathbf{f}\left(\mathbf{u}''\right)\right)^T\left(\mathbf{c}'-\mathbf{c}''\right)\end{aligned} \tag{A.13}$$

The monotonicity of function (8) implies that

$$\left(\mathbf{q}\left(\boldsymbol{\tau}'\right)-\mathbf{q}\left(\boldsymbol{\tau}''\right)\right)^T\left(\boldsymbol{\tau}'-\boldsymbol{\tau}''\right)>0 \tag{A.14}$$

According to (A.13) and (A.14), we thus obtain

$$\left(\mathbf{f}\left(\mathbf{u}'\right)-\mathbf{f}\left(\mathbf{u}''\right)\right)^T\left(\mathbf{c}'-\mathbf{c}''\right)>0 \tag{A.15}$$

As $\mathbf{c}' = \Delta^T\widehat{\mathbf{t}}(\mathbf{v}(\mathbf{u}'),\mathbf{u}')$ and $\widehat{\mathbf{c}}'' = \Delta^T\widehat{\mathbf{t}}(\mathbf{v}(\mathbf{u}''),\mathbf{u}'')$, (A.15) can be equivalently written as

$$\begin{aligned}&\left(\mathbf{f}\left(\mathbf{u}'\right)-\mathbf{f}\left(\mathbf{u}''\right)\right)^T\\&\quad\cdot\left(\Delta^T\widehat{\mathbf{t}}\left(\mathbf{v}\left(\mathbf{u}'\right),\mathbf{u}'\right)-\Delta^T\widehat{\mathbf{t}}\left(\mathbf{v}\left(\mathbf{u}''\right),\mathbf{u}''\right)\right)>0\end{aligned} \tag{A.16}$$

Rearranging the left side of (A.16), it follows that

$$\begin{aligned}&\left(\mathbf{v}\left(\mathbf{u}'\right)-\mathbf{v}\left(\mathbf{u}''\right)\right)^T\\&\quad\cdot\left(\Delta^T\widehat{\mathbf{t}}\left(\mathbf{v}\left(\mathbf{u}'\right),\mathbf{u}'\right)-\Delta^T\widehat{\mathbf{t}}\left(\mathbf{v}\left(\mathbf{u}''\right),\mathbf{u}''\right)\right)>0\end{aligned} \tag{A.17}$$

According to the definition of generalized link travel time function, (20), we have

$$\left(\mathbf{v}\left(\mathbf{u}'\right)-\mathbf{v}\left(\mathbf{u}''\right)\right)^T\left(\mathbf{u}'-\mathbf{u}''\right)>0 \tag{A.18}$$

Together with Assumption 1, $\mathbf{g}(\mathbf{u})$ is a strictly monotone function with respect to link flow solutions $\mathbf{v}(\mathbf{u})$; it follows that

$$\left(\mathbf{g}\left(\mathbf{u}'\right)-\mathbf{g}\left(\mathbf{u}''\right)\right)^T\left(\mathbf{u}'-\mathbf{u}''\right)>0 \tag{A.19}$$

which implies $\mathbf{g}(\mathbf{u})$ is strictly monotone with respect to $\mathbf{u}$ on $\overline{\Omega}$. □

B. Proof of Proposition 6

Vector function $\mathbf{g}(\mathbf{u})$ is Lipschitz continuous on $\overline{\Omega}$; namely,

$$\left\|\mathbf{g}\left(\mathbf{u}'\right)-\mathbf{g}\left(\mathbf{u}''\right)\right\|_2\le L\left\|\mathbf{u}'-\mathbf{u}''\right\|,\quad \forall \mathbf{u}',\mathbf{u}''\in\overline{\Omega} \tag{B.1}$$

Proof. Following the approach presented by Meng, Liu, and Wang [28], it is easily to conclude that $\mathbf{v}(\mathbf{u})$ is continuously differentiable in a neighbourhood of $\mathbf{u}_0$ according to the implicit function theorem [41].

Since $\mathbf{v}(\mathbf{u})$ is continuously differentiable on $\overline{\Omega}$, its Jacobian matrix $\nabla_{\mathbf{u}}\mathbf{v}(\mathbf{u})$ is thus continuous on $\overline{\Omega}$. The two-norm of $\nabla_{\mathbf{u}}\mathbf{v}(\mathbf{u})$ is therefore bounded from the nonempty closed and convex set $\overline{\Omega}$; namely, there is a positive constant such that

$$\left\|\nabla_{\mathbf{u}}\mathbf{v}\left(\mathbf{u}\right)\right\|_2\le L_1,\quad \forall \mathbf{u}\in\overline{\Omega} \tag{B.2}$$

According to the Theorem 3.2.4 of Ortega and Rheinboldt [41] (mean-value theorem), it can be seen that

$$\left\|\mathbf{v}\left(\mathbf{u}'\right)-\mathbf{v}\left(\mathbf{u}''\right)\right\|_2 \leq L_1\left\|\mathbf{u}'-\mathbf{u}''\right\|_2, \quad \forall \mathbf{u}',\mathbf{u}'' \in \overline{\Omega} \quad (B.3)$$

Similarly, according to **Assumption 1**, $\mathbf{g}(\mathbf{v})$ is a continuous differentiable function w.r.t. $\mathbf{v}$. The Jacobian matrix $\nabla_{\mathbf{v}}\mathbf{g}(\mathbf{v})$ is thus continuous on Ω which is also nonempty closed and convex. Thus, there is a positive constant such that

$$\left\|\nabla_{\mathbf{v}}\mathbf{g}(\mathbf{v})\right\|_2 \leq L_2, \quad \forall \mathbf{v} \in \Omega \quad (B.4)$$

and we have

$$\left\|\mathbf{g}\left(\mathbf{v}'\right)-\mathbf{g}\left(\mathbf{v}''\right)\right\|_2 \leq L_2\left\|\mathbf{v}'-\mathbf{v}''\right\|_2, \quad \forall \mathbf{v}',\mathbf{v}'' \in \Omega \quad (B.5)$$

According to (B.3) and (B.5), we thus have

$$\begin{aligned}\left\|\mathbf{g}\left(\mathbf{u}'\right)-\mathbf{g}\left(\mathbf{u}''\right)\right\|_2 &\leq L_2\left\|\mathbf{v}'-\mathbf{v}''\right\|_2 \\ &\leq L_2 \cdot L_1\left\|\mathbf{u}'-\mathbf{u}''\right\|, \quad (B.6) \\ &\qquad \forall \mathbf{u}',\mathbf{u}'' \in \overline{\Omega}\end{aligned}$$

This completes the proof. □

C. Proof of Proposition 7

$\mathbf{u}^* = (u_a^*, \forall a \in E)$ is a solution of $VI(\mathbf{g}, \mathfrak{R}_+^{|E|})$ if and only if $\mathbf{u}^*$ and $\mathbf{v}(\mathbf{u}^*)$ fulfil the generalized CMSTA conditions (15)-(18).

Proof.

Necessary Condition. Suppose $\mathbf{u}^*$ is a solution of $VI(\mathbf{g}, \mathfrak{R}_+^{|E|})$, and we now show that $\mathbf{u}^*$ fulfil the generalized CMSTA conditions.

We first demonstrate that $u_a^* < \overline{M}, \forall a \in E$. Assume that there is at least on element $b \in E$ with $u_b^* = \overline{M}$. Then, we define a feasible vector $\mathbf{u}' \in \mathfrak{R}_+^{|E|}$.

$$\mathbf{u}' = \left(u_a' = u_a^*, \ \forall a \in E \setminus \{b\}, \ u_b' = 0.5\overline{M}\right)^T. \quad (C.1)$$

Substituting $\mathbf{u}$ in the VI model $VI(\mathbf{g}, \mathfrak{R}_+^{|E|})$ with vector $\mathbf{u}'$, it follows that

$$\left(-g_b\left(v_b\left(\mathbf{u}^*\right)\right)\right) \cdot \left(0.5\overline{M}-\overline{M}\right) \geq 0; \quad (C.2)$$

namely,

$$g_b\left(v_b\left(\mathbf{u}^*\right)\right) \geq 0 \quad (C.3)$$

Now, we proceed to demonstrate $u_a^* < \overline{M}, \forall a \in E$ by using an apagogical approach. Note that, according to the monotony increasing of $\mathbf{g}(\mathbf{v})$, there exists a upper bound of link flow $\overline{v}_b$ that, for any flow $v_b < \overline{v}_b$, we have

$$g_b\left(v_b\left(\mathbf{u}^*\right)\right) < 0 \quad (C.4)$$

Then, we proceed to show $v_a(\mathbf{u}^*) < \overline{v}_a, \forall a \in E$. Suppose at least one flow loaded path on the road network uses link b. According to the DUE conditions, for any flow loaded paths, they have identical path travel time. Denote one of these paths as k_o and its $\lambda_{k_0}^{m,od}(\mathbf{u}^*)$ follows that

$$\lambda_{k_0}^{m,od}\left(\mathbf{u}^*\right) = \sum_{a \in E} u_a^* \zeta_{e,k_0}^{m,od} > \overline{M} \quad (C.5)$$

Therefore, its generalized path travel time and the corresponding modal travel time fulfil the condition:

$$c_{k_0}^{m,od}\left(\mathbf{u}^*\right) + \lambda_{k_0}^{m,od}\left(\mathbf{u}^*\right) > \overline{M} \quad (C.6)$$

$$c_m^{od}\left(\mathbf{u}^*\right) + \lambda_m^{od}\left(\mathbf{u}^*\right) \geq c_{k_0}^{m,od}\left(\mathbf{u}^*\right) + \lambda_{k_0}^{m,od}\left(\mathbf{u}^*\right) > \overline{M} \quad (C.7)$$

According to the monotonicity of function (8), we have

$$q_m^{od}\left(c_m^{od}\left(\mathbf{u}^*\right) + \lambda_m^{od}\left(\mathbf{u}^*\right)\right) \leq q_m^{od}\left(\overline{M}\right) < \frac{\overline{v}_b}{|W| \cdot |M|} \quad (C.8)$$

In other words,

$$v_b < \sum_{od \in W} \sum_{m \in M_{od}} q_m^{od} < \overline{v}_b \quad (C.9)$$

$$g_b\left(v_b\left(\mathbf{u}^*\right)\right) < 0 \quad (C.10)$$

However, (C.10) contradicts (C.3), then the hypothesis $u_b^* = \overline{M}$ is incorrect. In other words,

$$u_a^* < \overline{M}, \quad \forall a \in E \quad (C.11)$$

To further prove the necessary conditions, we construct the following two specific vectors for any link $a \in E$:

$$\mathbf{u}' = \left(u_a' = \left(1-\gamma_1\right)u_a^* + \gamma_1\overline{M}, \ \forall a \in E\right)^T \quad (C.12)$$

$$\mathbf{u}'' = \left(u_a'' = \gamma_1 u_a^*, \ \forall a \in E\right)^T \quad (C.13)$$

where $\gamma_1 \in (0,1)$ is a given parameter. Substituting vector $\mathbf{u}$ of VI model $VI(\mathbf{g}, \mathfrak{R}_+^{|E|})$ with these two particular vectors, respectively, yields that

$$g_a(\mathbf{v})\left(\gamma_1\left(\overline{M}-u_a^*\right)\right) \leq 0 \quad (C.14)$$

$$g_a(\mathbf{v})\left(\left(\gamma_1-1\right)u_a^*\right) \leq 0 \quad (C.15)$$

Since $M - u_a^* > 0$, (C.14) implies that

$$g_a(\mathbf{v}) \leq 0, \quad \forall a \in E \quad (C.16)$$

In accordance with the fact that $u_a^* \geq 0$, we thus have

$$\left(-g_a(\mathbf{v})\right) \cdot u_a^* \geq 0, \quad \forall a \in E \quad (C.17)$$

Hence, (C.17) in conjunction with (C.15) implies that

$$\left(-g_a(\mathbf{v})\right) \cdot u_a^* = 0, \quad \forall a \in E \quad (C.18)$$

In other words, $\mathbf{u}^*$ and $\mathbf{v}(\mathbf{u}^*)$ fulfil the generalized CMSTA conditions.

Sufficient Condition. Suppose $\mathbf{u}^*$ fulfil the generalized CMSTA conditions (15)-(18); we have the following conditions:

$$-\mathbf{g}\left(\mathbf{v}\left(\mathbf{u}^*\right)\right) \geq \mathbf{0} \quad (C.19)$$

$$\mathbf{g}\left(\mathbf{v}\left(\mathbf{u}^*\right)\right)^T \cdot \mathbf{u}^* = 0 \quad (C.20)$$

In addition, we have $\mathbf{u} \geq 0$. Therefore, according to (C.19) and (C.20), it follows that

$$\mathbf{g}\left(\mathbf{v}\left(\mathbf{u}^*\right)\right)^T \cdot \mathbf{u} \geq \mathbf{g}\left(\mathbf{v}\left(\mathbf{u}^*\right)\right)^T \cdot \mathbf{u}^* \quad (C.21)$$

$$\mathbf{g}\left(\mathbf{v}\left(\mathbf{u}^*\right)\right)^T \cdot \left(\mathbf{u}-\mathbf{u}^*\right) \geq 0 \quad (C.22)$$

This completes the proof. □

Table 9

ID	T	H	Fft	Capa	Type
1	8	9	3	400	Auto
2	9	8	3	400	Auto
3	9	10	3	400	Auto
4	10	9	3	400	Auto
5	10	16	3	400	Auto
6	16	10	3	400	Auto
7	8	16	3	400	Auto
8	16	8	3	400	Auto
9	1	3	3	800	Auto
10	3	4	4	1200	Auto
11	4	3	4	1200	Auto
12	4	5	5	800	Auto
13	5	4	5	800	Auto
14	5	9	5	800	Auto
15	1	2	8	600	Auto
16	3	12	6	1200	Auto
17	4	11	7	600	Auto
18	5	6	3	800	Auto
19	6	5	3	800	Auto
20	12	11	5	600	Auto
21	2	6	3	800	Auto
22	6	8	6	800	Auto
23	2	1	8	600	Auto
24	13	24	3	600	Auto
25	24	23	4	600	Auto
26	23	14	4	600	Auto
27	14	15	3	1200	Auto
28	15	14	3	800	Auto
29	15	10	6	800	Auto
30	13	12	7	800	Auto
31	24	21	3	600	Auto
32	21	24	3	600	Auto
33	23	22	3	600	Auto
34	22	23	3	600	Auto
35	21	22	4	600	Auto
36	22	15	4	800	Auto
37	21	20	3	600	Auto
38	20	19	5	600	Auto
39	19	17	3	1200	Auto
40	17	16	3	600	Auto
41	17	10	5	400	Auto
42	20	22	3	400	Auto
43	15	19	3	600	Auto
44	7	8	3	400	Auto
45	19	15	3	600	Auto
46	11	10	5	600	Auto
47	14	11	4	400	Auto
48	18	16	3	600	Auto
49	18	7	4	600	Auto
50	20	21	3	600	Auto
51	20	18	7	600	Auto
52	24	13	3	600	Auto
53	9	10	3	1500	Train
54	9	8	3	1500	Train
55	10	9	3	1500	Train
56	10	16	3	1500	Train
57	8	9	3	1500	Train
58	8	16	3	1500	Train
59	16	10	3	1500	Train
60	16	8	3	1500	Train
61	20	30	7	1000	Train
62	30	16	6	1000	Train
63	13	24	2	1000	Train
64	24	23	4	1000	Train
65	23	28	4	1000	Train
66	28	27	4	1000	Train
67	27	10	4	1000	Train
68	2	29	10	1000	Train
69	29	8	2	1000	Train
70	1	26	3	1000	Train
71	26	25	9	1000	Train
72	25	9	7	1000	Train
73	3	26	4	1000	RPR
74	6	29	4	1000	RPR
75	14	28	3	1000	RPR
76	19	30	3	1000	RPR

*T is tail node; H is head node; Fft denotes free flow time; Capa denotes capacity.

Table 10

ID	Origin	Destination	Demand
1	1	8	220
2	1	9	180
3	1	10	200
4	1	16	220
5	2	8	240
6	2	9	190
7	2	10	180
8	2	16	180
9	13	8	160
10	13	9	130
11	13	10	170
12	13	16	130
13	24	8	140
14	24	9	150
15	24	10	140
16	24	16	100
17	23	8	130
18	23	9	150
19	23	10	150
20	23	16	170
21	20	8	180
22	20	9	160
23	20	10	140
24	20	16	170

D. Attributes of Study Network

See Table 9.

E. OD Travel Demand of Study Network

See Table 10.

Conflicts of Interest

The authors declare that they have no conflicts of interest.

References

[1] X. Xu, A. Chen, and L. Cheng, "Reformulating environmentally constrained traffic equilibrium via a smooth gap function," *International Journal of Sustainable Transportation*, vol. 9, no. 6, pp. 419–430, 2015.

[2] S. Kitthamkesorn, A. Chen, X. Xu, and S. Ryu, "Modeling mode and route similarities in network equilibrium problem with go-green modes," *Networks and Spatial Economics*, vol. 16, no. 1, pp. 33–60, 2016.

[3] W. Y. Szeto, X. Jaber, and S. C. Wong, "Road network equilibrium approaches to environmental sustainability," *Transport Reviews*, vol. 32, no. 4, pp. 491–518, 2012.

[4] X. Chen, Z. Liu, I. Kim, and J. Leng, "Optimal Sitting and Sizing of the Remote Park-and-Ride Scheme in the Multimodal Transport Network," in *Proceedings of the Transportation Research Board 97th Annual Meeting*, Washington DC, USA, 2018.

[5] Z. Liu and Q. Meng, "Bus-based park-and-ride system: a stochastic model on multimodal network with congestion pricing schemes," *International Journal of Systems Science*, vol. 45, no. 5, pp. 994–1006, 2014.

[6] C. E. Wallace, K. Courage, D. Reaves, Schoene. G., and G. Euler, TRANSYT-7F user's manual.In.

[7] L. R. Rilett and C. M. Benedek, "Traffic assignment under environmental and equity objectives," *Transportation Research Record*, no. 1443, pp. 92–99, 1994.

[8] C. M. Benedek and L. R. Rilett, "Equitable traffic assignment with environmental cost functions," *Journal of Transportation Engineering*, vol. 124, no. 1, pp. 16–22, 1998.

[9] S. Sugawara and D. Niemeier, "How much can vehicle emissions be reduced? Exploratory analysis of an upper boundary using an emissions-optimized trip assignment," *Transportation Research Record Journal of the Transportation Research Board*, no. 1815, pp. 29–37, 2002.

[10] A. Chen, Z. Zhou, and S. Ryu, "Modeling physical and environmental side constraints in traffic equilibrium problem," *International Journal of Sustainable Transportation*, vol. 5, no. 3, pp. 172–197, 2011.

[11] Z.-C. Li, W. H. K. Lam, S. C. Wong, and A. Sumalee, "Environmentally sustainable toll design for congested road networks with uncertain demand," *International Journal of Sustainable Transportation*, vol. 6, no. 3, pp. 127–155, 2012.

[12] A. Nagurney, P. Ramanujam, and K. K. Dhanda, "A multimodal traffic network equilibrium model with emission pollution permits: complicance vs noncompliance," *Transportation Research Part D: Transport and Environment*, vol. 3, no. 5, pp. 349–374, 1998.

[13] E. Fernandez, J. de Cea, M. Florian, and E. Cabrera, "Network equilibrium models with combined modes," *Transportation Science*, vol. 28, no. 3, pp. 182–192, 1994.

[14] R. García and A. Marín, "Network equilibrium with combined modes: models and solution algorithms," *Transportation Research Part B: Methodological*, vol. 39, no. 3, pp. 223–254, 2005.

[15] Z.-C. Li, W. H. K. Lam, S. C. Wong, D.-L. Zhu, and H.-J. Huang, "Modeling park-and-ride services in a multimodal transport network with elastic demand," *Transportation Research Record: Journal of the Transportation Research Board*, no. 1994, pp. 101–109, 2007.

[16] W. H. Lam, Z. Li, S. C. Wong, and D. Zhu, "Modeling an elastic-demand bimodal transport network with park-and-ride trips," *Tsinghua Science and Technology*, vol. 12, no. 2, pp. 158–166, 2007.

[17] C. Pineda, C. E. Cortés, P. Jara-Moroni, and E. Moreno, "Integrated traffic-transit stochastic equilibrium model with park-and-ride facilities," *Transportation Research Part C: Emerging Technologies*, vol. 71, pp. 86–107, 2016.

[18] Z. Song, Y. He, and L. Zhang, "Integrated planning of park-and-ride facilities and transit service," *Transportation Research Part C: Emerging Technologies*, vol. 74, pp. 182–195, 2017.

[19] S. Kitthamkesorn and A. Chen, "Alternate weibit-based model for assessing green transport systems with combined mode and route travel choices," *Transportation Research Part B: Methodological*, vol. 103, pp. 291–310, 2017.

[20] Z. Liu, X. Chen, Q. Meng, and I. Kim, "Remote park-and-ride network equilibrium model and its applications," *Transportation Research Part B: Methodological*, vol. 117, pp. 37–62, 2018.

[21] S. Wang, Q. Meng, and H. Yang, "Global optimization methods for the discrete network design problem," *Transportation Research Part B: Methodological*, vol. 50, pp. 42–60, 2013.

[22] Z. Liu, S. Wang, and Q. Meng, "Optimal joint distance and time toll for cordon-based congestion pricing," *Transportation Research Part B: Methodological*, vol. 69, pp. 81–97, 2014.

[23] T. Liu, H. Huang, H. Yang, and X. Zhang, "Continuum modeling of park-and-ride services in a linear monocentric city with deterministic mode choice," *Transportation Research Part B: Methodological*, vol. 43, no. 6, pp. 692–707, 2009.

[24] D. Z. W. Wang and B. Du, "Reliability-based modeling of park-and-ride service on linear travel corridor," *Transportation Research Record: Journal of the Transportation Research Board*, vol. 2333, pp. 16–26, 2013.

[25] B. Du and D. Z. W. Wang, "Continuum modeling of park-and-ride services considering travel time reliability and heterogeneous commuters – A linear complementarity system approach," *Transportation Research Part E: Logistics and Transportation Review*, vol. 71, no. 0, pp. 58–81, 2014.

[26] Z. Liu, Q. Meng, and S. Wang, "Variational inequality model for cordon-based congestion pricing under side constrained stochastic user equilibrium conditions," *Transportmetrica A: Transport Science*, vol. 10, no. 8, pp. 693–704, 2014.

[27] T. Larsson and M. Patriksson, "Side constrained traffic equilibrium models— analysis, computation and applications," *Transportation Research Part B: Methodological*, vol. 33, no. 4, pp. 233–264, 1999.

[28] Q. Meng, Z. Liu, and S. Wang, "Asymmetric stochastic user equilibrium problem with elastic demand and link capacity constraints," *Transportmetrica A: Transport Science*, vol. 10, no. 4, pp. 304–326, 2014.

[29] C. F. Daganzo, "Stochastic network equilibrium with multiple vehicle types and asymmetric, indefinite link cost Jacobians," *Transportation Science*, vol. 17, no. 3, pp. 282–300, 1983.

[30] G. E. Cantarella, "A general fixed-point approach to multimode multi-user equilibrium assignment with elastic demand," *Transportation Science*, vol. 31, no. 2, pp. 107–128, 1997.

[31] A. Chen, X. Xu, S. Ryu, and Z. Zhou, "A self-adaptive Armijo stepsize strategy with application to traffic assignment models and algorithms," *Transportmetrica A: Transport Science*, vol. 9, no. 8, pp. 695–712, 2011.

[32] A. Chen, Z. Zhou, and X. Xu, "A self-adaptive gradient projection algorithm for the nonadditive traffic equilibrium problem," *Computers & Operations Research*, vol. 39, no. 2, pp. 127–138, 2012.

[33] Z. Zhou, A. Chen, and S. C. Wong, "Alternative formulations of a combined trip generation, trip distribution, modal split, and trip assignment model," *European Journal of Operational Research*, vol. 198, no. 1, pp. 129–138, 2009.

[34] S. Ryu, A. Chen, and K. Choi, "Solving the combined modal split and traffic assignment problem with two types of transit impedance function," *European Journal of Operational Research*, vol. 257, no. 3, pp. 870–880, 2017.

[35] B. S. He, H. Yang, Q. Meng, and D. R. Han, "Modified Goldstein-Levitin-Polyak projection method for asymmetric strongly monotone variational inequalities," *Journal of Optimization Theory and Applications*, vol. 112, no. 1, pp. 129–143, 2002.

[36] D. Han and W. Sun, "A new modified Goldstein-Levitin-Polyakprojection method for variational inequality problems," *Computers Mathematics with Applications*, vol. 47, no. 12, pp. 1817–1825, 2004.

[37] Y. Yin and S. Lawphongpanich, "Internalizing emission externality on road networks," *Transportation Research Part D: Transport and Environment*, vol. 11, no. 4, pp. 292–301, 2006.

[38] A. Nagurney, Q. Qiang, and L. S. Nagurney, "Environmental impact assessment of transportation networks with degradable links in an era of climate change," *International Journal of Sustainable Transportation*, vol. 4, no. 3, pp. 154–171, 2010.

[39] X. Yang, X. J. Ban, and R. Ma, "Mixed equilibria with common constraints on transportation networks," *Networks and Spatial Economics*, vol. 17, no. 2, pp. 547–579, 2017.

[40] S. Ryu, A. Chen, and K. Choi, "Solving the stochastic multiclass traffic assignment problem with asymmetric interactions, route overlapping, and vehicle restrictions," *Journal of Advanced Transportation*, vol. 50, no. 2, pp. 255–270, 2016.

[41] J. M. Ortega and W. C. Rheinboldt, *Iterative Solution of Nonlinear Equations in Several Variables*, Academic Press, Thailand, 1970.

16

Developing a Collaborative Planning Framework for Sustainable Transportation

Okan Örsan Özener

Department of Industrial Engineering, Ozyegin University, 34794 Istanbul, Turkey

Correspondence should be addressed to Okan Örsan Özener; orsan.ozener@ozyegin.edu.tr

Academic Editor: X. Zhang

Currently, as being the highest petroleum consuming sector in the world, transportation significantly contributes to the total greenhouse gas emissions in the world. Road transportation not only is responsible for approximately 20% of the total emissions of carbon dioxide in the EU and in the US but also has a steadily increasing trend in contributing to global warming. Initiatives undertaken by authorities, such as *Emission cap and trade in the EU,* limit the emissions resulted from the actions of the companies and also give economic incentives to companies to reduce their emissions. However, in logistics systems with multiple entities, it is difficult to assess the responsibilities of the companies both in terms of costs and emissions. In this study, we consider a delivery network with multiple customers served by a single carrier, which executes a delivery plan with the minimum transportation cost, and allocate the resulting costs and the emissions among the customers in a fair manner. We develop allocation mechanisms for both costs and emissions. In order to develop a mechanism that provides further reduction of the emissions, we study a setting where the carrier takes the responsibility of the emissions and reflects the resulting inefficiencies while charging the customers.

1. Introduction and Literature Review

Transportation sector is facing formidable challenges and recently environmental issues lie in the heart of these challenges. Currently, as being the highest petroleum consuming sector in the world, transportation significantly contributes to the total greenhouse gas emissions and hence to the global warming. In the EU, road transportation is responsible for approximately 20% of the total emissions of carbon dioxide. According to European Commission Directorate-General for Energy and Transport [1], in the EU in no other sector has the growth rate of greenhouse gas (GHG) emissions been as high as in transport, which amounted to 23% in 2010 [2] compared to 1990 levels. In the US, transportation is responsible for approximately 28% of the GHG emissions, second to the electricity sector, and compared to 1990 levels it has been increased by 18% [3]. This increasing trend in both the EU and the US is mainly due to the increased demand for travel of individuals and increased freight transportation activities.

On the other hand, road transportation can be considered as the most important mode for passenger and freight transportation. In 2010, total goods transport activities in the EU27 (only intra-EU) have amounted to 3,831 billion tonne kilometers. Road freight transport accounts for 45.8% of the total freight volume in the EU [4]. In the US, road transportation is responsible for hauling 70% of the total freight in 2012. Even though road transportation plays such a crucial role in passenger and freight transportation, it is also the largest contributor to global warming. Over the next 50 years, the GHG emissions of the transportation sector could grow 80% compared to current levels due to increased transportation activities [5].

Recently, there have been several initiatives to reduce the environmental damage caused by transportation activities. According to European Commission Directorate-General for Energy and Transport [1], the environmental suitability of transport activities is a major concern and immediate action should be taken to mitigate the negative effects of transportation on the environment. According to European

Commission Directorate-General for Energy and Transport [1], the EU has recently adopted a climate and energy package that sets a target of reducing GHG emission in the EU by 20% with respect to 1990. The US transportation sector also considers three potential approaches that could reduce GHG emissions as well as fuel consumption: (i) improvements in vehicle technology, (ii) switching to environmentally friendly fuels, and (iii) transportation/travel demand management (TDM), which corresponds to better utilization of the transportation resources [5].

Based on these facts and figures, it is clear that governments and international organizations should play an active role in achieving sustainable transportation. Several governments have already passed legislations to support sustainable transportation including the EU's Transport Protocol of the Alpine Convention [6]. An initiative for reducing GHG emissions is to limit the amount of pollutant emitted by the direct and the indirect actions of a company. This is referred to as "emission cap and trade," which allows controling pollution using economic means. European Union Emission Trading Scheme [7] is an example to these cap and trade initiatives. In this context, companies have certain limits either allocated or sold by the central authority. Companies must limit the emissions resulted by their actions to their thresholds, which are set by the permits they have. If the thresholds are not sufficient, then companies must acquire additional permits from other companies whose emissions are below their limit. The allowance of acquiring additional permits is referred to as "trade," which gives economic incentives to the companies to limit their GHG emissions. However, it is quite difficult to assess the companies' emission responsibilities resulted from their indirect actions.

Having this in mind, we develop mechanisms for the fair assessment of the emission responsibilities of the customers in a logistics system. In this context, a single logistics provider (carrier) may serve several customers along a delivery route in order to minimize the transportation costs. Allocating cost among the customers is required to cover the expenses plus the mark-up of the carrier and allocating emissions is required for emission cap and trade purposes discussed above. The GHG emissions are allocated among the customers in units of grams of CO_2 (gCO_2). Even though the transportation cost and the emissions are correlated as they both depend on the fuel consumption of vehicle, the relationship is not trivial as there are several other factors contributing to each of these values. For instance, the operating cost of a truck is estimated as 49 cents per mile on average; however, only 24.2 cents of this cost is resulted from the fuel consumption and the rest is due to maintenance, repairs, and depreciation [8]. On the other hand, the emissions are largely affected by the payload of the vehicle, as it varies considerably along a delivery route. The cost of emissions depends on the final emission responsibilities of the customers, which determine the total number of permits required by the company as well as the trading cost of a permit on the market. The cost of the permit on the market varies considerably mostly due to the available number of permits. The standard reporting unit is tones of carbon dioxide equivalent (tCO_2-e) and there exist several studies to forecast the price per ton of CO_2 [9]. In this study, we assume the price as $42 per ton of CO_2 even though this value might fluctuate over time. Note that both per mile cost of transportation and per ton cost emission represent just two parameter values and do not affect the solution methodology at all. On the other hand, the allocation of responsibility is important and affects the cost/emission allocation mechanisms. In the first setting we consider the customers are responsible for both the transportation cost and the emissions directly and the carrier minimizes the transportation cost to serve the customers and allocates both the cost of transportation and the total emissions to the customers individually. In the second setting, the emission responsibility is diverted to the carrier and the carrier decides the minimum total cost plan to serve the customers and allocates a single cost figure, which includes the transportation cost and the emission cost, to the customers. Our motivation for considering the second setting is that it might favor social welfare as the total overall emission might be reduced when the carrier's objective includes this as well.

Unfortunately, neither allocating costs nor allocating emissions are a simple task. The main reason is even though the total cost and the total emission can be easily calculated, it is not trivial to determine which customer is responsible for how much of these totals. We illustrate this on a simple example. Consider a delivery route of a carrier (Figure 1) serving three customer locations. Suppose that the carrier uses a truck with capacity Q. Suppose that Customers *B* and *C* are requesting deliveries with weights equal to 0.1Q and Customer *A* is requesting a delivery with a weight equal to 0.8Q. Suppose that the cost of traveling between any two locations except from *A* and *B* is 1 and negligible between *A* and *B*. The optimal delivery route is trivial as all demand can be consolidated in one truck (Figure 1). Note that this route also results in the minimum GHG emission, since the largest bulk delivery spends as less time as possible on the vehicle. If we were to allocate both the costs and the emissions proportional to the customers' distance to the depot, then all customers will have equal allocations in both costs and emissions. It is clear that this allocation is not acceptable since Customer *A* should be allocated a higher emission (due to its load weight) and Customer *C* should be allocated a higher cost value (due to its location). If we were to allocate both the costs and the emissions proportional to their individual solution values, respectively, then Customers *B* and *C* will be allocated the same cost and emission. Again, this is not an acceptable allocation as the cost and the emission contribution of Customer *B* on this delivery route is less than Customer *C* and hence should be allocated less in each category (as Customer *B* has a synergy with Customer *A*). Finally, if we were to reverse the route, then the contributions of Customer *C* seem relatively lower as the load delivered to Customer *C* now spends less time on the vehicle. However, delivery route determination decision is an exogenous decision of the carrier. Hence, the allocation decisions should be independent of the routing decisions. These issues remain even when the carrier takes the responsibility of the emissions and only allocates costs to the customers.

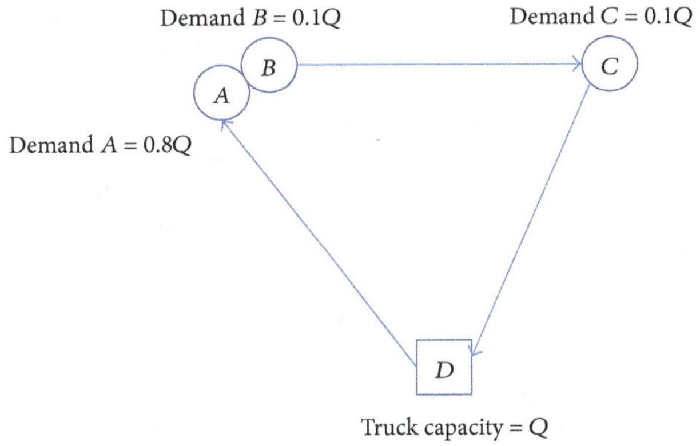

Figure 1: A delivery route.

Before discussing the related works in the literature, we present a list of desirable properties for an allocation method. First, an allocation method should allocate the entire costs/emissions among the participants, which is referred to as "budget balancedness." Second, each individual should be allocated at most its individual solution value, which is called "individual rationality." Third, if we take this property one step further, the same principle should apply to each subset (coalition) of the members, which is referred to as "stability." Fourth, the carrier's routing decisions should not have an effect on the cost/emission allocations. Otherwise, it would not be proper to charge a customer a higher cost/emission value just because the routing decision did not favor that particular customer. Finally, an allocation mechanism should treat each participant "fairly". In our context, a fair allocation method should satisfy budget balanced and stability conditions (referred to as "core allocation" in the literature) and should be independent of the routing decision of the carrier. Faigle et al. [10], Pál and Tardos [11], Engevall et al. [12], and Geomans and Skutella [13] are some examples from the literature which adopt the same fairness definition. In most cases, a core allocation may not exist; hence, our objective is to identify a good approximation to the core.

Allocating the costs/benefits in a logistics system have been studied in the literature under various settings. Tamir [14], Potters et al. [15], and Kuipers [16] consider a traveling salesperson problem (TSP) and discuss the existence of core allocations. Faigle et al. [10] propose a moat-packing based allocation method for TSP to obtain fair allocations. On the other hand, Göthe-Lundgren et al. [17] and Engevall et al. [12] analyze a vehicle routing problem (VRP) setting from a more practical perspective and propose a computational approach to compute nucleolus. Finally, Özener and Ergun [18] consider a vendor managed inventory setting and propose methods to compute cost-to-serve values under this setting.

Recent papers study the carbon emission and transportation relationship from different perspectives. Bektaş and Laporte [19] study the pollution-routing problem, an extension to VRP that takes other factors besides travel distance into account such as greenhouse emissions, fuel, and travel times. Pitera et al. [20] consider a case study on an urban transportation system and propose local search algorithm for the underlying emissions minimization problem. Suzuki [21] proposes a solution approach for a truck-routing problem where the objective is to minimize the fuel consumption and pollutants emission. Xiao et al. [22] study the capacitated VRP with a load-dependent fuel consumption rate and propose a solution approach to this problem using a simulated annealing algorithm with a hybrid exchange rule. Gajanand and Narendran [23] formulate a multiple-route VRP problem to minimize the fuel consumption and show that taking alternative routes may offer reduction in fuel consumption. Xue et al. [24] develop a model of the relationship between traffic flow and vehicle emission under road capacity constraints. Wang et al. [25] focus on the carbon emission rights and propose an auction mechanism which converges to a unique equilibrium.

Finally, Sichwardt [26], Leenders [27], and Naber [28] consider CO_2 allocation among the customers under a VRP setting. Sichwardt [26] analyzes the problem based on the empirical data of a Swedish medium-sized transportation company. The author assumes that the transportation decisions are given based on a combination of the sweep and savings algorithm. After the CO_2 emission calculations on a specific route, the emissions are allocated to the customers based on their distance to depot. Leenders [27] assumes a set of properties as the fairness criteria: (i) individual rationality, (ii) marginality, (iii) efficiency, and (iv) no negative allocation. Based on a comparative analysis, the selected allocation method is the one based on the individual rationality of the shipments. Naber [28] also discusses the vagueness of the term fairness and comes up with a long list of fairness concepts including stability and budget balancedness. The author compares methods such as Shapley Value and nucleolus as well as a proportional-based method. However, the analysis is on a per-route basis and hence the carrier's routing decision may have a huge impact on the final allocation of the emissions. More importantly, as the computations are on an individual route level, the number of customers, which significantly affects the computational complexity of the generic allocation methods, is less than 15. Hence, other than the proportional based method, these approaches are not scalable for real-life sized problems.

Contrary to the studies reviewed above, we propose methods to simultaneously allocate both the transportation costs and the emissions among the customers. We develop an allocation mechanism based on duality and propose an approximation for the Shapley Value. We refer the reader to Owen [29], Kalai and Zemel [30], Pál and Tardos [11], and Geomans and Skutella [13] as examples of the application of the duality-based allocation methods. Our fairness criteria are a more restrictive one compared to the studies discussed above. We computationally show that our methods outperform the proportional methods proposed in those studies. Finally, to the best of our knowledge, this is the first study that considers the setting where the carrier takes over the emission responsibility and allocates a joint cost to the customers.

The remainder of the paper is organized as follows. In Section 2, we provide a formal definition of the problem and list our assumptions. In Section 3, we present solution approaches for the carrier's routing problem under two different settings. In Section 4, we present our proposed solution approaches, dual LP based and Shapley Value, respectively. In Section 5, we computationally demonstrate how our proposed mechanisms perform in comparison with the other allocation schemes. Concluding remarks are provided in Section 6.

2. Problem Definition

In this section, we provide a formal definition of the problem, explain our emission calculation approach in detail, and finally list our assumptions.

We consider a set of customers $N = \{1, \ldots, n\}$ that are served by a common carrier from a single depot location. Including the depot location, the vertex set is $V = \{0, 1, \ldots, n\}$, where 0 represents the depot and edge set is $E = \{(i, j) : i, j \in V\}$. The distance along an edge (i, j) is denoted by d_{ij} and these figures are assumed to satisfy the triangle inequality. Note that the proposed solution approaches are valid under an asymmetric distance setting and can be used with minor modifications. The carrier serves customers via a set of delivery routes and each route is assigned to one vehicle from a finite number, K, of homogenous vehicles with capacity Q. Customer i is to be delivered a shipment with a weight denoted by q_i, $q_i \leq Q$ for all i. Note that a feasible delivery route corresponds to an ordered customer list with total shipment weight less than equal to the truck capacity.

In this context, delivery routes executed by the carrier determine the total transportation costs and the total emissions. It is important to note that the delivery routes determined by the carrier are not necessarily the optimal set of delivery routes to serve the customers. This is the case in many real life delivery systems; nevertheless, the cost to be allocated among the customers corresponds to the total cost of the selected delivery routes. We assume that the carrier is operating a commercial delivery truck with an operating cost of 49 cents per mile or equivalently 30 cents per kilometer [8]. Note that these values are just parameters and do not actually affect neither the proposed methodology nor the comparative results. We denote the transportation cost along an edge (i, j) by c_{ij}. Finally, there are two distinct approaches for calculating CO_2 emissions of transportation: energy-based approach and activity-based approach. Similar to the studies discussed above, we employ an activity-based approach. However, the difference of our approach is that we use a piecewise linear function with several break points rather than a linear function in order to better approximate the CO_2 emission values. The general formula is as follows:

$$CO_2 \text{ emissions} = \text{Transport volume} \times \text{Distance} \times \text{Ave. } CO_2\text{-emission factor per ton-km}. \quad (1)$$

Note that the transport volume is the payload of the truck covering that particular distance. The calculation of the average CO_2-emission factor per ton-km is done assuming a 40–44 ton truck and emission factors presented in Table 1, which is acquired from McKinnon and Piecyk [31].

Based on the data above, the emission value along an edge (i, j) with a payload of t, which is denoted by $g_{ij}(t)$, is calculated by identifying the average CO_2-emission factor per ton-km with respect to the payload along the edge and multiplying this factor by the payload volume and the distance. We assume that the truck itself weighs around 14-15 tons and hence the payload volume varies between 0 and 30 excluding the truck's own weight.

In this paper, we consider two different settings based on the emission responsibility (hence the objective function) of the carrier. Our motivation for analyzing two different settings is to assess the performance of these distinct policies from the perspective of fairness in allocating costs and emissions. The settings are as follows.

(i) First setting: the emission responsibility is on the customers and thus the objective of the carrier is to minimize the total transportation costs.

(ii) Second setting: the emission responsibility is on the carrier and thus the objective of the carrier is to minimize the total cost including the transportation costs and the emission costs.

In the first setting, the carrier does not have an incentive to execute low-emission routes and hence treats the emission caused by the delivery routes as a byproduct of the transportation activities. As a result, after the execution of the routes both the total transportation costs and the total emissions are to be allocated among all the customers.

In the second setting, the carrier wants to execute cost-effective and low-emission routes. As a result, after the execution of the routes the carrier has to allocate a joint cost among the customers in a fair manner. For this setting, we assume a unit cost of \$42 per ton of CO_2 emission and one permit is equivalent to one ton of carbon dioxide. Hence, the emission cost is a step function in terms of tons of emission. However, as the carrier might probably have other activities besides serving these customers and will pay an overall emission cost annually for all its activities, it will be an overestimate to use a step function with intervals of tons. Instead, we assume a linear relationship while estimating the emission cost of the delivery routes of the carrier.

3. Solution Method for the Carrier's Routing Problem

In this section, we present a solution method that determines the best routing of shipments from the perspective of the carrier. Note that, in the first setting, the carrier's objective is to minimize the cost of transportation. Hence, the routing decision of the carrier does not take into account the emission caused by the resulting delivery routes. The underlying routing problem is a well-studied problem in the literature, VRP.

In order to solve this VRP, we first iteratively generate clusters using a modified K-means clustering approach, and then based on these clusters we generate a number of feasible delivery patterns. Next, we compute the transportation cost and CO_2 emission of the generated patterns, and finally we

TABLE 1: Carbon emission factors (gCO_2/ton-km) for 40–44 ton trucks with varying payloads and levels of empty running.

Payload in tonnes	% of truck-kms run empty										
	0%	5%	10%	15%	20%	25%	30%	35%	40%	45%	50%
10	81	84.7	88.8	93.4	98.5	104.4	111.1	118.8	127.8	138.4	151.1
11	74.8	78.2	81.9	86.1	90.8	96.1	102.1	109.1	117.3	127	138.6
12	69.7	72.8	76.2	80	84.3	89.2	94.7	101.1	108.6	117.5	128.1
13	65.4	68.2	71.4	74.9	78.9	83.4	88.5	94.4	101.3	109.5	119.3
14	61.7	64.4	67.3	70.6	74.2	78.4	83.2	88.7	95.1	102.7	111.8
15	58.6	61	63.8	66.8	70.3	74.2	78.6	83.7	89.7	96.8	105.3
16	55.9	58.2	60.7	63.6	66.8	70.5	74.6	79.5	85.1	91.7	99.7
17	53.5	55.7	58.1	60.8	63.8	67.2	71.2	75.7	81	87.2	94.7
18	51.4	53.5	55.8	58.3	61.2	64.4	68.1	72.4	77.4	83.3	90.4
19	49.6	51.5	53.7	56.1	58.8	61.9	65.4	69.5	74.2	79.8	86.5
20	48	49.8	51.9	54.2	56.8	59.7	63	66.9	71.4	76.7	83
21	46.6	48.3	50.3	52.5	54.9	57.7	60.9	64.5	68.8	73.9	80
22	45.3	47	48.8	50.9	53.3	55.9	59	62.5	66.5	71.4	77.2
23	44.2	45.8	47.6	49.6	51.8	54.3	57.2	60.6	64.5	69.1	74.7
24	43.2	44.7	46.4	48.3	50.5	52.9	55.7	58.9	62.7	67.1	72.4
25	42.3	43.8	45.4	47.3	49.3	51.7	54.3	57.4	61	65.2	70.3
26	41.5	42.9	44.5	46.3	48.3	50.5	53.1	56	59.5	63.6	68.5
27	40.8	42.2	43.7	45.4	47.3	49.5	52	54.8	58.1	62.1	66.8
28	40.2	41.5	43	44.6	46.5	48.6	51	53.7	56.9	60.7	65.3
29	39.7	41	42.4	44	45.7	47.8	50.1	52.7	55.8	59.5	63.9

solve a set partition problem (SPP) to select the delivery patterns to serve the customers with the minimum cost possible. As the number of customers along a route is limited due to the truck capacity restrictions, we are able to solve these TSPs using exact methods and commercial solvers. Our motivation for choosing this particular solution method among all possible VRP solution methodologies is that we are able to generate a large set of delivery routes, all of which will be used in our allocation methodology.

We use a variant of the K-means clustering algorithm, which is discussed in [32, 33]. In this approach, we partition the customers into a certain number of clusters. In our problem, this number would be equal to the number of available trucks, which is denoted by K. After initiating the first K clusters randomly, we implement a variant of the set partition problem (VSPP) to assign customers to these cluster bases. VSPP ensures that the total weight of the shipments assigned to a cluster does not exceed the truck capacity. The details of the overall solution algorithm are as Procedure 1. Note that H represents the iteration number of the outer loop and F represents the iteration number in the inner loop in generating the delivery patterns. Before presenting the formulation of PS, we introduce the notation used in the formulation. Let x_p be the binary variable indicating whether a pattern is selected in the final solution or not. Let P_i be the set of delivery patterns that visit customer i. The formulation of PS is as follows:

$$PS: z_\alpha = \min \sum_{p \in P} \alpha_p x_p, \tag{2}$$

$$\text{s.t.} \quad \sum_{p \in P_i} x_p = 1, \quad \forall i \in N, \tag{3}$$

$$\sum_{p \in P} x_p \le K, \tag{4}$$

$$x_p \in \{0,1\}, \quad \forall p \in P. \tag{5}$$

The objective is to minimize the total transportation costs. Constraints (3) ensure that each customer is served by exactly one delivery route. Constraint (4) ensures that the number of delivery routes executed should be less than or equal to the total number of vehicles. The last constraints are the binary restrictions of the variables.

Solving the carrier's routing problem would determine the minimum cost solution of delivering to all the customers. Note that, in applying Procedure 1, we calculate not only the transportation costs but also the CO_2 emissions of the delivery routes. Here, z^*_α corresponds to the optimal objective function value of PS, which may not correspond the optimal solution to the problem as Procedure 1 does not guarantee optimality. We assume that z^*_α denotes the total cost to be allocated among the customers. We also need to compute the total emissions to be allocated among the customers. Let z^*_β denote that value, which is equal to $\sum_{p \in P} \beta_p x^*_p$. Finally, we are able to generate a large set of feasible delivery patterns, not only the ones executed in the final solution. This is very crucial because, for our proposed allocation methods, we require a range of feasible delivery routes, especially

```
Initiate a set of delivery patterns P = {}
For h = 1,...,H
    Initiate the cluster centers by randomly choosing K locations out of all customer locations
    For f = 1,...,F
        Assign customers to cluster centers by solving VSPP
        For k = 1,...,K
            Generate a delivery pattern p for cluster k and add this pattern to set P
            Calculate the centroid for all locations assigned to this cluster center
            Relocate the cluster center to the location closest to center of gravity of the cluster
        End For
    End For
End For
Remove the duplicated patterns in P
For p = 1,...,|P|
    Solve a TSP to calculate the cost and the emission of the route α_p and β_p, respectively
End For
Solve PS to select the final K delivery patterns
```

PROCEDURE 1: Solving the carrier's routing problem.

to determine a fair allocation method independent of the routing solution preferred and executed by the carrier.

In the second setting, the carrier's objective is to minimize the total cost of the delivery routes, which includes the transportation costs and the emission costs. Under this setting, we modify Procedure 1, specifically the step "solving the TSPs to calculate the cost and the emission of the route." Unfortunately, this modification is a major one. In the original procedure, we solve a mathematical model for TSP with regular assignment constraints as well as subtour elimination constraints. However, when the emission costs are also included in the objective function and since the emission costs depend on the payload of the vehicle on a certain link, we need to keep track of the route traveled and hence the payload on the vehicle at any given time. Therefore, we need an entirely different model to handle this situation. The problem resembles the delivery man problem (DMP) studied by Fischetti et al. [34]. In addition to this modification, we calculate the emission values using a piecewise linear approximation with several break points. Hence, we need a further modification of the flow-based model proposed by Fischetti et al. [34]. Unfortunately incorporating the piecewise approximation into this flow formulation makes the model too time consuming to be used for large number of delivery patterns. Therefore, we use a step function for assessing the emission cost in this flow-based model's objective function and calculate the emission cost by using our original piecewise approximation after solving the model. The formulation of our flow-based TSP model, FTSP, is as follows:

$$\text{FTSP}: \gamma_P = \min \sum_{i\in\widetilde{N}} \sum_{j\in\widetilde{N}} c_{ij} r_{ij} + \sum_{i\in\widetilde{N}} \sum_{j\in\widetilde{N}} o_{ij}, \tag{6}$$

$$\text{s.t.} \quad \sum_{i\in\widetilde{N}} r_{ij} = 1, \quad \forall j \in \widetilde{N}, \tag{7}$$

$$\sum_{j\in\widetilde{N}} r_{ij} = 1, \quad \forall i \in \widetilde{N}, \tag{8}$$

$$\sum_{i\in\widetilde{N}\setminus\{0\},} s_{i0} = 1, \tag{9}$$

$$\sum_{i\in\widetilde{N}} s_{ik} - \sum_{j\in\widetilde{N}} s_{kj} = q_k, \quad \forall k \in \widetilde{N}, \tag{10}$$

$$\sum_{i\in\widetilde{N}} s_{i0} - \sum_{j\in\widetilde{N}} s_{0j} = -\sum_{k\in\widetilde{N}} q_k, \tag{11}$$

$$s_{ij} \le M r_{ij}, \quad \forall i, j \in \widetilde{N}, \tag{12}$$

$$\sum_{l=0}^{30} t_{ijl} = 1, \quad \forall i, j \in \widetilde{N}, \tag{13}$$

$$\sum_{l=0}^{30} t_{ijl} (l-1) \le \frac{s_{ij}}{1000}, \quad \forall i, j \in \widetilde{N}, \tag{14}$$

$$\sum_{l=0}^{30} t_{ijl} (l) \ge \frac{s_{ij}}{1000}, \quad \forall i, j \in \widetilde{N}, \tag{15}$$

$$42 \left[\sum_{l=1}^{30} t_{ijl} g_{ij} (l - 0.5) \right] = o_{ij}, \quad \forall i, j \in \widetilde{N}, \tag{16}$$

$$r_{ij} \in \{0, 1\}, \quad \forall i, j \in \widetilde{N}, \tag{17}$$

$$s_{ij} \in \mathbb{Z}^+, \quad \forall i, j \in \widetilde{N}, \tag{18}$$

$$o_{ij} \ge 0, \quad \forall i, j \in \widetilde{N}, \tag{19}$$

$$t_{ijl} \in \{0, 1\}, \quad \forall i, j \in \widetilde{N}, \ \forall l \in \{0, 1, \dots, 30\}. \tag{20}$$

In the formulation above $\widetilde{N}$ represents the set of customers to be visited by the given TSP pattern p plus the carrier's depot location. The binary variable r_{ij} denotes whether edge (i, j) is selected in the final TSP tour or not. The integer variable s_{ij} represents the flow volume (weight) along edge (i, j). Note that as q_k values are in kilograms, the flow values along the edges are represented in integer multiples of kilograms and also the weight of the truck is not included in this flow volume. The variable o_{ij} represents the cost of emission along the edge (i, j). Finally, the binary variable t_{ijl} denotes whether on edge (i, j) the flow volume falls within an interval between $l - 1$ and l or not.

The objective function is to minimize the total transportation costs and the emission costs. Constraints (7) and (8) ensure that each customer is visited only once by the delivery pattern. Constraints (9), (10), and (11) ensure that the flow conservation within the network is achieved while satisfying the volume delivery requirements of the customers. In order to represent a positive flow on the vehicle along the selected edges of the pattern, we increase the outgoing flow value from the depot by one kilogram, which is negligible in emission calculations. Constraints (10) also guarantee that there will be no subtours in a feasible solution. Suppose that there is a subtour that does not include the depot (otherwise there is only one grand tour, and hence no subtours). For all the nodes in that particular subtour, we sum up the corresponding constraints of type (10) and this would yield a contradiction. Hence, any feasible solution to FTSP does not have subtours. Constraints (12) guarantee that if an edge has a positive flow then that edge has to be covered by the pattern. Constraints (13), (14), (15), and (16) make sure that the flow on an edge falls within a certain interval between 0 and 30 and the emission cost along an edge is computed according to that specific interval. In Constraints (16), the constant, 42, represents unit cost of emission per ton of CO_2 and $g_{ij}(l - 0.5)$ represents the total amount of emission in tons along edge (i, j) if this edge is covered with a payload of $l - 0.5$ tons. Note that $g_{ij}(l - 0.5)$ values are parameters that can be computed in a preprocessing step. Also, $g_{ij}(l - 0.5)$ is an approximation as even though the real flow value, $(s_{ij} - 1)/1000$, falls within the interval $[l-1, l]$, it may not be exactly equal to $l-0.5$. The last four constraints are the binary, integer, and nonnegativity restrictions of the variables. Recall that, after solving the model above, we calculate the total emission cost using our original piecewise approximation and update the total cost of the TSP pattern accordingly. The rest follows by replacing α_p values with γ_p values and solving a modified *PS* to obtain $z^*_\gamma = \sum_{p \in P} \gamma_p x^*_p$.

4. Cost/Emission Allocation Methods

In this section, we describe our proposed allocation methods both for cost and emission allocation. Before describing each method, note that the cooperative game under a VRP setting may have an empty core, which follows directly from the proof on the existence of the core allocation under a TSP setting, and hence both the cost and the emission allocation problems are not guaranteed to have a budget balanced and stable cost allocation. Having this in mind, our objective is to find the most fair allocation method, which we define as being exactly budget balanced and as stable as possible (also note that stability implies individual rationality). Hence, for all the allocation methods, we calculate the percent instability values for all possible groups of customers and decide which allocation method yields better solutions based on these instability values. Finally, ease of computation is another criterion in choosing the best allocation method as the applicability of the proposed methods directly depends on their computational times and the related works in the literature mostly focus on the theoretical contribution and ignore this aspect of the problem.

4.1. An Allocation Method Based on the Distance to Depot. First, we describe an allocation method, which allocates the costs and the emissions proportional to the customers' distances to depot. It is natural to assume that the contribution of a customer to both transportation cost and CO_2 emission of a deliver route is correlated to the customer's distance to the depot, even though this correlation might be stronger in cost contribution than that of emission contribution as the weight of the shipment is ignored. The obvious advantages of such a cost/emission allocation method are as follows: (i) it is a simple/fast allocation method that requires no complex calculations. (ii) it provides individually rational allocations, and (iii) obtained allocations do not depend on the routing solution of the carrier. The disadvantage of this method is that it completely ignores the synergies among the customers and it is very unlikely to obtain approximate core, hence fair, allocations using this proportional method. A particular disadvantage of this method in allocating CO_2 emission is that this method completely ignores the shipment weights of the customer, which is a very important factor in determining the CO_2 emissions. We compute θ_i, the cost allocated to customer i, and σ_i, the emission allocated to customer i, as Procedure 2.

Note that for our second setting, we replace z^*_α value with z^*_γ value in Procedure 2 to obtain cost allocations.

4.2. An Allocation Method Based on the Individual Solutions. Second, we describe an allocation method, which allocates the costs and the emissions proportional to the solutions where the customers are served individually by the carrier. It is clear that the cost allocations will be exactly the same as the proportional allocation based on the distance to the depot. However, the emission allocation will be different as the weight of the shipments becomes a factor in this allocation method. Hence, we expect this allocation method to perform similar to the one discussed above except that the emission allocations might be better in terms of fairness criteria. The details of the procedure are as Procedure 3.

Again, for our second setting, we replace z^*_α value with z^*_γ value in Procedure 3 to obtain cost allocations.

4.3. An Allocation Method Based on Duality. Third, we introduce our first proposed method for allocating the costs and the emissions, which is based on duality. The relationship between core allocation and LP duality is a well-established one since Owen [29]. In this allocation method, we use the

For $i = 1, \ldots, N$

Compute $\theta_i = \frac{d_{i0}}{\sum_{j=1}^{N} d_{j0}} z_\alpha^*$

Compute $\sigma_i = \frac{d_{i0}}{\sum_{j=1}^{N} d_{j0}} z_\beta^*$

End For

PROCEDURE 2: Allocation method based on the distance to depot.

For $i = 1, \ldots, N$

Compute $\theta_i = \frac{2(d_{i0})}{\sum_{j=1}^{N} 2(d_{j0})} z_\alpha^*$

Compute $\sigma_i = \frac{g_{0i}(q_i) + g_{i0}(0)}{\sum_{j=1}^{N}(g_{0j}(q_j) + g_{j0}(0))} z_\beta^*$

End For

PROCEDURE 3: Allocation method based on the individual solutions.

dual of the LP relaxation of PS to obtain cost allocations. Note that, from the optimal values of the corresponding dual variables, we obtain weights and by multiplying with z_α^* and z_β^* we obtain the actual allocations.

There are several advantages of the duality-based allocation method: (i) it is a simple/fast allocation method that only requires an LP solution, (ii) obtained allocations do not depend on the routing solution of the carrier, (iii) and it assesses the synergies among the customers better than other allocation methods discussed and hence is more likely to yield fair allocations.

The LP relaxation of PS is as follows:

$$PS_{\mathrm{LP}}: \bar{z}_\alpha = \min \sum_{p \in P} \alpha_p x_p, \quad (21)$$

$$\text{s.t.} \quad \sum_{p \in P_i} x_p = 1, \quad \forall i \in N, \quad (22)$$

$$\sum_{p \in P} x_p \le K, \quad (23)$$

$$x_p \le 1, \quad \forall p \in P, \quad (24)$$

$$x_p \ge 0, \quad \forall p \in P. \quad (25)$$

Let y_i be the dual variables associated with constraints (22), let v be the dual variable associated with constraint (23), let u_p be the dual variables associated with constraints (24), and finally let I_p be the set of customers visited by pattern p, and then the dual of the PS_{LP} is as follows:

$$\mathrm{DPS}_{\mathrm{LP}}: \hat{z}_\alpha = \max \sum_{i \in I} y_i + \sum_{p \in P} u_p + vK, \quad (26)$$

$$\text{s.t.} \quad \sum_{i \in I_p} y_i + u_p + v \le \alpha_p, \quad \forall p \in P, \quad (27)$$

$$y_i \; u.r.s, \quad \forall i \in I, \quad (28)$$

$$v \le 0, \quad (29)$$

$$u_p \le 0, \quad \forall p \in P. \quad (30)$$

Solving the dual LP above in polynomial time yields the optimal values for the dual variables y_i, which are the variables to be used in determining the cost allocations. Note that we need to solve another, modified dual LP in order to obtain dual variables for allocating the emissions. In that case, we just replace α_p values with β_p values to obtain $\tilde{y}_i^*$. The details of the procedure are as Procedure 4.

To obtain duality-based cost allocations for our second setting, we replace z_α^* value with z_γ^* value and α_p values with γ_p values.

4.4. An Approximation for the Shapley Value. Finally, we introduce an allocation method based on an approximation for the Shapley Value. The Shapley Value is a generic allocation method introduced by Shapley [35]. In a nutshell, Shapley Value is the weighted average of the marginal contribution of each member to each subset of the collaboration. Let $m^i(S)$ be the marginal cost of adding customer i to the subset S and let $l^i(S)$ be the marginal emission of adding customer i to the subset S, and then the Shapley Value calculations are as follows:

$$\theta_i = \sum_{S \subseteq N \setminus i} \frac{|S|!\,|N \setminus (S \cup i)|!}{|N|!} m^i(S), \quad (31)$$

$$\sigma_i = \sum_{S \subseteq N \setminus i} \frac{|S|!\,|N \setminus (S \cup i)|!}{|N|!} l^i(S). \quad (32)$$

The advantages of the Shapley Value allocation method are as follows: (i) obtained allocations do not depend on the

Solve PS_{LP} to obtain $y_i^* \quad \forall i$
Solve $\widetilde{PS}_{LP}$ to obtain $\widetilde{y}_i^* \quad \forall i$
For $i = 1, \dots, N$

$$\text{Compute } \theta_i = \frac{y_i^*}{\sum_{j=1}^{N} y_i^*} z_\alpha^*$$

$$\text{Compute } \sigma_i = \frac{\widetilde{y}_i^*}{\sum_{j=1}^{N} \widetilde{y}_i^*} z_\beta^*$$

End For

PROCEDURE 4: Allocation method based on duality.

For $i = 1, \dots, N$
Let N_i be the set of the customers including the nearest 5 customers and customer i

$$\text{Compute } \theta_i = \sum_{S \subseteq N_i \setminus i} \frac{|S|!|N_i \setminus (S \cup i)|!}{|N_i|!} m^i(S)$$

$$\text{Compute } \sigma_i = \sum_{S \subseteq N_i \setminus i} \frac{|S|!|N_i \setminus (S \cup i)|!}{|N_i|!} l^i(S)$$

End For

PROCEDURE 5: Approximation for the Shapley Value.

routing solution of the carrier, (ii) it assesses the synergies among the customers, and (iii) it satisfies additional properties such as equal treatment of equals, additivity, and dummy properties. A major disadvantage of the Shapley Value is that it requires an exponential effort to compute the allocation unless an implicit method is developed (as the number of subsets grows exponentially). Hence, in its current form Shapley Value is not a suitable method for the large instances of the problem. Therefore, we propose an approximation for the Shapley Value, in which we calculate the weighted average of the marginal contribution of each customer to the subsets that include only the nearest 5 customers. By doing that, we limit the number of subsets and hence limit the computational effort that is required to calculate the Shapley Value. The details of the procedure are as Procedure 5.

To obtain the Shapley Value cost allocations for our second setting, we update $m^i(S)$ values and repeat Procedure 5.

5. Computational Study

We carried out a computational study to demonstrate the performance of our proposed allocation methods. We perform our experiments on randomly generated instances with varying characteristics. These instances are all generated on a region of $1{,}000 \times 1{,}000$ unit square and the Euclidian distances between two locations are divided by 10 to obtain the distances in kilometers. Note that, for calculating the transportation cost, we multiply the distances in kilometers by the operating cost per kilometer multiplier, which is taken as \$0.30.

We generate a total of 20 instances using different parameter configurations. Half of these instances have 25 customers and the rest has 50 customers. We generate a certain portion of the customer locations in clustered regions, representing metropolitan areas. There are 3 clusters in 25-customer instances and 4 clusters in 50-customer instances. Among all the customers, 40% of them are generated within the cluster regions and the remaining ones are randomly placed over the entire map. The load-carrying capacity of the trucks is assumed to be equal to 30 tons. In half of the instances, low volume instances, the shipment weight is uniformly distributed between 5% and 25% of the truck capacity. In the rest of the instances, high volume instances, the shipment weight is uniformly distributed between 10% and 30% of the truck capacity. Low volume and high volume instances are equally distributed among 25- and 50-customer instances.

A major decision in our computational analysis is the number of iterations in our iterative VRP solution procedure. Recall that H represents the iteration number of the outer loop and F represents the iteration number in the inner loop in generating the delivery patterns. Our objective is to generate as many patterns as possible; however, the computational time of the procedure is also a concern. In our computations, we perform two different sets of runs with $H = 50$ and $H = 100$, whereas F is always set to 20. In solving PS, we set the time limit to 3600 seconds (1 hour) and the optimality gap to 1% for the termination of the run. Finally, all the computational experiments are carried out on a 64-bit Windows Server with two 2.4 Ghz Intel Xeon CPU's and 24 GB RAM. The algorithms are implemented using C++ and CPLEX Concert Technology.

TABLE 2: Instability values of the proposed methods on 25- and 50-customer instances when $H = 50$.

		25-customer				50-customer			
		PR	IS	DU	SV	PR	IS	DU	SV
Cost	Subset	608.9	608.9	178	286.1	1063.1	1063.1	169.2	636.5
	Ave	7.26	7.26	3.59	5.64	10.14	10.14	3.48	6.36
	Max	39.83	39.83	17.43	27.56	51.34	51.34	18.61	30.2
CO_2	Subset	883.5	894.4	304.2	272.7	1020.9	1036.5	279	433.4
	Ave	9.43	9.3	5.27	6.63	11.74	11.89	4.38	6.16
	Max	52.02	51.36	25.81	40.03	57.64	58.83	30.39	32.81

We test the performance of the proposed allocation methods with respect to both solution quality and computational time. Recall that the quality of an allocation method is assessed using our fairness criteria, budget balanced, and stability. As we cannot allow any budget deficiency or surplus, all the allocation methods have to satisfy the budget balanced property. Hence, we need to compute the maximum percent deviation from stability of each allocation method over all the subsets of the customers. Then, we call this maximum percentage deviation as "percent instability" of allocation method and assume that the higher this value is, the less fair the allocation is. Unfortunately, the problem comes with the fact that there are 2^N subsets and this number is huge when $N = 25$ or $N = 50$. Therefore, we use the "approximate stability assessment" procedure developed by Özener et al. [18] to assess the stability of the allocation methods. The purpose of this method is to generate only a fraction of all the possible subsets and, however, at the same time make sure that these subsets are the important ones. The definition of the "importance of a subset" is having the highest possibility of being instable as well as having a high percentage instability value. In that procedure, the inherent assumption is that the customers that are closer to each other have the highest synergy potential and hence constitute an important subset. In their procedure, they pick a random point on the map and assign probabilities to the customers based on their distance to that base point. Then, they generate a subset based on these probabilities and check the percent instability of that subset. We modify this procedure for our application. In our modified procedure, we discard the generated subsets if the customers cannot be served by a single delivery route (total shipment volume exceeds the truck capacity) or their total shipment volumes are less than 75% of the truck capacity. In that sense, we assume that the highest synergy customers subsets are the ones that correspond to the efficient delivery routes. We generate 25,000 subsets for each of the 25-customer instances and to 50,000 for each of the 50-customer instances. Note that the procedure does not allow duplicate subsets.

Table 2 summarizes the performance of the allocation methods when $H = 50$. The first four columns correspond to 25-customer instances and the next four columns correspond to 50-customer instances. Within each four-column group, each column represents an allocation method: proportional allocation method PR, individual solutions based allocation method IS, duality based method DU, and Shapley Value SV, respectively. The first three rows summarize the performance of the methods on allocating costs and the next three rows summarize the performance of the methods on allocating CO_2 emissions. The first row, "Subset," presents the average number of instable subsets (over all the generated subsets) with each cost allocation method. The row "Ave" presents the average of the average percent instability of each cost allocation method over all the generated subsets whereas the next row "Max" presents the average of the maximum percent instability of cost allocation methods over all the generated subsets. The next three rows present the same performance measures for the CO_2 emission allocations.

The computational experiments reveal that the duality-based method provides significantly better allocations especially compared to the proportional and individual solution based allocation methods. The average instability of duality-based cost allocation method over all instances is equal to 3.59% on 25-customer instances and 3.48% on 50-customer instances. The same values for proportional, individual solution based and Shapley Value allocation methods are 7.26%, 7.26%, and 5.64%, respectively, for 25-customer instances and 10.14%, 10.14%, and 6.36%, respectively, for 50-customer instances. Hence, we conclude that the superior performance of the duality-based method is even more significant in larger instances. When we compare the number of instable subsets, again duality-based method outperforms the other methods. When the CO_2 emission allocations are considered, we observe the same pattern, the superiority of the duality-based method, except that proportional and individual solution based allocation methods do not yield the same values.

Table 3 presents the same results except that the classification of the 20 instances is now based on the load volumes. Hence, the first four columns present the results for the low volume instances and the next four columns present the results for the high volume instances. Even though the ranking of the allocation methods does not change with respect to the instance characteristics based on the volume of the loads, we observe that the overall performances of all the allocation methods are slightly worse in low volume instances. The reason for such a difference in performance is due to the fact that, in the low volume instances, the synergies among customers are higher, which lead to more challenging instances in terms of stability.

Tables 4 and 5 present the results when $H = 100$, whereas in the former table the classification is based on the number of customers; in the latter one, the classification

TABLE 3: Instability values of the proposed methods on low and high volume instances when $H = 50$.

		Low volume				High volume			
		PR	IS	DU	SV	PR	IS	DU	SV
Cost	Subset	863.9	863.9	218	309.2	808.1	808.1	129.2	613.4
	Ave	9.29	9.29	4.04	5.75	8.1	8.1	3.03	6.25
	Max	46.72	46.72	21.52	24.32	44.45	44.45	14.52	33.44
CO_2	Subset	1033.5	1045.8	321.1	307.5	870.9	885.1	262.1	398.6
	Ave	11.32	11.35	5.35	6.42	9.85	9.84	4.3	6.37
	Max	57.16	57.44	33.21	32.49	52.5	52.75	22.99	40.35

TABLE 4: Instability values of the proposed methods on 25- and 50-customer instances when $H = 100$.

		25-customer				50-customer			
		PR	IS	DU	SV	PR	IS	DU	SV
Cost	Subset	598.9	598.9	98	282.6	1044.3	1044.3	127.5	584.7
	Ave	7.3	7.3	2.92	5.48	10	10	2.93	6.08
	Max	39.7	39.7	13.65	27.26	48.33	48.33	14.87	30.74
CO_2	Subset	914.8	925.2	270.8	280.8	1103.2	1125.5	281.3	510.7
	Ave	9.61	9.52	4.94	6.68	11.85	11.92	4.05	6.16
	Max	54.4	53.68	26.37	40.89	56.65	57.41	20.19	32.32

TABLE 5: Instability values of the proposed methods on low and high volume instances when $H = 100$.

		Low volume				High volume			
		PR	IS	DU	SV	PR	IS	DU	SV
Cost	Subset	867.4	867.4	134.6	312	775.8	775.8	90.9	555.3
	Ave	9.22	9.22	3.36	5.45	8.08	8.08	2.48	6.1
	Max	44.05	44.05	16.11	23.95	43.97	43.97	12.4	34.04
CO_2	Subset	1059.7	1077.4	259.6	304.8	958.3	973.3	292.5	486.7
	Ave	11.38	11.36	4.55	6.49	10.07	10.07	4.44	6.35
	Max	56.24	56.3	23.26	31.75	54.81	54.8	23.3	41.46

is based on load volumes. From both tables, we observe that the performance of the duality-based method improves whereas the performances of the other allocation methods do not change significantly. Here, one might expect the exact same performance from all the allocation methods except the duality-based method as their procedures are not affected. The differences are due to selecting a different seed in random variable generation in each of these computations.

Table 6 summarizes the computational times of the proposed allocation methods. The proportional method and individual solution based method clearly do not take even a second to compute and hence their computational times are not presented. The first row presents the computational times of the methods on 25-customer instances when $H = 50$, the second row presents the same values for 50-customer instances when $H = 50$, the third row for 25-customer instances when $H = 100$, and finally the last row for 50-customer instances when $H = 100$. We observe that duality-based method does not take longer than a second to compute even on the 50-customer instances when $H = 100$. Even though the Shapley Value method has the highest computational time of all the allocation methods, it is still within the acceptable limits. Finally, the total computational time values also include the stability assessment procedure run times, which are significant as expected. However, we should stress that the assessment computations should not be included in the running time of the allocation methods. We conclude that our proposed methods are not only effective but also computationally efficient.

Under our second setting, we provide additional incentives to decrease the emissions by switching the emission responsibility to the carrier. From Table 7, we observe that the second setting in fact decreases the total emission values. The average percentage improvement is 5.39% and the maximum percentage improvement is 14.60%.

Tables 8 and 9 present the results for the joint cost allocation methods under both classification rules, whereas in the former table $H = 50$ and in the latter $H = 100$. From both tables, we observe that the relative performances of the methods are similar with respect to each other; however, the instability values of all the methods are worse compared to their performance in allocating cost and emissions individually. This is inline with our expectations as in the joint cost setting opportunities for cost saving for the subsets might be higher, which leads to more restrict stability conditions.

TABLE 6: Computational times of the proposed methods in CPU seconds.

	DU	SV	Total
25 C—50 I	0	10.4	2363.2
50 C—50 I	0.2	27.9	6621.8
25 C—100 I	0	9.9	2278.1
50 C—100 I	0.2	28.5	6841.8

TABLE 7: Percentage improvement values in the emission values under the second setting.

Instance number	Improvement
1	8.33
2	0.97
3	6.36
4	7.10
5	8.69
6	2.05
7	3.31
8	4.01
9	5.62
10	5.37
11	0.85
12	8.21
13	2.27
14	3.10
15	0.52
16	1.35
17	9.13
18	14.51
19	14.60
20	1.44
Ave	5.39
Max	14.60

TABLE 8: Instability values of the proposed methods for joint cost allocation when $H = 50$.

		25-customer				50-customer			
		PR	IR	DU	SV	PR	IR	DU	SV
	Subset	983.8	986.2	392.7	511.5	1158.5	1163.7	852.5	608.9
Cost	Ave	8.04	8.04	4.57	5.85	10.32	10.33	5.7	6.05
	Max	45.61	45.63	23.4	32.66	51.6	51.72	32.02	30.92
		Low volume				High volume			
		PR	IR	DU	SV	PR	IR	DU	SV
	Subset	1254.2	1257.2	460.1	551.7	888.1	892.7	785.1	568.7
Cost	Ave	10.09	10.1	4.87	6.14	8.27	8.27	5.4	5.76
	Max	52.75	52.81	27.76	28.98	44.46	44.54	27.65	34.59

Finally, Table 10 summarizes the computational times of the joint cost allocation methods. Compared with the previous performance of the allocation methods, we observe that there is an increase in terms of CPU seconds. This due to the additional complexity of the TSP procedure. Fortunately, the increase is not significant except the stability assessment procedure run time, which is not included in the actual running time of the algorithms.

Based on our detailed computational study, we conclude that the Shapley Value approximation performs better than proportional and individual solution based allocation methods. The duality-based method significantly outperforms all

TABLE 9: Instability values of the proposed methods for joint cost allocation when $H = 100$.

		25-customer				50-customer			
		PR	IR	DU	SV	PR	IR	DU	SV
Cost	Subset	965.8	968.1	241.1	501.9	1106.1	1110.6	457.2	531
	Ave	8.02	8.03	3.78	5.85	10.54	10.57	4.62	5.74
	Max	45.34	45.36	17.27	32.23	46.3	46.43	46.37	29.49
		Low volume				High volume			
		PR	IR	DU	SV	PR	IR	DU	SV
Cost	Subset	1219.4	1222.9	313.8	534.6	852.5	855.8	384.5	498.3
	Ave	10.12	10.13	4.09	6.03	8.45	8.47	4.31	5.56
	Max	49.31	49.37	19.81	28.14	42.34	42.41	43.83	33.57

TABLE 10: Computational times of the proposed methods for joint cost allocation in CPU seconds.

	DU	SV	Total
25 C—50 I	0	13.9	8471.4
50 C—50 I	0.1	40.1	20404.7
25 C—100 I	0	14.1	8523.4
50 C—100 I	0	35.8	21301.9

other methods both in allocating costs and CO_2 emissions with respect to the average and the maximum percent instability as well as the number of instable subsets. Finally, the average instability value of the duality-based allocation method is always below 6% regardless of difference instance characteristics.

6. Concluding Remarks

In this paper, we study the cost and emission allocation problem of a logistics system. Road transportation is considered as the largest contributor to the world's GHG emissions and consequently to global warming. In order to mitigate the negative effects of the transportation activities to the environment, certain initiatives are undertaken, such as emission cap and trade in the EU. These initiatives try to give incentives to companies to reduce their emissions resulted from their direct and indirect actions. However, it is quite difficult to assess the companies emission responsibilities resulted from their indirect actions.

Having this in mind, we develop mechanisms for the fair assessment of the emission responsibilities of customers (recipients) in a logistics system. We consider a delivery network with multiple customers served by a single carrier and develop allocation mechanisms for both costs and emissions. Our assessment criterion is the fairness of the allocations. We also analyze the setting where the carrier takes over the emission responsibility and allocates a joint cost to the customers, which provides additional incentives to decrease the emissions. Based on a detailed computational study, we observe that our proposed methods, a duality-based allocation method and an approximation for the Shapley Value, perform significantly better than the allocation methods discussed in the literature. Our computational analysis also establishes that the proposed methods are computationally efficient and can be implemented for real-life sized problems.

Conflict of Interests

The authors declare that there is no conflict of interests regarding the publication of this paper.

References

[1] European Commission Directorate-General for Energy and Transport, A sustainable future for transport, 2009, http://ec.europa.eu/transport/media/publications/doc/2009_future_of_transport_en.pdf.

[2] European Commission. Road transport: Reducing CO2 emissions from vehicles, 2014, http://ec.europa.eu/clima/policies/transport/vehicles/indexen.htm.

[3] US Environmental Protection Agency, "Sources of Greenhouse Gas Emissions," 2014, http://www.epa.gov/climatechange/ghgemissions/sources/transportation.html.

[4] European Commission, "EU transport in figures," 2012, http://ec.europa.eu/transport/facts-fundings/statistics/doc/2012/pocketbook2012.pdf/.

[5] S. Mui, J. Alson, B. Ellies, and D. Ganss, "A wedge analysis of the U.S. transportation sector," 2014, http://www.epa.gov/otaq/climate/420r07007.pdf.

[6] Council of the European Union, "The Transport Protocol of the Alpine Convention," 2006, http://europa.eu/legislation_summaries/transport/transport_energy_environment/l24468_en.htm.

[7] European Commission, "The EU emissions trading system (EU ETS)," 2014, http://ec.europa.eu/clima/policies/ets/indexen.htm.

[8] G. Barnes and P. Langworthy, "The per-mile costs of operating automobiles and trucks," State and Local Policy Program, University of Minnesota, 2014, http://conservancy.umn.edu/bitstream/11299/909/1/200319.pdf.

[9] R. Wilson, P. B. Luckow, F. Biewald, E. Ackerman, and E. Hausman, "2012 carbon dioxide price forecast," http://www.synapse-energy.com/Downloads/SynapseReport.2012-10.0.2012-CO2-Forecast.A0035.pdf.

[10] U. Faigle, S. P. Fekete, W. Hochstattler, and W. Kern, "On approximately fair cost allocation in euclidean tsp games," *OR Spectrum*, vol. 20, no. 1, pp. 29–37, 1998.

[11] M. Pál and É. Tardos, "Group strategyproof mechanisms via primal-dual algorithms," in *Proceedings of the 44th Annual IEEE*

Symposium on Foundations of Computer Science (FOCS '03), pp. 584–593, October 2003.

[12] S. Engevall, M. Göthe-Lundgren, and P. Värbrand, "The heterogeneous vehicle-routing game," *Transportation Science*, vol. 38, no. 1, pp. 71–85, 2004.

[13] M. X. Geomans and M. Skutella, "Cooperative facility location games," *Journal of Algorithms*, vol. 50, no. 2, pp. 194–214, 2004.

[14] A. Tamir, "On the core of a traveling salesman cost allocation game," *Operations Research Letters*, vol. 8, no. 1, pp. 31–34, 1989.

[15] J. A. M. Potters, I. J. Curiel, and S. H. Tijs, "Traveling salesman games," *Mathematical Programming*, vol. 53, no. 2, pp. 199–211, 1992.

[16] J. Kuipers, "A note on the 5-person traveling salesman game," *Mathematical Methods of Operations Research*, vol. 38, no. 2, pp. 131–139, 1993.

[17] M. Göthe-Lundgren, K. Jörnsten, and P. Värbrand, "On the nucleolus of the basic vehicle routing game," *Mathematical Programming*, vol. 72, no. 1, pp. 83–100, 1996.

[18] O. O. Özener, Ö. Ergun, and M. Savelsbergh, "Allocating cost of service to customers in inventory routing," *Operations Research*, vol. 61, no. 1, pp. 112–125, 2013.

[19] T. Bektaş and G. Laporte, "The pollution-routing problem," *Transportation Research B: Methodological*, vol. 45, no. 8, pp. 1232–1250, 2011.

[20] K. Pitera, F. Sandoval, and A. Goodchild, "Evaluation of emissions reduction in Urban pickup systems," *Transportation Research Record*, vol. 2224, no. 1, pp. 8–16, 2011.

[21] Y. Suzuki, "A new truck-routing approach for reducing fuel consumption and pollutants emission," *Transportation Research D: Transport and Environment*, vol. 16, no. 1, pp. 73–77, 2011.

[22] Y. Xiao, Q. Zhao, I. Kaku, and Y. Xu, "Development of a fuel consumption optimization model for the capacitated vehicle routing problem," *Computers and Operations Research*, vol. 39, no. 7, pp. 1419–1431, 2012.

[23] M. S. Gajanand and T. T. Narendran, "Green route planning to reduce the environmental impact of distribution," *International Journal of Logistics Research and Applications*, vol. 16, no. 5, pp. 410–432, 2013.

[24] H. Xue, S. Jiang, and B. Liang, "A study on the model of traffic flow and vehicle exhaust emission," *Mathematical Problems in Engineering*, vol. 2013, Article ID 736285, 6 pages, 2013.

[25] M. Wang, C. Dang, and S. Wang, "A Pareto optimal auction mechanism for carbon emission rights," *Mathematical Problems in Engineering*, vol. 2014, Article ID 438104, 7 pages, 2014.

[26] A. Sichwardt, *CO_2 allocation in road transportation for Alwex Transport AB [M.S. thesis]*, University of Linnaeus, 2011.

[27] B. P. J. Leenders, *Carbon allocation in travelling salesman problems: an application in retail distribution [M.S. thesis]*, Eindhoven University of Technology, 2012.

[28] S. K. Naber, *Allocating CO_2 emissions to customers on a distribution route [M.S. thesis]*, Erasmus Universiteit Rotterdam, 2012.

[29] G. Owen, "On the core of linear production games," *Mathematical Programming*, vol. 9, no. 3, pp. 358–370, 1975.

[30] E. Kalai and E. Zemel, "Generalized network problems yielding totally balanced games," *Operations Research*, vol. 30, no. 5, pp. 998–1008, 1982.

[31] A. McKinnon and M. Piecyk, "Measuring and managing CO_2 emission," The European Chemical Industry Council, 2011, http://cefic-staging.amaze.com/Documents/Media%20Center/News/McKinnon-Report-Final-230610.pdf.

[32] K. Ganesh and T. T. Narendran, "CLOVES: a cluster-and-search heuristic to solve the vehicle routing problem with delivery and pick-up," *European Journal of Operational Research*, vol. 178, no. 3, pp. 699–717, 2007.

[33] B. Kim, S. Kim, and S. Sahoo, "Waste collection vehicle routing problem with time windows," *Computers and Operations Research*, vol. 33, no. 12, pp. 3624–3642, 2006.

[34] M. Fischetti, G. Laporte, and S. Martello, "The delivery man problem and cumulative matroids," *Operations Research*, vol. 41, no. 6, pp. 1055–1064, 1993.

[35] L. S. Shapley and L. S. Shapley, "A value for n-person games. Contributions to the Theory of Games, II," *Annals of Mathematical Studies*, vol. 28, pp. 307–317, 1953.

17

Mixed Transportation Network Design under a Sustainable Development Perspective

Jin Qin,[1] Ling-lin Ni,[2] and Feng Shi[1]

[1] *School of Traffic and Transportation Engineering, Central South University, Changsha 410075, China*
[2] *Dongfang College, Zhejiang University of Finance and Economics, Hangzhou 310012, China*

Correspondence should be addressed to Jin Qin; csu-qinjin@hotmail.com

Academic Editors: B. Uy and J. S. Zhang

A mixed transportation network design problem considering sustainable development was studied in this paper. Based on the discretization of continuous link-grade decision variables, a bilevel programming model was proposed to describe the problem, in which sustainability factors, including vehicle exhaust emissions, land-use scale, link load, and financial budget, are considered. The objective of the model is to minimize the total amount of resources exploited under the premise of meeting all the construction goals. A heuristic algorithm, which combined the simulated annealing and path-based gradient projection algorithm, was developed to solve the model. The numerical example shows that the transportation network optimized with the method above not only significantly alleviates the congestion on the link, but also reduces vehicle exhaust emissions within the network by up to 41.56%.

1. Introduction

Sustainable development, which is generally defined as "development that meets the needs of the present without compromising the ability of future generations to meet their own needs" [1], has in recent years become a widely appreciated concept with strong political and popular support. There are many components to this notion, such as economy, society, resources, and the environment. In particular, followed with the improvement of urbanization, the conflicts between transportation and the environment are becoming more and more serious. Therefore sustainable transport, which is defined as "satisfying current transport and mobility needs without compromising the ability of future generations to meet these needs" [2], has in recent decades become a very important goal in the field of transportation engineering.

The transportation network is one of the most critical infrastructures in urban structures and plays an essential role in meeting normal city operation. It is regularly improved to cope with the ever-growing demands in travel and problems of congestion. In order to optimize some given performance measures (e.g., total travel time or generalized cost) of the network, a crucial decision in the improvement of the network is the allocation of the limited resources to the capacity expansion of the existing links and/or the addition of new candidate links. Such decision problem is always referred to as a transportation network design problem (TNDP).

Morlok firstly proposed the quantitative TNDP in 1973, and since then it has been widely studied by researchers over the last 40 years. The TNDP is always formulated as a mathematical programming model with traffic equilibrium assignment, which can be categorized into three classes: continuous transportation network design problem (CTNDP), discrete transportation network design problem (DTNDP), and mixed transportation network design problem (MTNDP). CTNDP determines the optimal continuous expansion of capacity to existing links [3–8]. DTNDP selects the optimal link additions from a set of candidate links [9–12]. MNDP is a mixture of CTNDP and DTNDP, which simultaneously incorporates both discrete and continuous decision variables [13, 14]. Obviously, MTNDP is more sensible than others in the description of the practical transportation design problem. Nevertheless, MTNDP deals with discrete variables and continuous variables at the same time, so it is very difficult to solve.

Although there is much literature on transportation network design, we can easily find that their common objective is to minimize related economic costs, and goals related to sustainable development are always not taken into account. Guided by this transportation network design theory, the development of transportation engineering has brought about serious consequences regarding the ecological environment in recent decades. Given this situation it is very important to develop a study on the methods and theories of sustainable transportation network design.

This paper mainly studies MTNDP in the consideration of sustainable development, in which the sustainable factors, such as car exhaust emissions or land used and link load are involved. The problem is described as a bilevel programming model and its solution method is proposed.

The remainder of this paper is organized as follows. Hypothesis and notations are given in Section 2. Section 3 formulates the bilevel programming model for the problem. Section 4 designs a solution method based on the annealing algorithm and path-based gradient projection method. In Section 5, numerical experiments are conducted to test the proposed approach. The final section concludes the paper and briefly discusses future research directions.

2. Notations

The complexity of the MTNDP lies mainly within the setting of link-state decision variables, because a single variable cannot describe the continuous link capacity and discrete link connection state at the same time.

Nonetheless, we know that in reality it is meaningless to subtly improve the link capacity. The improvement of a link always comprises of adding new lanes or significantly increasing its capacity; namely, in practice the link capacity expansion is not continuous but within a limited status. Thus the link status can be divided into different discrete grade variables according to the link capacity, which could discretize the mixed problem and lay a foundation for the mathematical model formulation.

For example, according to the Road Commission of China Association for Engineering Construction Standardization [15] the link states are usually divided into 5 grades: high-speed link, 1st-level link, 2nd-level link, 3rd-level link, and 4th-level link. Each link grade represents different lanes, different unit construction costs, different design speeds, and different capacities. On this occasion it is possible to discretize the mixed problem by using the link grade decision variables.

It can be seen from Table 1 that we can set the link grade decision variable u with 10 integers, from 0–9 (in which 0 indicates that the link is not built) respectively representing the different link grades, meanwhile corresponding to the different link capacities and number of lanes, as shown in Table 1. So the link states in the mixed transportation network can be uniformly described by this link grade decision variable. The numbers of the decision variables need not to be restricted to the range of 0 to 9, and in practice can be set properly according to the needs.

In formulating the optimization model, the following notation is used:

$\mathbf{A_1}$: set of existing links in the network,

$\mathbf{A_2}$: set of candidate links to be added,

$\mathbf{A_3}$: set of links that are determined to be added,

$\mathbf{A}$: set of all links in the network, $\mathbf{A} = \mathbf{A_1} \cup \mathbf{A_3}$,

$\mathbf{R}$: set of original nodes,

$\mathbf{S}$: set of destination nodes,

r: origin node index, $r \in \mathbf{R}$,

s: destination node index, $s \in \mathbf{S}$,

rs: origin-destination (OD) pair index,

a: link index, $a \in A$,

K_{rs}: set of the paths connecting OD pair rs,

u_a: state variable of link $a \in \mathbf{A}$; we assume that there are $n+1$ grade states from 0 to n in the network, so $u_a \in \{0, 1, 2, \ldots, n\}$, in which $u_a = 0$ means link a will not be built,

u_a^0: the initial grade of link $a \in \mathbf{A}$, $u_a^0 \in \{0, 1, 2, \ldots, n\}$,

l_a: length of link $a \in \mathbf{A}$ (in kilometers),

q_{rs}: travel demand between OD pair rs,

x_a: total traffic flow on link $a \in \mathbf{A}$,

c_a: capacity of link $a \in \mathbf{A}$ is determined by the link grade u_a,

v_a: design speed of link $a \in \mathbf{A}$ (kilometers/hour) is determined by the link grade u_a,

t_a^0: free-flow travel time on the link $a \in \mathbf{A}$ (in minutes). And if $u_a = 0$, $t_a^0 = +\infty$,

t_a: travel time function of link $a \in \mathbf{A}$ (in minutes) we use the BPR function as $t_a = t_a^0 \cdot (1 + \alpha(x_a/c_a)^{\beta})$ (α, β are all given positive parameters),

$R_{\max}$: maximum values of allowed link load degree for all links,

$p_a(x_a)$: CO emission function of vehicles on the link $a \in \mathbf{A}$,

$I_a(u_a)$: construction cost function per unit length of link $a \in \mathbf{A}$ when its link grade is u_a,

E: upper limit value of land amount which can be used

$e(u_a, u_a^0)$: land-use scale function of the link $a \in \mathbf{A}$, namely the land-use scale of the link a is a function on the initial state u_a^0 and final state u_a,

$I_a(u_a, u_a^0)$: the construction cost function per unit length of the link $a \in \mathbf{A}$, that is to say, the construction cost of link a is a function of its initial state u_a^0 and final state u_a,

f_k^{rs}: traffic flow on path k connecting OD pair rs,

$\delta_{a,k}^{rs}$: path/link incidence variables, if path k connecting OD pair rs passes link $a \in \mathbf{A}$, then $\delta_{a,k}^{rs} = 1$, otherwise $\delta_{a,k}^{rs} = 0$.

TABLE 1: Decision variables of link grade and its corresponding information in China.

Link status variable u	0	1	2	3	4	5	6	7	8	9
Link grade	No-way	4th-level	3rd-level	2nd-level	2nd-level	1st-level	1st-level	Highway	Highway	Highway
Lane	0	2	2	2	4	4	6	4	6	8
Design speed	0	40	60	80	100	100	100	120	120	120
Capacity	0	1500	3000	4000	6000	15000	20000	30000	45000	60000
Construction cost (tens of thousands of Yuan/km)	0	80	200	500	800	1500	3000	4000	6000	7000

3. Model Formulation

The TNDP consists in seeking a transportation network configuration that minimizes some objectives, subject to the equilibrium constraint.

3.1. Sustainable Development in Transportation Network Design. From the perspective of sustainable transport, we should take the sustainability factors into consideration in the design of the transportation network. Moreover, according to the definition of sustainable transport we could conclude that the sustainability factors in the transportation network should mainly focus on the pollution caused by vehicle exhaust emissions, resources utilization, and link load.

3.1.1. Vehicle Exhaust Emissions. Vehicles continuously generate hazardous exhaust gases when they are running, which is the major cause of air pollution. In order to protect the ecological environment and human health, reducing vehicle exhaust emissions has been an ongoing endeavor of many governmental authorities over the past few decades.

Vehicle exhaust emissions contain a range of pollutants, such as carbon monoxide (CO), nitric oxide (NO), and nitrogen dioxide (NO_2). Because it is almost solely emitted by vehicles, CO is considered as the indicator for the level of atmospheric pollution generated by vehicular traffic in the transportation network [16]. Thus if we can reduce one pollutant from vehicle emissions, other pollutants in the emissions will also be reduced. In this regard we only calculate CO as a control subject in this paper.

Obviously, vehicle exhaust emissions on the link depend on the link traffic flow. Yin and Lawphongpanich [17] proposed the following function to estimate vehicular CO emissions:

$$p_a(x_a) = 0.2038 \cdot t_a \cdot e^{0.7962(l_a/t_a)}, \quad (1)$$

where l_a is the length of link $a \in \mathbf{A}$ (in kilometers) and $t_a(x_a)$ is the travel time for link $a \in \mathbf{A}$ (in minutes), and $p_a(x_a)$ is the vehicular CO emissions on link a (in g/hour).

So the total CO emissions in the network are:

$$\sum_{a\in A} p_a(x_a) x_a = 0.2038 \times \sum_{a\in A} t_a \cdot x_a \cdot e^{0.7962(l_a/t_a)}. \quad (2)$$

From (2) it can be found that the total CO emission is associated with the vehicle travel time and the average travel speed on the link, so the CO emission in the network is difficult to determine. We could regard it as an environmental cost of the transportation network system, which should be minimized in the optimization objectives of the TNDP.

3.1.2. Link Load. The service level of the transportation network perceived by travelers is generally the average travel speed or travel time. Obviously, the higher the average speed of the vehicles, the shorter the travel time, and the greater the perceived level of service.

It is easy to know that the average speed, $\overline{v}_a$, on the link, a, should be

$$\overline{v}_a = \frac{l_a}{t_a} = \frac{l_a}{t_a^0 \cdot \left(1 + \alpha(x_a/c_a)^\beta\right)} = \frac{l_a}{t_a^0} \cdot \frac{1}{1 + \alpha(x_a/c_a)^\beta}. \quad (3)$$

Because l_a, t_a^0, α, and β are all the parameters given, the average travel speed on the link, a, only depends on the value of x_a/c_a, which is usually known as the link load or congestion degree. Thus, to reach a certain service level, as well as to reserve some link capacity for future development, it is necessary to impose restrictions on the maximum link load.

According to the definition, $R_{\max}$ is the maximum load allowed for all links in the transportation network, then the link load constraint can be expressed as

$$\frac{x_a}{c_a} \le R_{\max}. \quad (4)$$

3.1.3. Land Resource Utilization. It is well known that urban land resources are limited and precious, which leads to the fact that they cannot be extravagantly used for traffic. A certain amount of land resource needs to be taken up when upgrading existing links or adding new links. To ensure sustainable development we should use as little land as possible on the premise that the expected goal is achieved. Therefore we should set an upper limit value for the amount of land used in the network design.

Obviously the amount of used land for the improvement of a link is relevant to the initial and final states of the link. According to the definition, the land-use scale function of link a is $e(u_a, u_a^0)$, thus the land-use scale constraint can be expressed as

$$\sum_{a\in A} e\left(u_a, u_a^0\right) \le E, \quad (5)$$

where E is a given upper limit for the amount of land available in the network design.

3.1.4. Construction Funds Utilization. That construction fund is also a limited resource. Therefore, with the premise of achieving the expected construction goal, the total construction cost should be as low as possible. Thus, the total construction cost $\sum_{a\in A} l_a \cdot I(u_a, u_a^0)$ should be minimized in the problem, which is also to be minimized in the model.

3.2. Model Formulation. Similar to a conventional TNDP, the mixed sustainable TNDP can be formulated as a bilevel programming model, in which the upper-level model is the network structure design model and the lower-level model is the user equilibrium assignment model.

Mathematically, the formulation of the following bi-level programming model is

$$\begin{aligned} \min C &= \sum_{a\in A} \left[\theta \cdot p_a(x_a) \cdot x_a + l_a \cdot I_a\left(u_a, u_a^0\right)\right], \\ \text{s.t.}\quad & R_{\min} \le \frac{x_a}{c_a} \le R_{\max}, \\ & \sum_{a\in A} e\left(u_a, u_a^0\right) \le E, \\ & u_a \ge u_a^0, \\ & u_a \in \{0, 1, 2, \ldots, n\}, \end{aligned} \tag{6}$$

where x_a is implicitly defined by the lower-level model:

$$\begin{aligned} \min T &= \sum_{a\in A} \int_0^{x_a} t_a\left(x_a, c_a\left(u_a\right)\right) d_x, \\ \text{s.t.}\quad & \sum_{k\in K_{rs}} f_k^{rs} = q_{rs}, \quad \forall s \in \mathbf{S},\ \forall r \in \mathbf{R}, \\ & x_a = \sum_{r\in R}\sum_{s\in S}\sum_{k\in K_{rs}} f_k^{rs}\delta_{a,k}^{rs}, \quad \forall a \in \mathbf{A}, \\ & f_k^{rs} \ge 0, \quad \forall s \in \mathbf{S},\ \forall r \in \mathbf{R},\ \forall k \in K_{rs}. \end{aligned} \tag{7}$$

In the upper-level model, the optimization objective is to minimize the total investment costs and vehicle emissions of CO at the same time, where parameter θ represents the unit conversion factor. Since the existing links generally will not be closed or degraded in the network improvement, it is possible to set the link grade restraint as

$$u_a \ge u_a^0. \tag{8}$$

4. Method of Solution

4.1. Simulated Annealing Algorithm. As mentioned previously, the MTNDP is NP-hard, and this nature makes heuristics the natural choice for it. In this paper we adopted the simulated annealing (SA) algorithm as the basic method to solve the model.

SA is a probabilistic heuristic for global optimization problems for finding a good approximation to the global optimum of a given objective function in the search space. The strength of SA is illustrated by the breadth of studies

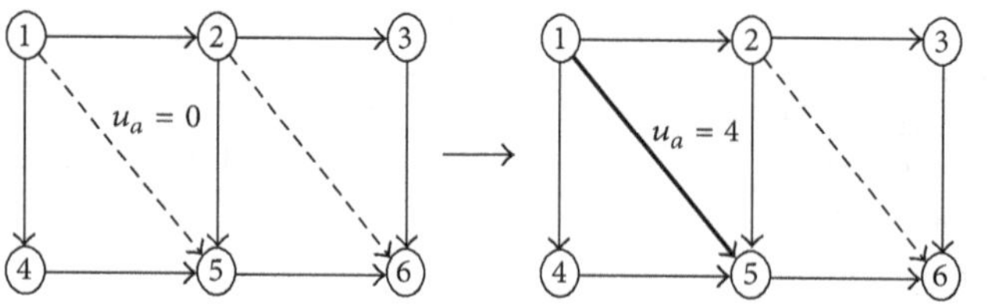

FIGURE 1: Neighboring function based on one link level being transformed.

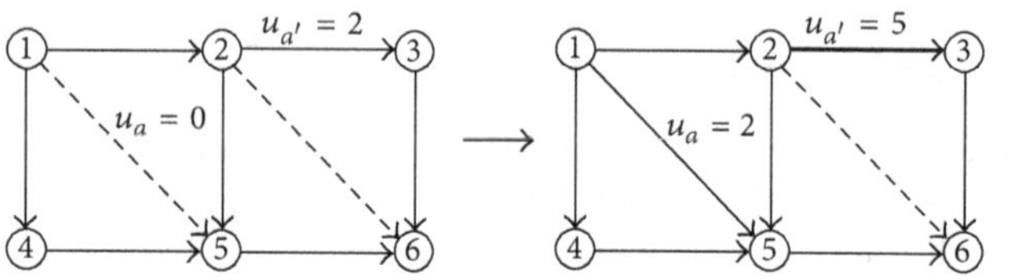

FIGURE 2: Neighboring function based on two link levels being transformed.

found in areas such as logistics network design, computer network design, and machine scheduling, in which SA has been proven as an effective tool for approximating globally optimal solutions to many NP-hard problems.

Although SA studies abound, the area of transportation network design has not been addressed in the literature. It is from this point that the current study embarks.

The neighborhood function plays a crucial role in the performance of SA. In order to improve the efficiency of local search we adopt neighborhood operations by simultaneously adjusting the grades of 1 or 2 links randomly in the neighborhood function. Figures 1 and 2 describe these two neighborhood operations respectively, which randomly select 1 or 2 links from the current network to conduct a link grade transformation operation (the solid lines represent the existing links, while the dashed lines represent the candidate links to be added). Note that the link grades can only be transformed into higher states.

In order to describe the procedure of the SA, $\mathbf{S}, \mathbf{S}', \overline{\mathbf{S}}$ are used to represent the different solutions to the model, $C(\mathbf{S})$ is used to represent the objective function value of the solution $\mathbf{S}$ in the upper-level model. Then the detailed steps of the SA algorithm can be described as follows.

Step 1 (initialization). Set the initial and final temperature values as T_0 and t_f, respectively. The cooling rate $\alpha(0 < \alpha < 1)$ is specified along with the maximum number of iterations N at each temperature value. Take the current network configuration as the initial solution $\mathbf{S}$ and the global optimal solution $\overline{\mathbf{S}}$. Define link set $\overline{\mathbf{A}} = \phi$. The iteration counter $i = 0$.

Step 2 (equilibrium assignment). Conduct equilibrium assignment in the initial network and obtain the equilibrium link flows. Check all link load constraints in the upper-level model. If link a violates the constraint then $\overline{A} = \overline{A} \cup a$, and the penalty cost in the objective function should be imposed, that is $C(\mathbf{S}) = C(\mathbf{S}) + n\sum_{a\in\overline{\mathbf{A}}} l_a I(u_a, 0)$ (where n is the given positive integer). Set the global optimal solution $\overline{\mathbf{S}} = \mathbf{S}$ and $C(\overline{\mathbf{S}}) = C(\mathbf{S})$.

Step 3 (generate a feasible neighboring solution). Generate a neighboring solution $\mathbf{S}'$ by altering the current solution $\mathbf{S}$ according to the given neighborhood function. Then check the feasibility of solution $\mathbf{S}'$. If $\mathbf{S}'$ is infeasible, it is necessary to repeat Step 3 to regenerate the neighboring solution; otherwise calculate $C(\mathbf{S}')$ and proceed to Step 4.

Step 4 (equilibrium assignment in a new network configuration). Conduct equilibrium assignment in the new network configuration (neighboring solution $\mathbf{S}'$). Get the new equilibrium link flows. Reset $\overline{A} = \phi$ and check the link load constraints in the upper model. If link a violates the link load constraints, then $\overline{A} = \overline{A} \cup a$ and $C(\mathbf{S}') = C(\mathbf{S}') + n\sum_{a\in\overline{\mathbf{A}}} l_a I(u_a, 0)$.

Step 5 (evaluate the current solution with the neighboring solution). If $C(\mathbf{S}') \le C(\overline{\mathbf{S}})$, $\overline{\mathbf{S}} = \mathbf{S}$ as well as $C(\overline{\mathbf{S}}) = C(\mathbf{S})$. If $C(\mathbf{S}') \le C(\mathbf{S})$, then set $\mathbf{S} = \mathbf{S}'$, $C(\mathbf{S}) = C(\mathbf{S}')$, proceed to Step 7, otherwise proceed to Step 6.

Step 6 (examine the metropolis condition). The probability at which the relatively inferior neighboring solution should be accepted is $P(\mathbf{S}') = \exp(-(C(\mathbf{S}') - C(\mathbf{S}))/T_i)$, where T_i is the current temperature. A random number ρ is then generated from the interval $(0, 1)$. If $\rho < P(\mathbf{S}')$, then $\mathbf{S} = \mathbf{S}'$ and $C(\mathbf{S}) = C(\mathbf{S}')$.

Step 7 (increment counters). $i = i + 1$. If $i \le N$, return to Step 3, otherwise proceed to Step 8.

Step 8 (adjust the temperature). Adjust the temperature value by the cooling rate. Mathematically this is $T_{i+1} = \alpha \cdot T_i$, where T_i is the temperature used to compute the acceptance probability at iteration i.

Step 9 (convergence check). If $T_{i+1} < t_f$, stop and output the optimal solution $\overline{\mathbf{S}}$ and the set $\overline{\mathbf{A}}$. Otherwise reset $i = 1$, and return to Step 4.

In the above algorithm, the global optimal solution $\overline{\mathbf{S}}$ is saved in Step 5, which is the best solution found so far. And only if the neighborhood solution $\mathbf{S}'$ is better than the optimal solution $\overline{\mathbf{S}}$, do we replace the optimal solution with the neighborhood solution.

It should be noted that if and only if $\overline{\mathbf{A}} = \phi$ after the algorithm terminates, the model has the optimal solution. Otherwise it can be deemed that the initial problem has no feasible solution, so then we should relax the link load constraints and/or land-use scale constraints appropriately and recalculate to obtain the optimal solution.

The method of solution for the lower-level traffic equilibrium model can be divided into two types, which are link-based and path-based. Compared to the link-based algorithm, the path-based algorithm can provide more traffic flow information with a higher computational efficiency [18, 19]. So we adopt the path-based gradient projection (pGP) algorithm to solve the low-level model. The following is a step-by-step description of the procedure for the pGP algorithm [18].

Step 1 (initialization). Generate an initial path for each OD pair.

Step 1.1. Set $x_a(0) = 0$, $t_a = t_a[x_a(0)]$, for all a and $K_{rs}(0) = \phi$.

Step 1.2. Set iteration counter $n = 1$.

Step 1.3. Solve the shortest problem and get $\overline{k}_{rs}(n)$. Update the path set:

$$K_{rs}(n) = \overline{k}_{rs}(n) \cup K_{rs}(n-1), \quad \forall r, s. \tag{9}$$

Step 1.4. Perform "All or No" assignment: $f^{rs}_{\overline{k}_{rs}} = q_{rs}$, for all r, s.

Step 1.5. Calculate the flows on roads according to the flow on the path:

$$x_a(n) = \sum_{r\in R}\sum_{s\in S}\sum_{k\in K_{rs}(n)} f^{rs}_k(n)\,\delta^{rs}_{ka}, \quad \forall a. \tag{10}$$

Step 2 (column generation). Generate the shortest path based on the current link travel times and augment the set of generated path if it is new.

Step 2.1. Increment iteration counter $n = n + 1$.

Step 2.2. Update link travel times: $t_a(n) = t_a(x_a(n-1))$, for all a.

Step 2.3. Solve the shortest problem: $\overline{k}_{rs}(n)$, for all r, s.

Step 2.4. Augment path $\overline{k}_{rs}(n)$ into path set $K_{rs}(n-1)$ if it is not already existed. If $\overline{k}_{rs}(n) \notin K_{rs}(n-1)$, then $K_{rs}(n) = \overline{k}_{rs}(n) \cup K_{rs}(n-1)$, for all r, s. Otherwise, if $\overline{k}_{rs}(n) \in K_{rs}(n-1)$, then tag the shortest path in the path set $K_{rs}(n-1)$ as $\overline{k}_{rs}(n)$ and set $K_{rs}(n) = K_{rs}(n-1)$.

Step 3 (equilibration). Solve the path-formulated traffic assignment problem over the restricted set of paths generated thus far.

Step 3.1. Calculate first derivative path costs $d^{rs}_k(n)$, $d^{rs}_{\overline{k}_{rs}}(n)$ and second derivative path costs $s^{rs}_k(n)$, where

$$s^{rs}_k(n) = \sum_{a\in A} t'_a(n)\left(\delta^{rs}_{ka} - \delta^{rs}_{\overline{k}_{rs}(n)a}\right)^2,$$

$$\forall k \in K_{rs}(n), \ \forall k \neq \overline{k}_{rs}(n), \ \forall r, s,$$

$$d^{rs}_k(n) = \sum_{a\in A} t_a(n)\,\delta^{rs}_{ka}, \quad \forall r, s,$$

$$d^{rs}_{\overline{k}_{rs}}(n) = \sum_{a\in A} t_a(n)\,\delta^{rs}_{\overline{k}_{rs}(n)a}, \quad \forall k \in K_{rs}(n),$$

$$\forall k \neq \overline{k}_{rs}(n),\ \forall r, s. \tag{11}$$

Step 3.2. Update the flows on the nonshortest paths:

$$f^{rs}_k(n+1) = \max\left\{\left[f^{rs}_k(n) - \alpha(n)\left(s^{rs}_k(n)\right)^{-1}\left(d^{rs}_k(n) - d^{rs}_{\overline{k}_{rs}(n)}(n)\right)\right], 0\right\},$$

$$\forall k \in K_{rs}(n), \forall k \neq \overline{k}_{rs}(n), \forall r, s, \tag{12}$$

where $\alpha(n)$ is the stepsize and it could be set as $\alpha(n) = \text{function}(n) = (n+1)^{-1}$.

Step 3.3. If $f^{rs}_k(n+1) = 0$, then drop the path k: $K_{rs}(n) = K_{rs}(n) \setminus k$.

Step 3.4. Update the flows on the shortest paths:

$$f^{rs}_{\overline{k}_{rs}(n)}(n+1) = q_{rs} - \sum_{k\in K_{rs}, k\neq \overline{k}_{rs}(n)} f^{rs}_k(n+1), \quad \forall r, s. \tag{13}$$

Step 3.5. Update link flows:

$$x_a(n+1) = \sum_{r\in R}\sum_{s\in S}\sum_{k\in K_{rs}(n)} f^{rs}_k(n+1)\,\delta^{rs}_{ka}, \quad \forall a. \tag{14}$$

Step 4 (termination). Terminate the algorithm if it satisfies the stopping criterion.

Step 4.1. If $\max_{r,s} \sum_{k\in K_{rs}, k\neq \overline{k}_{rs}(n)} f^{rs}_k(n)(d^{rs}_k(n) - d^{rs}_{\overline{k}_{rs}}(n))(q_{rs} d^{rs}_k(n))^{-1} \leq \varepsilon$, then terminate, output the link flows; otherwise, proceed to Step 2.

5. Numerical Examples

In this section a transportation network, illustrated in Figure 3, is used to show the numerical results of the proposed model and solution method, which consists of 12 nodes and 5 OD pairs: $q_{1,11} = 4900$, $q_{1,12} = 10000$, $q_{2,8} = 8000$, $q_{2,12} = 9000$, and $q_{5,12} = 8000$. The 17 solid lines represent the existing links and the 6 dashed lines represent the candidate links. The length and initial grade of each link are given as the first and second values in parentheses, respectively. The available link grade s and its related information are shown in Table 1.

As mentioned previously, the construction cost of the link is determined by its initial and final link grade. We could regard the link construction cost function $I_a(u_a, u^0_a)$ as

$$I_a\left(u_a, u^0_a\right) = I_a\left(u_a\right) - I_a\left(u^0_a\right). \tag{15}$$

Table 2: Comparison of OD traveling times.

OD	Travel time in initial network (hour)	Travel time in optimized network (hour)	Gap (%)
(1, 11)	11.690	2.513	78.50
(1, 12)	12.664	3.018	76.17
(2, 8)	9.973	1.934	80.61
(2, 12)	11.720	2.354	79.92
(5, 12)	4.677	2.608	44.24

From Table 1 we could conclude that if the link grade u_a is determined, the free-flow speed on the link, as the designed speed v_a, can also be found (kilometers/hour), and thus the free-flow time on link a is

$$t^0_a = \frac{l_a}{v_a}. \tag{16}$$

For the travel time function, $t_a = t^0_a \cdot (1 + \alpha(x_a/c_a)^\beta)$, the parameters are uniformly set as $\alpha = 0.15$, $\beta = 4$, thus the travel time on link a is

$$t_a = t^0_a \cdot \left(1 + \alpha\left(\frac{x_a}{c_a}\right)\right)^\beta = \frac{l_a}{v_a} \cdot \left(1 + 0.15\left(\frac{x_a}{c_a}\right)\right)^4. \tag{17}$$

The land-use amount function can be set specifically as

$$e\left(u_a, u^0_a\right) = 8\left(u_a - u^0_a\right) + 30\left(L\left(u_a\right) - L\left(u^0_a\right)\right), \tag{18}$$

where $L(u_a)$ refers to the number of lanes owned by link a at the link grade of u_a.

For the link load constraint, set $R_{\max} = 0.85$ for all links. The upper limit value of land amount $E = 1000$, the unit matching coefficient is $\theta = 100$.

The heuristic parameters are fixed as follows: cooling rate $\alpha = 0.7$, and the maximum number of iterations at each temperature value $N = 10|R||S|$.

For these mentioned models and solution algorithms, the results can be achieved with the help of Visual C# 2005 programming. A personal computer with Intel DUO Core 2.0 Ghz CPU, 4 G RAM and Windows XP Professional operating system was used for the test. The optimal solution for the problem can be obtained within about 4 seconds.

The travel times for each OD pair in the initial network and optimized network are listed in Table 2. Compared with the initial network, the OD travel times in the optimized network fell by 44.24–80.61%. The reconstruction of the transportation network significantly improves its performance.

Figure 4 shows the information on the grades and loads of various links under the flow equilibrium state before and after optimization, where the information marked above the line is the initial link grade and load, and the information below the line is the corresponding information after optimization. From the figure we can determine that most link loads before optimization exceed the allowable range. The congestion is obvious in the links, especially in links such as (1,5), (2,3),

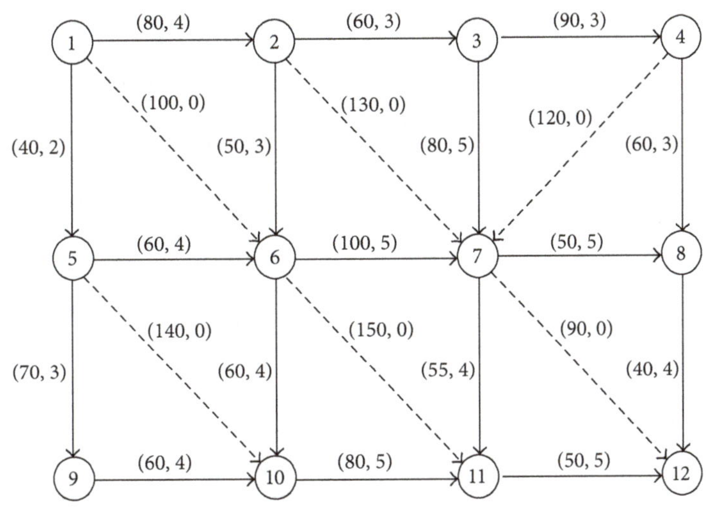

FIGURE 3: 12-node transportation network.

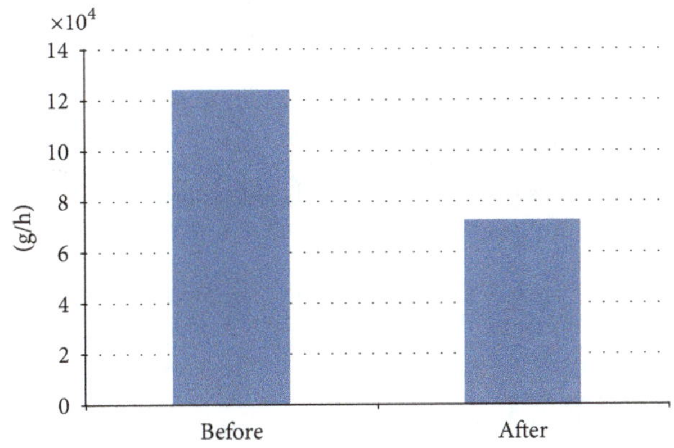

FIGURE 4: CO emission amount in network before and after optimization.

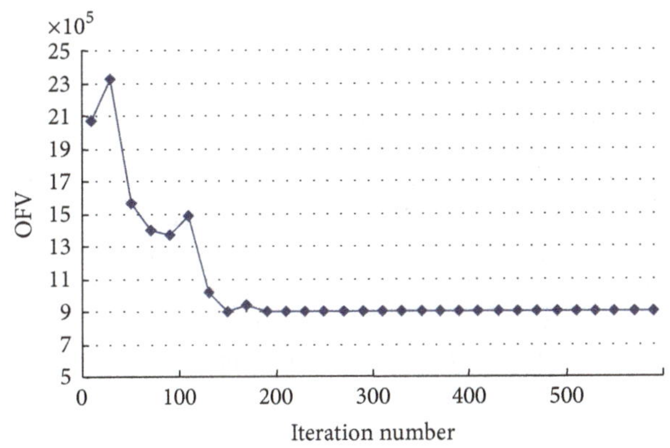

FIGURE 5: Objective value of upper model versus iteration number of SA.

(2,6), (5,9), and (6,10). These links have less flexibility and low elasticity for demand, which is not conducive to sustainable development of the transportation network. However, in the optimized network links (1,5), (2,3), (2,6), (5,6), (6,7), and (6,10) are upgraded, whilst at the same time constructing new links (1,6), (2,7), and (7,12). The loads of links with flows are in the range (0.29, 0.84), which all meet the related constraints. Thus, the new links play a good role in diverting the flows on the congestion links in the network and effectively improve the gridlock of the transportation network.

The total investment of the network improvement project is 192.8 million Yuan, and the land-use amount is 862. The total CO emission in the transportation network drops by 41.56%, reduced from 123,729.00 g/hour before optimization to 72,301.74 g/hour after optimization. It shows a significant improvement in environmental protection in the transportation network.

In addition it can be seen from Figure 5 (OFV is the objective function value of the upper-level model) that that the SA-pGP algorithm determines the optimal solution after about 200 iterations. The algorithm convergence speed and convergence effect are really good, which also indicates that the SA algorithm proposed in this paper is effective. Figure 6 shows that the objective function of the lower model decreases quickly with the number of iterations in the computation process of the GP algorithm (OFV is the objective function value of the lower-level model) and converges to an optimal solution by the 25–30th iteration.

6. Conclusions

This paper studied the sustainable MTNDP. Link-grade decision variables were used to discretize the mixed problem, based on which the bi-level programming model is developed. The upper model is built to minimize the CO emissions of vehicles on the links and also the total investment, in terms of land-use scale and link load constraints, the lower one is the deterministic traffic assignment model. A heuristic algorithm is proposed to solve the model, in which the SA algorithm and the path-based GP algorithm are used to solve the upper and lower models, respectively.

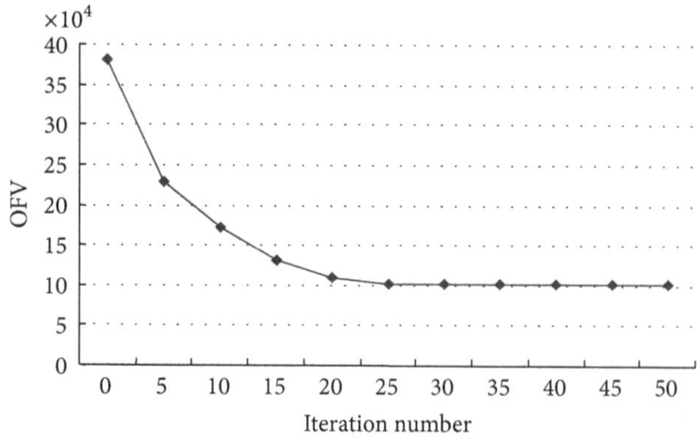

FIGURE 6: Objective value of lower model versus iteration number of GP.

The numerical examples demonstrated that the optimization model and the solution method are effective, in which the SA is not only more efficient and robust with regard to the calculations, but it also rapidly converges to the optimal solution. The path-based GP algorithm calculates quickly and converges steadily. Compared to the initial transportation network, the network optimization significantly improved congestion and pollution in the network. Thereby, the bi-level programming model of the sustainable mixed transportation network design and its solution algorithm can be used to solve real transportation network design problems, and it could provide some scientific evidence for decision making during transportation network planning when considering sustainable development.

Acknowledgment

The work was supported by the National Natural Science Foundation of China (Grant no. 71101155).

References

[1] World Commission on Environment and Development, *Our Common Future*, Oxford University Press, Oxford, UK, 1987.

[2] W. R. Black, "Sustainable transportation: a US perspective," *Journal of Transport Geography*, vol. 4, no. 3, pp. 151–159, 1996.

[3] M. Abdulaal and L. J. LeBlanc, "Continuous equilibrium network design models," *Transportation Research B*, vol. 13, no. 1, pp. 19–32, 1979.

[4] J. X. Ban, H. X. Liu, M. C. Ferris, and B. Ran, "A general MPCC model and its solution algorithm for continuous network design problem," *Mathematical and Computer Modelling*, vol. 43, no. 5-6, pp. 493–505, 2006.

[5] S. W. Chiou, "Bilevel programming for the continuous transport network design problem," *Transportation Research B*, vol. 39, no. 4, pp. 361–383, 2005.

[6] H. Sun, Z. Y. Gao, and J. C. Long, "The robust model of continuous transportation network design problem with demand uncertainty," *Journal of Transportation Systems Engineering and Information Technology*, vol. 11, no. 2, pp. 70–76, 2011.

[7] T. Xu, H. Wei, and Z. D. Wang, "Study on continuous network design problem using simulated annealing and genetic algorithm," *Expert Systems with Applications*, vol. 36, no. 2, pp. 2735–2741, 2009.

[8] G. Zhang and J. Lu, "Genetic algorithm for continuous network design problem," *Journal of Transportation Systems Engineering and Information Technology*, vol. 7, no. 1, pp. 101–105, 2007.

[9] D. E. Boyce and B. N. Janson, "A discrete transportation network design problem with combined trip distribution and assignment," *Transportation Research B*, vol. 14, no. 1-2, pp. 147–154, 1980.

[10] H. Farvaresh and M. M. Sepehri, "A single-level mixed integer linear formulation for a bi-level discrete network design problem," *Transportation Research E*, vol. 47, no. 5, pp. 623–640, 2011.

[11] Z. Gao, J. Wu, and H. Sun, "Solution algorithm for the bi-level discrete network design problem," *Transportation Research B*, vol. 39, no. 6, pp. 479–495, 2005.

[12] H. Poorzahedy and M. A. Turnquist, "Approximate algorithms for the discrete network design problem," *Transportation Research B*, vol. 16, no. 1, pp. 45–55, 1982.

[13] Y. Sun, R. Song, S. W. He, and Q. Chen, "Mixed transportation network design based on immune clone annealing algorithm," *Journal of Transportation Systems Engineering and Information Technology*, vol. 9, no. 3, pp. 103–108, 2009.

[14] P. Luathep, A. Sumalee, W. H. K. Lam, Z. C. Li, and H. K. Lo, "Global optimization method for mixed transportation network design problem: a mixed-integer linear programming approach," *Transportation Research B*, vol. 45, no. 5, pp. 808–827, 2011.

[15] *Road Commission of China Association for Engineering Construction Standardization*, The People Jiaotong Press, Beijing, China, 2004.

[16] A. Alexopoulos, D. Assimacopoulos, and E. Mitsoulis, "Model for traffic emissions estimation," *Atmospheric Environment B*, vol. 27, no. 4, pp. 435–446, 1993.

[17] Y. Yin and S. Lawphongpanich, "Internalizing emission externality on road networks," *Transportation Research D*, vol. 11, no. 4, pp. 292–301, 2006.

[18] A. Chen, D. H. Lee, and R. Jayakrishnan, "Computational study of state-of-the-art path-based traffic assignment algorithms," *Mathematics and Computers in Simulation*, vol. 59, no. 6, pp. 509–518, 2002.

[19] D. H. Lee, Y. Nie, A. Chen, and Y. C. Leow, "Link- and path-based traffic assignment algorithms computational and statistical study," *Transportation Research Record*, vol. 1783, pp. 80–88, 2002.

18

Estimation of Performance Indices for the Planning of Sustainable Transportation Systems

Alexander Paz,[1] Pankaj Maheshwari,[1] Pushkin Kachroo,[2,3] and Sajjad Ahmad[1]

[1] *Department of Civil and Environmental Engineering, University of Nevada, Las Vegas, NV 89154-4015, USA*
[2] *Transportation Research Center, Harry Reid Center for Environmental Studies, University of Nevada, Las Vegas, NV 89154-4007, USA*
[3] *Department of Electrical and Computer Engineering, University of Nevada, Las Vegas, NV 89154-4026, USA*

Correspondence should be addressed to Alexander Paz; apaz@unlv.edu

Academic Editor: Ching-Hung Lee

In the context of sustainable transportation systems, previous studies have either focused only on the transportation system or have not used a methodology that enables the treatment of incomplete, vague, and qualitative information associated with the available data. This study proposes a system of systems (SOS) and a fuzzy logic modeling approach. The SOS includes the Transportation, Activity, and Environment systems. The fuzzy logic modeling approach enables the treatment of the vagueness associated with some of the relevant data. Performance Indices (PIs) are computed for each system using a number of performance measures. The PIs illustrate the aggregated performance of each system as well as the interactions among them. The proposed methodology also enables the estimation of a Composite Sustainability Index to summarize the aggregated performance of the overall SOS. Existing data was used to analyze sustainability in the entire United States. The results showed that the Transportation and Activity systems follow a positive trend, with similar periods of growth and contractions; in contrast, the environmental system follows a reverse pattern. The results are intuitive and are associated with a series of historic events, such as depressions in the economy as well as policy changes and regulations.

1. Introduction

1.1. Background. With the rapid increase in economic development throughout the world, there is stress on the resources used to support global economy, including petroleum, coal, silver, and water. Currently, the world is consuming energy at an unprecedented rate never seen before. Based on data from 2005, about 30.6 billion barrels of petroleum are used annually worldwide [1]. The estimates indicate that the availability of total world reserves is in the vicinity of 1.3 trillion barrels and will be depleted by 2047 [2]. The finite nature of such nonrenewable natural resources as petroleum and coal puts pressure on the environmental system and ultimately reduces the availability of resources for future generations. Hence, it is critical to develop planning and operational strategies that seek to achieve a sustainable use of existing natural resources.

The development of a sustainable system and its corresponding planning strategies requires an adequate definition of sustainability as well as mechanisms to quantify, qualify, and assess sustainability. The quantification of sustainability poses considerable challenges, ranging from data availability to adequate methods to process information. Numerous studies have established different measures to quantify sustainability [3]. According to Bell and Morse [4], sustainability primarily is measured by means of three components: (i) time scale, (ii) spatial scale, and (iii) system quality. The time and spatial scale corresponds to the analysis period and the geographical region of interest, respectively. On the other hand, system quality corresponds to the quantification of the overall system performance or state. In order to quantify system quality, Sustainability Indicators (SIs) have been developed in a diverse range of fields, including biology and the life sciences, hydrology, and transportation. Harger and Meyer [5] argued that SIs should be simple, diverse, sensitive, timely, quantifiable, and accessible. Bossel [6] proposed a system-based approach for developing 21 SIs for environmental characteristics. The approach suggested that a system cannot exist independently, and several external factors can intrude on its boundaries. Some studies argue about the various dimensions associated with sustainability considerations [7, 8].

It is clear that a truly sustainable state for a system requires all the relevant interdependent subsystems/sectors and components, at levels so that the consumption of and the impact on the natural and economic resources do not deplete or destroy those resources. Hence, the assessment of a system state requires a holistic analysis in order to consider all the relevant sectors and impacts. However, existing approaches used to study the sustainability of a transportation system are not comprehensive enough to include key interactions with other systems such as the environment, the economy, and society in general. For example, the current planning of transportation systems is limited in terms of the number, accuracy, length, and approaches used to consider simultaneously important characteristics, including energy consumption, emissions, accidents, congestion, reliability, economic growth, and such social impacts as human health. That is, the existing practices only consider some effects, the estimations are approximate [9], and the analysis period is relatively short, in the order of 30 years [10]. In addition, these effects are synthetized only on the basis of approximated monetary considerations that are unlikely to capture the full extent of the effects, for instance, the financial cost of emissions or greenhouse gases [11, 12]. For example, Zheng et al. [3] described various system indicators by primarily considering economic aspects. Although the study provided valuable insights about the quantification of the economic domain of transportation sustainability, it is primarily focused on the transportation sector.

Among several studies that focused on different sectors, impacts, and aspects of sustainability, the following key characteristics have emerged as fundamental for a sustainable system:

(i) continuity through time [13, 14];

(ii) development of the needs of current generations without compromising the needs of future generations [15];

(iii) utilization of resources without compromising their health and productivity [16];

(iv) development that improves quality of life [17]; and

(v) assimilation of economic, ecological, social, and biophysical components of resource ecosystems [18, 19].

In terms of the methodologies available to estimate SIs, numerous studies have proposed different approaches. For example, Multi-criteria Decision Making (MCDM) and Analytical Hierarchy Process (AHP) techniques have been proposed to consider multiple criteria and estimate relevant SIs [20–26]. The MCDM approach selects or ranks different predetermined alternatives and is based on making discrete decisions [23]. Traditional MCDM techniques assume that the criteria are well-defined, certain (deterministic rather than stochastic), and independent. In reality, the criteria usually involve stochasticity and interdependence. In addition, some aspects in MCDM models are subjective in nature. The weights used in MCDM always include some uncertainty. The basic idea behind the AHP is to convert subjective assessments of relative importance to a set of overall weights or scores. The scale suggested by Saaty [27] is used to compute the weights, using linear algebra. These weights are the elements in the eigenvector associated with the maximum value of the matrix. The eigenvalue-based method has been criticized by researchers on the grounds of lack of prioritization and consistency [28]. In addition, there is an issue of rank reversal possibly arising when a new criteria is added. Due to the above reasons, the theoretical foundation of a rigid scale used in the methods is also questionable [29]. There have been attempts to address some of these limitations. The computation of the weights in MCDM and AHP requires significant amounts of data and a priori or expert knowledge of the system under study. Furthermore, different regions may require different weights to capture local conditions.

Given the complexities, interdependencies, nonlinearities, vagueness, and incomplete information associated with the various factors that are generally involved when considering the sustainability of a system, some studies have adopted concepts from fuzzy set theory for the development of SIs [30–32]. Awasthi et al. [33] applied a fuzzy Technique for Order Preference by Similarity to Ideal Situation approach, to evaluate the sustainability of transportation systems using partial or incomplete information. Opricovic and Tzeng [34] used a fuzzy multi-criteria model to evaluate post-earthquake land use planning. The modeling approach was developed to deal with qualitative or incomplete information. Mendoza and Prabhu [35] applied fuzzy logic for assessing criteria and indicators for sustainable forest management. In addition, linear aggregation techniques were used to combine multiple indicators. Liu [36] tried to integrate MCDM and fuzzy logic techniques to evaluate environmental sustainability. The environmental sustainability of 146 countries was calculated, ranked, and clustered. The study was extensive in dealing with multiple variables and indicators. However, only the environment aspects of sustainability were evaluated without considering any other SIs related to the transportation or activity system. Similarly, Prato [37] discussed a fuzzy logic approach for evaluating ecosystem sustainability. Data needs as well as the lack of technical expertise were important issues in this study. Marks et al. [38] used fuzzy logic techniques to develop a theoretical framework for the evaluation of sustainable agriculture. The study argued about the advantages of fuzzy logic over conventional MCDM techniques. An important characteristic in these studies is their limited scope in terms of the system(s) considered in the analysis.

1.2. Motivation. It is clear that sustainability analysis of transportation systems requires a broad perspective including various systems, such as the economic, and the political, social, and environmental systems. This perspective enables the consideration of such relevant aspects as biodiversity, human health, quality of life, and life expectancy. Such analysis requires significant amounts of data as well as methods to develop adequate SIs. Although not all data that one may want to use is available, there is a vast amount of relevant information that can be obtained from such organizations as The World Bank, the United Nations, the Bureau of Transportation Statistics, and the U. S. Environmental Protection Agency.

Although fuzzy logic has been used in the context of sustainability to handle key characteristics of the relevant data, its use has not been coupled with a broad perspective considering multiple systems. To consider, explicitly, important broad effects and the characteristics of the associated data, this study proposes a system of systems (SOS) [39] and a fuzzy logic modeling approach. The SOS includes the Transportation, Activity, and Environment systems. The fuzzy logic modeling approach enables the treatment of the vagueness associated with some of the relevant data. Performance Indices (PIs) are computed for each system using a number of performance measures. The PIs illustrate the aggregated performance of each system as well as the interactions among them. The proposed methodology also enables the estimation of a Composite Sustainability Index to summarize the aggregated performance of the overall SOS.

The PIs are calculated with an emphasis on transportation systems, while explicitly considering and calculating the PIs for the other two relevant and affected systems. The PIs are calculated based on multiple performance measures with various degrees of resolution and units. These multi-resolution, multi-unit characteristics are intrinsic to the systems under consideration.

The paper is organized as follows. Section 2 describes three interdependent systems: the Transportation, Activity, and Environmental systems. Section 3 summarizes the fuzzy logic methodology used in this study. Section 4 provides information about the study region and data. Results and analysis are presented in Section 5. Some policy perspectives are illustrated in Section 6. Section 7 provides conclusions and recommendations for future work.

2. Interdependent Systems

In the context of sustainability, it is difficult to isolate systems or narrow the analysis to a particular region. Different systems such as Transportation have interdependencies with other systems including the economy and the environment. For example, energy resources, which are part of the environmental system, are required by both the transportation sector and the economy. Hence, any policy or strategy affecting the consumption or production of energy has effects at least on the transportation, the economy, and the environment. This research explicitly considers and defines three major interdependent systems, the transportation system, the activity system, and the environmental system.

2.1. The Transportation System. The transportation system includes all the infrastructure facilities, vehicles, operators, and control strategies used to provide transportation services to people and to move products. Thus, the overall transportation system includes all modes of transportation, including highways, transit, and fluvial and air modes. Existing literature uses a number of measures to describe or assess transportation system performance. Lomax et al. [40] identified several measures of congestion, such as travel time, total segment delay, corridor mobility index, delay ratio, and relative delay rate. The Roadway Congestion Index uses volume and capacity to provide a measure of congestion [41]. A Roadway Congestion Index exceeding 1.0 denotes an average congestion level that is undesirable during the peak period. Black [42] uses principal component analysis to examine the relationships among multiple performance measures, including Vehicle Miles Traveled (VMT), travel time, mobility, crashes, fuel consumption, and emissions. The results indicate that VMT is the single most important factor in the context of sustainability. High VMT values do not necessarily mean high congestion; therefore, similar to the Roadway Congestion Index, VMT needs to be used in conjunction with the corresponding capacity. Thus, VMT per lane mile is a desirable performance measure. In addition, transit passenger miles and the number of intersections per capita can be important performance measures depending on the geographic location. Thus, both the demand and supply side should be taken into account for the selection of performance measure.

The Transportation Service Index (TSI) is a performance measure that seeks to quantify the movement of passenger and freight by the for-hire transportation sector [43]. This index, which is reported every month, can be used in conjunction with economic indicators to analyze the relationships between the economy and the transportation sector. Another interesting performance measure is the amount of personal money spent on transportation; this includes motor vehicles and parts, gasoline, and such transportation services as transit. The public investment on infrastructure is another important performance measure. Depending on the available data, some or all of the above performance measures can be used to develop the Transportation System PI (TSPI). The proposed modeling framework is modular and very flexible to enable the seamlessly incorporation of additional performance measures.

2.2. The Activity System. Previous studies have described the activity system as the combination of social, economic, political, and other transactions taking place over time and space [44, 45]. These transactions create and determine the demand for transportation. For example, changes in such economic policies as gas taxes or VMT fees create changes in the demand for transportation. In this research, the activity system consists of the social, cultural, health-related, and economic/financial aspects. A commonly used indicator for the socio-economic development of any country is its Gross Domestic Product. However, the United Nations Development Program (UNDP) [46] recommends using the Human Development Index because it incorporates all the basic and necessary dimensions for economic prosperity. This index measures the average achievements in a country by considering (i) a long and healthy life, or life expectancy; (ii) access to knowledge, or the education index; and (iii) a generous standard of living, measured by gross national income per capita. Life expectancy is the average number of years a child is expected to live, assuming that the mortality rate will remain constant [46]. The Education index includes the average number of years of education received in a lifetime and the expected number of years a child will attend school, assuming constant enrollment rates. The gross national income combines the gross domestic product of a country with its income received from other countries, less similar payments made to other countries. Some of these indices or indicators are used in this study to develop the Activity System PI (ASPI).

2.3. The Environmental System. The environmental system includes the air, water, soil, and all other natural resources as well as all living organisms that are affected and/or used by the transportation and activity systems. In the United States, data from the Federal Highway Administration and the Environmental Protection Agency suggests that emissions from the transportation system has been reduced drastically over the last 30 years, despite substantial gains in VMT, gross domestic product, population, and employment [47]. This has been attributed to the introduction of the Clean Air Act in 1973 and the emergence of fuel-efficient vehicles. However, such other sectors as industrial and chemical have generated increased carbon dioxide emissions over the years, thereby affecting climate change.

The Environmental Sustainability Index (ESI) was created by the end of the 1990s by Yale and Columbia Universities [48]. This index, which is a single indicator that provides insight into human health and the environment, was promoted by the World Economic Forum. This index currently is considered the most powerful tool available to measure environmental sustainability. The ESI uses 76 variables, including air pollution, emissions related to human health, environmental factors, water pollution, and resource minimization. In addition, it incorporates response factors relating to international agreements, such as the preservation of extinct species, limitations to the use of natural resources, limitations to the release of pollutants, and biodiversity conservation.

In 2006, the ESI became the Environmental Performance Index (EPI). Since then, the EPI has been published every two years. The primary constituents of the EPI are environmental health and ecosystem vitality. Policy weights used to calculate the EPI are approximate percentages that can be summarized as follows: environmental burden of disease, 25%; climate change, 25%; air pollution, 17%; water pollution, 17%; biodiversity and habitat, 4%; forestry, 4%; fisheries, 4%; and agriculture, 4%.

3. Methodology

This section provides a detailed framework of the modeling approach used in this study.

The concept of Fuzzy Logic was introduced by Lotfi Zadeh in 1965. It is a way of processing data by allowing partial set membership rather than crisp set membership or non-membership [49, 50]. Fuzzy logic provides a simple and efficient way to arrive at a definite conclusion based upon vague, ambiguous, imprecise, noisy, or missing input information. In the current study, multiple performance measures are combined and corresponding PIs are computed using fuzzy logic for the Transportation, Activity, and Environmental Systems. The PIs are calculated independently for each of the three systems. Their interdependencies are inherent in the data and are illustrated later in the results and discussion section. Considering a vector of performance measures X for system J as the inputs, the following three steps are used to calculate the corresponding PI: (1) an inference step (2), an aggregation step, and (3) a defuzzification step.

3.1. Inference Step. The inference step uses "*if-then*" rules and associated membership functions to develop and capture logical relationships between the different performance measures (inputs) and the PI (output).

3.1.1. If-then Rules. "*If-then*" rules are logical statements developed based on observation and expert knowledge of the system. The "*if*" part, left-hand side (LHS) or antecedent, is used with the inputs. The "*then*" part, the right-hand side (RHS) or consequent, is related to the output. An example of an "*If-then*" rule is as follows.

If [the VMT per lane mile is High and the TSI is Medium and the personal spending on transportation is Low], then [the TSPI is High].

As illustrated in this rule, in order to build the logical relationships between inputs and output, both the LHS and RHS are related to three fuzzy sets, High (H), Medium (M), and Low (L). Table 1 shows the set of "*if-then*" rules used in this study to calculate the TSPI. Here, three performance measures are used, namely. (i) the VMT per lane mile (v), (ii) the TSI, and (iii) the personal spending on transportation (ps) per year. If required, and if the relevant data is available, additional performance measures can be used; the corresponding rules are added to the table. Similar rules have been developed for each of the PIs in order to cover all possible relationships between the chosen system performance measures and the corresponding PI. Thus, the Transportation and Activity Systems each have three inputs and 27 rules while the environmental system has four inputs and 81 rules.

3.1.2. Membership Functions. The quantitative estimation of a PI requires knowledge about the interdependencies between the system performance measures and the corresponding PI. Considering the complexity of the Transportation, Activity, and Environmental Systems, this required knowledge is limited, vague, and sometimes ambiguous. Fuzzy logic provides a mathematical construct to combine all the available knowledge and develop meaningful PI estimates. The "*if-then*" rules are used in conjunction with sets of membership functions to relate the performance measures to the PIs, based on the available knowledge and data. Membership functions are used to define the grade or degree associated with every input and output. In this study, three membership functions are associated with each input and output, as illustrated in

Table 1: "*If-then*" rules for TSPI.

Rules	1	2	3	4	5	6	7	8	9	10	11	12	13	14	15	16	17	18	19	20	21	22	23	24	25	26	27		
		L	x	x	x	x	x	x	x	x	x																		
	v	M										x	x	x	x	x	x	x	x	x									
		H																			x	x	x	x	x	x	x	x	x
		L	x	x	x							x	x	x							x	x	x						
LHS	TSI	M				x	x	x							x	x	x							x	x	x			
		H							x	x	x							x	x	x							x	x	x
		L	x			x			x			x			x			x			x			x			x		
	ps	M		x			x			x			x			x			x			x			x			x	
		H			x			x			x			x			x			x			x			x			x
RHS	TSPI		L	L	L	L	L	L	M	M	M	L	L	L	M	M	M	M	H	H	M	M	M	H	H	H	H	H	H

Figure 1. Triangular membership functions are used in this study because they are easy to define; only three parameters are required: a modal point, the upper width, and the lower width. In addition, due to their conceptual and computation simplicity, triangular fuzzy numbers are commonly used in practical applications [30, 51, 52]. The domain for the membership functions corresponding to the LHS is defined based on the absolute value of the associated performance measures; the domains for the PIs corresponding to the RHS are normalized so as to use a simple $[0, 1]$ range. Figure 1 shows the membership functions for the calculation of the TSPI. Similar functions are defined for the other two PIs.

Once the "*if-then*" rules and the membership functions are defined, the Mamdani max-min composition operator and the Mamdani min implication operator are used for the fuzzy inference step [50]. For example, the three inputs for the calculation of TSPI, v, TSI, and ps are matched against the membership functions by using the "*if-then*" rules to determine the degree of activation. The degree at which each rule α is activated (δ^{α}) is obtained by using v, TSI, and ps as well as the max-min operator, as shown by (1)

$$\delta^{\alpha} = \max_{z \in Z} \min \left(\mu_{v}{}^{\alpha}(z), \mu_{\mathrm{TSI}}{}^{\alpha}(z), \mu_{\mathrm{ps}}{}^{\alpha}(z) \right), \quad (1)$$

where Z represents the universe of domains of the fuzzy sets v, TSI, and ps; and μ is a membership function. Equation (2) represents the membership functions of the fuzzy outcomes for the TSPI obtained, using the min implication operator

$$\mu_{\mathrm{TSPI}^{\alpha *}} = \min \left(\delta^{\alpha}, \mu_{\mathrm{TSPI}^{\alpha}} \right). \quad (2)$$

3.2. Aggregation Step. The Aggregation Step represents the aggregation of all the fuzzy output sets obtained after matching all the inputs to the membership functions by using all the "*if-then*" rules. A total of R rules for the calculation of TSPI are defined. The aggregation step is given by (3):

$$\mu_{\mathrm{TSPI}^{*}} = \sum_{\alpha=1}^{R} \mu_{\mathrm{TSPI}^{\alpha *}}. \quad (3)$$

3.3. Defuzzification Step. The output from the Aggregation Step combines all the available information by using all the defined rules. However, this output needs to be defuzzified to obtain a single crisp value for the corresponding PI, in this case, TSPI. The Center of Gravity method [50], illustrated in (4), is used for the Defuzzification Step:

$$\mathrm{TSPI} = \frac{\sum_{\alpha=1}^{R} \overline{\theta}^{\alpha} \cdot S\left(\mu_{\mathrm{TSPI}^{\alpha *}}\right)}{\sum_{\alpha=1}^{R} S\left(\mu_{\mathrm{TSPI}^{\alpha *}}\right)}, \quad (4)$$

where $\overline{\theta}^{\alpha}$ is the centroid of the fuzzy set for the TSPI, given by the RHS of rule α; and $S(\cdot)$ calculates the area of the membership function for a fuzzy set.

4. Study Region and Data

Sustainability considerations make it difficult to isolate systems and narrow the analysis to a particular transportation system or region. It is clear that impacts on the Environmental System, the Activity System, and even the Transportation System extend across regions and boundaries. In addition, the level of resolution of the available data may limit localized analyses. Hence, to illustrate the proposed method, without loss of generality, the United States is used as the study area. Similar analyses can be conducted for other regions and, ideally, the entire globe. In this case, the analysis was conducted for a period of 20 years between 1990 and 2010.

The three performance measures used in the examples in Section 3 for the estimation of the TSPI in this study were obtained from the Bureau of Transportation Statistics [43]. The ASPI includes the following performance measures provided by the United Nations [46]:

(i) Gross national income (*gni*);
(ii) The Education Index (*ei*);
(iii) Life expectancy (*le*).

The Environmental System Performance Index (ESPI) is based on the following performance measures:

(i) Carbon dioxide emissions (*ce*) [53];
(ii) Air pollutants (*ap*) [54];
(iii) Water pollutants (*wp*) [55];
(iv) Energy consumption (*ec*) [56].

5. Results and Discussion

Figure 2 shows the normalized performance measures and performance index for the Transportation System. The historic trend for the VMT per lane mile (in thousands) covers a period from 1990 to 2008. It is clear that the trend is increasing except between 1990 and 1991. This could be attributed to the recession during each of those time periods [57, 58]. During 2005-2006, the VMT started decreasing probably as a consequence of the rising oil prices [59]. The trend for the TSI covers from 1990 to 2010. The base year for TSI = 100 is taken as the year 2000. The figure shows the decrease in TSI between the years 2000 and 2002, when the terrorist attack on September 11 occurred. In 2001, there was less freight and passenger travel. Between years 2008 and 2010, the financial crisis resulted in a severe recession with consequences on TSI, as illustrated in Figure 2. Personal spending on transportation is included during 1995–2010. It is evident that spending increases from 1995–2005 as a result of economic development. However, in 2006, spending started decreasing as a result of a rise in gas prices, which hit $4 a gallon. Later, the financial crisis during 2007–2010 resulted in decreased spending for transportation-related activities.

Figure 2 also shows the historic trend of the Transportation System performance index from 1990 to 2009. The crisp value in the y-axis is obtained by using the fuzzy approach discussed in earlier sections. Here, the closer the TSPI to 1, the better the performance of the Transportation System; if its value is close to 0, then performance is poor. The crisp values can only be used as a relative measure to compare between alternatives and scenarios. It cannot be used to assess

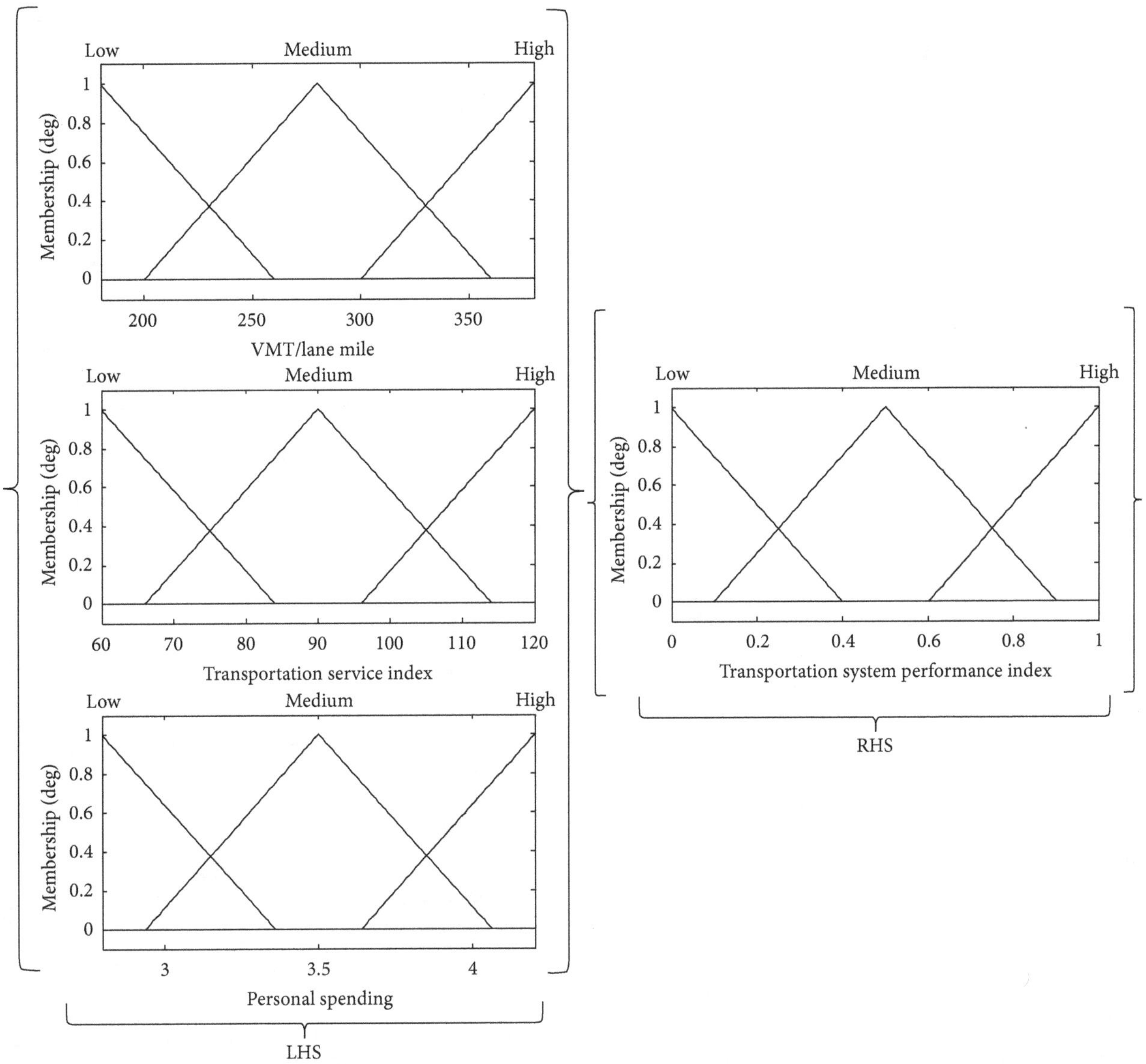

Figure 1: Membership functions for the calculation of the Transportation System Performance Index.

the absolute value of the sustainability of the system. It is evident that TSPI has the best value between years 2005 and 2006, when the economy was booming, and the least value between years 1990 and 1991. The curve for the TSPI follows a pattern consistent with VMT/lane mile and TSI. That is, the TSPI increases with the increase in VMT/lane mile and TSI. According to Genier [59], rising oil prices during 2005-2006 has led to reduced VMT and promoted alternate modes of transportation, such as transit and car-pooling, as well as the use of more efficient vehicles.

Figure 3 shows the normalized performance measures and performance index for the Activity System. The trend of the average annual income in Gross National Income per capita is provided from 1990 to 2010. The trend increased, with a high growth rate until 1999. The rate started decreasing in 2000 following the technology bust, also known as the Dot-Com Bubble, and later in 2006, following the housing crisis. It is noted that the rate of growth in income is less in the past decade as compared to earlier decades.

The trend of the average annual education index is provided from 1990 to 2010. This index started increasing from 1990 to 2000, the primary reason being the invention of new technologies and innovations that kept the United States in the forefront of education. In addition, a new wave of technological revolution was seen in the form of startups. Also, science, engineering, and medical disciplines saw a new era of growth and development. The reason for a slight decline in the education index between 2000 and 2004 is not clear yet. The trend of the average annual life expectancy is provided from 1990 to 2010. The average life expectancy has increased from 74 years in 1990 to 80 years in 2010. This increase can be attributed to the technological advancement in medical facilities and billions of dollars spent on research and the development of new and effective drugs.

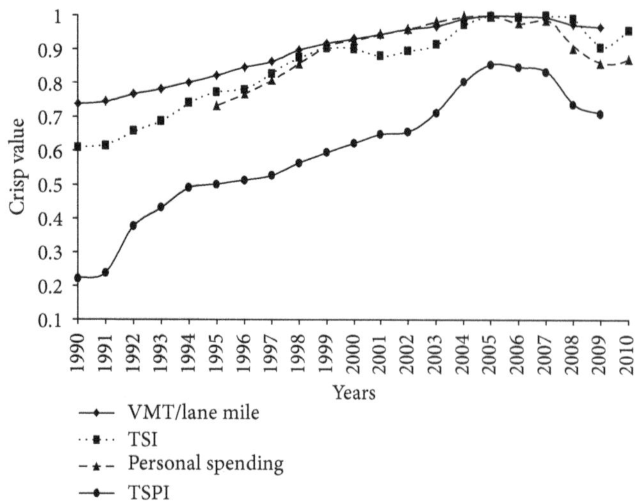

FIGURE 2: Historical Trend of Transportation System Performance Index and its Performance Measures.

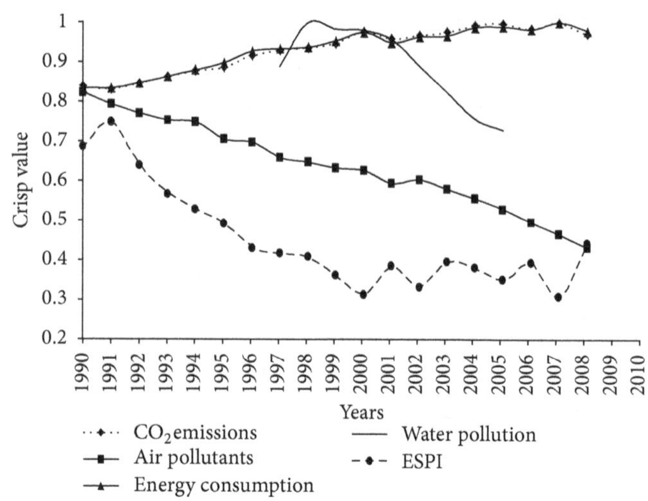

FIGURE 4: Historical Trend of Environmental System Performance Index and its Performance Measures.

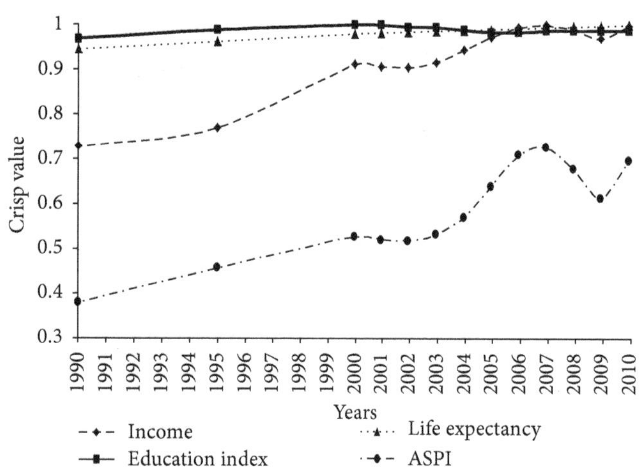

FIGURE 3: Historical Trend of Activity System Performance Index and its Performance Measures.

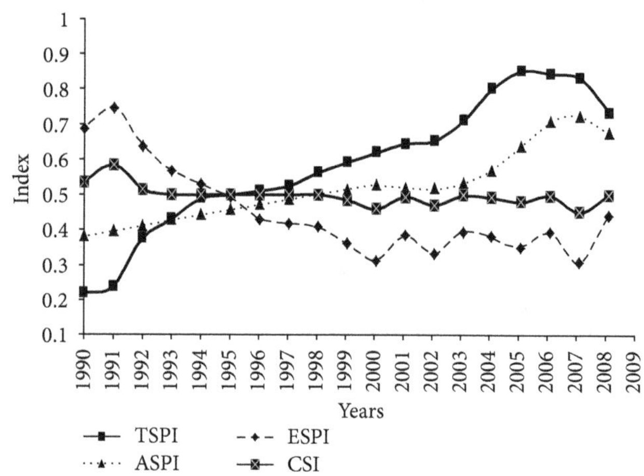

FIGURE 5: Historical Trend of Performance Indices and the Composite Sustainability Index for the Transportation, Activity, and Environmental systems.

Figure 3 also shows the trend for the Activity System's performance index from 1990 to 2010. This index started increasing from year 1990 until the year 2000 as a result of economic development. Starting with the technology bust in 2000 and terrorist attacks in 2001, the economic activity started to decrease and did not recover until the end of the year in 2003. Since 2003, the Activity System started an upward trend before hitting a peak in 2007. The financial crisis from 2007 to 2009 resulted in a dramatic decrease in economic activity, often compared as equivalent to the Great Depression of 1930s. The year 2009 marks the period of "official recovery" from the recession.

Figure 4 shows the normalized performance measures and performance index for the Environmental System. The trend of carbon dioxide emissions is provided from years 1990 to 2008. This is an increasing trend except during 1990-1991, a time of global political unrest and high inflation; 2000-2002, due to the technology bust and September 11 attacks; 2005-2006, due to high gas prices; and 2007-2008, with the financial crisis. The trend of air pollutants is provided from 1990 to 2008. With the introduction of the Clean Air Act in 1973, there has been a dramatic reduction in air pollution. In addition, the introduction of innovative technologies, such as hybrid and battery powered vehicles, have led to reduced air pollution over the years.

The trend for water pollutants is provided from 1997 to 2005. This trend decreases with time as a result of innovative and efficient waste management techniques. The trend for the average annual energy consumption in quadrillion British Thermal Units is provided from 1990 to 2008. This trend indicates that energy consumption decreased during the financial crisis of 1990-1991. After 1991, energy consumption started an upward trend and finally peaked in 2007. However, there were short periods of decline in energy consumption

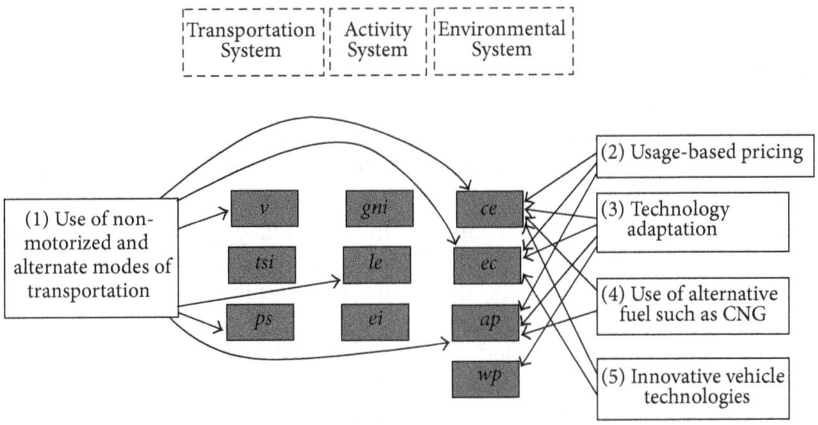

FIGURE 6: Direct effects of policy options on Performance Measures.

both in 2001, attributed to the September 11 terrorist attacks, and 2006, due to high oil prices. The terrorist attack resulted in decreased travel and less economic activity, while the exorbitant high oil prices promoted the use of new battery-powered and hybrid vehicle technologies.

Figure 4 also shows the trend of the Environmental System's performance index from 1990 to 2008. If the value for ESPI is close to 1, then the environmental system is excellent; if its value is close to 0, then the system quality is very poor. The best value for ESPI occurred during 1990–1995, when economic development was slow as a result of global political unrest and the first gulf war. Since 2000, the quality started to improve, probably as a consequence of multiple periods of economic contractions. Again, the year 2007 marked the beginning of a slight uptrend in the index as a result of a global financial crisis. In general, the environment improves during periods when economic activity is down and/or oil prices are high. In addition, the curve for the ESPI follows a pattern consistent with carbon-dioxide emissions and energy consumption. That is, the ESPI decreases with the increase in carbon-dioxide emissions and energy consumption.

Figure 5 shows the three performance indices from 1990 to 2009. In this figure the Transportation and Activity Systems follow an increasing trend over the years, with similar periods of growth and contractions; on the other hand, the Environmental System follows a reverse pattern. These trends seem intuitive, as growth in the economy and the transportation sectors are expected to happen simultaneously; this growth requires resources from the environment, thereby increasing emissions and energy consumption.

Figure 5 also illustrates a Composite Sustainability Index (CSI), an index used to access the overall sustainability of the SOS used this research. It is calculated using the proposed fuzzy logic approach and the performance index for the Transportation, Activity, and Environmental Systems. The CSI shows an overall increasing trend from year 1990 to 1995. However, considering the overall negative slope and corresponding decrease on the ESPI, the CSI does not continue increasing after 1995 presenting some negative periods and increases only when there is a significant improvement on the ESPI. Based on these observations and the chosen performance measures, negative impacts to the environment seem to be associated with negative consequences on the overall sustainability of the SOS. In general, under the proposed framework, a system is sustainable if the slope of the corresponding PI curve presents a nonnegative slope. Similarly, the overall SOS is sustainable if the slope of the CSI is nonnegative. The axioms presented in this paper are an attempt to summarize our observations based on chosen performance measures. There is a vast literature with similar observations. For example, Young et al. [60] as well as Lahiri and Yao [61] have observed that the transportation and activity system follows a lead-lag phase pattern and environment system is inversely related to the other two. The following axioms can be postulated to assess the sustainability or the unsustainability of our SOS.

(1) The SOS is sustainable when the overall slopes for the TSPI, ASPI, and ESPI have a positive trend. This is an ideal scenario with positive growth in all systems, and implies that there is no need of nonrenewable natural resources to sustain growth in the transportation and the economy.

(2) The SOS is unsustainable when the slopes of TSPI and ASPI have a positive trend but the slope of ESPI has a negative trend. This is the scenario that we have been observing in the USA. In general, the SOS is unsustainable when the overall trend of at least one of the three slopes is negative.

(3) The SOS is sustainable when the overall slopes for the TSPI, ASPI, and ESPI have a nonnegative trend. This scenario is sustainable because all the transportation and other social activities can continue in perpetuity without degradation of the environmental system. Although this is a scenario preferred over an unsustainable situation, it may represent an unstable equilibrium that can easily become unsustainable.

6. Policy Perspectives

This section discusses some policy options for the sustainability of the SOS studied in this research. Some of these options have been implemented in the past revealing some of their effects. Other options are currently under consideration by multiple stakeholders. Figure 6 illustrates five policy options that can be used to improve performance and support the sustainability of the SOS considered here. The dashed boxes correspond to the three major systems, the grey boxes represent the performance measures within each system, and the suggested policies are depicted by rectangular boxes. These policies have direct and indirect effects on some performance measures and systems. Only the direct effects of the proposed policies are shown through the arrows in Figure 6. Conclusion regarding indirect effects will be immature at this point; hence are not discussed here. Each policy is described as follows.

Use of Nonmotorized and Alternate Modes of Transportation. This policy consists of the promotion of non-motorized modes of transportation, such as bicycles, and the use of alternatives for driving alone, such as transit and carpooling. The success of this policy depends on multiple factors, including land use. It may require the establishment of commuter-friendly and transit-friendly development near the central business district. In addition, changes in travel and demand patterns depend on the users' preferences and attitudes as well as convenience. Expected consequences of implementing this policy, among others, include reductions on (i) VMT [62, 63], (ii) air pollution, (iii) carbon dioxide emissions, (iv) energy consumption, (v) health issues, and (vi) out-of-pocket cost. The money and resources saved can be used to improve such sectors as education and research with further impacts on the gross domestic product.

Usage Based Pricing. Currently, the implementation of a VMT fee is being considered as a viable alternative to replace the current fuel tax for collecting the required resources for highway maintenance [64]. This policy also can promote the reduction of VMT, along with all the other associated consequences. However, this policy faces a number of deployment as well as acceptance issues.

Technology Adaptation. The rapid industrialization and technological revolution has resulted in reduced emissions over the years. For example, the use of efficient boilers in coal-fired plants will help reduce carbon dioxide emissions, pollution, and energy consumption [65, 66]. Health related issues will be reduced as a consequence of less pollution.

Use of Alternative Fuels Such As Compressed Natural Gas (CNG). The use of alternative fuels in the form of CNG will reduce carbon-dioxide emissions and pollution [67, 68]. This will lead to a green and cleaner environment [69] with all the associated benefits to health, the economy, and the quality of life. In the United States, the reserves of natural gas are larger than those of petroleum [70]. Hence, this policy seems plausible from an environmental and economic perspective. The only drawbacks are the time and cost associated with retrofitting vehicles and the supply chain.

Innovative Vehicle Technologies. Replacement of conventionally powered vehicles with hybrid and electric vehicles will reduce carbon-dioxide emissions and nonrenewable fuel consumption [71]. Automakers are particularly interested in this policy [72]. In addition, the federal government provides tax incentives to develop and manufacture lithium ion batteries in the United States.

Ideally, each of these policies is evaluated before deployment and adoption. Some of them are currently under analysis by multiple agencies and sectors. The proposed framework in this study is descriptive rather than normative. Hence, it can only be used to appreciate the effectiveness and benefits of past policies. Currently, the proposed framework is been extended to enable a normative analysis in order to evaluate potential policy alternatives such as those described earlier.

7. Conclusions

Previous studies about sustainable transportation have either focused only on the transportation system, or have not used a methodology that enables the treatment of incomplete, vague, and qualitative information present in the problem context. This study adopted a holistic approach to compute Performance Indices for a SOS including the Transportation, Activity, and Environmental systems. The Performance Indices are synthetized to calculate a Composite Sustainability Index to evaluate the sustainability of the overall SOS. Considering the complexity, vagueness, nonlinearity, qualitative, and incomplete information characterizing the quantification of the Performance and Composite Sustainability Indices, a fuzzy logic approach was used to calculate these indices. Historic events and policy changes indicated that fuzzy logic provided an adequate approach to estimate both the Performance Indices and the Composite Sustainability Index.

Results of the analysis for the US showed that the Transportation and Activity Systems both follow a positive trend over the years, with similar periods of growth and contractions. In contrast, the environmental system follows a reverse pattern. This seems intuitive, as periods of economic and transportation growth is expected to have a negative effect on the environment, leading to increased emissions and energy consumption. In general, the performance of the environmental system has decreased significantly over time. Policies adopted to protect the system have shown positive effects. However, the current performance of the Environmental System is questionable, and long-term policies need to be developed.

The conclusions provided here are based on the results obtained using a limited number of performance measures. Adding or removing performance measures are expected to change the results and conclusions. In general, following a holistic approach, it is expected that the more relevant performance measures are used, the more comprehensive and accurate the analysis is. Planning and operational policies for the sustainability of the Transportation, Activity, and Environmental systems can be developed using the proposed approach. Considering the current practice of making planning decisions at the regional and jurisdictional level, the framework used in this study is currently been extended to enable the analysis of regional systems including large metropolitan areas. A simulation-based approach has been developed to estimate multiple performance measures required to calculate adequate performance indices.

References

[1] EIA (Energy Information Administration), *International Energy Outlook*, EIA, Washington, DC, USA, 2006.

[2] J. J. MacKenzie, "Alternative fuels to reduce petroleum consumption, global warming gases, and urban air pollution," in *Symposium on Challenges and Opportunities for Global Transportation in the 21st Century*, John A. Volpe Transportation Systems Center, Cambridge, Mass, USA, 1995.

[3] J. Zheng, C. Atkinson-Palombo, C. McCahill, R. O'Hara, and N. W. Garrick, "Quantifying the economic domain of transportation sustainability," in *Proceedings of the Annual Meeting of the Transportation Research Board—CDROM*, Washington, DC, USA, 2011.

[4] S. Bell and S. Morse, *Sustainability Indicators: Measuring the Immeasurable*, Earthscan, London, UK, 2nd edition, 2008.

[5] J. R. E. Harger and F. M. Meyer, "Definition of indicators for environmentally sustainable development," *Chemosphere*, vol. 33, no. 9, pp. 1749–1775, 1996.

[6] H. Bossel, "Assessing viability and sustainability: a systems-based approach for deriving comprehensive indicator sets," *Conservation Ecology*, vol. 5, no. 2, article 12, 2001.

[7] C. M. Jeon, A. A. Amekudzi, and R. L. Guensler, "Evaluating plan alternatives for transportation system sustainability: Atlanta metropolitan region," *International Journal of Sustainable Transportation*, vol. 4, no. 4, pp. 227–247, 2010.

[8] T. Litman, "Developing indicators for comprehensive and sustainable transport planning," *Transportation Research Record*, no. 2017, pp. 10–15, 2007.

[9] J. A. Paravantis and D. A. Georgakellos, "Trends in energy consumption and carbon dioxide emissions of passenger cars and buses," *Technological Forecasting and Social Change*, vol. 74, no. 5, pp. 682–707, 2007.

[10] S. Huzayyin and H. Salem, "Analysis of thirty years evolution of urban growth, transport demand and supply, energy consumption, greenhouse and pollutants emissions in Greater Cairo," *Research in Transportation Economics*, vol. 40, no. 1, pp. 104–115, 2013.

[11] T. Litman, *Climate Change Emission Valuation for Transportation Economic Analysis*, Victoria Transport Policy Institute, 2012.

[12] E. J. Zolnik, "Estimates of statewide and nationwide carbon dioxide emission reductions and their costs from cash for clunkers," *Journal of Transport Geography*, vol. 24, pp. 271–281, 2012.

[13] G. R. Conway, "Sustainability in agricultural development: trade-offs with productivity, stability and equitability," *Journal of Farming Systems Research and Extension*, vol. 4, no. 2, pp. 1–14, 1994.

[14] R. Gray, "Economic measures of sustainability," *Canadian Journal of Agricultural Economics*, vol. 39, no. 4, pp. 627–635, 1991.

[15] WCED (World Commission on Environment and Development), *Our Common Future*, Oxford University Press, Oxford, UK, 1987.

[16] R. Costanza, B. Norton, and B. D. Haskell, *Ecosystem Health: New Goals for Environmental Management*, Island Press, Washington, DC, USA, 1992.

[17] IUCN (World Conservation Union), UNEP (United Nations Environment Programme), and WWF (World Wide Fund for Nature), *Caring for the Earth: A Strategy for Sustainable Living*, IUCN, Gland, Switzerland, 1991.

[18] K. Rennings and H. Wiggering, "Steps towards indicators of sustainable development: linking economic and ecological concepts," *Ecological Economics*, vol. 20, no. 1, pp. 25–36, 1997.

[19] G. Munda, *Fuzzy Information on Multi-Criteria Environmental Models*, Physika, Heidelberg, Germany, 1995.

[20] J. Zietsman, L. R. Rilett, and S. J. Kim, "Transportation corridor decision-making with multi-attribute utility theory," *International Journal of Management and Decision Making*, vol. 7, no. 2-3, pp. 254–266, 2006.

[21] R. Islam and T. L. Saaty, "The analytic hierarchy process in the transportation sector," in *Multiple Criteria Decision Making for Sustainable Energy and Transportation Systems*, vol. 634 of *Lecture Notes in Economics and Mathematical Systems*, Springer, Physica, Berlin, Heidelberg, 2010.

[22] G. A. Mendoza and R. Prabhu, "Multiple criteria decision making approaches to assessing forest sustainability using criteria and indicators: a case study," *Forest Ecology and Management*, vol. 131, no. 1-3, pp. 107–126, 2000.

[23] H. J. Zimmermann, *Fuzzy Set Theory and Its Applications*, Kluwer Academic, Boston, Mass, USA, 4th edition, 2001.

[24] S. Yedla and R. M. Shrestha, "Multi-criteria approach for the selection of alternative options for environmentally sustainable transport system in Delhi," *Transportation Research A*, vol. 37, no. 8, pp. 717–729, 2003.

[25] D. Tsamboulas and G. Mikroudis, "EFECT—evaluation framework of environmental impacts and costs of transport initiatives," *Transportation Research D*, vol. 5, no. 4, pp. 283–303, 2000.

[26] A. Awasthi and H. Omrani, "A hybrid approach based on AHP and belief theory for evaluating sustainable transportation solutions," *International Journal of Global Environmental Issues*, vol. 9, no. 3, pp. 212–226, 2009.

[27] T. L. Saaty, *The Analytic Hierarchy Process*, McGraw-Hill, New York, NY, USA, 1980.

[28] G. Crawford and C. Williams, "A note on the analysis of subjective judgment matrices," *Journal of Mathematical Psychology*, vol. 29, no. 4, pp. 387–405, 1985.

[29] J. Barzilai, "Consistency measures for pairwise comparison matrices," *Journal of Multi-Criteria Decision Analysis*, vol. 7, no. 3, pp. 123–132, 1998.

[30] G. J. Klir and B. Yuan, *Fuzzy Sets and Logic: Theory and Application*, Prentice Hall, Upper Saddle River, NJ, USA, 1995.

[31] R. R. Yager, "Aggregation operators and fuzzy systems modeling," *Fuzzy Sets and Systems*, vol. 67, no. 2, pp. 129–145, 1994.

[32] W. Silvert, "Ecological impact classification with fuzzy sets," *Ecological Modelling*, vol. 96, no. 1–3, pp. 1–10, 1997.

[33] A. Awasthi, S. S. Chauhan, and H. Omrani, "Application of fuzzy TOPSIS in evaluating sustainable transportation systems,"

Expert Systems with Applications, vol. 38, no. 10, pp. 12270–12280, 2011.

[34] S. Opricovic and G. H. Tzeng, "Fuzzy multicriteria model for postearthquake land-use planning," *Natural Hazards Review*, vol. 4, no. 2, pp. 59–64, 2003.

[35] G. A. Mendoza and R. Prabhu, "Fuzzy methods for assessing criteria and indicators of sustainable forest management," *Ecological Indicators*, vol. 3, no. 4, pp. 227–236, 2004.

[36] K. F. R. Liu, "Evaluating environmental sustainability: an integration of multiple-criteria decision-making and fuzzy logic," *Environmental Management*, vol. 39, no. 5, pp. 721–736, 2007.

[37] T. Prato, "A fuzzy logic approach for evaluating ecosystem sustainability," *Ecological Modelling*, vol. 187, no. 2-3, pp. 361–368, 2005.

[38] L. A. Marks, E. G. Dunn, J. M. Keller, and L. D. Godsey, "Multiple criteria decision making (MCDM) using fuzzy logic: an innovative approach to sustainable agriculture," in *Proceedings of the 3rd International Symposium on Uncertainty Modeling and Analysis and Annual Conference of the North American Fuzzy Information Processing Society (ISUMA-NAFIPS'95)*, pp. 503–508, IEEE Computer Society, Washington, DC, USA, September 1995.

[39] R. L. Ackoff, "Towards a system of systems concepts," *Management Science*, vol. 17, no. 11, pp. 661–671, 1971.

[40] T. Lomax, S. Turner, and G. Shunk, *Quantifying Congestion*, vol. 1, National Cooperative Highway Research Program, Transportation Research Board, National Academy Press, Washington, DC, USA, 1997.

[41] D. L. Schrank and T. J. Thomas, *Urban Mobility Report*, Texas Transportation Institute, College Station, Tex, USA, 2009.

[42] W. R. Black, "Sustainable transport and potential mobility," *European Journal of Transport and Infrastructure Research*, vol. 2, no. 3-4, pp. 179–196, 2002.

[43] BTS (Bureau of Transportation Statistics), "Key Transportation Indicators," 2011, http://www.bts.gov/publications/key_transportation_indicators.

[44] M. L. Manheim, *Fundamentals of Transportation Systems Analysis, Volume 1: Basic Concepts*, MIT Press, Boston, Mass, USA, 1979.

[45] E. Cascetta, *Transportation Systems Analysis: Models and Applications*, Springer, New York, NY, USA, 2009.

[46] UNDP (United Nations Development Programme), "Human Development Report," 2010, http://www.hdr.undp.org/en/statistics/.

[47] ARTBA (American Road and Transportation Builders Association), 2011, http://www.artba.org/mediafiles/regulatorylegalartbacafecommentsjanuary2011.pdf.

[48] ESI (Environmental Sustainability Index), Benchmarking National Environmental Stewardship, Yale Center for Environmental Law and Policy, Yale University, 2005, http://www.yale.edu/esi/ESI2005_Main_Report.pdf.

[49] R. R. Yager, S. Ovchinnikov, R. M. Tong, and H. T. Nguyen, *Fuzzy Sets and Applications: Selected Papers By L.A. Zadeh*, John Wiley & Sons, New York, NY, USA, 1987.

[50] L. H. Tsoukalas and R. E. Uhrig, *Fuzzy and Neural Approaches in Engineering*, John Wiley & Sons, New York, NY, USA, 1997.

[51] W. Pedrycz, "Why triangular membership functions?" *Fuzzy Sets and Systems*, vol. 64, no. 1, pp. 21–30, 1994.

[52] C. H. Yeh and H. Deng, "A practical approach to fuzzy utilities comparison in fuzzy multicriteria analysis," *International Journal of Approximate Reasoning*, vol. 35, no. 2, pp. 179–194, 2004.

[53] EIA (Energy Information Administration), *Emissions of Greenhouse Gases in the United States*, Washington, DC, USA, 2009, http://www.eia.gov/environment/emissions/ghg_report/pdf/tbl6.pdf.

[54] EPA (Environmental Protection Agency), Clearinghouse for Inventories and Emissions Factors (CHIEF), Current Emission Trends Summaries, 2009, http://www.epa.gov/ttn/chief/trends/index.html.

[55] World Databank, World Development Indicators and Global Development Finance, 2010, http://www.databank.worldbank.org/ddp/home.do?Step=2&id=4&hActiveDimensionId=WDI_Series.

[56] EIA (Environmental Information Administration), Monthly Energy Review, 2011, http://www.eia.doe.gov/emeu/mer/contents.html.

[57] M. Mussa, "U.S. Macroeconomic policy and third world debt," *Cato Journal*, vol. 4, no. 1, 1984, http://www.cato.org/sites/cato.org/files/serials/files/cato-journal/1984/5/cj4n1-5.pdf.

[58] R. H. Kamery, "A brief review of the recession of 1990-1991," *Allied Academies International Conference, Proceedings of the Academy of Legal, Ethical and Regulatory Issues*, vol. 8, no. 2, 2004.

[59] B. Genier, "Peak Demand—U.S. Gasoline Demand Likely Peaked in 2007," Cambridge Energy Research Associates (CERA), 2008, http://www2.cera.com/news/details/1,2318,9568,00.html.

[60] P. Young, K. Notis, G. Feuerberg, and L. Nguyen, "BTS Technical Report: transportation services index and the economy," 2007.

[61] K. Lahiri and W. Yao, "A dynamic factor model of the coincident indicators for the US transportation sector," *Applied Economics Letters*, vol. 11, no. 10, pp. 595–600, 2004.

[62] T. Litman and R. Steele, *Land Use Impacts on Transport: How Land Use Factors Affect Travel Behavior*, Victoria Transport Policy Institute, 2011.

[63] Nelson/Nygaard, *Crediting Low-Traffic Developments: Adjusting Site-Level Vehicle Trip Generation Using URBEMIS, Urban Emissions Model*, California Air Districts, 2005.

[64] D. S. Kim, D. Porter, and R. Wurl, *Technology Evaluation for Implementation of VMT Based Revenue Collection Systems*, Oregon Department of Transportation, Road User Fee Task Force, Salem, Ore, USA, 2002.

[65] K. Jordal, M. Anheden, J. Yan, and L. Strömberg, "Oxy-fuel Combustion for coal-fired power generation with CO_2 capture—opportunities and challenges," in *Proceedings of the 7th International Conference on Greenhouse Gas Control Technologies (GHGT-7)*, Vancouver, Canada, 2004.

[66] M. B. Toftegaard, J. Brix, P. A. Jensen, P. Glarborg, and A. D. Jensen, "Oxy-fuel combustion of solid fuels," *Progress in Energy and Combustion Science*, vol. 36, no. 5, pp. 581–625, 2010.

[67] M. P. Hekkert, F. H. J. F. Hendriks, A. P. C. Faaij, and M. L. Neelis, "Natural gas as an alternative to crude oil in automotive fuel chains well-to-wheel analysis and transition strategy development," *Energy Policy*, vol. 33, no. 5, pp. 579–594, 2005.

[68] P. Goyal and Sidhartha, "Present scenario of air quality in Delhi: a case study of CNG implementation," *Atmospheric Environment*, vol. 37, no. 38, pp. 5423–5431, 2003.

[69] S. Yeh, "An empirical analysis on the adoption of alternative fuel vehicles: the case of natural gas vehicles," *Energy Policy*, vol. 35, no. 11, pp. 5865–5875, 2007.

[70] M. Q. Wang and H. S. Huang, "A full fuel-cycle analysis of energy and emissions impacts of transportation fuels produced from natural gas," Report No. ANL/ESD-40, Center for Transportation Research, Energy Systems Division, Argonne National Laboratory, IL, US Department of Energy, 2000.

[71] M. S. Dresselhaus and I. L. Thomas, "Alternative energy technologies," *Nature*, vol. 414, no. 6861, pp. 332–337, 2001.

[72] S. G. Wirasingha, N. Schofield, and A. Emadi, "Plug-in hybrid electric vehicle developments in the US: trends, barriers, and economic feasibility," in *Proceedings of the IEEE Vehicle Power and Propulsion Conference (VPPC'08)*, Harbin, China, September 2008.

Permissions

All chapters in this book were first published in JAT, by Hindawi Publishing Corporation; hereby published with permission under the Creative Commons Attribution License or equivalent. Every chapter published in this book has been scrutinized by our experts. Their significance has been extensively debated. The topics covered herein carry significant findings which will fuel the growth of the discipline. They may even be implemented as practical applications or may be referred to as a beginning point for another development.

The contributors of this book come from diverse backgrounds, making this book a truly international effort. This book will bring forth new frontiers with its revolutionizing research information and detailed analysis of the nascent developments around the world.

We would like to thank all the contributing authors for lending their expertise to make the book truly unique. They have played a crucial role in the development of this book. Without their invaluable contributions this book wouldn't have been possible. They have made vital efforts to compile up to date information on the varied aspects of this subject to make this book a valuable addition to the collection of many professionals and students.

This book was conceptualized with the vision of imparting up-to-date information and advanced data in this field. To ensure the same, a matchless editorial board was set up. Every individual on the board went through rigorous rounds of assessment to prove their worth. After which they invested a large part of their time researching and compiling the most relevant data for our readers.

The editorial board has been involved in producing this book since its inception. They have spent rigorous hours researching and exploring the diverse topics which have resulted in the successful publishing of this book. They have passed on their knowledge of decades through this book. To expedite this challenging task, the publisher supported the team at every step. A small team of assistant editors was also appointed to further simplify the editing procedure and attain best results for the readers.

Apart from the editorial board, the designing team has also invested a significant amount of their time in understanding the subject and creating the most relevant covers. They scrutinized every image to scout for the most suitable representation of the subject and create an appropriate cover for the book.

The publishing team has been an ardent support to the editorial, designing and production team. Their endless efforts to recruit the best for this project, has resulted in the accomplishment of this book. They are a veteran in the field of academics and their pool of knowledge is as vast as their experience in printing. Their expertise and guidance has proved useful at every step. Their uncompromising quality standards have made this book an exceptional effort. Their encouragement from time to time has been an inspiration for everyone.

The publisher and the editorial board hope that this book will prove to be a valuable piece of knowledge for researchers, students, practitioners and scholars across the globe.

List of Contributors

Kezhen Hu, Jianping Wu and Mingyu Liu
Department of Civil Engineering, Tsinghua University, Beijing 100084, China

Kezhen Hu
Transport Studies Unit, School of Geography and the Environment, University of Oxford, Oxford, UK

Xiaohong Chen, Jiaqi Cheng and Jianhong Ye
Key Laboratory of Road and Traffic Engineering of the Ministry of Education, Tongji University, 4800 Cao'an Road, Shanghai 201804, China

Yong Jin, Xi Li and Fei Zhang
Global Carsharing & Rental Co., Ltd., 888 South Moyu Road, Shanghai 201805, China

Peng Chen, Cong Yan and Yunpeng Wang
Beijing Key Laboratory for Cooperative Vehicle Infrastructure Systems and Safety Control, School of Transportation Science and Engineering, Beihang University, Xue Yuan Road No. 37, Hai Dian District, Beijing 100191, China

Peng Chen and Yunpeng Wang
Beijing Advanced Innovation Center for Big Data and Brain Computing, Beihang University, Xue Yuan Road No. 37, Hai Dian District, Beijing 100191, China

Jian Sun and Keping Li
Department of Traffic Engineering and Key Laboratory of Road and Traffic Engineering, Ministry of Education, Tongji University, Shanghai 201804, China

Shenyang Chen
Department of Civil and Environmental Engineering, University of California, Davis, CA 95616, USA

Mirosław Targosz, Wojciech Skarka and Piotr Przystałka
Silesian University of Technology, Institute of Fundamentals of Machinery Design, 18A Konarskiego Street, 44-100 Gliwice, Poland

Quan Yuan, Haotian Su, Guanwen Bing and Xinyuan Gui
State Key Laboratory of Automotive Safety and Energy, Tsinghua University, Beijing 100084, China

Wei Hao
Hunan Key Laboratory of Smart Roadway and Cooperative Vehicle-Infrastructure Systems, Changsha University of Science and Technology, Changsha, Hunan 410114, China
Key Laboratory of Safety Design and Reliability Technology for Engineering Vehicle, Changsha University of Science and Technology, Changsha, Hunan 410114, China

Abolfazl Safikhani
Department of Statistics, Columbia University in the City of New York, NY 10031, USA

Francesco Galante, Luigi Pariota, Luigi Biggero and Gennaro N. Bifulco
Department of Civil, Architectural and Environmental Engineering, University of Naples Federico II, 80125 Naples, Italy

Fabrizio Bracco and Carlo Chiorri
Department of Educational Sciences, University of Genoa, 16128 Genoa, Italy

Noelia Caceres
Transportation Engineering Unit, AICIA, Camino de los Descubrimientos, s/n, 41092 Seville, Spain

Francisco G. Benitez
Transportation Engineering, Faculty of Engineering, University of Seville, Camino de los Descubrimientos, s/n, 41092 Seville, Spain

Wei-Hsun Lee and Jiang-Yi Li
Department of Transportation and Communication Management Science, National Cheng Kung University, No. 1, University Road, Tainan City 701, Taiwan

Wanjing Ma and Lei Wang
The Key Laboratory of Road and Traffic Engineering of the Ministry of Education, Tongji University, Shanghai 201804, China

Lin Li
Shanghai SH Intelligent Automotive Technology Co., Ltd., Shanghai 201805, China

Yiping Wu, Xiaohua Zhao, Ying Yao and Jian Rong
Beijing Key Laboratory of Traffic Engineering, Beijing University of Technology, 100 Ping Le Yuan, Beijing 100124, China

Michelle S. F. Arcúrio and Talita Armborst
National Civil Aviation Agency, AVSEC Certification Technical Branch, Brasília, Brazil

Eliane S. Nakamura
National Civil Aviation Agency, AVSEC Quality Control Technical Branch, Brasília, Brazil

Sina Torabi and Mattias Wahde
Department of Mechanics and Maritime Sciences, Chalmers University of Technology, 412 96 Göteborg, Sweden

Ranjit R. Desai, Roger B. Chen and William Armington
Golisano Institute for Sustainability, Rochester Institute of Technology, One Lomb Memorial Dr., Rochester, NY 14623-5608, USA

Shan Chang, Rui Song, Shiwei He and Guo Qiu
MOE Key Laboratory for Urban Transportation Complex Systems Theory and Technology, Beijing Jiaotong University, Beijing, China

Jun Zeng and Zhenjun Zhu
School of Transportation, Southeast University, Nanjing 211189, China

Jun Zeng
Department of Civil, Architectural and Environmental Engineering, The University of Texas at Austin, Austin, TX 78712,USA

Chaoxu Sun
Zhejiang Provincial Natural Gas Development Co. Ltd., Hangzhou 310052, China

Zhenjun Zhu
Department of City and Regional Planning, University of California, Berkeley, Berkeley, CA 94720, USA

Jiangling Wu
School of Civil Engineering and Architecture, Henan University, Kaifeng 475004, China

Hongsheng Chen
School of Architecture, Southeast University, Nanjing 210096, China

Zhanbo Sun and Yu Tan
National Engineering Laboratory of Integrated Transportation Big Data Application Technology, Chengdu, China
School of Transportation and Logistics, Southwest Jiaotong University, Chengdu, China

Rui Ma
Department of Civil and Environmental Engineering, University of California, Davis, CA, USA

Xia Yang
Department of Civil Engineering, SUNY Polytechnic Institute, Utica, NY, USA

Jie Zhang
School of Information Science and Technology, Southwest Jiaotong University, Chengdu, China

Ziyan Luo
State Key Laboratory of Rail Traffic Control and Safety, Beijing Jiaotong University, Beijing 100044, China

Xiaoyu Li and Naihua Xiu
Department of Mathematics, Beijing Jiaotong University, Beijing 100044, China

Xinyuan Chen and Inhi Kim
Monash Institute of Transport Studies, Australia

Index

www.ingramcontent.com/pod-product-compliance
Lightning Source LLC
LaVergne TN
LVHW080918210926
848485LV00004B/673